THE PRACTICE OF SILVICULTURE: APPLIED FOREST ECOLOGY

THE PRACTICE OF SILVICULTURE: APPLIED FOREST ECOLOGY

Ninth Edition

DAVID M. SMITH

Morris Jesup Professor Emeritus of Silviculture
Yale University

BRUCE C. LARSON

Lecturer in Forestry
Director of Yale Forests
Yale University

MATTHEW J. KELTY

Associate Professor of Forestry
University of Massachusetts

P. MARK S. ASHTON

Associate Professor of Silviculture
Yale University

JOHN WILEY & SONS, INC.
NEW YORK • CHICHESTER • BRISBANE • TORONTO • SINGAPORE • WEINHEIM

Acquisition Editor   Eric Stano
Marketing Manager   Rebecca Herschler
Senior Production Editor   Elizabeth Swain
Cover Design   Harry Nolan
Manufacturing Manager   Mark Cirillo
Illustration Coordinator   Gene Aiello
Cover photos: Photo in circle: Photo by U.S. Forest Service. Photo in rectangle: Photo by Yale School of
Forestry and Environmental Studies. Background photo: U.S. Forest Service.

This book was set in 10/12 Times Roman by University Graphics, Inc. and
printed and bound by Hamilton. The cover was printed by Phoenix Color.

Recognizing the importance of preserving what has been written, it is a
policy of John Wiley & Sons, Inc. to have books of enduring value published
in the United States printed on acid-free paper, and we exert our best
efforts to that end.

The paper in this book was manufactured by a mill whose forest management programs include
sustained yield harvesting of its timberlands. Sustained yield harvesting principles ensure that
the number of trees cut each year does not exceed the amount of new growth.

Library of Congress Cataloging-in-Publication Data

The practice of silviculture / David M. Smith . . . [et al.].—9th ed.
          p.   cm.
      Rev. ed. of: The practice of silviculture / David M. Smith. 8th ed.
   1986.
      Includes bibliographical references and index.
      ISBN 0-471-10941-X (cloth : alk. paper)
         1. Forests and forestry.   I. Smith, David M. (David Martyn)
   II. Smith, David M. (David Martyn) Practice of silviculture.
   SD391.S57   1996
   634.9—dc20   96-25883
                                                                      CIP

Printed in the United States of America

10 9 8 7 6 5 4 3 2 1

PREFACE

The first five editions of this book, starting in 1921, were written by the late Ralph C. Hawley, one of the pioneers of American forestry. He had to base it all on knowledge imported from Europe and what he and a few hundred foresters had learned from managing the limited tracts of forest on which true long-term forestry was being conducted. Since American society of the times regarded forests only as timber mines and knew that money did not grow on trees, the main focus of the book was on kinds of timber-production silviculture that would be financially sound in the long run. When David M. Smith joined as an author in 1954, increased emphasis was placed on presenting the scientific basis for silvicultural practice; more attention was given not only to such fields of natural science as forest ecology but also to the social sciences, especially economics.

Addition of the phrase, "Applied Forest Ecology," to the title of this edition simply describes scientific silviculture and previous editions of this book for what they have always been. The basic purpose is to call attention to the fact that foresters should design forests just as machinery should be designed on the basis of scientific knowledge by engineers. The kind of ecology that is applied is that which is concerned with the mechanisms of interactions among organisms and their environment, regardless of the degree to which the forests involved resemble ones devoid of human influence.

There are several ways in which this ninth edition responds to increased concern about non-timber uses of forests. Separate chapters deal with the silviculture of management of watersheds and wildlife. The chapter on wildlife management is almost entirely new and reflects that fact that scientifically based knowledge of the species of animals that go with certain kinds of forest vegetation is much stronger than it was previously. Some material that involves mainly timber management is consolidated in a new single chapter and the same is true of a chapter on silvicultural control of damaging agencies. However, we have tried to avoid so much consolidation of such information as to obscure the idea that all silvicultural solutions must include some sort of consideration of water, animal life, recreation, aesthetics, social concerns, and the fact that trees inevitably produce wood.

Part of the purpose in having separate chapters on wildlife management, watersheds, protection, and timber management is to respond to concerns about ecosystem management. In varying degrees these particular chapters deal with building whole forests out of different kinds of stands. In previous editions attention was limited to treatment of individual stands out of deference to the view that making forests out of stands was in the academic field of "forest management." That field seems to remain limited to concerns about dollar-sign economics and scheduling of timber harvests, so we have moved to fill part of a vacuum. Forest management is not exclusively the realm of economists, nor is that of ecosystem management exclusively the realm of ecologists.

Previous editions have had chapters labelled with the names of methods of regeneration and silvicultural systems. This lends too much credence to the idea that the practice of silviculture can be reduced to choosing among a few ready-made programs of treatment and then following them in something like the mindless conformity to orders of military drill. Therefore, we have reduced emphasis on treatments and focused more directly on the kinds of stands and stand-development processes that are the results and objectives of treatments. Some of this has been accomplished by adding a chapter on the dynamics of stand development and some simply by changing the names of certain chapters. The treatments themselves are still covered and those who preferred the previous approach will find it easy to put the old names on Chapters 12–16.

The book is still intended as a collection of ideas about silviculture and analytical approaches to its practice and not as a cookbook or manual of practice. It is written primarily for use in North America. However, that continent has so many different kinds of forests that it is not a long step from them to the rest of the world. The principles of silviculture are independent of geography, even though natural and socioeconomic diversity makes the application of the principles highly variable.

The list of references at the end of each chapter is intended to lead the reader to publications that bring together as much other literature, chiefly North American, as possible. Heavy reliance is placed on the published proceedings of symposia; these have the desirable feature of including the ideas of practicing foresters as well as those of researchers. The amount of silvicultural literature has become so vast that we make no pretense of presenting a review of it. The original sources of most ideas are so elusive that they have gone uncited.

Since it is no longer possible for one person to keep abreast of all developments in silviculture, this edition has four authors. Each chapter has been written primarily by one author but critically edited by each of the other three. We specifically thank Mary Ann Fajvan, Lauren Fins, Michael Messina, Kevin O'Hara, Edmund Packee, David Wm. Smith, Kathleen Schomaker, and Sarah Warren for the useful comments that they have made in reviews that they made at various stages on the writing of this edition. More generally we thank the many students and other readers for all of the suggestions that they have contributed over the years. This book is based on the thoughts, research, and experience of many generations of foresters and and forest scientists; we respectfully acknowledge our debt to them.

The opportunity to make selections of illustrations from the collections of the U. S. Forest Service, American Forests, and other organizations credited in the text has been invaluable.

We thank the members of our families for their patience with us about the time and attention that has been diverted to the writing of this book.

David M. Smith
Bruce C. Larson
Matthew J. Kelty
P. Mark S. Ashton

New Haven, Connecticut and Amherst, Massachusetts

CONTENTS

PART 2   Tending and Intermediate Cutting

CHAPTER 3   THE RESPONSE OF
                INDIVIDUAL TREES TO
                THINNING AND PRUNING   47

CHAPTER 4   MANAGEMENT OF GROWTH
                AND STAND YIELD BY
                THINNING   69

CHAPTER 5 METHODS AND APPLICATION OF THINNING 99

CHAPTER 6 RELEASE OPERATIONS AND HERBICIDES 131

P A R T 3   R e g e n e r a t i o n

C H A P T E R 7   E C O L O G Y O F
                   R E G E N E R A T I O N   **161**

CHAPTER 8   PREPARATION AND TREATMENT OF THE SITE   195

PART 4   Stand Development and Structure

CHAPTER 11   DEVELOPMENT OF SILVICULTURAL SYSTEMS AND METHODS OF REGENERATION   301

CHAPTER 12   THE SILVICULTURE OF PURE EVEN-AGED STANDS   316

CHAPTER 16   STANDS OF MIXED SPECIES   391

PART 5   Silvicultural Management Objectives

CHAPTER 17   TIMBER MANAGEMENT   423

CHAPTER 18 SILVICULTURAL MANAGEMENT OF WATERSHED ECOSYSTEMS 449

CHAPTER 19 SILVICULTURAL CONTROL OF DAMAGING AGENCIES 464

PART 1

INTRODUCTION TO SILVICULTURE

CHAPTER *1*

SILVICULTURE AND ITS PLACE IN FORESTRY

Silviculture has been variously defined as the art of producing and tending a forest; the application of knowledge of silvics in the treatment of a forest; or the theory and practice of controlling forest establishment, composition, structure, and growth. Since silvicultural practice is applied forest ecology, it is also a major part of the biological technology that carries ecosystem management into action. This chapter provides a preliminary review of all of silviculture and an account of how it relates to the rest of forestry.

Silvicultural practice consists of the various treatments applied to forests to maintain and enhance their utility for any purpose. The duties of the forester are to analyze the natural and social factors bearing on each stand and then to devise and conduct the treatments most appropriate to the objective of management.

Silviculture is to forestry as agronomy is to agriculture in that it is concerned with the technology of growing vegetation. Like the rest of forestry itself, silviculture is an applied science that rests on the more fundamental natural and social sciences. The immediate foundation of silviculture in the natural sciences is **silvics**, which deals with the growth and development of single trees and other forest biota as well as of whole forest ecosystems. Among the sources of information about silvics are books by Packham, et al. (1992); Kozlowski, Kramer, and Pallardy (1991); Lassoie and Hinckley (1991); Oldeman (1990); Burns and Honkala (1990); Whitmore (1990); Kimmins (1987); Spurr and Barnes (1980); and Daniel, Helms, and Baker (1979).

The competent practice of silviculture, whether it be crude or elaborate, demands as much knowledge of such fields as ecology, plant physiology, entomology, and soil science as a forester can acquire. It is through silviculture that a major part of the growing store of knowledge about trees and forests is applied. This knowledge is not learned once for a lifetime. The forestry practitioner must keep abreast of new information and ideas through communication with other members of the profession and familiarity with the results of formal research. Although formal research is indispensable, it does not lead to total knowledge, nor does it relieve the forester of responsibility for additional thought and continual observations in the forest. In applied science, in the absence of total knowledge we are always condemned to act on the basis of thoughtful judgment. Skillful practice itself is a continuing, informal kind of research in which understanding is sought, new ideas are applied, and old ideas are tested for validity. The observant forester, who is wise to seek to explain what is observed, will find answers to many silvicultural questions in the woods by examining the results of accidents of nature and earlier treatments of the forest.

The Purpose of Silviculture

Silviculture is designed to create and maintain the kind of forest that will best fulfill the objectives of the owner and the governing society. The production of timber, though the most common objective, is neither the only nor necessarily the dominant one. It is a mistake for foresters to assume that timber production is or should be the sole objective of silviculture. Frequently, especially with public forests, such benefits as water, wildlife, grazing, recreation, or aesthetics may be more important; water and wildlife always have to be taken into account.

Most silvicultural practices are applied in the course of timber harvesting if only because the value of the products removed greatly reduces the cost of the operations; this is true even if timber production is only a secondary or tertiary objective of management. Silviculture for timber production is also the most intricate kind because the species and quality of trees are of greater concern than they would be with other forest uses. Designing silviculture for wildlife management is also complicated, but mainly because of the difficulty of determining the kinds of vegetation that mobile and elusive animal populations require. Once the kind of habitat required is known, the silviculture is not difficult to design. If one understands silviculture for timber production, that for other uses is comparatively easy to formulate.

Some of the biggest problems in silviculture are getting owners and society to define their management objectives and, especially, the degree of priority attached to various uses. Foresters can determine how much of what uses are feasible, but the owners determine actual policies about allocations. It is the responsibility of foresters to work out the details, which include the design and conduct of silvicultural treatments. Management is hamstrung if owners, particularly legislative bodies, fail to provide for the making of hard choices about allocations of forest uses and leave them to be fought over by single-minded user groups. Even worse problems can be caused by amateur prescription of silvicultural practice through simplistic rules ordained by legislatures, courts, or accountants.

Improving on Nature Through Silviculture
The most magnificent forests that are ever likely to develop were present before the dawn of civilization and grew without human assistance. Furthermore, under reasonably favorable conditions, forests may remain productive even after long periods of mistreatment. Therefore, it is logical to consider why foresters should attempt to direct any of the powerful natural forces at work in the forest.

The purely natural forest is governed by no purpose unless it be the unceasing struggle of all the component plant and animal species to perpetuate themselves. Human purpose is introduced by preference for certain tree species, stand structures, or processes of stand development that have desirable characteristics. Where fine forests have developed in nature, they are usually found to have been the result of fortuitous disturbances followed by long periods of growth. In silviculture, natural processes are deliberately guided to produce forests that are more useful than those of nature, and to do so in less time.

The productivity of the managed forest is improved through attainment of the objectives listed in the following sections.

Control of Stand Structure and Process

Silviculture is a kind of process engineering or forest architecture aimed at creating structures or developmental sequences that will serve the intended purposes, be in harmony with the environment, and withstand the loads imposed by environmental influences. Because the stands grow and change with time, the designing is more sophisticated and difficult to envision than that of static buildings. Furthermore, the stands alter their own environment enough that the forester is partly creating a new ecosystem and partly adapting to the one that already exists.

As will be described in more detail in Chapter 2, the possible variations in stand structure and process are almost infinite. The shapes and sizes of stands can be altered for many purposes. A few among these are expediting silvicultural treatments and harvesting, creating attractive scenery, governing animal or pest populations, trapping snow, and reducing wind damage. The shapes of stands should be fitted to the immutable patterns dictated by soils and terrain. While the arrangement of stands in checkerboard patterns has a certain administrative appeal, the natural characteristics of land are not arranged in such ways.

The internal structure of a stand is determined by considerations such as variation in species and age classes (or lack of it), the arrangement of different layers or stories of vegetation (usually differing as to species), and the distribution of diameter classes. Much of this book is concerned with the purpose and means of achieving these kinds of variations in structure and developmental process.

Control of Composition

Undesirable species or inferior individuals of otherwise desirable species appear in almost any forest. One objective of silviculture is to restrict the composition of stands to that most suited to the location from economic and biological standpoints. This frequently means that the total number of species in a managed forest is less than that of the natural forest of the same locality. Unwanted plants commonly flourish at the expense of the desirable, so every reasonable effort is usually made to keep them in check.

Species composition is controlled basically through regulating the kind and degree of disturbance during periods when new stands are being established. In this way environmental conditions can be adjusted to favor desirable vegetation and exclude undesirable species. Regulation of the regeneration process by itself is not always enough to provide adequate control over stand composition. It is often necessary to supplement this approach by direct attack on the undesirable vegetation during or after periods of stand establishment. Cutting, poisoning, controlled burning, or regulated feeding by various forms of animal life may be used to restrict the competition and regeneration capacity of undesirable vegetation.

Desirable species and genotypes can be favored in a more positive fashion by planting or artificial seeding. By this means, it is also possible to improve on nature through the introduction of species that do not occur in the native vegetation, provided that they are

Figure 1.1 Much of silviculture has always consisted of rehabilitation efforts and of knowing what will happen as a result of treatments of the forest. This sequence of pictures from 1938, 1949, and 1969 shows a planted stand, at a National Forest in northern Idaho, in three stages of development. The tract had been logged over from a logging railroad in 1930–1931 and both burned and acquired just before the first picture was taken. Planting of western white pine and Engelmann spruce was done in 1939 and 1940. The subsequent pictures show the development, to age 30, of the mixture of planted trees and other conifers that seeded-in naturally. *(Photographs by U. S. Forest Service.)*

Figure 1.1   (continued)

adequately adapted to the environment and do not get out of control. The goal of controlling stand composition is to achieve the best fit between the purposes of management and the natural constraints imposed by the site.

Control of Stand Density

Inadequately managed forests are often too densely or too sparsely stocked with trees. If stand density is too low, the trees may be too branchy or otherwise malformed, and the unoccupied spaces are likely to be filled with unwanted vegetation. This condition arises from inadequate provision for regeneration; it is most common in the early life of a stand but its consequences may linger after the trees have grown to occupy all of the space available. Excessively high stand density causes the production to be distributed over so many individual trees that none can grow at an optimum rate and too many may decline in vigor. Unless stand density is controlled at the time a stand is established or during its development, it is almost sure to depart from optimum density at some stage of its life.

Restocking of Unproductive Areas

Without proper management many areas of land potentially suited to growth of forests tend to remain unstocked with trees (Fig. 1.1). Fires, destructive logging, grazing, agricultural clearances, and other kinds of forest devastation have already created many large, open areas that can be reforested only by planting.

Protection and Reduction of Losses

In unmanaged stands, severe losses are commonly caused by damaging agencies such as insects, fungi, fire, and wind, as well as by the loss of merchantable trees through competition. Substantial increases in production may be achieved merely by salvaging material that might otherwise be lost. Proper control of damaging agencies can result in further increases in production. Forest protection often involves modification of silvicultural techniques. Adequate protection should be extended to all forests, even the poorest, because fire and pest outbreaks developing in stands where they may be of little concern can sometimes spread to those deemed worthy of protecting. Those areas set aside for wilderness, scenery, or scientific study require some protection; sound policies about their care and use inevitably involve something other than leaving them absolutely alone.

Control of Rotation Length

Stands of trees are not immortal. In most situations, there is an optimum size or age to which trees should be grown. The period of years required to grow a stand to this specified condition of either economic or natural maturity is known as the **rotation**. Controlled reductions of stand density or such measures as fertilization and drainage can shorten rotations by making the final-harvest trees grow to the desired sizes at earlier ages. Premature cutting is a common type of mismanagement. Trees allowed to grow beyond the optimum do not continue to increase in value at rates sufficient to provide an acceptable return on either the costs of growing them or the investment represented by their own value. The risk of decay or other damage may increase the possibility that they will decline in value, be lost, or become a hazard. However, the reservation of overmature trees or even of dead trees may be necessary for some wildlife species or simply for scenery.

Facilitating Harvesting

In the unmanaged forest, timber, like gold in the hills, is where one finds it: the greater the amount extracted, the more difficult and expensive it becomes to find and extract more. In the managed forest, the growth of stands can be planned so that any use of them is on a more efficient, economical, and predictable basis. It helps to create good stands that are so located that the cost of transporting timber from them is kept under control. Planned reductions in the number of trees on an area not only makes them reach merchantable size more quickly but also leaves more space between trees for extraction of logs during partial cutting.

Conservation of Site Productivity

Paramount among the objectives of forestry in general and of silviculture in particular is the maintenance of the productivity of the living forest. The site is the total combination of the factors, living and inanimate, of a place that determine this productivity. The site factors that are most subject to long-lasting harm are those of the soil, which is the most nearly nonrenewable of the resources used in silviculture. The basic supply of solar energy, the most vital site factor, is beyond silvicultural control. Silviculture rests heavily on manipulation of the microclimate of a site, but its effects on the macroclimate are limited to those caused by photosynthetic removal of carbon dioxide from the atmosphere. Forests are the result rather than the cause of geographical precipitation patterns.

The living organisms of a place are site factors themselves; however, they can reproduce themselves and are thus the epitome of the renewable resource. If none are rendered extinct, damage to these living components of the site is not likely to be permanent, even though it can be serious and long-lasting. There are always uncertainties over the extent to which silviculture should discriminate against "undesirable" forms of life.

The most obvious and least reparable kind of damage to the soil is physical erosion. Careless treatment, especially that associated with roads and trails used for timber extraction, can cause several years of accelerated erosion that may negate the soil formation processes of a thousand years. A more subtle kind of chemical erosion can result if the remarkable capacity of forest vegetation to recycle nutrients in place is so impaired that large amounts of vital chemicals are irretrievably lost to surface runoff or leaching. These two kinds of erosion cause harm twice because they reduce not only the productivity of the soil but also the quality of the water that flows from it. Soil damage impairs the capacity of the site to yield all of the primary tangible benefits of the forests—vegetation, animals feeding on plants, and good water.

It is entirely within the realm of possibility to conduct forestry permanently without the degradation that is almost inevitable in most agriculture and in other "higher" uses of land. However, realization of this potentiality is not automatic.

Silviculture as Applied Ecology

Silviculture is the oldest conscious application of the science of ecology and is a field that was recognized before the term *ecology* was coined. Many of the ways of governing the development of forest stands rest heavily on cuttings and other treatments that alter or modify the factors of the stand environment that regulate the growth of the vegetation. The reliance on ecological knowledge is all the firmer for resting on the virtue of necessity and not simply on philosophical principle. The economic returns from forestry are not high enough to make it feasible to shield forests from all the vicissitudes of nature. Therefore, silviculture is usually far more the imitation of the natural processes of forest growth and development than of substitution for them. These processes may be improved upon, channelized; and limited; however, excessive disruption leads to severe losses, high costs, and other unfortunate consequences, immediate or delayed.

The necessity that nature should be understood and emulated does not mean that silviculture should slavishly follow either the reality of natural processes or abstract theories about them. Most forests live longer than people. It is, therefore, not easy to recognize that the natural disturbances that renew forests, often after intervals of centuries, are usually catastrophes such as fires, windstorms, and insect outbreaks (Kimmins, 1987; Oliver, 1981). There are also forests that are slowly and continuously renewed by minor disturbances, but these are far from being any universal or typical form of nature. The various patterns in the development of forest vegetation over time and after disturbance are discussed in Chapter 2 on stand dynamics.

Silviculture is concerned with the building and control of forest ecosystems. The continuity of life and the health of ecosystems ultimately depend on completion of ecological cycles of vital materials. Some are closed in seconds within the space of a few millimeters, whereas those that involve the natural loss of water and certain gases to the atmosphere will not be closed within a single continent or year. Some purely natural cycles are not completed, if ever, until nutrients lost to streams are heaved up from the ocean floor millions of years later.

It is desirable that these cycles be closed as nearly as possible within distances of time and space that are not needlessly large. However, this does not mean that they need to be closed and contained within single hectares on an annual basis; they were not in nature, and it is not feasible to do so in practice. In other words, the concept of the mixed, all-aged stand of stable climax vegetation with some ancient trees and no net gain or loss of materials is more of an abstraction than a reality. Diagrams depicting such forests are useful for conveying ideas succinctly, but the concept that they depict may be to forest ecology and silviculture what the perfect black body is to physics and engineering.

Ecological Technology

Silviculture is not conducted in a pure state of nature, and this state ceased to be pure when society advanced beyond the food-gathering stage. To the extent that civilization is partly artificial, it can be argued that forestry must also be partly artificial in ways calculated to keep the world ecosystem in healthy dynamic equilibrium. However, the farther that any artificially induced process departs from nature, the more perilous it is if only because it becomes harder to predict the results.

The web of life is so intricate that it is easy to argue that one should do nothing to the forest for fear of doing something wrong. This is the indictment pure science continually brings against the technology of applied science. The charge can never be entirely refuted, but society requires practitioners of applied science to act in the absence of full knowledge. The best that they can do is to proceed by **adaptive management** in which one takes action based on the most complete knowledge available. The results are monitored, and the management practices are changed or otherwise adapted as one learns from the results.

Silviculture is conducted on the basis of ecological principles and not in spite of them. There is virtually no way of using land resources so heavily with so little risk of depleting their productive capacity. The goods and benefits that flow from forests under sound, long-term management depend on living processes and are thus renewable to the extent that basic productive site factors are maintained; they can even be increased if these factors are permanently improved.

The wood produced by forests is, apart from concrete, the most important structural substance in human use. Unlike alternative mineral or agricultural materials, its production requires much less energy and does little that would damage or pollute. In fact, the growing of it increases the stock of resources even as it cleans both air and water. If forest vegetation were more efficient in yielding human food and in concentrating sources of fuel, the future of the world ecosystem would be much brighter for the human race. It is ecologically ignorant to assume that "saving forests" by substituting wood with substances produced with fossil fuel from mineral resources benefits any ecosystem that includes human civilization.

Economic and other social factors also affect the silvicultural policy for any given area; the objective is to operate so that the value of benefits derived from a forest should exceed the value of efforts expended. The most profitable forest type is not necessarily the one with the greatest potential growth or the one that can be used or harvested at lowest cost. One must also consider the silvicultural costs of growing the crop or maintaining the stand and the prospective losses to the damaging agencies. In fact, it is usually the inroads of insects, fungi, and atmospheric agencies that ultimately show where silvicultural choices have run afoul of the laws of nature. The majority of the best choices are imitations of those natural communities. These are not necessarily or even commonly old-growth climax communities.

It is not entirely safe to accept the success of modern agriculture as justification for highly artificial kinds of silviculture. The environment of a cultivated field is much more thoroughly modified and readily controlled than that of a forest stand. Furthermore, forest crops must survive winter and summer over a long period of years, whereas most agricultural crops need survive only through a single growing season. One disastrous year harms the production of just one year with an annual crop, but it can destroy the accumulated production of many years in a stand of trees. Neither economic nor ecological principles permit the forester to engage in the wholesale, routine use of pesticides and fertilizer on which intensive agriculture often rests. Any silvicultural application of refinements borrowed from agriculture must be combined with all the kinds of measures appropriate to the intensity of agriculture imitated. Forestry can profitably borrow much more than it ever has from the science on which modern agriculture is based, but there is little place for uncritical imitation.

Some silvicultural measures depart very far from natural precedent. These usually involve the introduction of exotic species or the creation of communities of native species unlike anything that might come into existence naturally. Departures of this sort cannot be thoughtlessly condemned but should be viewed with reservations until they have been tested over long periods. Otherwise, most of the choices can be thought of in terms of the degree to which natural processes are accepted or arrested, pursued or reversed. Any departure from nature that still looks good at the end of a rotation is probably sound.

Relationship with Forest Management and the Social Sciences

The decisions made in silvicultural practice are based fully as much on economic constraints and social objectives as on the natural factors that govern the forest. Recognition of objectives and limitations set by society in any given case reduces the silvicultural alternatives that need be considered. Even though intelligent application of silviculture can make a positive contribution to the tactics, the strategy for solving problems associated with social sciences is not academically part of silviculture. Those matters that involve economic considerations are dealt with in the field of forest management, which is concerned with forest planning, economic analysis, harvest scheduling, and the administrative aspects of forestry (Davis and Johnson, 1987; Leuschner, 1984). The field of forest policy deals more indirectly with the effects of sociological and political phenomena, as well as economics, on the use and treatment of forests.

Silviculture and forest management are interdependent, and not parallel, alternative approaches to the same problem. Because of its dominant concern for efficient application of the natural sciences, silviculture is no less "practical" than forest management, with its tendency toward preoccupation with economic considerations. No management plan is better than the silviculture it stipulates, nor is any silvicultural treatment better than the usefulness of the results it produces.

The Stand and the Forest

A **stand** is a contiguous group of trees sufficiently uniform in species composition, arrangement of age classes, site quality, and condition to be a distinguishable unit. The internal structure of stands varies mainly with respect to the degree that different species and age classes are intermingled. The simplest kind of structure and developmental pattern is that of the pure, even-aged plantation. The range of complexity can extend to a wide variety of combinations of age classes and species in various vertical and horizontal arrangements. The development of stands over time, or **stand dynamics**, is considered in Chapter 2.

From the standpoint of forest management, the term **forest** has a special meaning and denotes a collection of stands administered as an integrated unit, usually under one ownership. Putting stands together into forests is especially important in regulating timber harvests as well as managing wildlife populations and large watersheds.

One objective of this type of planning for timber is usually the achievement of a sustained yield of products. The forest, and not the stand, is the unit from which this sustained yield is sought. Management studies of prospective growth and yield determine the volume of timber to be removed from the whole forest in a given period; silvicultural principles should govern the sequence and manner in which individual stands are cut to produce the required volume. The tendency to treat large groups of dissimilar stands as if they conformed to a uniform, hypothetical average should be studiously avoided. However, an arbitrary decision must always be made regarding the minimum size of stand that can be regarded as a separate entity.

The silviculture that is concerned with natural processes involving mobile animals, flowing water, and whole landscapes also involves the arrangement and juxtaposition of stands. Differences between adjacent stands and the distribution of stands across landscapes are taken into account as part of the management of forests or ecosystems.

The size and number of stands recognized depend on the intensity of practice, the value of the stands, the diversity of site conditions, and the ease of mapping. Where intensive forestry is feasible, stands as small as a quarter-hectare may be recognized; under crude, extensive practice, the same forest might be divided into units no smaller than several hundred hectares. The best policy is to recognize the smallest stands that can be conveniently delineated on the maps of forest types and age classes used in administration. Even after stand maps have been put on paper, the forester must still deal with variations that actually exist within each stand. From the technical standpoint, each portion is best treated separately, although acceptance of too many variations would eventually create a mosaic of conditions that would be awkward for most operations.

Scope and Terminology of Silvicultural Practice

Silvicultural practice encompasses all treatments applied to forest vegetation. Although there is much more to the understanding of these treatments than their definitions and nomenclature, the terminology must be understood and used carefully and precisely. Sloppy use of the terms causes all manner of misunderstanding within the forestry profession and in dealings with the general public. For example, some categorize all cutting as either "clearcutting" or "selective cutting." This not only stunts the development of their own understanding of forestry practice and causes blunders but also keeps everyone else confused. The terminology in this book generally adheres to that promulgated by the Society of American Foresters Silviculture Instructors Sub-Group (1994) and the Commonwealth Forestry Bureau (Ford-Robertson, 1978); it departs only where further improvement in clarity or precision seems imperative.

The act of replacing old trees, either naturally or artificially, is called **regeneration** or **reproduction;** these two words, which are synonymous in this usage, also refer to the new growth that develops.

Kinds of Stands

The production of benefits by forest stands is controlled by the developmental processes of those stands whether these benefits be wood, wildlife, water, forage, or scenery. The processes are all of those that start with the birth of the stands, continue with competition between trees, and end with the death of old trees and their replacement. The simplest kind of stand development process is that of the **pure even-aged stand** in which the trees are "pure," that is, all of one species, and start together after the previous stand is removed; such stands are often ones that have been planted. Slightly more complicated are pure stands that start from so-called **advance regeneration** that is already in place before the old stand is removed. **Uneven-aged stands** have trees or, more commonly, groups of trees of different ages and much more complicated developmental patterns. **Mixed stands** have more than one tree species, and the interaction between them makes their development even more complicated, especially if they also have more than a single age-class of trees. The development of these different kinds of stands is discussed in Chapter 2 and Chapters 11–16.

Kinds of Silvicultural Treatments

There is also a terminology of silvicultural treatments, of which there are two categories. (1) **Methods of reproduction** refer to treatments of stand and site during the period of regeneration or establishment while (2) **tending** or **intermediate cutting** refers to treatments at other times during the rotation (Fig. 1.2).

    Reproduction or **regeneration cuttings** are made with the twin purposes of removing the old trees and creating environments favorable for establishment of regeneration. The period over which such regeneration treatments extend is the **reproduction** or **regeneration period**. Regeneration cuttings range from one to several in number, and the regeneration period may extend from several years to several decades. In truly uneven-aged stands, regeneration is almost always under way in some part of the stand. The regeneration period begins when preparatory measures start, and it ends when young trees, free to grow, are dependably established in acceptable numbers. The **rotation** is the period during which a single crop or generation is allowed to grow. The act of replacing old trees, either naturally or artificially, is called **regeneration** or **reproduction** and these two words, which are synonymous in this usage, also refer to the new growth that develops.

    The names of the various methods of regeneration (see Chapter 11) denote the patterns of cutting in time and space that determine the structure of the stands created or maintained by the process. They distinguish between reliance on reproduction from seeds or that from sprouts and may tell a little about the degree of shading of new seedlings. Later chapters describe how clearcutting is associated with pure, even-aged stands; shelterwood or the

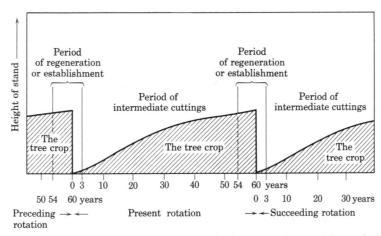

Figure 1.2   The relationship between the period of regeneration and the period of intermediate cuttings is shown for a sequence of even-aged stands managed on a 60–year rotation according to the shelterwood system. In this system the new stand is started before the older one is completely removed.

seed-tree method, with advance regeneration; the selection system, with uneven-aged stands; and coppice methods, with sprout regeneration. The names of the methods, systems, and kinds of stand usually only begin to describe the details of treatment programs.

Silvicultural treatments are not limited to securing regeneration. Other treatments may be applied after the stand is established and during the long period that elapses while the crop grows through various stages until it is ready for replacement. Various **intermediate cuttings** or **tending operations** are conducted to improve the existing stand, regulate its growth, and provide for early financial returns, without any effort directed at regeneration. Sometimes these treatments are referred to as **stand improvement operations** or **timber stand improvement (=TSI)**, when they yield no products.

Intermediate cuttings that are aimed primarily at controlling the growth of stands by adjusting stand density or species composition are called **thinnings** (Chapters 3–5). Treatments conducted to regulate species composition and improve very young stands are **release operations** (Chapter 6). Those that involve only the branches are **pruning** (in Chapter 3). Those tending operations that involve outright investment without any harvest of wood are sometimes referred to as **stand improvement**. Many kinds of intermediate cutting or tending can now be accomplished without actually cutting down trees. Therefore, usage of the word "cutting" is now often correct only if viewed in a traditional and figurative sense.

Protection against injury is as much a part of silviculture as harvesting, regenerating, and tending of forests. It is so important that it has led to fields of specialization in forestry such as entomology, pathology, and fire control and is no longer included as a separate section of this book, although Chapter 19 is devoted to the silvicultural aspects. The details of almost any successful silvicultural system include significant modifications designed to reduce injuries. Where such measures fail or are inadequate it is sometimes desirable to conduct **salvage cuttings** to recover the values represented by damaged trees or stands.

A program for the treatment of a stand during a whole rotation is called a **silvicultural system**. The silvicultural system is usually given the same name as the method of regeneration that is used at the stage of stand replacement. This is because these regeneration methods determine the kinds of stands and stand developmental processes that go on during a whole rotation.

Role of Cutting in Silviculture

The techniques of silviculture proceed on the basic assumption that the vegetation on any site tends to extend itself aggressively to occupy the available growing space. The limit on growing space is usually set by the availability of light, water, inorganic nutrients, or carbon dioxide. In a general way, the available amount of growing space is set by the most limiting of these factors, although an abundant supply of one factor can partially offset deficiency of another. If the vegetation nearly fills the growing space, the only way that the forest can be altered or controlled is by killing trees and other plants. In reproduction cutting, this is done to provide room for the establishment of new trees; in intermediate cutting, to promote the growth of desirable trees already in existence. Paradoxical as it may seem, useful forests are created and maintained chiefly by the destruction of judiciously chosen parts of them. One of the characteristics of life is death; if there were no death, there would be no space for new life.

The ax and other means of killing trees can, in other words, be used for the construction as well as the destruction of the forest. What is left or what replaces what is harvested is more important silviculturally than what is cut. Unfortunately, much of the general public as well as some loggers have eyes only for what is cut and regard the harvests as simply the mining of a nonrenewable resource.

Preoccupation with the trees should not cause foresters to overlook the lesser vegetation and the animals that are a part of the forest community. The animals ultimately depend on the vegetation for food and thus do not compete directly for the growing space. However, whether they be defoliating insects or carnivores that feed on herbivorous mammals, they exert major influence on the nature of the vegetation even as they are, in turn, controlled by it. The fauna and nonarborescent vegetation of the forest are as likely to respond to cutting as the trees.

Effect of Cutting on Growing Stock

Cutting controls not only the composition and form of forest stands but also the relationship between trees reserved for growth and the amount of growth available for cutting. It is important to understand the long-term, cumulative effect of cutting operations in building, or degrading, a forest.

The growing of timber and animals is one of the few kinds of creative processes in which both product and productive machinery are the same thing (Heiberg, 1945). The wood of the tree stem cannot be removed without destroying the machinery that produced it, and the same is true of the animals. A clear distinction must, therefore, be drawn between the individuals that must be left to produce more wood or animals and the surplus that can be regarded as product and harvested.

The trees that must be reserved somewhere in the forest to continue production are the **growing stock** or **forest capital**. The volume of wood that is grown in the future depends on the quantity and condition of growing stock that is maintained. Cuttings regulate the amount of this growing stock and its distribution within individual stands or among the various stands that comprise the forest. The regulation of growing stock is of most crucial importance in silviculture when partial cuttings are applied within stands.

Silviculture and the Long-Term Economic Viewpoint

It is said that money does not grow on trees; the bane of forestry is the popular view that trees exist but do not grow. The short-term outlook of conventional economic theory holds, in effect, that the silviculturist cannot win while certain naturalistic ecological theories warn against trying. The economic time scale of forestry is so vast and unique that to many investors and in terms of ordinary economic theory it really is not economic at all.

There is scarcely any part of forestry in which this issue must be faced more squarely than in silviculture, especially when investments in establishing or treating young stands are considered. It takes a certain kind of ambivalence to keep the economics of forestry in perspective. The decision to practice forestry is usually a matter of ethics, politics, and social concern for posterity but not basically one of conventional economics. However, once the decision is made, it becomes logical to apply economic analysis to determine how best to execute the details. Any conflict is not between ''silviculture'' and ''economics'' but between the long-term economic viewpoint of forestry itself and customary short-term outlooks on financial matters. In the long run, short-sighted silviculture and poor environmental management become uneconomic.

The holding of land for future production of wood or other benefits involves silviculture, even if nothing more is done than to let nature take its course and to harvest trees occasionally. Ownership incurs costs, and these constitute investments in the future even if nothing is invested in treatments to increase future production. The question is not whether owners are investing in silviculture because they are already doing so, even if unintentionally; it is whether the capacity for long-term investments in silviculture and other parts of the forest enterprise is being used to optimum advantage.

Funds are rarely available for all the silvicultural work likely to prove profitable; thus the forester must ensure that those that are invested are expended on the most remunerative lines of work possible. In any situation, it is logical to apply first those treatments that will yield the greatest increase in value of benefits per dollar of investment.

The first stage in the evolution of silvicultural practice is that in which continued production is actively sought but without any cash investments in it (Spurr, 1979). This "no-investment" silviculture places emphasis on treatments that can be accomplished by removing merchantable timber without significant increase in harvesting costs. The removals cannot exceed the productive capacity of the forest. Some forests are sufficiently easy to control that treatments made on this basis give reasonably good results. This kind of silviculture is practiced over wide areas and will doubtless continue to be for a long time. The idea of taking values out of the forest without really reinvesting anything in future production has a powerful appeal. It almost completely dominated American silviculture for decades, and there are still many instances in which it is consciously or unconsciously regarded as the only economically feasible alternative.

Orderly policies of long-term investment in silviculture emerge if economic conditions and natural productivity are favorable, provided that adequate experience has accumulated. The kind and amount of investment are limited only by the economic law of diminishing returns. The actual amount expended on this kind of silviculture varies widely but is not likely to be trivial.

Variations in Intensity of Practice

The amount of effort expended on the treatment and care of stands—that is, the **intensity** of silviculture—varies widely, depending chiefly on economic circumstances. The converse of intensive silviculture is **extensive** silviculture. The degree of intensity is usually estimated in terms of such things as the amount of money invested in cultural treatment, the frequency and severity of cuttings during the rotation, and the amount of concern accorded to future returns relative to immediate returns.

The appropriate intensity of silviculture varies with accessibility, markets, site quality, management objective, and nature of ownership. The proper level often must be chosen for each stand because the application of a single degree of intensity will not give optimum results throughout a given forest, unless it is exceedingly small and uniform. The more favorable the combined economic effect of all factors, the higher the appropriate level of intensity of silviculture. The place for extensive silviculture is found in remote areas, on poor sites, or where owners are not willing or able to make more than minimum investments. It often plays a role where timber production is secondary to other purposes of forest management.

In the past, American forests have been exploited in such a manner that the poorest and most ill-treated stands are often found on the best sites and in the most accessible areas, such as those along permanent roads. This situation arises because the best and most conveniently located stands have been exploited first, most heavily, and most frequently. Ultimately, silviculture of the highest intensity should be practiced in many of these situations rather than in the remote areas where the best forests are now often found. Permanent roads and good markets for a diversity of forest products do not automatically ensure optimum practice, but they are essential to intensive management.

The intensity of timber-production silviculture depends in large measure on the nature and objectives of ownership. Variations in the species and sizes of trees desired may necessitate different procedures on adjoining lands that are fundamentally similar. Stability or longevity of ownership also controls intensity of silviculture. Large corporations and public agencies, which are relatively immortal, are in a far better position to practice intensive silviculture than individuals or small corporations of uncertain stability.

The intensity of silviculture often depends on the extent to which the owner processes the wood grown in his forest. The closer the product is carried to the ultimate consumer, the greater is the ability to capture the values added by increases in intensity of practice in the woods. Prices for **stumpage**, that is, standing trees, do not necessarily reflect all the values added to the product by silvicultural measures to improve the quality of wood. Therefore, the owner who is not in a position to do more than sell stumpage may not be able to practice silviculture as intensive as that suitable for owners who also harvest, manufacture, or sell the final product. This relationship is modified, however, by the ability and willingness to make long-term investments. For example, public forestry agencies usually confine their operations to producing stumpage; they may, however, practice intensive silviculture without concern for profit on their investments in order to discharge their long-term responsibilities to the national economy.

Silviculture in Application

The practice of silviculture does not consist of rigid adherence to any set of simple or detailed rules of procedure. For example, no part of this book can be wisely used as a manual of operations. Many of the techniques of cutting are described in simplified form shorn of the myriad of refinements and modifications necessary to accommodate the special circumstances and local variations encountered in practice. Each procedure described is merely an illustration intended to demonstrate the application of a set of treatments designed to meet a uniform set of circumstances. Even though uniform stands have important advantages that make them worthy of creation, the stands encountered in the field may lack uniformity and thus call for variation in treatment.

Any consideration of silviculture covers a variety of treatments wider than is likely to be practiced in any locality at a particular time. In times when all the forests of a locality are immature, silvicultural practice may be limited to intermediate cuttings, and anything connected with regeneration may be limited to the reforestation of vacant areas. In localities where it is customary to secure regeneration by planting, the forester may regard methods of natural regeneration only as matters of intellectual exercise. Conversely, where planted stands are an anathema or owners are not ready to invest in them, only natural regeneration may seem important. Where attention is for the time being concentrated on replacing old-growth forests with regeneration, consideration of intermediate cuttings may seem quite unessential. At times and places where economic conditions support only the crudest kind of extensive silviculture, intensive treatments may seem visionary indeed.

The subject matter of this book reflects a wide variation in intensity of silvicultural practice because an attempt is made to describe all known techniques that seem applicable in any significant forest area, especially of North America, within the foreseeable future. The procedures characteristic of the more intensive kinds of silviculture cannot be described as briefly as those associated with extensive silviculture and so get more attention. This does not mean that a management program must include a long series of different treatments to be silviculture. Some of the most astute silviculture is the kind conducted at low intensity in which much is accomplished with a limited amount of treatment.

The student forester interested in only one particular region cannot safely restrict attention to the kind of silviculture currently practiced there. Foresters move, times change, and ideas from other places are often as fruitful as the indigenous ones. Scientific knowledge and technology also grow at an accelerating pace. The demands that society places on forests continually increase even as that same society places increasing restrictions on the ways of meeting the demands.

In many places today, the impractical or impossible of 20 years ago is now the routine—yet may prove to be the naive, illegal, or inadequate of a decade hence. Because of cutting and growth, the forests of a locality often change, and this calls forth new methods of treatment. This is especially true in North America, where the forests of many localities tend to be in uniform condition either because they have been scarcely touched or because they were all cut over in a short space of time. This book may seem to contain more techniques and ideas than a forester might need in a professional lifetime. Although some may go unused or go quickly out of date, there are really only enough to provide a start.

It is not enough for the forester to know what to do and how to do it. The important questions in silviculture, whether they involve matters of natural or social science, start with the word ''why.'' As in other applied sciences, action proceeds from the knowledge represented by the answer, or sometimes the merest inkling of an answer. The forester can find as many solutions in the woods as in the printed word. However, it is necessary to ask oneself the questions that generate the solutions and also to be ready to take the time to observe how the flora and fauna of the forest develop. Growing knowledge calls for those changes in procedure that are part of adaptive management.

Silvicultural Literature

Each chapter of this book ends with a listing not only of the references cited in that chapter but also of others chosen to lead the way to as much literature as possible on topics covered in the chapter. It often helps to find a recent article on the chosen topic and then follow it backward in time through use of the references in that article. Any sort of literature review is useful.

Abstracting journals can be very valuable, especially *Forestry Abstracts,* a British publication that covers significant forestry literature on a worldwide basis. The professional and scientific journals of forestry of one's locality can also be helpful, especially if they have annual or cumulative indexes that are used with appropriate imagination. The use of computerized information retrieval systems is growing rapidly. More detailed information and many additional literature references about silviculture in the United States can be obtained from at least three consolidated publications. In *Regional Silviculture in the United States* (Barrett, 1994), various silviculture professors have written about their localities. Research scientists of the U.S. Forest Service (Burns and Honkala, 1990; Burns, 1983) have summarized information about the ecological characteristics of tree species and about the silviculture of the important forest types. One advantage of these three sources is that they will help locate many of the large numbers of publications issued by research and extension agencies of governments and universities.

The *Forestry Handbook* of the Society of American Foresters (Wenger, 1984) presents much information about silviculture and closely associated topics, as do similar compendia designed to help the practicing foresters of a locality. The written word can bring the forester ideas from distant places. Not all of the problems of growing loblolly or ponderosa pine have to be solved exclusively by study of these individual species. Much has also been learned about the silviculture of pines in Finland and Australia; knowing about teak in Asia may also help. In fact, new and useful insights often come faster from distant sources. Most of the world literature of forestry is in English, although English-speaking forestry students should be more ambitious about mastering other languages.

A forester should not read about silviculture just to absorb information. Reading should be a stimulus to thought, a way of synthesizing new patterns of understanding, and of both expanding and testing ideas. It can make comprehension of processes seen in the woods surer and more serviceable.

BIBLIOGRAPHY

Barrett, J. W. (ed.). 1994. *Regional silviculture of the United States.* 3rd ed. Wiley, New York. 643 pp.

Burns, R. M. (ed.) 1983. Silvicultural systems for the major forest types of the United States. *USDA Agr. Hbk.* 445. 191 pp.

Burns, R. M., and B. H. Honkala (eds.). 1990. *Silvics of North America.* 2 vols. *USDA, Agric. Hbk.* 654. 1552 pp.

Burschel, P., and J. Huss. 1987. *Grundriss des Waldbaus.* Verlag Paul Parey, Hamburg. 352 pp.

Daniel, T. W., J. A. Helms, and F. S. Baker. 1979. *Principles of silviculture.* 2nd ed. McGraw-Hill, New York. 500 pp.

Davis L. S., and K. N. Johnson. 1987. *Forest management: regulation and valuation.* 3rd ed. McGraw-Hill, New York. 790 pp.

Dengler, A. 1990. *Waldbau, auf ökologischer Grunglage. II. Baumartenwahl, Bestandesgrundung und Bestandspflege.* 6th ed. revised by E. Rohrig and H. A. Gussone. Verlag Paul Parey, Hamburg. 314 pp.

Ford-Robertson, F. C. (ed.). 1978. *Terminology of forest science, technology, practice and products.* Society of American Foresters, Washington, D.C. 349 pp.

Heiberg, S. O. 1945. Is forestry unique? *J. For.* 43: 294–295.

Kimmins, J. P. 1987. *Forest ecology.* Macmillan, New York. 531 pp.

Kozlowski, T. T., P. J. Kramer, and S. G. Pallardy. 1991. *The physiological ecology of woody plants.* Academic, San Diego. 657 pp.

Lassoie, J. P., and T. M. Hinckley (eds.). 1991. *Techniques and approaches in forest tree ecophysiology.* CRC Press, Boca Raton, Fla. 599 pp.

Leuschner, W. A. 1984. *Introduction to forest resource management.* Wiley, New York. 298 pp.

Nyland, R. D. 1996. *Silviculture: concepts and applications.* McGraw-Hill, New York. 576 pp.

Oldeman, R.A.A. 1990. *Forests: elements of silvology.* Springer-Verlag, New York. 624 pp.

Oliver, C. D. 1981. Forest development in North America following major disturbances. *FE&M* 3:169–182.

Packham, J. R., D.J.L. Harding, G. M. Hilton, and R. A. Suttard. 1992. *Functional ecology of woodlands.* Chapman & Hall, New York. 208 pp.

Smith, W. K., and T. M. Hinckley. 1995. *Resource physiology of conifers: acquisition, allocation, and utilization.* Academic Press, San Diego. 396 pp.

Society of American Foresters, Silviculture Instructors Sub-Group. 1994. *Silviculture terminology.* Society of American Foresters, Bethesda, Md. 14 pp.

Spurr, S. H. 1979. Silviculture. *Scientific American* 240:76–82, 87–91.

Spurr, S. H., and B. V. Barnes. 1980. *Forest ecology.* 3rd ed. Wiley, New York. 687 pp.

Wenger, K. F. (ed.). 1984. *Forestry handbook.* 2nd ed. Wiley, New York. 1335 pp.

Whitmore, T. C. 1990. *An introduction to tropical rain forests.* Oxford University Press, New York. 226 pp.

CHAPTER *2*

STAND DYNAMICS

A forest stand may seem static but is really a dynamic, ever-changing, living structure. The choices made in silviculture are usually among the stands and processes to start or to alter after they have started. **Stand dynamics** is the study of changes in forest stand structure with time, including stand behavior after disturbances (Oliver and Larson, 1996; Means, 1982; Reichle, 1981). In practicing silviculture, the forester must be able to predict what kind of vegetation will follow regenerative disturbances and what patterns of development should be anticipated in *all* the vegetation as the stands grow older. This chapter introduces the different kinds of stands and their developmental processes.

Initiating Disturbances and Sources of Regeneration

All silvicultural procedures are, at least to some degree, simulations of natural processes in which stands start, develop, and are replaced gradually or suddenly. No plant starts or accelerates its development unless something dies or is killed to provide vacant growing space for it. Established plants often exclude others; vigorous individuals can expand and suppress or kill weaker ones; some can endure beneath taller neighbors and thus share growing space with them. Silviculture imitates and regulates the processes involved.

The lethal natural disturbances that initiate new stands include fires, pest attacks, landslides, windstorms, and various atmospheric agencies. Their effects vary greatly in intensity. For example, some fires burn so hot that virtually all preexisting plant life is killed, whereas others do little more than reduce the litter that often inhibits seed germination. Decisions about the regeneration of stands usually involve choices of the kinds of disturbance to simulate. An important distinction can be made between lethal disturbances (Fig. 2.1) such as (a) fire that can be said to "kill from the bottom up" because they are

Figure 2.1 (a) Fire-killed stand of lodgepole pine in western Montana. Natural regeneration of pines has started, and the fire eradicated the serious dwarf-mistletoe infestation that caused the witches brooms in the pines. (*Photograph by U.S. Forest Service.*) (b) Stand of loblolly pine in east Texas destroyed by tornado; some understory hardwoods remain and will dominate the site unless a fire in the dense fuels kills their roots. (*Photo by Texas Forestry Service.*)

more likely to kill small plants than large and (b) windstorms or pests that kill more large than small trees and thus "kill from the top down."

New trees can be recruited from many sources. Among these are seeds, nursery-grown plants, vegetative sprouts, and various forms of stored advance growth that have previously started beneath old stands. The sources and processes of regeneration are covered in much more detail later in this book.

Figure 2.1   (continued)

Cohorts and Age Classes

Regenerative disturbances, whether naturally or artificially induced, determine when new trees appear or start active development on any given unit of ground area. Each aggregation of trees that starts as a result of a single disturbance is a single **cohort.** If the range of ages of trees within the cohort is very narrow, the new aggregation is regarded as a single **age class** which is also **even-aged**. An age class may be a single cohort, but a cohort is not necessarily an age class.

For purposes of planning for cuttings and forecasting the future growth and yield of stands, it is necessary to ascribe ages to stands or components of stands that have arisen at different times in the past. This is no problem if the trees all germinated or were planted during the same year because they are clearly of the same even-aged class.

Quandaries develop when the effects or characteristics of the disturbances and the sources of the regenerating trees are so variable that the true ages vary widely. Confusion can be reduced by recognizing the difference between **chronological age**, which is the true age of the plant, and **effective age**, which is the number of years since the trees were free to start rapid growth.

A cohort has an effective age even if the chronological age varies widely. It may include trees that germinated or were planted in a single year, those that sprouted from stumps or roots that were really hundreds of years old, advance regeneration of many different heights that had accumulated over many decades, or new seedlings that slowly appear for several decades after a severe disturbance. In such cases, the effective age of the whole is best dated arbitrarily from the time of the regenerative disturbance. This does not mean that one may blithely adopt any assumption that the trees of the cohort are all the same because they were all put in the same pigeonhole.

In this book, the term *cohort* will not be used except in cases in which chronological and effective might often differ. Terms such as *age class*, *even-aged*, and *uneven-aged* will be used not only where the range of chronological age within a class is very small but also where it simplifies discussion to refer to a cohort as an age class. The most notable examples of the latter exception involve long-established distinctions between "even-aged" and "uneven-aged" stands that are used in developing management for sustained yield (as in Chapters 14 and 17).

Differences in the timing of regenerative events create various spatial patterns of age classes or cohorts. The area occupied by a given cohort can be of any size, provided that it is large enough that some new trees can continue to grow in height without being arrested by expansion of the crowns of older adjacent trees. Only those truly regenerative events that leave new or small trees free to grow really affect the arrangement of age classes. Intermediate cuttings such as thinnings do not leave new trees free to grow and thus have no effect on age-class arrangement.

There are three general types of age-class structure within stands: even-aged, stands with two age classes, and uneven-aged. They are most easily distinguished in pure stands, so these will be considered first, and the complexities of mixed stands will be taken up later. In an **even-aged** or **single-cohort stand** (Fig. 2.2), all trees are the same age or at least of the same cohort. An **uneven-aged stand** or **multiple-cohort stand** contains at least three age classes intermingled intimately on the same area. **Double-cohort stands** or **stands with two age classes** represent an intermediate category in which the presence of both cohorts may be temporary or continuous. All gradations of age distribution may be found in nature or created by cuttings designed to make way for new age classes or cohorts.

For some management purposes, a distinction is made between balanced and irregular uneven-aged stands. A **balanced uneven-aged stand** consists of three or more different age classes (or cohorts), each of which occupies an approximately equal area. The age classes are also spaced at uniform intervals all the way from newly established reproduction to trees near rotation age. Such stands, once created, may function as self-contained, sustained yield units. **Irregular uneven-aged stands** do not contain all the age classes necessary to ensure that trees will arrive at rotation age at short intervals indefinitely. Uneven-aged virgin stands and stands that have been partially cut without plan are almost always irregular in age distribution. In fact, irregular uneven-aged stands are common and may be highly desirable, as long as they are recognized and treated for what they are.

Identification of Age Classes

The profile of the top of a stand is a good criterion of age distribution because trees of the same age grow in height at roughly the same rate, provided site conditions are uniform. An even-aged stand tends to be almost smooth on top. An uneven-aged stand is distinctly irregular in height; the greater the number of age classes or cohorts, the more uneven the canopy.

There are two exceptional kinds of cases in which stands with more than one cohort can become rather smooth on top: (1) in very old stands, all of the trees, even those of very different age classes, may have culminated in height growth at a common level; (2) in some cases, isolated older trees that remain after cutting or other disturbance may have

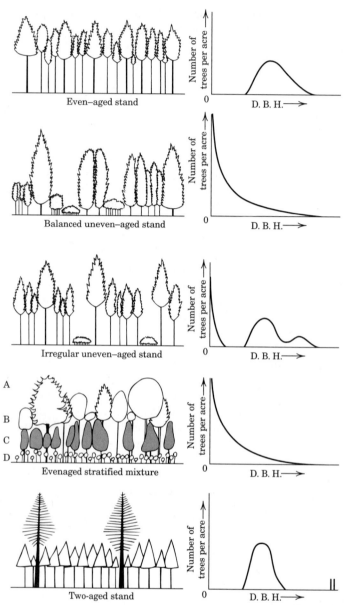

Figure 2.2   Typical examples of five different kinds of stand structure show the appearance of stands in vertical cross section and corresponding graphs of diameter distribution in terms of numbers of trees per unit of area. The trees of the first three stands are all of the same species. The fourth stand consists of several species, but all of the same age, and the fifth stand has trees of two different age classes.

decelerated in height growth sufficiently that more numerous younger trees around them catch up, and both age classes continue growing slowly in one smooth-topped stand.

Although it might seem that fat trees are always older than thin ones, diameter is not a very good criterion of age and must be used as such with caution. The diameter growth of trees is much more variable than that in height. Therefore, the trees in an even-aged stand are not as uniform in diameter as they are in height. If one plots the number of trees in each diameter class over diameter for a given *pure* even-aged stand, the distribution approximates the normal, bell-shaped curve (Fig. 2.2). The continuing loss of small trees from competition accounts for the typically abrupt slope of the left-hand side of the curve. It should be borne in mind that even-aged or single-cohort stands typically have a wide range of diameter classes; the age-class structure of a stand cannot be determined merely from the *range* of diameter classes present.

Uneven-aged stands are composed essentially of small even-aged groups of different ages. The distribution of diameters within each group also fits a bell-shaped curve, provided that the group consists of only one species or species that grow at the same rate in height and diameter. However, as each little even-aged group grows older, competition reduces its number of trees, rapidly at first and more slowly later on; the point may even be reached where only one tree remains from 100 or more. Therefore, if each age class occupies the same area, the composite diameter distribution curve for a balanced uneven-aged stand (Fig. 2.2) follows an asymptotic relationship commonly referred to as "reverse-J-shaped," or simply "J-shaped."

If the age classes or cohorts of an irregular uneven-aged stand differ widely in age, they are revealed as humps on the diameter distribution curve. The diameter distribution of each even-aged component broadens with age and will also be modified if the age class is composed of different species that grow at varying rates.

The most accurate assessment of the age-class structure of a stand comes from actual counts of annual rings. It is seldom reliable to depend on the criteria illustrated in Fig. 2.2 until direct age determinations have been made in representative stands typical of a locality. When such counts are made, consideration should be given to the fact that many species start as suppressed advance regeneration beneath older trees. In such instances, the effective age (i.e., the period since the trees were released) is more important than the chronological age. In other words, any core of fine growth rings around the pith is best discounted in assigning a tree to its proper cohort. With species in which this phenomenon is common, the number of annual rings at breast height is a good approximation of effective age if there is no tight core of rings at that level.

Differences in age distribution are most easily recognized in pure stands and in mixed stands composed of species with rates of height growth so nearly identical that the trees of a single cohort are aggregated into a single stratum in the crown canopy. However, even-aged mixtures of tree species usually segregate into different canopy strata and exist as **stratified mixtures** (Fig. 2.2) in which species of differing ecological status occupying different strata. The structure and development of these are considered later in this chapter.

Stages of Stand Development

A given aggregation of trees of a single age class or cohort proceeds from birth to death through a sequence of developmental steps (Oliver and Larson, 1996). These must be recognized if understanding of stand dynamics is to be used to achieve management objectives by imitating, guiding, or altering natural processes in silvicultural treatments.

The first stage in stand development (Fig. 2.3) is called the **stand initiation stage.** After a lethal disturbance has created a unit of vacant growing space, the trees that become established in it (or preexisting smaller ones that expand into it) do not fully occupy the space. Until they do so, there is opportunity for additional plants to fill the empty spaces. Often the plants that fill the newly vacant spaces are herbaceous annuals or other short-lived species that may come and go quickly.

Figure 2.3 The same mixed stand at advancing stages of development starting at the top with the stand initiation stage, then proceeding through the stem exclusion stage to the beginning of the understory reinitiation stage. Numbers and crown shapes identify the size of different species in the stand. Species 1 is a short-lived pioneer; 2 is a fast-growing emergent; 3 is an emergent with delayed ascent to top of canopy; 4 is another species with slow initial development that reaches the main canopy; 5 is a species with initial rapid development that is overtaken by species 4; and 6 is a very shade-tolerant species that usually remains in the lower strata. Figure 2.4 shows the same stand developing into old growth.

Ultimately, the above- and below-ground growing space is filled with plants, mostly woody perennials in the case of forests. The tree crowns become closed in the horizontal dimension, and some of their lowermost foliage starts to die because of shading by the upper foliage. This event starts the second stage of stand development, the **stem exclusion stage**. During this stage, the trees start to compete with each other; the more vigorous usurp the growing space of weaker ones that die, usually from lack of light or soil moisture, in a process called **suppression**. Establishment of additional regeneration of tree species is also prevented.

Unless some disturbance wipes out the stand and starts a new stand initiation stage, an aging stand will gradually enter the **understory reinitiation stage**. In this stage, scattered trees that have previously been successful in competition with other trees begin to be lost to pests, other damaging agencies, or cutting operations, and their crowns do not fully close again. The small vacancies thus created allow the establishment of new plants beneath the old stands. These are often advance regeneration of shade-tolerant species.

Unless something happens to replace most of the stand, the gradual process of death of overstory trees and replacement by younger age classes depicted in Fig. 2.4 leads gradually into an **old-growth stage.** This stage commences when a majority of the original trees are gone and one or more of the new age classes or cohorts compose parts of the top canopy. It may continue on past the death of the last original trees. The number of different age classes increases, although it would be most unlikely that any balanced uneven-aged stands would come into existence without cutting deliberately done to create such an outcome. The stands would have many dead trees both standing and fallen. Under purely natural conditions, production of wood and other organic matter would tend to be balanced by losses to death and decay. The total number of species of plants may increase. Stratified mixtures typically develop, and foliage often extends from the ground to the top of the stand at least in some parts of the stands.

Pure, Even-aged, Single-Canopied Stands

Many ideas about silviculture are dominated by the view that all stands are or should be perfectly even-aged and of only one species. Such stands are indeed very common. They typically arise in nature after severe disturbances such as hot fires or, artificially, from programs of clearcutting and planting. Usually, the species composition remains pure only if soil moisture or some other site factor is so restrictive that only one species can endure. The uppermost sketch of Fig. 2.2 depicts this kind of simple single-canopied stand. The two sketches beneath show uneven-aged stands composed of the same kind of pure, even-aged aggregations of trees; each constituent unit of a pure uneven-aged stand goes through essentially the same developmental stages as a pure even-aged stand of the same species.

During the stem exclusion stage, the trees of pure stands compete fiercely with each other, mainly because they all have crowns in the same stratum. The typical pure stand starts life with a relatively large number of small trees, usually thousands or tens of thousands per acre. The number of trees decreases as they grow larger, at first rapidly but more slowly with each passing decade. By the time the understory reinitiation stage starts, this number has been reduced to a few hundred trees per acre or to even less than one hundred.

This continual diminution in numbers is the result of competition and rigorous natural selection, and is the expression of one of the most fundamental biological laws of silviculture. Those trees that are most vigorous or best adapted to the environment are most

Figure 2.4   Continuation of the stand-development sequence of Figure 2.3, starting late in understory reinitiation stage and extending into the old growth stage. The five different species have the same numbers as in Figure 2.3. Species 1 was too intolerant to become re-established in the understory.

likely to survive the intense competition for light, moisture, and nutrients. Growth in height is the most critical factor in competition, although those trees that increase most rapidly in height are almost invariably the largest in all dimensions, especially in size of crown. As the weaker trees are crowded by their taller associates, their crowns become increasingly misshapen and restricted in size. Unless freed by random accidents or deliberate

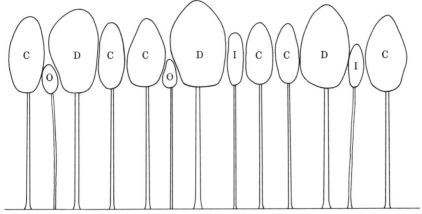

Figure 2.5 The relative positions of trees in different crown classes in an even-aged, pure stand. The letters D, C, I, and O denote dominant, codominant, intermediate, and overtopped crown classes, respectively.

thinning, such trees gradually become overtopped and ultimately die. In this constant attrition, the weaker members of an age class are progressively submerged and the strongest forge ahead. Very few trees ever recover a leading position after they have fallen behind in the race for the sky. This process by which the rich get richer and the poor get poorer is known as **differentiation into crown classes**.

Four crown classes are generally recognized, as illustrated in Fig. 2.5; the process of differentiation is depicted in Fig. 2.6. This classification of tree crowns is named the Kraft Crown Classification after the nineteenth-century German forester who devised it. *It ap-*

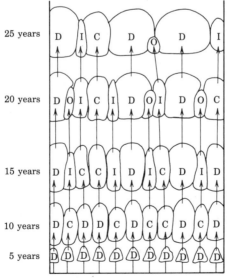

Figure 2.6 The process of differentiation into crown classes as a result of competition in a pure, single-canopied, even-aged stand, showing how some trees that were initially dominants lose in the race for the sky.

plies only to pure stands or stands composed of different species with identical regimes of height growth. The four classes are as follows:

> **Dominant:** Trees with crowns extending above the general level of the crown cover and receiving full light from above and partly from the sides; larger than the average trees in the stands and with crowns well developed but possibly somewhat crowded on the sides.
>
> **Codominant:** Trees with crowns forming the general level of the crown cover and receiving full light from above but comparatively little from the sides; usually with medium-sized crowns more or less crowded on the sides.
>
> **Intermediate:** Trees shorter than those in the two preceding classes but with crowns extending into the crown cover formed by codominant and dominant trees; receiving a little direct light from above but none from the sides; usually with small crowns considerably crowded on the sides.
>
> **Overtopped:** Trees with crowns entirely below the general level of the crown cover, receiving no direct light either from above or from the sides. Synonym: "suppressed."

This qualitative classification is simple and can be applied without measuring tree heights or crown widths. The intermediate and overtopped classes are well defined and easily distinguished from the two superior categories. However, it is difficult to draw a sharp distinction between the dominant and codominant classes. It would be desirable if the codominant class, which usually includes the majority of the main canopy trees, could be objectively subdivided into vigorous and less vigorous categories. Anything growing in a stratum definitely below the main crown canopy is better considered part of an understory rather than as part of the overtopped crown class. The development of pure, even-aged stands is presented in more detail in the chapters on thinning, and the use of clearcutting in their establishment is described in Chapter 12.)

Double-Cohort Stands

It is possible to have stands in which a new cohort or age class starts beneath an old one that is entirely or partially eliminated, allowing the new one to continue its development (Fig. 2.2). Natural fires may create such stands. They can be of pure or mixed species composition. Chapter 12 describes the use of shelterwood and seed-tree cutting methods in establishing these kinds of stands. If the stands are pure, each new cohort develops with the same differentiation into crown classes (described in the previous section).

Pure, Uneven-aged or Multicohort Stands

When small gaps are created in stands by cutting or destructive events, new cohorts may start to develop in the small openings thus created. The creation and development of such stands, as discussed in Chapter 15, is easiest to understand if they are introduced as pure stands without the complications found in mixtures of species. The development processes of the little pure groups of trees that constitute pure uneven-aged stands (Fig. 2.2) are essentially the same as those of pure even-aged stands except where older groups impede the younger ones at their interfaces. Natural stands of this kind most commonly exist in habitats where soil moisture deficiencies or occasional light fires allow only one tree spe-

Figure 2.7   Multi-cohort stand of ponderosa pine in southern Oregon after a group of mature trees has been removed to make a vacancy for establishment of re-generation. *(Photograph by U.S. Forest Service.)*

cies to grow. Many ponderosa pine forests of the western interior epitomize this condition (Fig. 2.7). Most multicohort stands are actually mixtures of species.

The management of uneven-aged stands is complicated even if they are pure, especially if attempts are being made to mold them into self-contained sustained-yield units. These efforts usually involve manipulation of diameter classes and efforts to create reverse-J-shaped diameter distributions such as those shown in Fig. 2.2. The management of such stands is much less complicated where the trees and groups of trees are thinned to enhance growth and harvested when mature without attempting to change age-class structure. These treatments involve what is aptly called selection cutting and are considered in much more detail in Chapter 15.

Mixed Single-Canopied Stands

It is uncommon, but not impossible, for two tree species of the same age to grow in height at the same rate for long periods. If they actually do, they can be thought of, and managed, in the context of the single-canopied structure almost as if they were pure stands (Guldin and Lorimer, 1985). Ordinarily, however, one species tends to suppress its associates; very small differences in height growth become greater as the leaders forge ahead and the laggards suffer. This can happen in mixtures of loblolly and shortleaf pine; loblolly gets ahead unless something happens to allow the shortleaf to take advantage of its superior adaptability to water shortages. Mixtures of beech, spruce, and fir, all tolerant species, in Central Europe must be thinned constantly to keep the beech or spruce from exterminating the fir. In fact, trees that stray out of the single canopy stratum are usually either eliminated or promoted into it by releasing operations.

Yet another concept about stand development patterns accommodates the idea of single-canopied structure to the tendency of different species to grow at different rates. This pattern is one in which mixed stands are mosaics composed of little pure stands, each arising from a small, pure patch of young trees. Because each patch develops independently (except at the edges), it is possible for each species to grow at its own rate. A process can be postulated in which each patch might be reduced to a single tree in a mixture that became stratified in some late stage of development. Patchwise mixtures of this kind are most likely to arise from deliberate planting. Patchy variations in soil or microenvironmental conditions may cause them to start from natural seeding, but the randomness imparted by seed dispersal and other factors usually causes species to be intermingled. Clumps of natural root-suckers are commonly pure, but they are not necessarily mixed with pure patches of other species.

Single-Cohort Stratified Mixtures

Single-cohort stratified mixtures (Fig. 2.2) develop when different species represented by advance regeneration, sprouts, new seedlings, or combinations of the three start off together upon release by some major disturbance such as a windstorm, insect outbreak, or heavy cutting. Although these mixtures are usually thought of as originating from natural regeneration, they can also start with the planting of mixtures of species.

The developmental processes of stratified mixtures of species are different from those of simple pure even-aged stands or cohorts. The differentiation of trees according to height is into different horizontal strata or stories, one above the other, with one species or a group of similar species in each stratum. The differentiation is not simply into crown classes within a single canopy stratum as it is in pure stands.

The sorting into strata begins in the stand initiation stage strata (Fig. 2.8) and becomes most pronounced in the stem exclusion stage. Competition is most intense among trees within a given species and stratum, but the lower strata are not excluded by the upper ones. The species of the lower strata are adapted to survive there without participating in the race for the sky that characterizes pure, single-canopied stands. The species that rose to the top of the total crown canopy during previous forest generations generally do so again, even though this development may not take place until the later stem exclusion stage is well underway.

The species groups of each stratum differ from the other groups in rate of height growth, tolerance of shade, rooting depth, and similar ecological characteristics (Fig. 2.9). It is rare that any two associated species grow in height at precisely the same rate throughout life, even when they are not actually competing with each other. If they are intimately intermingled, the species with the most rapid rate of juvenile growth in height will gain ascendancy over the slower growing species, which will lag even farther behind because of deficiency of light.

In the simplest kind of stratified mixture, each species ultimately tends to occupy a different stratum of the total crown canopy. In general, there will be as many strata as there are groups of species that differ from one another in height growth and tolerance. The stratification is not always perfect or readily apparent. Even when the stratification is well differentiated, a few individuals of species that go with one stratum may by chance have grown upward into a higher stratum or have been left behind in a lower one. Fur-

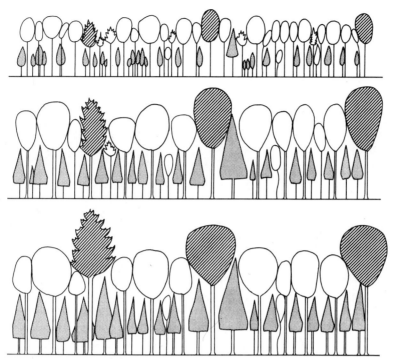

Figure 2.8 Stages in the natural development of an untreated stratified mixture in an even-aged stand of the eastern hemlock–hardwood–white pine type. The upper sketch shows the stand at 40 years with the hemlock (gray crowns) in the lower stratum beneath an undifferentiated upper stratum. By the seventieth year (middle sketch) the emergents (hatched crowns) have ascended above the rest of the main canopy, except for the white pine, which has only started to emerge. The lower sketch shows the stand as it would look after 120 years with the ultimate degree of stratification developed.

thermore, the observer standing beneath a stratified mixture will have to look closely and exercise some imagination to perceive the different strata.

The different strata can be designated A, B, and C downward, a terminology first applied in the forests of the moist tropics, where the concept of the stratified mixture originated (Fig. 2.2). Stratum A may be continuous but is more often composed of scattered and isolated **emergents** that either grow faster or continue growing longer than their associates. Sometimes the emergents are simply called that and Stratum A is regarded as the uppermost fully closed canopy layer. Each stratum can also be designated by the name of the most characteristic species within it as when a pure Douglas-fir stratum is above one of pure western hemlock or an oak stratum is over a "sugar maple stratum" in which that species predominates over beech.

An understanding of the structure and development of stratified mixtures provides a way of dealing with many kinds of complex stands. These are usually found where soil moisture and other site factors are so favorable that many species can grow, although they can also occur on less favorable sites. On sites that do not restrict the entry of many species, it is very difficult to maintain pure, single-canopied stands. Frequently, a single species is

Figure 2.9　A single-cohort mixture of northern hardwoods, about 90 years old, in the Adirondack Mountains of New York. The emergents are the white pines at the left and some white ashes in the middle (which are still nearly leafless in this spring picture). Sugar maples and yellow birches form the main canopy stratum with American beeches in the understory stratum. (*Photograph by Yale University School of Forestry and Environmental Studies.*)

not capable of fully occupying such sites and stands turn into stratified mixtures unless costly treatments are applied to reduce the invaders. It is sometimes useful to encourage some productive lower-stratum species and to use them to exclude the unwanted.

If the soil factors are not seriously limiting, the vegetation collectively intercepts more of the photosynthetically functional light than could a pure stand of some shade-intolerant upper-stratum species (Kelty, 1989). The vertically distributed foliage in a stratified mixture represents a sequence of sun- and shade-leaves that are more fully adapted to do this than the rather similar array of leaves within a single species. However, in some habitats the lower stratum species can be undesirable or not permanently adapted to the site and so a single species would be better adapted. For example, the exclusion of fire on some dry, fire-prone sites has sometimes allowed establishment of lower-stratum species that cause excessive buildup of forest-fire fuels, rob the upper stratum of water, or die of drought themselves.

Plants follow many different strategies for exposing their leaves to solar radiation and claiming growing space. Most annual plants are designed to expend all their carbohydrates on roots and tops that fill a small amount of growing space quickly, with no provision made for woody stems and roots and none for the future except for seeds. Perennial herbaceous plants tend to develop enduring control of small amounts of the soil space but avoid investing substance in the woody stems needed for races toward the sky. Trees and other woody perennials have a wide variety of strategies. Shrubs invest so little in building supportive stems that the sizes of their crowns are limited. The stems of woody vines are designed almost entirely for conduction of water and dissolved substances.

Most pioneer tree species have weak stems that are designed to expose large amounts of foliage to sunlight rapidly. Usually, they either collapse of their own weight when young or are overtaken by species with stronger stems that start height growth more slowly. However, some quick starters, such as yellow-poplar and certain hard pines, continue height growth for long periods and dominate the top stratum for whole rotations. It should be noted that every tree species has a limiting total height, usually greater on good sites than on poor, which should guide decisions about how to handle mixtures of species.

Many ideas about silviculture are based on the view that all useful species of trees grow quickly and steadily in height for whole rotations. This neglects the fact that many are not so adapted. A very large number grow slowly, or even not at all, in height until they have built root systems adequate to supply water to the crowns. When they have done that they may initiate rapid height growth immediately upon release from trees that have been taller. However, if not released, they may remain stunted but retain potential for rapid growth for many years or even for decades. Occasionally, there is the further complication that these kinds of shade-tolerant species may grow in height for a time and then again lapse into quasi-dormancy after being overtaken by faster-growing species.

As a result of these phenomena, there can be remarkable reversals in the position of different species in the different strata. Northern red oak, for example, grows at a steady, moderate rate that cannot be accelerated; after several decades, associates such as red maple and black birch that had previously gone ahead slow down and lapse into the lower strata (Oliver, 1978). Some species, such as the white pines and spruces (Fajvan and Seymour, 1993), may linger below the top of the canopy and ultimately emerge above it simply because they survive or continue growing in height longer than their associates.

Stratified mixtures were first recognized by Richards (1952) in the wet evergreen and moist deciduous tropical forests, which are practically incomprehensible without this means of analyzing their structure. The fact that stratified mixtures can be even-aged has been learned in the temperate zone where trees have annual rings (Kelty, Larson, and Oliver, 1992). From these it has been possible to reconstruct the patterns of height growth that lead to whatever structural arrangement exists at any developmental stage (Oliver, 1981). The myriad of such patterns that must exist in moist tropical forests is almost unknowable until ways are invented to determine ages of the trees. In one case, Terborgh and Petren (1991) were able to reconstruct the development of stratified mixtures in a datable series of even-aged stands that became established annually as new soil was laid down while a tropical river shifted its course sideways each year.

Efforts to apply the Kraft Classification of crown dominance to stratified mixtures usually cause confusion about the past and future development of individual trees without adding much that is useful to tree description. However, trees of the *same* species do compete strongly with each other (Kittredge, 1988), and sometimes the different pure-stand crown classes can be discerned *within* a stratum.

Although the lowermost strata of shrubs or herbaceous vegetation are usually not counted among the arborescent strata, they may play an important role in the total structure and function of the forest ecosystem. Subordinate strata are very common even beneath stands of single tree species that are casually thought of as absolutely pure.

In mixed stands, diameters tell almost nothing about ages when different species are compared. A slender tree of a C-stratum species may be as old or even older than a fat emergent of another species above it. Large gaps in diameter distribution *within* a species normally denote truly different age classes. It is common for stratified mixtures composed of a single cohort to have varying degrees of semblance of the ''J-shaped'' diameter distribution that would be a plausible indicator of the balanced, *uneven-aged* structure if the trees were all of the same species. In such cases, the different diameter classes generally represent species with different schedules of height growth rather than different age classes.

The species of even-aged stratified mixtures are ordinarily arranged with intolerants in the upper strata, with species of increasing tolerance in each successively lower stratum (Fig. 2.9). There may be more than one species in a stratum. The order of vertical arrangement of species is not necessarily the same at all stages or ages. Not only do some species drop out, but also some may even exchange their positions. Some intolerants race quickly ahead but soon slow down and die; other intolerants grow rapidly and steadily to endure to old ages. The more shade-tolerant species almost always start slowly and accelerate later. The very shade-tolerant ones may endure as practically dormant seedlings or saplings and shoot for the sky only when some tree above is eliminated. Some species grow rather rapidly at first and then slow down so that they lapse into the understory, while another that had lagged forges ahead. In other words, the development is not like that of a pure even-aged stand in which laggards in the race for the sky are usually doomed.

Collectively, the various strata often fill the growing space so long that the stem exclusion stage may extend for long periods of time. Even if one of the upper layers is completely eliminated a lower one will usually take over any vacant space. If there are no major lethal disturbances, vacancies gradually develop in the lower strata; the understory reinitiation stage starts and ultimately leads into a very complex, uneven-aged, old-growth stage.

Mixed, Multicohort Stands

Mixed stands with a history of partially effective or patchy lethal disturbances develop into complex mixtures of species, fragments of stratified mixtures, age classes or cohorts (Fajvan and Seymour, 1993). All stages of stand development are likely to be going on simultaneously in some part of the stand. If there are truly different age classes or cohorts, there will be real variations in the height of the top of the main canopy in different parts of the stand. If shade-intolerant species normally found only in the upper strata exist in very different diameter classes within a stand, the stand is very likely to be composed of more than one cohort.

Where site conditions are conducive to mixed stands, this chaotic kind of structure is characteristic of undisturbed old-growth stands. The same is true of **"high-graded"** **stands** from which the best trees have been cut, except that the best species and largest trees may have been eliminated.

The intermingling of trees of very different age often obscures and complicates any processes by which different species sort themselves into different strata. Older individuals of species normally found in some lower stratum often reach the uppermost levels and the stands become irregularly uneven-aged. *The best silvicultural solution for managing such stands lies in diagnosing the processes taking place in each separate part and treating it accordingly.* Even though they may exhibit some semblance of a J-shaped curve of diameter distribution, they almost never approach the balanced all-aged condition. That condition is actually artificial; it is created by deliberate silvicultural action and not by random events of nature.

Relationship to Other Interpretations of Vegetational Development

Vegetation changes over time by many different patterns, and various attempts have been made to systematize them (Shugart, 1984; MacIntosh, 1980; Cattelino et al.,1979). The concept of natural succession formulated by Clements (1936) once dominated American ecology. An oversimplified version of this concept holds that pioneer vegetation of an

Figure 2.10   Successional development in which an overmature stand of western larch in western Montana is dying and being replaced by subalpine fir and a few Engelmann spruce. *(Photograph by U. S. Forest Service.)*

initial Stage 1 colonizes an area from which all preexisting vegetation has been eliminated. Stage 1 soon dies and is replaced by a Stage 2 composed of other species that start under Stage 1 and are relatively shade-tolerant (Fig. 2.10). Ultimately, Stage 2 is similarly replaced by Stage 3, and there may be additional stages leading to a stable, endlessly self-replacing stage called the **climax**. The originator of the concept made it far more complicated and sophisticated, but many of the disciples have oversimplified it.

Climax vegetation has been regarded by some as an ultimately perfect state of nature in which all organisms are represented and all physical and biotic factors are in perpetual balance. It was once postulated that this perfect condition required freedom from disturbance, although it has been conceded more recently that perpetuation of the essentially uneven-aged state depended on the occasional creation of small gaps in the growing space. In fact, the terms **gap** and **patch dynamics** have been coined to describe the study of patterns of establishment and subsequent development of vegetation in all vacancies of *any* size in the growing space (Pickett and White, 1985). In this sense, the term *forest stand dynamics*, as used in this book, is a kind of gap or patch dynamics, except that gaps or patches filled by single cohorts are called stands.

The Clementsian natural succession model does describe at least the early stages of development after most vegetation has been killed by severe disturbances, but it does not cover other sequences very well. According to the stand dynamics principles proposed by Oliver and Larson (1996), the stages of stand initiation, stem exclusion, and understory reinitiation represent early stages of the natural succession just described, and the old-growth stage covers all of the subsequent steps leading to the climax stage.

Enthusiasm for the simple Clementsian concept of natural succession and the climax stage as the sole pattern of vegetational development has declined mainly because of two things. First, it was observed that fires and other major disturbances were so common, even in nature, that in many localities the theoretical climax vegetation never had time to develop. Second, it became apparent that more complex ideas were necessary to account for the behavior of intimate mixtures of species and the effects of lethal disturbances that killed only some of the plants growing in a unit of space.

The concepts of stratified mixtures and the principles of stand dynamics covered earlier in this chapter represent attempts to deal with the effects of partial disturbances and the interaction of different species. Each of the individual strata of a single-cohort stratified mixture often represents one of the sequential stages envisioned in the Clementsian concept of succession, with the species of the earliest stage being those of the A Stratum.

Some ecologists have applied the term **initial floristics** to the simultaneous appearance of many different categories of plant species in what is here termed the stand initiation stage. These ecologists call Clementsian succession **relay floristics**. The members of a single cohort that develop into a stratified mixture arise through initial floristics. Each one of these terms describes one of the various ways in which vegetation develops; the two concepts are complementary rather than conflicting because each fits a different disturbance pattern. Relay floristics usually fits developments following disturbances such as hot fires that wipe out nearly all of the vegetation. Initial floristics is associated more with regeneration from sprouts and advance growth as well as new seedlings that start development following an initiating disturbance by wind or other agencies that kill stands from the top downward.

Another way of viewing stand development over time emphasizes the accumulation of biomass and chemical nutrients after a destructive regenerating disturbance (Bormann and Likens, 1981). The regeneration step is called **organization**; the buildup of biomass is referred to as **aggradation**; and the state in which the accumulation of biomass and nutrients comes into equilibrium with losses is the **steady state.** These stages are analogous to steps of stand initiation, stem exclusion, understory reinitiation, and old growth that relate to changes in tree populations.

There is no single universal pattern; in fact, there are more patterns than terms to describe them. In managing any category of stands, the forester should know as much as possible about their past and future development. It is well to be wary of preconceived ideas about standard patterns because there are more developmental sequences than are described in this book. Oliver and Larson (1996) cover the topic in detail.

The Ecosystem Concept

It was recognized long ago that both study and application of ecology suffered from excessive compartmentalization. The total flora of interacting forest plants is far more than just trees; the total biota of a place includes not only plants but all the orders of the animal kingdom that are present. An ecosystem, however, is more than just the living organisms. It also includes the nonliving physical and chemical factors that interact with the living organisms (Tansley, 1935).

In applying silvicultural treatments one is, in some degree, manipulating many sizes of ecosystems simultaneously (Perry, 1994; Reichle, 1981). At one extreme are the world cycles of carbon, oxygen, and water; the microenvironment around a pine seedling in the shade of a log is at another; and in between are the cycles of mineral nutrients and the combination of different kinds of forest stands on a hillside. **Forest ecosystem management** includes the design and application of silvicultural solutions that are based on analysis of all the ecological factors known to operate in the system involved. In other words, silviculture has always been ecosystem management, provided that it is conducted on the basis of such analysis. Ecosystem management is one of the keys to maintaining biodiversity for it requires consideration of the interaction and habitat requirements of all living organisms.

Silvicultural treatments achieve their results through deliberate manipulation of the forces represented by physical, chemical, and biological processes that alter ecosystems in somewhat the same manner that purely natural forces produce changes in ecosystems. In ecosystem management, it is necessary to consider the spatial arrangement of stands of differing ages and species composition. In other words, it is best if silvicultural planning and forest management are not restrained by boundaries of stands and ownerships. Insofar as possible, such planning should consider the landscape scale in which ecosystem boundaries are defined by watersheds, climate, topography, and the ranges of plant and animal species.

All good ideas survive to be overdone, and the ecosystem concept is no exception. It is too easily translated into the philosophically attractive concept that each biotic community is a superorganism in which each constituent species is indispensable and somehow depends on every other species. Although it can perhaps be said that every part of an ecosystem has some *effect*, even if minuscule, on every other part, each part does not *depend* on every other part. There are many important interactions between particular species, such as symbiosis and competition, predation and parasitism, or simply shading of one plant species by another. Although these interactions need to be recognized (and used) in silviculture, they do not mean that all parts are like essential cogs in a whole engine. The vast majority of species are adaptable to many different conditions and often move around independently of their associates. If two different species are dependent on each other, they are usually adapted to move or respond to change together.

Natural disturbances and subsequent development processes commonly lead to the development of particular combinations of species of trees, lesser plants, animals, and other forms of life on particular kinds of sites in a given climatic region. These are called **communities**. Some species within them are dependent on others. For example, pines depend on mycorrhizal species of fungi; herbivores, on the foliage; and bark beetles, on dead or dying trees. However, most of the trees and other organisms are not dependent on each other, and the weight of evidence is against the idea that they have, as is often claimed, lived in association and been dependent on each other for millions of years.

Evidence from pollen deposits shows that most tree species have moved around quite independently of each other since the continental glaciers started to shrink about 15,000 years ago. Most modern plant communities, including those in the tropics, are less than 8000 years old and have responded to climatic changes that have taken place even more recently (Davis, 1983; Delcourt and Delcourt, 1987; Hunter, Jacobson, and Webb, 1988).

There does, therefore, seem to be some entirely natural precedent for silvicultural changes in species composition of forest ecosystems. However, this does not mean that all changes, subtractions, or additions are safe or desirable. Changes should be made only in the light of the best knowledge available about relevant mutual relationships between species in natural forests of a locality. Foresters who deal with any managed forests should always watch for undesirable (and desirable) consequences of departures from more natural conditions and be ready to act on the knowledge.

In considering silvicultural manipulations of ecosystems, it should be recognized that the same degree of ''naturalness'' cannot be maintained in all forests. Not all silviculture should mimic the old-growth stage of development, even if there was enough freedom from disturbance to allow it. Just as there are variations in intensity of silvicultural practice, so should there be variations in ''naturalness.'' It is also necessary to recognize that most forests situated where silvicultural management is feasible are already significantly modified by human action (Whitney, 1994), so it may be virtually impossible to return to a pure state of nature.

Natural preserves should be extensive enough to maintain all native species and represent all natural habitats; it is not enough to confine them to forests, such as wilderness areas, that are merely difficult of access. Many stands from which wood is harvested can be expected to maintain most of the biodiversity of an area if appropriate attention is given to the relative dominance of species and the characteristics of silvicultural disturbances. Intensively managed forests, such as plantations, do not necessarily maintain a high degree of natural diversity; however, even these maintain most of the basic ecosystem equilibria, such as high biological productivity, uptake of carbon dioxide, retention of nutrients, control of erosion, and regulation of hydrologic processes.

Choice of Developmental Patterns

Silvicultural choices can be thought of as determining what kind of stand developmental process or stage of natural succession is most desirable in a given situation. In the Pacific Northwest, for example, the forester must often decide whether to perpetuate pure stands of Douglas-fir or allow them to be succeeded by stratified mixtures of Douglas-fir, western hemlock, and red-cedar. In the Lake Region, the choice may be between the pioneer aspen association and a climax stage such as the spruce–fir association. In the South, decisions must be made about whether to let old-field stands of loblolly pine revert to pine-hardwood mixtures. In almost every kind of forest, it must be recognized that some wildlife species may depend on stands that have some dead trees and other features of old-growth stands.

Several generalizations of wide, but not universal, application may be introduced at this point. In the first place, the most valuable commercial species tend to be relatively intolerant but comparatively long-lived trees representative of the early or intermediate stages in natural succession. Species such as pines, Pacific Coast Douglas-fir, yellow-poplar, and white ash definitely fall in this category. It is no coincidence that intolerant species are important commercially because they are the ones most likely to lose their lower branches through natural pruning. It is of significance that some of them are adapted to reproduce mostly after major disturbances that happen infrequently. If these are to survive from one major disturbance to another, they must be long-lived; as a result, they are likely to develop the economically desirable attributes of large size and resistance to decay or other relatively minor sources of damage.

Late successional forest types, characterized by species such as hemlock, true firs, and beech, are frequently composed of branchy trees that produce less valuable wood. Because of their shade tolerance, they can reproduce almost continuously; thus the ability of individuals to endure for long periods is not so crucial in the survival of the species. Many pioneer species have even less capacity for individual longevity. However, they usually exhibit good natural pruning, and the necessity that they grow rapidly to seed-bearing age is an economically desirable attribute, although they usually have weak wood of low density.

Natural succession proceeds most rapidly and vigorously on the better sites, that is, on soils that are both moist and well aerated. Here it is sometimes impossible to resist the invasion of additional species without expensive silvicultural treatments. Furthermore, good sites are hospitable to the growth of so many species that silvicultural treatment becomes complicated and difficult. These considerations often have the paradoxical effect of making silviculture most efficient on sites of intermediate quality where uncomplicated stands can be maintained without strenuous effort. In fact, on poor sites occasionally it may be virtually impossible for succession to proceed beyond an intermediate stage, which is sometimes referred to as a **physiographic climax**. For example, pure stands of jack or red pine occasionally represent valuable physiographic climaxes on certain dry, sandy soils in the Lake Region.

It has been claimed that late successional or old-growth types may be more resistant to, and more cheaply protected from, fire, insects, fungi, wind, and weather than earlier stages. In those cases where this advantage exists, it results more from the diversity of species and age classes than from age or position in the successional scale. Similar advantages sometimes prevail in mixed stands with a variety of age classes that are still typical of earlier successional stages.

It is often postulated that natural climax or old-growth communities are in a stable and favorable equilibrium with the physical and biological environment. Perfect stability and complete favorability do not exist, so one must think in terms of relative degrees of each quality. For example, the balance achieved by long-continued natural processes, operating more or less at random, is not necessarily more favorable to the trees than to the organisms that feed upon them. The more artificial dynamic equilibrium produced by prudent silviculture may be less stable but ought to be more favorable from the standpoint of the integrated effect of all socioeconomic factors. If the dynamic equilibrium created by treatment ultimately balances at some disastrous condition, the silviculture was hardly prudent.

The naturalistic doctrine of silviculture did not arise from any clearly demonstrated disadvantages of early or middle stages of natural forest succession. It developed largely from disappointments with attempts to create unnatural types, particularly with exotic species or native ones not adapted to the sites involved. In more recent times it has been advanced as a result of concern for wildlife diversity and need for perpetuation of natural areas. The most extreme manifestations of the naturalistic viewpoint are usually unwarranted extensions of otherwise sound observations.

BIBLIOGRAPHY

Attiwill, P. M. 1994. The disturbance of forest ecosystems: the ecological basis for conservative management. *FE&M* 63:247–300.

Barnes, B. V., Z. Xü, and S. Zhao. 1992. Forest ecosystems in an old-growth pine–mixed hardwood forest in the Changbai Shan Preserve in northeastern China. *CJFR* 22:144–160.

Bormann, F. H., and G. E. Likens. 1981. *Pattern and process in a forested ecosystem.* Springer-Verlag, New York. 253 pp.

Cattelino, P. S., I. R. Noble, R. O. Slayter, and S. R. Kessell. 1979. Predicting the multiple pathways of plant succession. *Environmental Management* 3(1): 41–50.

Clements, F. E. 1936. Nature and structure of the climax. *Jour. Ecol.* 24:252–284.

Cobb, D. F., K. L. O'Hara, and C. D. Oliver. 1993. Effects of variation in stand structure on development of mixed-species stands in eastern Washington. *CJFR* 23:545–552.

Davis, M. B. 1983. Quaternary history of deciduous forests in eastern North America. *Annals, Missouri Bot. Garden* 70:550–563.

Deal, R. L., C. D. Oliver, and B. T. Bormann. 1991. Reconstruction of mixed hemlock—spruce stands in coastal southeast Alaska. *CJFR* 21:643–654.

Delcourt, P. A. and H. R. Delcourt. 1987. *Long-term forest dynamics of the temperate zone: a case study of late-quaternary forests in eastern North America.* Springer-Verlag, New York. 439 pp.

Djailany, V. R. 1986. Natural regeneration of dipterocarp forests: the secondary forest types. In: Forest regeneration in Southeast Asia. *Biotrop Special Publ.* 25. Pp. 129–138.

Fajvan, M. A., and R. S. Seymour. 1993. Canopy stratification, age structure, and development of multicohort stands of eastern white pine, eastern hemlock, and red spruce. *CJFR* 23:1799–1809.

Glenn-Lewin, D. C., R. K. Peet, and T. T. Veblen. 1992. *Plant succession: theory and prediction.* Chapman & Hall, New York. 352 pp.

Guldin, J. W., and C. G. Lorimer. 1985. Crown differentiation in even-aged northern hardwood forests of the Great Lakes Region, USA. *FE&M* 10:65–86.

Kelty, M. J. 1989. Productivity of New England hemlock-hardwood stands as affected by species composition and canopy structure. *FE&M* 28:237–257.

Kelty, M. J., B. C. Larson, and C. D. Oliver (eds.). 1992. *The ecology and silviculture of mixed-species forests.* Kluwer Academic Press, Norwell, Mass. 287 pp.

Kittredge, D. B. 1988. The influence of species composition on the growth of individual red oaks in mixed stands in southern New England. *CJFR* 18:1550–1555.

Hunter, M. L, Jr., G. L. Jacobson, Jr., and T. Webb, III. 1988. Paleoecology and the coarse-filter approach to maintaining biodiversity. *Conservation Biology* 2:375–385.

Leopold, D. J., G. R. Parker, and W. T. Swank. 1985. Forest development after successive clearcuts in the southern Appalachians. *FE&M* 13:83–120.

Longpré, M.-H., Y. Bergeron, D. Paré, and M. Bélard. 1994. Effect of companion species on the growth of jack pine. *CJFR* 24:1846–1853.

McIntosh, R. P. 1978. *Phytosociology.* Academic, New York. 416 pp.

McIntosh, R. P. 1980. The relationship between succession and the recovery process in ecosystems. Pp. 11–62. In: J. Cairns (ed.), *The recovery process in damaged ecosystems.* Ann Arbor Science Publ., Ann Arbor, Michigan.

Means, J. E. (ed.) 1982. *Forest succession and stand development research in the Northwest.* Oreg. State Univ., For. Res. Lab., Corvallis. 171 pp.

O'Hara, K. L. 1995. Early height development and species stratification across five climax series in the eastern Washington Cascade Range. *New Forests* 9:53–60.

Oliver, C. D. 1978. Development of northern red oak in mixed species stands in central New England. *Yale Univ. Sch. For. & Env. Studies Bul.* 91. 63 pp.

Oliver, C. D. 1981. Forest development in North America following major disturbances. *Forest Ecol. and Mgmt.* 3:169–182.

Oliver, C. D., and B. C. Larson. 1996. *Forest stand dynamics.* Wiley, New York. 520 pp.

Perry, D. A. 1994. *Forest ecosystems.* Johns Hopkins University Press, Baltimore. 549 pp.

Pickett, S.T.A., and P. S. White (eds.). 1985. *The ecology of natural disturbance and patch dynamics.* Academic, New York. 472 pp.

Reichle, D. E. 1981. *Dynamic properties of forest ecosystems.* Cambridge, New York. 683 pp.

Richards, P. W. 1952. *The tropical rain forest, an ecological study.* Cambridge Univ. Press, Cambridge, UK. 450 pp.

Richards, P. W. 1992. *The tropical rain forest, an ecological study.* 2nd ed. Cambridge Univ. Press, New York. 600 pp.

Rowe, J. S., and B. V. Barnes. 1994. Geo-ecosystems and bio-ecosystems. *Bull. Ecol. Soc. Amer.* 75:40–41.

Runkle, J. R. 1982. Patterns of disturbance in some old-growth mesic forests in eastern North America. *Ecology* 63:1533–1546.

Shugart, N. H. 1984. *A theory of forest dynamics, the ecological implications of forest succession models.* Springer-Verlag, New York. 304 pp.

Spurr, S. H., and B. V. Barnes. 1980. *Forest ecology.* 3rd ed. Wiley, New York. 687 pp.

Sutton, S. L., T. C. Whitmore, and A. C. Chadwick (eds.). 1983. *Tropical rain forest, ecology and management.* Oxford, Boston. 498 pp.

Tansley, A. G. 1935. The use and abuse of vegetational concepts and terms. *Ecology* 16:284–307.

Terborgh, J., and K. Petren. 1991. Development of habitat structure through succession in Amazonian floodplain forest. In: E. D. McCoy and H. R. Mushinsky (eds.), *The structure of habitats.* Chapman & Hall, New York.

Waring, R. H., and W. H. Schlesinger. 1985. *Forest ecosystems.* Academic Press, Orlando, Fla. 340 pp.

West, D. C., H. H. Shugart, and D. B. Botkin (eds.). 1981. *Forest succession: concepts and application.* Springer, New York. 517 pp.

Whitney, G. G. 1994. *From coastal wilderness to fruited plain: a history of environmental change in temperate North America.* Cambridge, New York. 451 pp.

PART 2

TENDING AND INTERMEDIATE CUTTING

CHAPTER *3*

THE RESPONSE OF INDIVIDUAL TREES TO THINNING AND PRUNING

Whether one grows trees by the thousands in forest stands or as shade trees on city streets, one must know how they develop as individuals. Among the attributes of trees that can be regulated are the size, shape, and structure of their stems as well as their branching characteristics. This chapter discusses the use of thinning to govern the amount of growing space allocated to trees. If a healthy tree is given more growing space by eliminating some of its neighbors (thinning), the tree grows faster, but it is more tapered and branchy. If left crowded from the sides, the lower branches will die and, although the tree will grow more slowly in diameter, the stem will be less tapered. The same effects can be produced artificially by pruning off living branches.

Trees, like all plants, capture sunlight energy through photosynthesis by producing carbohydrate. Some of this photosynthate is used for respiration to meet the basic metabolic needs of the living cells. The remaining energy is used for growth. The silviculturist tries to influence the size and shape of each tree to achieve the desired stand structure and value. The amount of growing space available to a tree will limit the amount of growth, and the genetic makeup of a tree will dictate how the growth is allocated to different parts. A most difficult tradeoff faced by the silviculturist is that rapid growth is almost always accompanied by large lower branches whose knots reduce the timber value of the tree. However, this contradiction in goals can sometimes be mitigated by pruning the lower branches.

RESPONSE OF TREES TO INCREASED GROWING SPACE

The amount of carbohydrate produced by a tree depends mainly on the size of the crown or leaf surface and the ability of the roots to supply the foliage. When a tree is released by cutting a competing tree, any prompt acceleration of growth is largely from an increase in water and nutrients supplied by the roots. The amount of foliage does not increase until there has been time for the crown to enlarge, although this delayed effect is ultimately the most important. Not all units of leaf surface are equally efficient in photosynthesis. Leaves that are severely exposed to sun and wind produce somewhat less than slightly protected ones; heavily shaded leaves do little more than supply themselves. The part of the crown above the point of horizontal crown closure, especially the upper middle portion, produces much more than anything below and more than the foliage at the very top of the tree.

The roots extend horizontally more widely and swiftly than the crown. Since they do not have to support themselves like aerial branches, their extension is not limited by structural necessities. They may extend throughout the soil even if this means intermingling with or going around the root systems of other trees. The root systems of healthy trees are much wider than the crowns. Many species, but not all, form intraspecific root grafts, so it is possible for a tree to incorporate part of the root system of an adjacent cut tree. In fact, in some single-species stands, the trees that are individuals above ground share some semblance of a common root system. There is some evidence from the tropics that closely related species can develop functional interspecific root grafts as well.

Priorities in Allocation of Carbohydrates

After many generations of natural selection, trees allocate carbohydrate in ways that will generally increase the tree's ability to survive. Because some functions are more crucial than others, a priority schedule for allocation is apparent.

Respiration is the highest priority because the living tissues would perish immediately without it. The greater the amount of crown surface per unit of living tissue, the greater will be the proportion of carbohydrate left for formation of new tissues. Very important is the indispensable renewal of foliage and fine roots, and this growth is the second highest priority. The extension of the crown and root system depends mainly on the room for expansion that thinning makes available or can be captured from less vigorous competitors. Part of this process is height growth, which is so important to the survival of competing trees that it commands the next highest priority. Being shaded by neighboring trees greatly lowers photosynthesis. Height growth varies surprisingly little with tree vigor unless vigor is so low that very little carbohydrate is left after respiration demands have been met. However, height growth differs greatly among species and is strongly affected by variation in site quality factors that control overall productivity.

Another vital function that must be fulfilled is conduction of materials between crown and root. Phloem must be renewed annually, and so ordinarily must xylem (wood). However, most trees lay down far more conductive tissue than is absolutely necessary. This can be verified by observing how small a "bridge" of tissues is necessary to maintain an imperfectly girdled tree in good vigor. Most of the wood laid down in the stem and larger roots serves to support the crown. Although this is a crucial function, it does not have enough immediate survival value to command high priority. The production of secondary compounds that protect the tree from insect and disease attack is of similar low priority.

Biologic success is basically the ability to reproduce. The priorities for carbohydrate are in direct relation to the importance in reaching reproductive size and then producing propagules. Seed production is so vital to species survival that whenever it takes place it does so at the expense of other functions.

In comparing the relative growth rates of individual trees, it can be seen that "the rich get richer.'' Trees that have sufficient carbohydrate after meeting basic respiratory needs are able to put more energy into larger crowns and root systems, stronger stems, and the production of secondary compounds that protect the foliage from herbivory and phloem from beetle attack. These factors all result in even greater photosynthesis, and the effect snowballs.

The extent to which a tree increases mechanical support depends on the vigor of the tree and the amount of carbohydrate that remains after provision for more vital functions. The part of the growth that is of greatest economic concern for timber production is thus low enough on the scale of biological priority that it is subject to great variation; anything that reduces the amount of photosynthesis will result in a reduction in stem growth. It is for this reason that growth in diameter is so readily controllable by thinning or other means of regulating stand density and tree vigor.

Stem Form

The tree stem as a whole tends to develop as a mechanical structure designed to bear the vertically acting load of its own weight and the horizontally acting load of wind. It seems possible that the better the carbohydrate supply the closer the stem approaches the engineered form of a cantilever beam of uniform resistance to bending throughout its length. This is the shape that the stem would have if an engineer designed it to meet the loads placed on it with the least amount of wood. If tree weight were the only load, the stem would be cone-shaped above the base of the crown. Because there is always some effect of the horizontally acting wind, the stem develops butt-swell and thus assumes the tapered form of the modified paraboloid recognized in mensurational work. At least in the lowest portions of the stem, the amount of wood formed appears to be controlled by the bending stress that is maximal at the ground line because that is the fulcrum of a long moment-arm of bending. This is the cause of the butt-swell at the base of the stem. However, if the tree loses vigor, the development of the butt-swell ceases, although it may remain obvious as an inheritance from an early period of more rapid growth.

The annual layer of wood deposited on the stem is not uniform in thickness, density, strength, or anatomical structure from top to bottom. Ring width is the parameter most frequently used to assess the stem growth rate, but care must be taken in its interpretation. Two primary factors affect ring width at a given height: (1) is the total volume of wood that is laid down at that height, and (2) the stem circumference around which the wood is deposited.

The volume of wood added to the stem generally increases from the top of the tree to the base of the live crown. That is to say, the volume of wood at any height is proportional to the amount of foliage above that height. For trees in the main canopy, the same relationship is true below the crown as well, so that the volume of wood added is nearly constant down to the ground. Very vigorous trees add more wood along the lower stem most likely in response to the mechanical stress of a heavy crown. On the other hand, trees low in vigor are subject to less mechanical bending and have less carbohydrate available for wood production. These trees have less growth in the lower stem portion. In fact, some years trees very low in vigor may add wood around only part of the circumference or may not add a ring at all along the lower stem. The additional wood near the foliage gives credence to the hypothesis that stem form is controlled by relative proximity to the carbohydrate source in the crown. Growth of wood in the lower part of the bole accords with the view that stem growth is a mechanically controlled response to bending. Both kinds

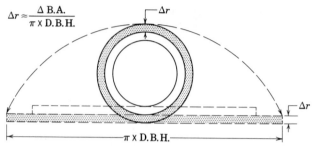

Figure 3.1 The relationship between annual increment in basal areas (ΔB.A.) and radial thickness (Δr) of growth rings. The annular rings of two successive years are straightened out into rectangles to show the large increase in basal area increment that is necessary to maintain constant ring width.

of response seem to be involved to some degree and are, in either case, governed by complex hormonal systems (Larson, 1963).

Even if the volume of wood laid down is constant along the stem, the thickness of the annual ring decreases below the live crown as the diameter of the bole increases. This is the result of spreading similar amounts of material over increasing circumference (Fig. 3.1). If the cross-sectional area of the ring is measured rather than radial thickness, it appears to be nearly uniform along the stem because cross-sectional area is the same as volume at a given height.

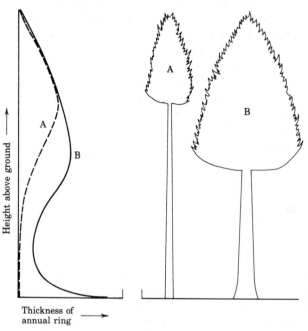

Figure 3.2 The variation in thickness of the sheath of wood annually laid down on the central stem for (A) a small-crowned tree that is barely surviving and (B) a vigorous one with well-developed crown and butt-swell. Note that there are peaks in ring thickness near the base of the crown of each tree.

Both factors together result in the widest rings being near the base of the live crown in all trees. In vigorous trees, a second widening occurs near the base, but in trees of poor vigor the thinnest rings will be near the ground. Figure 3.2 shows the difference in ring widths between very vigorous and less vigorous trees. When this pattern continues over time, dominant trees develop stems that are quite tapered (and very strong); trees that are overtopped have very cylindrical (and weaker) stems.

The general effect of thinning is that the crown expands. The lower branches live longer and become thicker in diameter, which degrades the value of lumber produced in the stem, but the increased foliage also produces much more carbohydrate. This in turn provides the structural material that is laid down as increased diameter growth. However, the diameter growth is increased much more in the lower than in the upper parts of the stem, so that it becomes more tapering.

The wide variations that can be induced in diameter and basal area growth in the zone of butt-swell can give exaggerated impressions of increases or decreases in growth of stem volume. Except in small trees, this zone definitely includes the breast-height level. Increment cores taken at breast height give good qualitative indications of growth trends but do not give good quantitative estimates of changes in volume without knowing something about the changes in stem taper. It is unfortunate that trees are tall and people are short because it is more reliable to assess diameter at some higher point.

Tree Growth as a Guide to Thinning

Patterns of annual-ring thickness are handy indicators of tree development in general, but they need to be interpreted with some mathematical sophistication. An ideal stem might have wide rings that were uniformly thick, but this can be produced only if the crown surface expands exceedingly rapidly and growth in basal area accelerates (Fig. 3.2). For example, a 10-inch ($\approx$25 cm) diameter tree must increase its basal area growth rate by 50 percent by the time it reaches 15 inches ($\approx$38 cm) in diameter in order to maintain a constant diameter growth rate. This may happen with an isolated shade tree, in trees that are very dominant, or after heavy thinning, but it is seldom possible with trees in closed stands.

It is erroneous to assume that successful thinning must cause dominant trees to show a shift from narrow rings to wide. The ring pattern of a timber crop tree that has been released in a series of timely thinnings is one in which ring thicknesses decrease slowly but steadily outward from the pith. Often, a tree actually accelerates in volume growth even while ring width slowly declines. Pronounced increases in growth can be welcome, but they really indicate that the thinning might better have been done earlier. Trees of the subordinate crown classes show the most spectacular response, provided they have not been suppressed so much that they fail to respond at all. Thinning is usually better directed at forestalling major decelerations of diameter growth than at remedying them after they have occurred. Ideally, there should be an ambitious but realistic plan for the regime of diameter growth of final crop trees.

Variations in stand density, such as those induced by thinning, cause very large variations in diameter growth but remarkably little in height growth. Very light thinning may have no effect on diameter growth, but, with drastic thinning, it is entirely possible to produce crop trees with twice the diameter that they would have attained in the same time without thinning. The height growth of the *dominant* trees, on the other hand, is changed scarcely at all; it is likely to be slightly decreased at the extremes of stand density. The height growth of the leading trees of a stand is, in fact, so independent of stand density and so closely controlled by the totality of growth-supporting factors of the site that it is used as an integrated expression of site quality. Thus **Site index**, the total height of the *leading* trees at 50 years or some other standardized age, is, in this context, a manifestation of the fact that thinning affects height growth very little.

The height growth of dominants is not absolutely and always unaffected by stand density or thinning. Trees that have become exposed, isolated, or open-grown may become somewhat stunted. The reasons are not clear but may involve growth of branches and lower bole at the expense of height. Very high stand density often causes reduced growth of trees in all dimensions, including height, probably because of the high respiration associated with a poor ratio of photosynthetic to aphotosynthetic tissue in each tree. Thinning these crowded stands might result in increased height growth if the trees are able to respond to the increase in growing space.

Values of average height calculated for *all* trees in stands are inversely related to stand density because the height growth of trees of the lower crown classes is stunted by competition. The greater the number of trees being submerged through competition within the crown canopy, the lower is the *average* height.

The common opinion that trees grow taller in dense stands is, in general, incorrect. The slender trees merely look taller than the more tapering ones of a less dense stand, and the small diameters lead one to underestimate stand age. The effects of this optical illusion vanish if the heights and ages are measured.

One rough but convenient index of the ability of the crown to nourish the remainder is the **live crown ratio**, which is the percentage of length of stem clothed with living branches. It approximates the ratio of photosynthetic to aphotosynthetic surface and is related to considerations of tree management such as the branch-free length and the taper of the stem. It is a parameter that is better measured than guessed at because the upper parts of a tree, being farther from the eye, always seem smaller than the lower parts. In closed stands, the bottom of the foliar canopy retreats upward with surprising rapidity. This is fine from the standpoint of natural pruning and reduction of stem taper but discouraging for the maintenance of good diameter growth. With most trees, it is desirable to plan to let the base of the crown retreat to a chosen height and then to halt or slow down the retreat. This means that the live crown ratio decreases from 100 percent to a certain amount, and then one attempts to make it increase again or at least to slow down any further decrease. This can be done only by thinning to keep the lower parts of the crown adequately illuminated. If the ratio is allowed to decrease to 30 percent or less, the general reduction in vigor will cause substantial loss of diameter growth. Ratios under 20 percent may start resulting in reduced height growth. If the ratio is very low, the recovery after thinning will, at best, be delayed or the tree may even succumb.

An indication of how well the crown has been able to nourish the stem in the past is the **height/diameter ratio**. This ratio is calculated by dividing the height by the diameter, making sure that the same units are used (such as centimeters). The preceding discussions of annual ring thickness throughout the height of the bole indicate that a dominant tree would have a low ratio because of a large diameter. An overtopped tree would have a much higher ratio because diameter is more sensitive to loss of vigor than height. A dominant tree with a low ratio will be much stronger than an overtopped tree with a high ratio, but the overtopped tree is exposed to much less wind force. If trees are overcrowded and the ratios of the main canopy trees become 100 or more, major portions of the stand may blow over.

The loss of trees with small crowns after thinning may be caused by insect attack, sunscald, or even the cutting of other trees that had formerly nourished the unthrifty through root grafts (Eis, 1972). The most important cause may be merely the increase in respiration induced by the sudden increase in temperature caused by exposure. (Respiration about doubles for every 10°C increase in temperature.) If the respiratory demand is great enough, very little carbohydrate will be left for renewal of vital tissues. However, unthrifty trees are usually eliminated by bark beetles or other biotic agencies before they actually starve to death. Such difficulties are avoided by selecting crop trees from the dominant and codominant classes with acceptable live crown ratios.

Biological Basis for Quantifying Tree Development

It is easy to demonstrate that the accelerated crown expansion induced by thinning increases diameter growth, but difficult to develop precise quantitative relationships necessary to predict how various thinning programs would affect the growth of individual crop trees. When such relationships have been established, it becomes possible to simulate alternative thinning regimes by computer. Calibrating a model requires data derived from thinning trials or other long-term observations necessitating decades of measurements. Most efforts to explain tree growth in quantitative terms seem to work best if tree diameter is bypassed in favor of direct relationships between amounts of foliage and wood volume growth. Many empirical models exist today, but without a direct functional relationship with the foliage, the model is limited by conditions that actually existed in the empirical data set used to parameterize and calibrate the model. These models are useful for interpolating between the experimental conditions used in the construction of the model, but are much less useful for extrapolating to conditions that were not measured. Models that predict stand averages rather than the response of individual trees are easier to construct and give more reliable estimates of yields.

Even though new wood is laid down on old wood by the vascular cambium, it is not the wood but the foliage of the crown that "grows" more wood. It is fairly obvious that wood grows in layers laid on the surface of core that gradually expands. It is not so obvious that the foliage that ultimately produces the wood is also borne on layers wrapped around an expanding core because, unlike the wood, the old layers of leaves disappear when they cease to perform their function. The total cubic volume of the tree is a cumulative expression of all wood ever produced by the main stem (Seymour and Smith, 1987). It is, at least theoretically, closely related to and controlled by the crown volume or, more precisely, by a kind of "historical" crown volume projected into the past as shown in Fig. 3.3; this is a cumulative or integrated approximation of all the foliage that the tree ever had.

According to the same line of reasoning, the growth of wood during any short period of years would be related to the expansion of the crown during that period. The growth of one year might have a close relationship to the amount of foliage or the living crown surface area of that same year. In trying to determine such quantitative relationships, it helps to distinguish between cumulative and current parameters and to draw relationships within these categories and not between them. For example, if things have to be simplified by the treacherous step of omitting the vertical dimension, the basal area of a tree, which is a cumulative parameter, can be related to the crown area, which is the area of ground directly under the crown and, in effect, a cumulative parameter. The changes in both basal area and crown area during a certain short period would be corresponding "current" parameters.

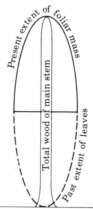

Figure 3.3   A two-dimensional representation of the relationship between the cumulative volumes of stemwood and space occupied by the foliage of a tree during all of its previous development. The present size of the crown is enclosed within solid lines and the space that has been successively filled with previous foliage is depicted by the dashed lines projected from the widest part of the crown down to the ground. This is intended to show how the total amount of foliage that the tree has produced can be related to the total amount of stemwood which that same foliage produced. The diameter of the tree stem is exaggerated.

Much of the practice of thinning rests heavily on the importance of crown relationships. However, trees do not live by light and carbon dioxide alone; water and soil nutrients often limit the amount of photosynthesis and growth. In fact, some sites are so deficient in available (oxygenated) water or some nutrient element that thinning will *not* improve the growth of the trees unless coupled with fertilization, drainage, or irrigation. Furthermore, there are many sites on which improvements in growth from thinning do not come directly from crown expansion but from allowing the roots to obtain more of the limited supply of soil water and nutrients. In such cases, the concomitant crown expansion is a necessary, but secondary, consequence and not a primary cause. Light and carbon dioxide become the main limiting factors only on sites that are "good" in terms of the supply of available water and nutrients in the soil.

Effect on Wood Quality

The structural requirements of tree stems are not met entirely by laying down wood of homogeneous density, structure, and strength. Instead, the strength of the wood in a stem usually increases from the pith outward as well as from the top downward. Resistance to breakage at any point is determined largely by the product of the strength of the outermost fibers and the distance by which they are separated (stem diameter). The strength of the wood between is of little consequence other than in making the stem more rigid and reducing the actual amount of bending caused by a given wind load. The overall strength of the stem is still further enhanced by the fact that the outermost fibers are under tension and the inner ones under compression. These internal growth stresses contribute to the tendency of lumber to warp. However, they make the stem stronger, just as prestressed concrete beams with stretched cables beneath the surface are stronger than reinforced concrete with inert rods in the same positions.

When a conifer is young and short, the loads placed on it are small and it usually produces the weak **juvenile** or **core wood**. However, during its period of active increase in size and height, the specific gravity and strength of wood laid down tend to increase. When a tree of almost any species reaches maturity and ceases to increase much in height or size of crown, the annual rings added to the bole become very thin and of low strength, representing little more than enough xylem to renew the water-conducting system. As a tree slows down in height growth and foliage expansion, because of the high energy demands of maintaining a large size, the lack of increase in size means that the mechanical stresses on the stem no longer increase; therefore, additional strength is not necessary for survival. Juvenile wood is normally produced only in those parts of tree stems that are still clothed with living branches.

Growth Rate—Not a Causative Factor

One erroneous view that long persisted was that the strength of wood is directly controlled by the rate of growth. According to this lore, rapid growth was reputed to produce strong wood in ring-porous hardwoods, weak wood in conifers, and no appreciable difference in diffuse-porous hardwoods. The main practical consequence of these ideas has been the persistent notion that thinning or any other measure that increased diameter growth of conifers weakens the wood.

The idea that fast-grown conifers produce weak wood is based on little more than the observation that young trees have weak wood and grow rapidly, whereas somewhat older trees grow more slowly and have stronger wood. It later became clear that the effects of growth rate and factors related to age were being confounded and that the age-related ones were more nearly the controls of wood strength and density. Correlation of two factors is not proof that one causes the other because both may be the result of some other more controlling factor. If a conifer is made to grow rapidly by thinning, it creates a greater volume of the same kind of wood that might have been laid down without release and does not shift to production of weak wood.

It is probably best to interpret differences in the strength, density, and structure of wood in terms of the main biological functions of xylem, which are mechanical support and water conduction. When trees are small and young, their stems do not have to be particularly strong. It is noteworthy that some fast-growing, weak-bodied, pioneer species never get out of this condition and topple over before they become very tall. As long as a tree continues to grow in height, the load of its own weight and that of the wind increase, and so does the strength of the stem. If the tree becomes very old, it shifts to the production of very thin annual layers of soft, weak wood of the kind prized for fine veneers, cabinet work, and finish grades of lumber. It may be suspected, but is not proven, that this kind of wood is produced by the necessary renewal of conductive xylem in trees that are no longer adding to the load on their stems because their height growth and crown expansion have nearly ceased. Such trees have already developed the stem strength necessary to support the crowns.

A more refined indicator of the strength and density of wood is the relative proportion of early-wood, which functions mainly for conduction, and of late-wood, which serves mainly for mechanical support. In conifers, the time of the annual change from formation of early-wood to that of late-wood appears to coincide with the end of the initial period of shoot and needle elongation (Larson, 1963). Any treatments, such as thinning, fertilization, or irrigation, or factors of soil and climate that prolong or increase diameter growth during the summer usually increase the proportion of late-wood, which is produced mainly at that time. Thinning in conifer stands in droughty areas can prolong the time of diameter growth later into the summer because water does not become limiting as quickly (Zahner and Whitmore, 1960). The amount of early-wood can be very similar between thinned and unthinned stands, and most of the additional ring width is late-wood. Because late-wood is denser and stronger, wide rings formed well below the base of the live crown will form wood that is heavier and stronger than wood with narrow rings from the same stem position.

Controlling Wood Properties by Thinning

The structure and anatomy of xylem vary tremendously between and within species as well as within tree stems; there are many important variations more subtle than those considered here (Haygreen and Bowyer, 1982). The choices that can be effected between good trees and poor ones in thinning improve the ultimate utility of the wood much more than they might be impaired by any vaguely suspected baneful effects of making the trees grow faster. In general, the larger the trees grow and the straighter their stems, the greater is the usefulness of the wood, at least until the heart-rots of old age start to cause deterioration.

At least with intolerant species, such as the two- and three-needled or ''hard'' pines, it is often best to allow the trees to grow unchecked when they are young so that they will develop crowns large enough to continue good health in later stages. The best way to deal with the juvenile wood that this produces is to develop suitable uses for it consistent with its shortcomings. This line of action will produce a greater amount of denser wood in later stages than if one attempts to restrict the amount of core wood and rebuild larger crowns later. Furthermore, it is desirable to avoid abrupt changes in the rate of diameter growth, such as those that can occur when trees are released by long-delayed thinning. This is one effect of rapid growth that can cause problems. There may be shake (splitting along the plane of annual rings) or warping in boards cut from the transition zone. This is one of the reasons why it is better to thin to forestall declines in diameter growth than to correct them after they have taken place.

Many shade-tolerant species are adapted to start as advance growth and grow very slowly in the early stages. They do not form any fast-grown juvenile wood, which can be an advantage. Once released, they generally commence rapid growth of wood with good properties. However, because of their ability to endure long periods of suppression, they are subject to the problems associated with sudden accelerations in growth. Ideally, they should be released in the sapling stage and be induced to grow steadily thereafter.

Some conifers, such as the spruces, true firs, and five-needled or ''soft'' pines, which have no pronounced variation between early- and late-wood, produce very homogeneous wood. They adjust to structural demands more by increasing taper than by laying down stronger wood in the lower and outer parts of the stems.

Thinnings that are heavy enough to cause major changes in the form and taper of boles may be detrimental as a consequence. If the logs are converted into lumber, there is increased waste in slabs, and the boards are more likely to be cross-grained. The trees may no longer meet the specifications for poles and piling if they taper too much or have excessively large branches high on the stems.

Thinning has a tendency to halt natural pruning and stimulate the development of large branches. If not remedied by artificial pruning, this effect will increase the size and number of knots in the wood. The only compensating effect is that the branches remain alive longer, thereby reducing the number of loose knots eventually produced. If a large amount of clear material is to be grown without artificial pruning, thinning should be delayed until natural pruning has proceeded to the extent ultimately desired. Regardless of how the pruning is accomplished, it is prudent to set some realistic goal as to the length of branch-free bole that will be developed; the remaining upper portion of the stem should ordinarily be kept clothed with living branches. The contemplated length of clear bole should be greater the better the site and the longer the rotation because it depends on the ultimate height and live crown ratio. If one plans to grow trees 100 feet ($\approx$33 m) tall with a live crown ratio of 40 percent, it is important to plan for the proper development of the 60-foot ($\approx$13 m) length below the live crown.

The criteria of quality for pulp depend on the pulping process and the characteristics desired in the product. If the paper needs to be strong, the procedures necessary to produce strong, dense wood are appropriate. Fibers that are long, strong, and narrow interlace to produce strong paper. However, if the paper must be smooth, it is necessary to have some short fibers to nestle in the gaps between the stronger ones. These differing requirements are better met by mixing fibers of different species than by trying to make spruce fibers like those of maple and vice versa. In general, the farther the wood is from the pith, the greater is the length, density, strength, cellulose content, and anatomical structural quality of the fibers. If the wood is of high density, the volume that must be processed for a given yield of pulp will be low. The wood of knots has anatomy that is different from that of the main stems, so it is usually an undesirable but unavoidable adulterant of pulpwood.

Not all wood is grown for pulp or structural material. Sometimes the most valuable wood is that which is sufficiently soft and uniform in texture to be shaped and finished easily for millwork, furniture, plywood, veneer, and similar products. Homogeneity of properties and freedom from defect are more crucial characteristics than strength and density. This kind of material now comes mainly from the very fine-ringed wood of the outer portions of large, ancient trees.

The problems of producing such wood in the future arise with species in which there is a sharp contrast between early- and late-wood. Such species include the oaks and other ring-porous hardwoods as well as the hard pines and Douglas-fir among the conifers. For these species, there is no good solution short of growing the trees to those advanced ages at which diameter growth becomes very slow. Ordinarily, the rotations are thought to be financially far too long. However, in the Spessart region of western Germany, oaks are grown on rotations of 300 to 400 years to produce such wood, and it brings some of the highest prices in the world. Sometimes the proportion of hard late-wood in such species is low when, as is the case with some interior ponderosa pine, the trees are subject to summer deficiencies of soil moisture.

These difficulties are less serious with diffuse-porous hardwoods, soft pines, true firs, and spruces, which show little contrast between early- and late-wood. Consequently, such species provide the best opportunities to grow soft or highly homogeneous wood on shorter rotations.

PRUNING

Trees must have branches, but the only good branches are the living ones. Part of the problem with branches is that they form knots, which are the most common defects of wood grown in managed forests.

Branches do not necessarily fall off when they cease to function. But, except to the extent that they act as infection courts for rotting fungi, their continued presence does not threaten tree survival. However, dead knots are more serious timber defects than ones from living branches and dead branches are unsightly on ornamentals.

Natural Pruning

Most of the branch pruning that takes place in forests is caused by physical and biotic agencies of the environment and is called **natural** or **self-pruning**. In most situations, the branches die from lack of light resulting from shading by higher branches. Wind can also cause the breakage and death of branches. As trees become taller, the crown edges hit adjoining crowns with increasing force during wind sway. This results in **crown shyness**, a phenomenon which leaves spaces between the crowns of adjacent trees. In alpine areas the buds are destroyed by wind and ice crystals leading to the stunted krumholz tree form. A similar process occurs near oceans where the buds are killed by salt spray. Crowns can also die back from water stress during drought periods or from the induced water stress after sudden increases in light levels.

In closed stands, natural pruning proceeds from the ground upward and starts with the killing of branches by the shade of those above. If the object is to keep the branches small and soon shed, then it is desirable to maintain high stand density and even to refrain from thinning until some desired branch-free length has been developed. The alternative, discussed later, is to grow stands at a lower density and prune living branches to create branch-free length. If pruning is not done, it often helps to distinguish between an initial period of "stem training" in which branch-free length is developed and a later one in which thinning is used to encourage crown expansion and to halt the dying of branches.

If branches die before they get large enough to contain heartwood, their bases are usually sealed by the formation of resins in conifers and gums in hardwoods. This tends to keep wood-deteriorating organisms out of the central stem. Fortunately, the fungi that literally rot dead branches off trees are saprophytes of dead sapwood and are rarely capable of attacking heartwood or living sapwood. The rotten branches ultimately break off through the action of precipitation, wind, or the whipping action of subordinate trees. However, they tend to persist if they have become large or if either chemical substances or merely excessive dryness inhibit fungal action. It is especially awkward if the initial stand density is so low that many branches become large and then die when an unthinned stand becomes tightly closed.

When new shoots form, they usually point vertically upward even if they are on the tips of old branches. If they sag from the vertical, it is mainly because they do not build enough supportive tissue to remain erect. This effect is influenced by growth regulating chemicals produced in the apical buds of the leaders. If the light reaching the branches is suddenly increased, they can form enough reaction wood to curve back toward the vertical. However, branches normally sag increasingly as they become older and more submerged in the crown. Their resistance to natural pruning is least when they are horizontal.

If a terminal shoot is killed and two or more branches turn upward to replace it, some deformities can result. Even if a single shoot quickly asserts dominance, it may remain curved enough to cause a crook in the main stem. As long as dominance is shared between two or more upturned branches, the main stem remains forked. Any upturned branch that competes for dominance for some years but is then subordinated may persist as a **rami-corn**. This is an abnormally large branch that projects at a small acute angle from the main stem; it tends to persist longer and is much more undesirable than an ordinary branch.

Ramicorns can also form if the terminal bud dies or merely if some hormonal abnormality causes it to fail to assert apical dominance over the branch buds.

One common occurrence in some conifers is **lammas growth**, the sudden expansion and growth from a bud late in the growing season. If the terminal bud of the leader expands, the single stem is maintained. However, if the leader remains dormant and a branch just below has this additional growth, the branch may contest the leader for apical control the next growing season. If the battle is even the stem will fork, but more often the battle will last for several growing seasons and the losing branch will form a ramicorn.

In most natural stands, the branches of surviving trees tend to be larger in the middle and upper parts of the bole than in the lower. This is because inter-crown competition is usually severe before differentiation into crown classes takes place. Trees on poor sites usually have small branches but the vertical distance between them is short.

The final step in natural pruning is the occlusion or covering with new tissues of the short stub left by the dead branch. Whether the pruning is natural or artificial, this process can be likened to the submergence of a post set in a rising stream of water. The rate of submergence depends on the rate at which the stream rises; that is, the faster the radial growth, the sooner the wound heals, but the diameter of the branch makes remarkably little difference. As with the post, the upstream edge of the top, that is, the upper side of the stub, is engulfed somewhat sooner than the lower or ''downstream'' side.

The new wood that ultimately covers the old branch stub is the most dependable seal for excluding water, oxygen, and fungal spores from the interior of the stem. As long as dead branches protrude from the stem, they can act as conduits for these materials. Even if rot has already started underneath a dead branch stub, it may cease to spread when water and oxygen are excluded.

Knots

The knots (Fig. 3.4) produced while branches are still alive are known as **live**, or **intergrown**, knots and are usually *red* in color. Those formed after the branches die are called **dead**, or **encased**, knots; in conifers they often become *black* because of heavy resin deposits. The conductive elements laid down around live knots are termed intergrown because they bend outward and *appear* to be continuous with those of the branches. Actually, this is true only of the xylem and phloem of the lower half of the live branch. The elements of the upper half actually sweep around the sides and top of the branch base without being continuous with it. The discontinuity can be observed by forcibly pulling a living branch out of the bole.

The elements laid down around dead branches bend inward and have no connection with those of the branches. In fact, the new xylem is separated from the old branch wood by an encasing layer of new bark. Dead knots are serious defects in lumber or veneer because they are apt to fall out when they dry, especially if they are long-dead parts of branches that had started to deteriorate when encased. In other words, live knots are tight knots, but dead knots have a strong tendency to be loose knots. Live knots remain tight because the attachment in the lower half of the former branch suffices to keep them in place.

Softwood lumber is usually graded visually on the basis of the size, soundness, number, and distribution of knots in the boards. Although it is best to have clear material with no knots at all, they are permitted in and characterize the medium and lower grades. Some softwood lumber to be used for structural support is stress graded. A force is applied to

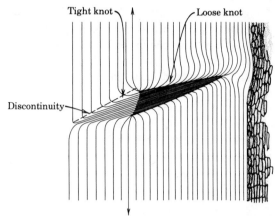

Figure 3.4   Radial section through a pine branch that persisted for many years after its death. The branch is thickest at the point where it joined the stem at time of death. Note that the annual rings laid down around the branch after death of the tree turn inward without joining the wood of the branch. Even when the branch was living the fibers of the upper part of the branch did not actually link with those of the main stem, though they appear to do so if one examines a radial section of this kind. The arrows at top and bottom mark the last annual ring formed before the branch died.

the center of the piece and the bend or deflection is measured. Silviculturally, the question of how grading is done makes little difference because knots always have the greatest influence on strength characteristics.

Most hardwood lumber is used, fortunately, in short lengths, for furniture and other highly discriminating purposes; it is therefore usually graded on the basis of the number of completely clear or defect-free pieces 2 feet long that can be cut from the boards being graded. A small, tight ''pin knot'' is as serious a defect by this reckoning as a very large rotten hole. The only reason why there is more concern about pruning with softwoods than with hardwoods is that softwood branches are generally more persistent. Knots in hardwood boards are actually more serious problems, and it is merely good fortune that they are less numerous.

As far as plywood is concerned, knots are scarcely permitted at all in the face veneer layers of high-grade material. Furthermore, one knot in rotary-cut veneer can degrade a wide sheet at many repetitive points. On the other hand, defective material can be either buried in the interior layers or laminated together into construction grades in which the structural weaknesses induced by knots are well compensated. If the face veneer has a small number of knots, they can be cut out and replaced with small wooden plugs. As a result, in most modern plywood manufacture, it is desirable to have as much clear material as possible, but there are also good ways of utilizing poor material.

Dead branches of maples and certain other diffuse-porous hardwoods constitute a special kind of detriment to wood quality (Shigo et al., 1979). Whenever any branch larger than about 1 inch (≈3 cm) in diameter dies, it becomes an entry point for bacteria that turn all of the existing stemwood below that point into undesirable, brown ''pathological heartwood.'' Fortunately, gums almost immediately form inside the stem in such manner that the bacteria are sealed off inside the tree, and the wood subsequently formed is desirable white sapwood at least until the next branch dies. This problem can be overcome by cutting the branches off before they die, making sure that the cut does not leave a stub. A more passive solution is to set some goal as to how tall and thick the pathological heartwood column is going to be and then thin hard enough to prevent branches from dying after the size of that column is determined.

Artificial Pruning

The training effects obtainable with manipulations of stand density are not always sufficient to eliminate branches early enough, and it is sometimes desirable to resort to artificial pruning. This is an expensive and labor-intensive operation, so it is necessary to be very discriminating about when and where it is done.

The most common tools (Fig. 3.5) are hand saws, usually with curved blades and handles of lengths appropriate to the height of pruning. Clubs are sometimes used to knock dead branches off, but they work best with the small-diameter branches produced in high density stands. They have become less common because of the greater emphasis on reducing rotation age by keeping stands at a lower density. Ax pruning is dangerous and for larger branches can be slow and ragged.

It is easy enough to prune up to about 10 feet ($\approx$3 meters), especially with hand saws mounted on ax handles, but difficulties and costs increase rapidly above that level, considering that pruning should be carried out as close to the base of the live crown as

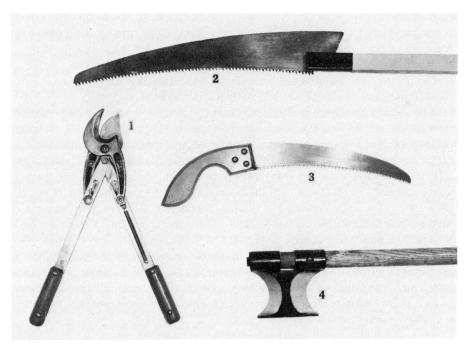

Figure 3.5 Typical instruments used in pruning forest trees: (1) A pair of shears that has both blades sharpened and is designed for close pruning. (2) The blade of pole saw. (3) A hand pruning saw. (4) The Rich pruning tool, typical of those that cut on impact. *(Photograph by Yale University School of Forestry & Environmental Studies.)*

possible. The most common tool is a saw or clipper on a long pole, which is effective to about 18 feet (≈6 meters) above ground but wobbles too much to enable one to prune higher. A variation is a power saw on the end of a long pole. The engine is at the end of the pole held by the operator. Many operators avoid such saws because they tend to be heavy and unwieldy, but improvements in design are making them more useful. It is also possible to use hand saws from ladders, but the moving of ladders is cumbersome enough that it is hard to prune higher with them also.

Several pruning machines have been developed. These machines power themselves up the tree, cutting branches as they go. They are usually controlled by either a wire or a radio control. They work best on cylindrical trees with only minor bumps or other irregularities. When they work well, trees can be pruned very high up the stem. So far most North American tree species have proven to have enough stem irregularities to render these machines of marginal utility. Expert tree climbers, with safety belts, can get to almost any height using climbing irons or ropes. As with the pruning machine, this method can be carried to almost any height, but is almost always is too costly for use except in parks and intensively managed settings.

Pruning for Timber Production

Artificial pruning is expensive enough that it must be limited to a small number of trees per acre. There must be a substantial premium for knot-free wood. This usually exists with species that have wood suitable for cabinet work, interior finish, furniture, or high-grade surface veneer. Ordinarily, the premium has been regarded as insufficient in the case of species that are used primarily for framing timbers and building construction. Even though knots substantially impair the load-bearing strength of wood, the construction industry usually has found it cheaper to compensate for knots by using thicker members than to pay higher prices for wood with fewer knots. In recent years, efforts to shorten rotations by growing trees at a lower density have changed this thinking, because short rotations without pruning produce much juvenile wood full of large knots.

Among the conifers, the species most worthy of pruning are those with low contrast between early- and late-wood that are soft and easily worked and finished because these milled products usually require knot-free wood. These include the five-needled pines, the ponderosa pine of certain localities, and sometimes spruces (O'Hara, Larvik, and Valappil, 1995). Most southern pines and Douglas-fir have wood that is strong but too coarse for finishing purposes, so that pruning of these species is for the surface veneer of plywood and for large construction timbers. Most hardwoods prune well naturally, but there can be great advantage in the timely removal of a few ramicorns or other large, persistent branches from otherwise branch-free stems.

Provided that there is an adequate premium for clear material, pruning is most advantageous for species such as eastern white pine and spruces, which have rot-resistant branches. The removal of such branches has the effect of converting what would be loose-knotted lumber of the lowest common grades and minimal stumpage value to clear material that can have very high value.

It is essential that pruned trees be carefully selected and rigorously limited in number. It is easy to forget that most of the trees in a young stand are doomed to die of suppression before the survivors reach sawtimber size. On one hand, it is undesirable to make the expensive choices involved until the trees are large enough to have expressed dominance; on the other, if the trees get too large, there will not be enough time for them to produce shells of clear wood thick enough to repay the high cost of the pruning. It also helps to postpone the first pruning at least until the terminal shoots have grown to the height of the first log, thus avoiding the risk that forks or crooks may develop before that height is attained. In typical cases, pruning might be limited to 65–120 trees per acre (150–290 per hectare) in the diameter class of 5–10 inches ($\approx$13–25 cm). If the object is to produce clear lumber, the clear shell must grow to at least 4 inches ($\approx$10 cm) in radial thickness to be of any advantage. Pruning is cheap enough up to about 10 feet ($\approx$3 m) that sometimes the pruning of more trees may be justified, but the number usually must be reduced when the time comes to carry the pruning higher.

Ordinarily, pruning is done after the branches are dead and in two or sometimes more stages of height. Different tools, usually saws with handles of differing lengths, are used for each stage. The object is often to keep the knotty core of uniform thickness. Commonly, the problem is not that the lower part of the knotty core becomes too thick but that the later stages of pruning are delayed so long that the upper part gets thicker than the lower part.

A very high proportion of the merchantable volume and value of a tree is often concentrated in this zone anyhow. In many instances, however, it would be fine if the pruning could be carried higher; sometimes it is. The branches that develop somewhat higher on the stem are usually the largest; they form the poorest kinds of knots and the worst sources of heart-rot infections if they die. Artificial pruning at these heights probably awaits the refinement of pruning machines. The problem can be alleviated to some extent by thinning hard enough to halt or slow down the dying of these upper branches. Large, live branches are better than dead ones but even large, tight knots can become so large that the only solution is to harvest the tree.

It is perhaps most important of all that artificial pruning be coupled with programs of thinning that are positively aimed at maintaining diameter growth (Fig. 3.6). It usually takes rapid growth of substantial amounts of clear material to repay the high compounded cost of an operation that takes 10 to 15 minutes per tree with a wait of 15 to 40 years for the returns. The chosen trees should initially be as vigorous as possible, although not so vigorous as to have excessively large branches. It is generally best not to invest in pruning in a stand until after the first thinning. This not only ensures that the chosen trees have been released, but it also reduces the risk of logging damage to them. Although it is possible to waste money on pruning, under some circumstances the combination of investments in pruning with thinning and good diameter growth can provide some of the highest long-term returns available in timber-production silviculture (Page and Smith, 1994).

The ideal program of growing trees for timber would be one in which either natural or artificial pruning caused branches to be gone as soon as they died up to a certain deliberately chosen level. After the chosen branch-free length was achieved, it would be ideal if thinning were prosecuted so vigorously that no more branches died and diameter growth was maintained at the highest possible rate. It also helps to keep track of the rate at which branch death causes the base of the live crown to retreat upward. During the crucial stages in the development of closed stands, this retreat is astonishingly and undesirably rapid. Any ideal thinning program should be aimed not only at controlling the diameter growth of the crop trees but also at regulating branch-free length, stem taper, and the sizes of branches.

It is often desirable to refrain from pruning large branches (Fig. 3.7). They are costly to remove and may also contain heartwood that will not seal itself adequately against rot. Their cut stubs may also project so far out from the cambial zone that it takes a long time for them to be submerged by new wood. It is this phenomenon rather than branch diameter itself that makes large pruning wounds slow to heal.

Figure 3.6   Cross-section through a node of an eastern white pine that was green-pruned 42 years ago when it was 3.5 inches (≈9 cm) in diameter. The average diameter growth of the tree was about 2.6 inches (≈6.5 cm) per decade in the period after pruning. The small wounds healed swiftly and almost perfectly. *(Photograph by Forest Products Laboratory, U.S. Forest Service.)*

It is usually best to delay pruning until a branch is dead or is about to die. The removal of living branches in what is called **green** or **live** pruning can cause wounding of the bole by tearing of bark from it. This risk can be reduced by making a cut on the lower side of the branch before completing the removal with a cut from above. It is also very desirable to try to confine pruning to dormant seasons when the bark is tight and fungal spores are presumably less numerous. In some species, the bases of branches begin to be sealed by resin and gum deposits when they die, so it is well to wait until this protection commences. Living branches below the level at which the crowns close can usually be cut without reducing the diameter growth of the trees. Height growth usually remains unaffected unless the live crown ratio is reduced to less than 25–30 percent (Långström and Hellquist,

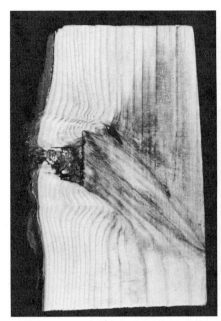

Figure 3.7 Typical radial sections through branches of red pines that were pruned 14 years ago. The branch at left was alive and 0.9 inch ($\approx$2.5 cm) in diameter when pruned; the stub was 0.7 inch ($\approx$2 cm) long, and the tree lapsed into a codominant position after pruning. The pitch pocket that formed as a result of the slow growth and rather large size of the stub would not have occluded for several years. The branch at right was 0.5 inch (1 cm) in diameter and dead when pruned; the stub was 0.5 inch (1 cm) long. The tree remained in a dominant position and grew rapidly so that healing was virtually complete 7 years after pruning. Note that both calluses grew more rapidly from above than from below. *(Photographs by Yale University School of Forestry & Environment Studies.)*

1991). Reductions in diameter growth slow down both wood production and healing of pruning wounds. Deceleration of height growth may cause a pruned tree to be suppressed by unpruned, faster growing neighbors.

There are cases in which it may be possible and desirable to substitute green pruning for the sort of natural pruning that is induced by crowding. It is, for example, possible to plant new stands at a very wide spacing with a small number of trees and to employ drastic green pruning not only to get rid of branches but also to prevent boles from becoming too tapering. It may be necessary to prune all of the trees in a stand, so that there will be no faster-growing, unpruned ones to claim dominance.

Excessive green pruning may result in sunscalding of trees with smooth, thin bark. It can also lead to the development of epicormic branches on hardwoods as well as on those conifers, such as Douglas-fir, spruce, and true firs, that have numerous dormant buds. On the other hand, green pruning usually halts production of juvenile wood at the levels from which the branches are removed.

To avoid ragged pruning cuts, it is not necessary to go to the cost and effort of sawing off any raised collars that form around branch bases. The reduction achieved in the diameter of the knotty core is too small to justify the work, and the practice may do more harm than good. In some diffuse-porous hardwoods, for example, cutting into the branch collar increases the amount of bacterially induced pathological heartwood (Shigo et al., 1979).

Bud pruning is a method of very early green pruning in which the lateral buds are rubbed or clipped off. The purpose is to produce logs with no knots at all. This can be an efficient system for producing small-diameter, short, plywood veneer bolts of high-quality hardwoods. It is important not to get carried away with the ease of this method because growth reductions and the serious results of injuries to the terminal shoots can make the treatment disappointing. The same effect, called "foxtailing," can develop naturally in some tropical pines planted where the dormant season is not long enough. The branch whorls that develop at the top or bottom of the long, straight, branch-free segment often produce large knots or other deformities.

Pruning of Ornamental Trees

Artificial pruning on open-grown shade trees is often desirable. Dead or broken branches may have to be removed so that they will not fall on people or become infection courts for fungi. Live branches also have to be removed before they touch electric wires, structures, or passing people and vehicles. It is also possible to prune in order to improve the appearance of the crown or to make the branches stronger by thinning them much as one might thin a stand of trees. Pruning can also be used to lengthen the branch-free bole of small ornamental trees; sometimes it is much cheaper, more dependable, and almost as fast to develop a good shade tree in this way starting with a small one than to transplant a large, expensive, preformed tree that is apt to succumb in the process.

Shade-tree pruning often involves the removal of large, green branches that have formed heartwood. The branches have to be severed with initial undercuts or supported during the removal so that they do not tear the remaining stem tissues. The work requires special equipment and well-trained workers because it can be quite dangerous, especially near electric wires. Putting paint or other coatings on the wounds provides more aesthetic benefit than any real protection against infections. In fact, the coatings may delay growth of protective callus tissue and slow recovery of the wound. Fungal spores are small enough to get through cracks in the coatings, although they may be killed by any fungicide chemicals applied with the coatings.

It probably helps to make the cuts smooth and to shape them, so that water flows off rather than onto the cut surfaces. If rotten cavities are too large to be sealed by growth of wood, they should be repaired in such ways that water will drain out of them through rust-proof pipes or through other means. Filling cavities with concrete or other substances may improve appearances but is as likely to speed decay as to slow it. Crotches that have started to split apart can sometimes be strengthened by screwing lag bolts into the inner side of each stem and connecting them with turnbuckles. Continuing diameter growth results in trouble with rods that go all the way through the stem or with modes of attachment that girdle the stems.

People get all kinds of erroneous ideas about the care of shade trees. Among these is the notion that it somehow helps to prune so that a projecting stub is deliberately left. This may stimulate the formation of wood gums or resins in a few species, but it helps only if the stub is soon removed. If it is not, it usually invites fungal or insect attack and delays healing for many years.

Shearing and Pruning of Christmas Trees

Conifers being grown for use as Christmas trees often require special pruning techniques to correct the effect of excessively rapid or asymmetrical growth and produce dense crowns of the proper shape. The most common practice, referred to as **shearing**, is the clipping of both terminal and lateral shoots. The objective is to develop crowns that are narrow and globose at the bottom and conical in the middle and upper portions. Shearing is usually confined to the removal of part of the shoots of the current or most recent growing season. Buds formed just below the point of cutting must be depended upon for renewed elongation of the decapitated shoots. Species that have dormant buds throughout the length of each internode, such as firs and spruce, can be sheared at almost any season except early summer. Pines rarely have any dormant buds along the internodes and are best sheared when the shoots are actively elongating. This stimulates the prompt development of vigorous adventitious buds just below the cut ends (Bramble and Byrnes, 1953; Brown, 1960).

Excessively tall and rapidly grown trees can sometimes be converted into good Christmas trees in one operation simply by pruning off the lower three-quarters or more of their crown. The remaining tuft of crown at the top then grows much more slowly and may soon develop into a properly compact form without any shearing. This technique has been successfully applied to fast-growing balsam firs and white spruces of natural origin.

BIBLIOGRAPHY

Bramble, W. C., and W. R. Byrnes. 1953. Effect of time of shearing upon adventitious bud formation and shoot growth of red pine grown for Christmas trees. *Penn. AES Progress Report* 91. 6 pp.

Brown, J. H. 1960. Fall and winter pruning of pines in West Virginia. *W. Va. AES Current Report* 26. 7 pp.

Eis, S. 1972. Root grafts and their silvicultural significance. *Can. J. For. R.* 2:111–120.

Esau, K. 1965. *Plant anatomy.* 2nd ed. Wiley, New York. 767 pp.

Chapman, A. G., and R. D. Wray. 1979. *Christmas trees for pleasure and profit.* Rutgers Univ. Press, New Brunswick, N.J. 212 pp.

Grey, G. W., and F. J. Deneke. 1978. *Urban forestry.* Wiley, New York. 279 pp.

Hallé, F., R. A. A. Oldemann, and P. B. Tomlinson. 1978. *Tropical trees and forests, an architectural analysis.* Springer, New York. 411 pp.

Haygreen, J. G., and J. L. Bowyer. 1982. *Forest products and wood science: an introduction.* Iowa State Univ. Press, Ames. 495 pp.

Holsoe, T. 1950. Profitable tree forms of yellow poplar. *W. Va. AES Bul.* 341. 23 pp.

Jacobs, M. R. 1939. A study of the effect of sway on trees. *Australia, Commonwealth For. Bur. Bul.* 26. 17 pp.

Kozlowski, T. T. 1971. *Growth and development of trees.* Academic, New York. 2 vols.

Långstrom, B., and Hellquist, C. 1991. Effects of different pruning regimes on growth and sapwood area of Scots pine. *FE&M*, 44:239–254.

Larson, P. R. 1963. Stem form development of forest trees. *For. Sci. Monogr.* 5. 42 pp.

Larson, P. R. 1969. Wood formation and the concept of wood quality. *Yale Univ. Sch. Forestry Bul.* 74. 54 pp.

O'Hara, K. L., D. A. Larvik, and N. I. Valappil. 1995. Pruning costs for four northern Rocky Mountain species with three equipment combinations. *W. J. App. For.* 10:59–65.

Oliver, C. D., and B. C. Larson. 1996. *Forest Stand dynamics.* Updated ed. Wiley, New York. 520 pp.

Ovington, J. D. 1956. The form, weights and productivity of tree species grown in close stands. *New Phytologist* 55:289–388.

Page, A. C. and D. M. Smith. 1994. Returns from unrestricted growth of pruned easter white pines. *Yale Univ. Sch. Forestry and Env. Studies Bul.* 97. 24 pp.

Seymour, R. S., and D. M. Smith. 1987. A new stocking guide formulation applied to eastern white pine. *For. Sci.* 33:469–484.

Shigo, A. L. 1989. *A new tree biology: facts, photos,, and philosophies on trees and their problems and proper care.* 2nd ed. Shigo and Trees Assoc., Durham, N.H. 618 pp.

Shigo, A. L., et al. 1979. Internal defects associated with pruned and non-pruned branch stubs in black walnut. *USFS Res. Paper* NE–440. 27 pp.

Stokes, B. J. 1992. An annotated bibliography of thinning literature. *USFS Gen. Tech. Rept.* 80–91. 178 pp.

Tomlinson, P. B., and M. H. Zimmermann (eds.). 1978. *Tropical trees as living systems.* Cambridge Univ. Press, London. 675 pp.

Wilson, B. F. 1984. *The growing tree.* Rev. ed. Univ. Mass. Press, Amherst. 152 pp.

Zahner, R. and F. W. Whitmore. 1960. Early growth of radically thinned loblolly pine. *J. For.* 58:628–634.

Zimmermann, M. H., and C. L. Brown. 1971. *Trees, structure and function.* Springer, New York. 505 pp.

CHAPTER 4

MANAGEMENT OF GROWTH AND STAND YIELD BY THINNING

One of the primary objectives of thinning is to manage the production of wood by individual trees and the aggregated yield of the forest stand. As unmanaged stands develop, the growing space on the site is reallocated to different trees mostly as a result of competition. Thinning is direct intervention in this reallocation process by eliminating some individuals and thereby adding to the competitive strength of other individuals. Removing weak competitors, small trees, will have little effect on the overall growth of the stand. Removing large trees will shift growing space to weak competitors that will not immediately, if ever, be able to use the additional growth factors efficiently. Removing large trees therefore reduces growth, resulting in lower yields over any fixed period of time. Yield is, in other words, regulated by thinning certain trees out of a stand and shifting growth to other trees.

FOREST PRODUCTIVITY

All living cells expend energy for their survival through respiration. Energy from the sun is captured by photosynthesis, which is the combination of water from the soil with carbon dioxide from the air to form glucose. This simple sugar contributes most of the basic building materials for all the compounds that provide trees with energy and structural material. Growth is simply the amount of glucose produced which exceeds the energy requirements of the tree, or the amount of photosynthesis minus respiration. Productivity is limited by the amount of photosynthesis that is limited by the amount of carbon dioxide in the air, the amount of sunlight, the availability of water in the soil, and temperature. The amount of foliage, which in turn controls the amount of photosynthesis, is limited by the amount of water and the availability of certain nutrients such as nitrogen, which are needed for chemical compounds in the leaves.

Forests are among the most productive plant communities, that is, in capacity for photosynthetic carbohydrate production as measured in terms of oven-dry organic substance, "dry matter," or "biomass." The high primary productivity of closed stands of tall, woody, perennial plants results from their deep canopies of sugar-producing foliage. The perennial woody stem is an efficient device both for displaying foliage in vertical depth and for conducting the upward movement that brings water and nutrients from the soil to the foliage. The vegetation of shallow waters and tidal marshes can be equally productive. Although it does not have to divert as much production on powerful supportive stems, it depends heavily on special vertical currents or the tides to deliver nutrients. Certain tall grasses, such as sugarcane or corn (maize) of agriculture, can be more productive but mainly because of fertilization, good water supply, or comparatively low losses to respiration.

The biological productivity of forests is so great that it is estimated that woody plant cover, occupying about a third of the land surface and a seventh of the total surface of the earth, accomplishes almost half of the world's annual photosynthetic fixation of carbon (Woodwell et al., 1978). Cultivated agricultural lands account for about a twentieth. Forests and the wood products produced from forests represent a significant proportion of the carbon fixed by the world's living plants (Burschel et al., 1993).

It is much more difficult to understand the production of material by woody perennials than that of typical agricultural annuals in which the production of a single year can be easily seen and assessed. Most of the comparatively greater production of a typical forest is laid down in forms such as the thin annual sheath of wood or new shoots that are not readily visible and can only be assessed with detailed measurements. At the very age when a stand is producing useful material most rapidly, the trees appear to change little in a year or even in a decade unless one measures their change in stem diameter.

Stand Development and Productivity

In the initial stages of development, a new stand of trees does not fully occupy the site. Herbaceous annuals and perennials or shrubs develop more rapidly and temporarily use most of the growth factors. Except on extremely droughty sites, the trees can overcome them because they can grow taller. Gradually the roots of woody plants assert occupancy of almost the whole stratum of soil in which the supply of water and oxygen is sufficiently ample and stable to allow them to persist.

The foliar surface tends to expand to form a complete crown canopy, unless severely limited by damaging agencies, water shortages, or other unfavorable conditions. The available crown space becomes fully occupied when the canopy has not only closed horizontally but also developed enough in the vertical so that the lower branches die and the insides of the crowns become vacant of foliage (Assmann, 1970). The stand then has as much foliage as it ever will have; the amount may remain stable or decline somewhat thereafter, but it is unlikely to increase unless the proportion of different species changes in mixed species stands.

After the crowns close, there may be a period in which the branches of adjoining trees interlace with one another. When they grow taller, however, the wind load on the crowns increases and the moment-arm lengthens, resulting in enough twig breakage from sway that the crowns become separated again and the gaps between them widen as trees sway and crown friction increases. This effect and more subtle ones gradually diminish the extent to which the original trees occupy the site; as more sunlight reaches the forest floor, new woody plants initiate in the understory to claim the unused growing space.

The natural, competitive suppression of trees by the shading effect of more aggressive trees does not create large vacancies in growing space. The death of such trees is merely a sign that their growing space has already been usurped by the others. At least until a stand is middle-aged, most silvicultural treatments modify a plant community that almost fully uses the available site factors. If scattered trees are removed, the remaining trees expand their crowns and root systems; the amount of foliage and effective root surface soon returns to that which existed before treatment. Root grafts (which are generally intraspecific) may enable the residual trees to immediately capture part of the root systems of cut trees.

The amount of sugar-producing foliage that a stand can maintain depends on species and, to a lesser extent, the ability of the soil factors to supply the foliage. Most evergreen species maintain more foliar surface than deciduous species, if only because the leaves persist longer than one growing season. Shade-tolerant species have more than the intolerants because their leaves can function at lower light intensity and thus form deeper crowns. Within a species, the total amount of foliage in a closed stand is much lower on a very poor site than on the best.

Poor sites produce less dry matter than good sites not so much because of major differences in the amount of foliage but because the foliage cannot function as efficiently (Assmann, 1970). Deficiencies in the supply of nutrients and water from the soil can slow photosynthetic rates and the creation of new tissues. If water becomes unavailable because of depletion, low temperature, or soil-oxygen deficiency, photosynthesis must cease. High temperature can also cause stomatal closure and cessation of photosynthesis if evaporation exceeds the rate of water uptake.

The proportion of growth concentrated below ground compared to stem and branchwood is influenced by both age and site quality (Keyes and Grier 1981; Mar-Møller, Müller and Nielson, 1954). Over time, the proportion going to above ground increases, but the diversion to below ground is much greater on poor sites. On very poor sites, more carbohydrate may go below ground for the life of the stand; on better sites more may go below ground at young ages and shift above ground in later years. On the best sites more carbohydrate may go above ground throughout the life of the stand. Although annual production of foliage is more apparent to the eye, as much or more carbohydrate is expended each year replacing fine roots (Gower, Vogt, and Grier, 1992).

Not all the carbohydrate produced by a stand is converted to permanent stem and root tissue. Some must be used for temporary tissues such as leaves, small twigs, fine roots, and fruits or cones. A major portion is expended in respiration to supply energy to the leaves, roots, meristematic tissues, and other living parts of the trees. The production of new tissues by a tree or stand of trees may be viewed as depending on the relationship between the photosynthetic or foliar surface and the aphotosynthetic surface (Satoo and Madgwick, 1982).

The aphotosynthetic surface is composed of the meristematic tissues of boles, branches, and roots; the growth of permanent tissues and respiration other than that of leaves and fine roots takes place at or near this surface. If the amounts of foliage and carbohydrate produced by a stand are fixed quantities and the fine roots completely occupy the available soil space, the amount of new stem and root tissue laid down will be determined by the amount of carbohydrate left for the aphotosynthetic surface after the requirements of respiration are met. From this standpoint, the goal of removing part of the trees in thinning could be viewed as reducing the aphotosynthetic surface substantially but in ways such that the foliage or photosynthetic surface quickly grows back to nearly its maximum extent. If the amount of foliar (and root) surface available to nourish each square centimeter of cambium is thus increased, the amount of carbohydrate left after respiration for wood formation should theoretically be greater and increased diameter growth should result. It is also important to remember that large solids have a lower ratio of surface to volume, so that the same stand volume made up of a small number of large trees will have less cambial area than the same stand volume made up of a larger number of smaller trees.

Quantification of Production

Estimates of the distribution of gross dry-matter production among the component uses have been prepared for whole stands (Kira and Shidei, 1967; Mar-Møller, Müller and Nielson, 1954). These studies indicate that respiration consumes more of the gross dry-matter production than is channeled into the permanent stem tissues. Respiration, being immediately essential to continued life, commands the highest priority in the carbohydrate metabolism of the trees.

Gross production is not constant with age. It does not approach its maximum value until the stand achieves full occupancy of the site and the foliage weight becomes constant several years after crown closure (Fig. 4.1). With further increase in age, the annual production of dry matter begins to dwindle, even though foliage weight remains constant with age. The decline may result from a combination of actual decreases in foliage and foliar efficiency and increases in respiration as stands grow taller. Taller trees will also suffer from additional water stress and close stomates more often. The slow increase of fungal, bacterial, and insect populations feeding on the tree also may make the system less efficient.

One of the most precise studies of energy transformation and dry-matter production by forest vegetation was conducted by Whittaker and Woodwell (1969) in a poor, 45-year-

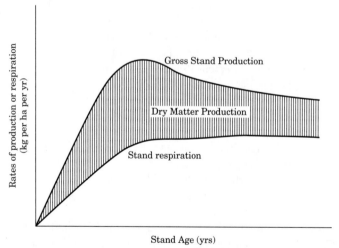

Figure 4.1   Rates of gross stand production, stand respiration, and dry-matter production of the tree component during the early life of a stand. Gross stand production reaches a maximum rate near the time of crown closure. Both the rates of gross stand production and stand respiration become nearly constant at the same time that foliar biomass becomes nearly constant.

old stand of oak and pitch pine on a droughty site on Long Island, New York. Although it would be easy to find stands that would be more productive, especially in economic terms, the study provides an example of the details of basic annual forest production:

Total sugar produced *(gross primary production)*	2647 (g per m²)
Respiration of the plants	1452
Remainder *(net primary production)*	1195
Consumed by insects, fungi, other organisms	653
Remainder *(net production of new plant material)*	542

Total solar radiation was 1,303,000 kcal per m², of which 560,030 were in photosynthetically active wavelengths. The 2647 g per m² of sugar produced is the energy equivalent of 12,440 kcal per m², which is only 0.95 percent of the total energy annually received. Forests on better sites have somewhat greater efficiency, but probably not much over 1.2 percent, even though they represent one of the most efficient ways of capturing energy from the sun. It can be seen in this case that respiration consumed more energy than that which went into tissue production. Nearly half of this production, including a major part of that which went into such temporary tissues as leaves, was consumed by insects, fungi, and other dependent organisms. This left a net production (of new plant material remaining in the stand) of 542 g per m² per year (or less than one-fourth of the gross production).

All of this production was the result of photosynthesis in only 384 g per m² of leaf tissue. The net annual primary production (g per m²) before the subtraction by consumption was distributed as follows:

	Above Ground	**Below Ground**	**Total**
Trees	796	260	1056
Shrubs	60	73	133
Herbs	2	4	6
Total	858	337	1195

The proportion of new root tissue on this dry site is probably greater than it would be in forests; as on most sites, it is noteworthy that the shrubs and herbs grew more below than above ground, but the trees did not. Only 149 g per m² of the annual tree production was in the form of main-stem wood; furthermore, the efficiency of conventional wood utilization is low in relation to the massive productivity of forests. Acorns, cones, and other reproductive structures accounted for 22 g per m² annually. As is the case with most forests, the tree stratum produced much more substance than the subordinate vegetation.

ANALYSIS OF INCREMENT

It is important to distinguish between two different ways of expressing the annual rates at which stands or individual trees add substance to themselves. Because they are perennial plants, some special bookkeeping is necessary to understand their production rates.

Current annual increment (C.A.I.) is the amount (in any unit of measure) that is actually added in a given year. The **periodic annual increment (P.A.I.)**, which is the increment of a short period (normally 5 or 10 years) divided by the number of years in that period, is usually taken as nearly equivalent to C.A.I. The effects of climatic variations and of errors in estimates of beginning and ending values are diluted or smoothed over by using a period longer than one year.

A basically different mode of expression is **mean annual increment (M.A.I.)**, which is the average amount of substance accumulated each year over the lifetime of the stand. Usually, this figure is easily determined by dividing the total accumulated substance and dividing by the age.

As shown in Fig. 4.2, these two different kinds of annual increment bear a definite relationship to each other. In an even-aged stand, C.A.I. or P.A.I. is equal to M.A.I. when M.A.I. culminates at its maximum. C.A.I. or P.A.I. can be spectacularly high early in the life of a stand but represents production rates that cannot be maintained. The M.A.I. provides the best estimate of the maximum production rate that can be continuously sustained by a given combination of species and site quality, provided that stands are not replaced much before or after the age of maximum M.A.I. The peak M.A.I. is a key value for determining how much can be harvested annually if a forest is managed to produce a sustained-yield of timber and how long the rotations should be to maximize production.

It is extremely difficult to determine M.A.I. for uneven-aged stands since they have no single age. However, every year the theoretical, balanced all-aged forest would have a C.A.I. virtually equal to the peak M.A.I. of an even-aged stand *of the same species grown to the same rotation age on a similar site*. This conclusion that age-class distribution does not alter average production rates in the long run emerged from long-term studies of forest production by Burger (1919–1953) in Switzerland.

The basic relationships between P.A.I. and M.A.I. curves shown in Fig. 4.2 are the same regardless of the units of measurement if each curve is for the same unit. Figure 4.2 is based on the merchantable cubic volume of stemwood; the curves do not originate at

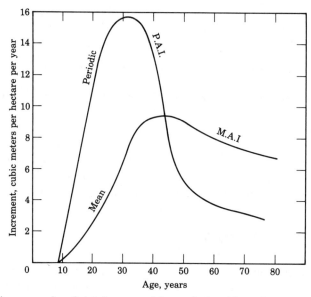

Figure 4.2   An example of the forms and interrelationships of curves of periodic and mean annual increment in an even-aged stand showing how the two rates are equal when M.A.I. is at its peak.

age zero because there is no merchantable volume until one tree reaches merchantable size. The curves would originate from zero and be shifted to the left if based on total dry-matter production. If they were based on board-foot volume with some large minimum top diameter for logs, the origins of the curves and their peaks would shift far to the right. If the measurement units define product objectives, the age of maximum M.A.I. is the optimum rotation length where the objective is to maximize production from a limited land base.

If the P.A.I. of a given stand exceeds the simultaneous M.A.I., this means that the stand has not yet reached the culmination of M.A.I.; one may also make an educated guess about the amount of M.A.I. and the time of its culmination. P.A.I. becomes negative after the age when decay and mortality become equal to growth, as in very old stands. M.A.I. would decrease to zero only when the very last tree of the initial stand was gone.

A distinction should also be made between gross and net increment. **Gross increment** includes material that formed earlier in the life of the stand or measurement period but was no longer present at the end; **net increment** includes only that present at the end. However, the values of any of these units can be quite different depending on whether the production of temporary tissues such as leaves and fine roots during the final year is included. One also needs to be clear about whether total increment is being considered or only the above-ground biomass.

Most of the estimates that have been made of the total dry-matter production of forests have been of P.A.I. or C.A.I. in immature stands. The rates of production by such stands are often impressively high but cannot be sustained over long periods. For example, the Long Island study involved a stand in which the culmination of P.A.I. appears to have been reached 5 to 10 years earlier but that of M.A.I. had not yet been attained.

Studies of total production elucidate the fundamental biology of the effects that differences in species, site factors, and silvicultural treatment have on production in forests. The much longer history of studies involving measurements of production of stemwood, often only in merchantable units, provides a large and useful body of information. However, very few studies have adequately addressed the biological and economic considerations simultaneously. The large number of careful measures that are needed to investigate primary productivity makes these studies useful in describing a biological mechanism. Most studies of merchantable units leave biological mechanisms unexplained but lend themselves to economic analysis. It is difficult to develop a study that can adequately investigate both issues with statistical rigor.

Parameters of Stand Density or Stocking

Thinning is the direct reduction in the number of trees in the stand. It is necessary to consider first the relationship between the number of trees in the stand and yield before the effects of thinning can become obvious. Growth is determined by the amount of foliage in the stand, but this is very difficult to measure. Many other measures are used to relate the foliage and the number of trees over which it is distributed.

The general term **stand density** is a measure of the amount of tree vegetation on a unit of land area. It can be the number of trees or the amount of basal area, wood volume, leaf cover, or any of a variety of less common parameters (West, 1983). **Stocking** is the proportion that any such measure of stand density bears to any of a wide variety of norms expressed in the same units and chosen for differing purposes. Density tells what is, and stocking how this density relates to someone's notion of what ought to be, usually in terms of percentages. Full stocking might, for example, be the basal area per acre of trees larger than 10 inches ($\approx$25 cm) D.B.H. that was deemed necessary to maximize the production of sawn boards. A thinning schedule is a guide to what stocking, expressed as some kind of stand density, should be at each stage of development.

No parameter of stand density has yet been devised that would, if held constant after each thinning of a whole rotation, define a series of thinnings that met any plausible set of objectives. Many of these parameters can be used to quantify stand density or stocking in thinning schedules, but neither they nor any other form of simple mathematical magic seem able to define a schedule by themselves. It may help to consider the utility of each of these parameters as measures of density or stocking.

The simplest parameter of all is **number of trees** per unit area. This takes no account of either the sizes of the trees or the space they occupy. However, there is no fundamental reason why a thinning schedule could not be presented in these terms, for a given species and site quality, if it had already been developed in terms of more meaningful units. Such a schedule would have to be adhered to rigidly because the trees would develop sizes different from those contemplated if there were significant departures from the program.

Basal area per unit of land is by far the most commonly used parameter, although there is probably more tradition than biological reason for this. Its main virtue is that it is a kind of integrated expression of numbers of trees and their sizes. It was originally devised not so much as a parameter of stand density but as a crude indicator of cubic volume of stemwood. Because it measures the cross-sectional area of physiologically dead wood, it has no direct biological significance to the current stand structure. However, the basal area of a tree is fairly well correlated with the cross-sectional area of the crown. This means that if one removed 25 percent of the basal area of a fully closed stand, one would also be making approximately 25 percent of the crown-level growing space vacant.

It is quite easy to measure basal area per unit area by various point-sampling techniques. However, there is seldom any virtue in continually thinning back to some constant basal area, even though it often seems convenient to think so. It does make some sense to reduce the basal area by some percentage and then let it grow to a level higher than existed previously before thinning again.

Theoretically, the cubic volume of stemwood per unit area should be a better indicator than basal area of the amount of foliage per acre, but it is more difficult to determine and there is no indication that the refinement would help. Because it is weighted by tree diameter, board-foot volume is a good way to assess the results of thinning programs, but for the same reason it would not be a good parameter for expressing them.

There is strong evidence that good correlations exist between the **cross-sectional area of sapwood** and the amount of foliage borne by a tree (Waring et al., 1977). This surrogate parameter for foliage is difficult to use for a thinning guideline except in experimental situations. All specific allometric relationships between tree parts depend on the individual growing situation of a tree. For example, the stem height for a given correlation depends on the amount of butt-swell that can be attributed to wind sway. However, like all surrogate measures, sapwood cross-sectional area is very useful conceptually and can be used within stands to compare foliar amounts of different trees. It is a difficult measure to determine because increment borings need to be made in the trees and the sapwood area must be identified.

The amount of stem or **bole surface** of trees on an acre or hectare has the virtue of approximating the amount of growing and respiring meristematic or aphotosynthetic surface of the above-ground portion of the stand and is proportional to the total aphotosynthetic surface. If the amount of foliage that a stand of given species and site can produce is fixed, it would seem that the average diameter growth of the trees could be predicted by knowing how much bole surface area was linked to that foliage.

One useful index of bole surface per unit area is the **sum of the diameters** of trees. This is because bole surface is the sum of the products of circumferences and heights of tree stems modified by some function of stem taper. If we make the sweeping assumption that the trees of an even-aged stand do not differ in height or taper, and note that *pi* is a constant too, then the sum of diameters becomes the chief variable function of bole surface. In this same way, we may deduce that basal area could be changed to the sum of the squares of the diameters and we would get just as good a surrogate for stem volume per unit area. As far as the present purpose is concerned, much analysis has been done in terms of basal area but little in terms of bole area.

One particular difficulty lurks behind the use of breast-high diameter in assessments of stand density or growth. The problem is that trees are taller than people, and breast height usually falls in the zone of the butt-swell where the effects of fast or slow growth of whole trees are greatly exaggerated. This is the point discussed in Chapter 3 and depicted in Fig. 3.2. What this means is that one unit of basal area stands for more crown area with a smallish tree in a stand than does that unit in a dominant.

This problem could be solved, at least in research work, by measuring diameters higher up on the trees. In American observations, the best fixed point would be at 17.5 feet, the level at which inside-bark diameter is measured to determine Girard Form Class. This mode of quantifying stem taper has been used in constructing many volume tables.

RELATIONSHIPS BETWEEN NUMBERS AND SIZES OF TREES

Studies of single-species plant populations of a variety of plants have shown a relationship between the maximum number of individuals that can occupy a site and the average size of the individuals. In these studies, the maximum number of plants of any given size that can exist on a site is correlated to the average biomass raised to the 3⁄2 power (Westoby, 1984). Many attempts have been made to quantify and apply this relationship to forest stands. The major problem is that it is almost impossible to know the living biomass of a tree. As well as being large, most of the tree is dead wood within the living sheath of the cambium. As a result, many surrogate variables for biomass are used such as volume, basal area, and diameter; all are imperfect and somewhat imprecise.

All thinning schedules derived from this type of straight-line relationship between the average sizes of trees and their total numbers are based on the fact that trees die from self-thinning when stands are near the maximum density for a given tree size. The relationship for Douglas-fir first derived by Drew and Flewelling (1977) is shown in Fig. 4.3. It is important to notice that a stand that starts at a certain stand density well below the maximum line will eventually reach the point of maximum density and trees will begin to die just because the individual trees grew in size. The relationship between the stand density and the maximum density that could occur at the same average tree size is referred to as the **relative density**. Thinning guidelines can be drawn up that indicate at what relative density thinning should take place based on the products desired and experience with the species. In such a guideline, many light low thinnings will have very little effect on the relative density of the stand because the average tree size will change little although the density will decrease. Mortality will be captured by the harvest without providing increased growth by the codominants (as will be discussed in Chapter 5).

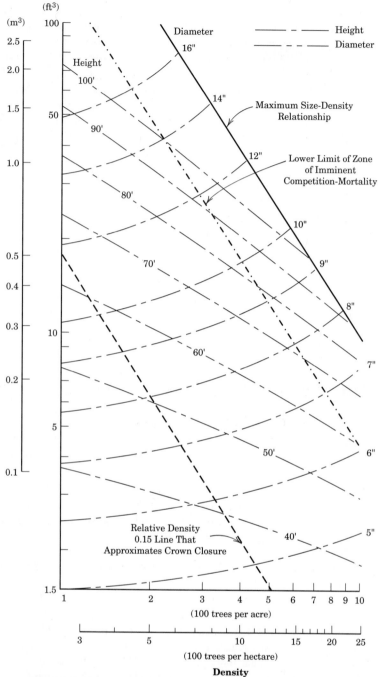

Figure 4.3 Relationship between maximum average tree size and number of trees per acre (hectare). As trees get bigger, the growing space can be occupied by fewer trees. Relative density is defined as the number of trees actually in a stand divided by maximum number of trees of that average size that could exist. When relative density exceeds 0.55, trees begin to die (Zone of Imminent Mortality); crown closure usually occurs at a relative density of about 0.15. Thinning regimes should be designed so that the relative density is kept below the 0.55 density line (———·———·———). (From Drew and Flewelling, 1977; reprinted from *Forest Science* with permission of Society of American Foresters, reproduction without permission is prohibited.)

Other measures of relative density have been developed based on other tree measures. The stand density index developed by Reinecke (1933) is based on inverse straight-line relationships between the logarithms of (1) average D.B.H. and (2) numbers of trees per acre. The index value is the number of trees for stands that have average D.B.H. of 10 inches (25 cm). This value is higher the more shade tolerant the species. This index is one of the easiest to use because it is not necessary to determine the tree volume.

It can be dangerous to deduce "laws" from artificially constructed models that are at least once removed from the original data. In this case, however, it is probably safe to infer that the relationships do exist in crowded stands and have biological basis.

A similar approach has been used for handling the problem from the opposite part of the possible range of stand density, the open-grown tree (Gingrich, 1967). In this it is postulated that the very lowest density to which a stand might logically be thinned is that in which each tree was left occupying the space that its crown could expand to cover if it were open-grown. The statistical relationships between the diameters and the numbers per acre of such trees are determined for hypothetical stands that consisted of open-grown trees of uniform stem sizes that were just barely closed. The upper limit, beyond which stand density should not be allowed to increase, is defined by similar relationships between numbers and sizes of trees in fully stocked stands. The results are usually plotted on nomograms like that shown in Fig. 4.4.

These relationships represent attempts to develop mathematical relationships between tree size and stand density. Because most of the diameter and volume of a tree is a reflection of past growth and not current status, parallel relationships do not exist between stands grown by thinning regimes that kept stand density away from the open-grown condition or that of overcrowding. Similarly, it is not clear that quantitative guidelines of this type should be consistent across thinning methods. The vigor of the trees left after thinning will determine how quickly the vacant growing space is reoccupied (see Chapter 5).

Modeling Growth and Yield

Many attempts have been made to refine thinning schedules by simplifying the stand growth processes and explicitly including time. The simplest approach is a yield table that indicates yield of stands at different ages on different sites. These tables are constructed by measuring stands of many different ages. Because not every age is equally represented, it is usually necessary to mathematically determine the yield for some ages to fill in the table. Different mathematical techniques can be used, but all represent some form of interpolation. The yields are assumed to be for fully stocked stands; stands less than fully stocked are simply reduced by a percent factor.

More sophisticated mathematical models can be constructed by using multiple regression equations (Clutter et al., 1983). The purpose of the equations is to predict the growth of trees or stands on the basis of measurable factors that are alterable by thinning. In this procedure, as many factors as possible that might govern growth, the single dependent variable, are tested as independent variables in multiterm equations. Whatever equation explains the greatest amount of variation in growth (of tree or stand) is taken as the best quantification of the phenomenon. Among the independent stand variables commonly used are age, basal area, numbers of trees, site index, average diameter, wood volumes, and various indexes of stand density based on the aforementioned relationships between average D.B.H. and numbers of trees.

When the goal is to explain the growth of individual trees, the variables often include the initial sizes of the trees themselves, site index, crown dimensions, distances to other

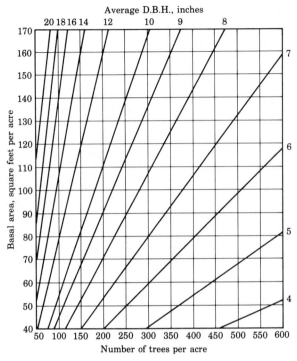

Figure 4.4 A nomogram for relating the average D.B.H. of stands to basal area and numbers of trees per acre in the guidance of thin ring schedules. Note that the average D.B.H. of a stand is, by convention, the quadratic mean or the D.B.H. of the tree of average basal area and not the arithmetic mean. The largest trees in the stand would usually be considerably larger than the average tree. Almost any thinning schedule could be described on this diagram by a saw-toothed line along which the stand "moved" over time from right to left. However, there is nothing inherent in this diagram that tells what the schedule should be or how rapidly stands could be made to "move" across it.

trees, live crown ratios, or parameters designed to quantify competitive effects. Each analysis is typically confined to data for some species in a particular locality and is designed to predict growth or yield rather than to explain the results.

This kind of analysis can be very useful in providing the basis for thinning schedules if it is done with due regard for both biological principles and those of statistical analysis. The results should not be used without critical appraisal and consideration of why the equations come out the way they do. The fact that the multiterm equations have a superficial resemblance to those of physics does not mean that they explain everything.

If such analyses are set up so that the goal is to maximize the production of cubic volume per unit area, the equations will tend to indicate that thinning should be very light or not done at all. If the initial diameters of trees in even-aged, single-species stands are used as variables to explain the subsequent growth of the trees, this method will obscure the importance of other more fundamental variables. Such analysis will always show that the trees that have grown the best in the past are likely to continue to grow the best in the near future. The factors that caused the tree to grow well are obscured. The resulting equation is mostly an extrapolation of the past. These analyses are important because they remove any subjective bias of the observer, but they need to be interpreted carefully. Many dependent variables are not truly independent of each other, and the data used in the analysis are not well balanced because of these interactions. All application of model results to combinations of conditions that did not exist in the original data set must be made with great caution.

Sometimes the equations developed by multiple regression can be used to form computer simulation models. The equations will be used to increment tree size and then the larger tree size will be run through the equations to increase tree size again. It is important to remember that simple repetition of the equations will result in errors being compounded; trees that grow too fast in the simulation will continue to grow faster. Random effects are often added to the regression equations to approximate natural variation in basic growth rates (Stage, 1973).

Regression-type simulation models can be developed for whole stands or for individual trees. In the latter case the stand yields are determined by summing the individual trees.

Another modeling approach is to grow each tree by calculating the effects of physiological processes and then determining the effects of competition between adjacent trees. These models can be either independent of tree location (Shugart and West, 1980) and competitive effects between trees averaged or individual crown growth can be simulated in location-dependent models (Mitchell, 1975). A great number of computer simulation models exist. Each model has characteristics that make it appropriate for certain management questions in a given situation. It is important to determine which model is based on the appropriate assumptions for the problem at hand rather than trying to rank models in general and having a "favorite" for all questions in all situations.

THINNING AND ITS OBJECTIVES

The yield of merchantable timber volume by stands can be optimized by judicious, temporary reductions in stand density that enhance diameter growth. This process can be regarded as a controlled and accelerated reduction of the number of trees or the amount of photosynthetic area left to regain full occupancy of the growing space. Artificial reductions in stand density cause temporary reductions in gross total production but enable the remaining trees to accelerate their occupancy of growing space and their diameter growth. Some of this same effect can also be achieved by control of initial stand density in the spacing of planted trees (see Chapter 11).

Surplus trees are removed in thinning to concentrate the potential wood production of the stand on a limited number of selected trees. Total yield is augmented largely by harvesting trees that would ultimately die of suppression. However, the value, utility, and health of the remaining trees are increased because of their more rapid growth.

It is usually well to keep thinnings light enough to restrict the growth of any shrubs, vines, or undesirable trees that will cause problems at the time of regeneration. The gaps created in the crown canopy by most thinnings are so small that they close before any understory trees can start the kind of rapid, sustained growth in height necessary to count as new age-classes of regeneration. Thinning is done to regulate the distribution of growing space for the benefit of the existing crop and not to vacate enough space to start a new one.

Attempts, either ambitious or casual, to obtain regeneration and secure the benefits of thinning simultaneously often wind up doing neither thing well. Desirable regeneration may appear under stands as a result of thinning or even without any treatment. If it is persistent enough or is deliberately sought to be a new crop of trees, then it is advance growth and is best regarded as shelterwood regeneration rather than as a result of what was intended as thinning. This semantic switch helps focus attention on the source and nature of the regeneration process, which is a more crucial step in the silvicultural system than thinning.

The general principles of thinning are most expeditiously considered in connection with aggregations of trees that are even-aged, of a single species, and thus all in the same crown-canopy stratum. Although it is simplest to think of such single-canopied aggregations as being whole stands, they could also be the even-aged groups that are immature components of uneven-aged stands. Thinnings can also be done in mixtures of species including those with differing rates of height growth that segregate into stratified mixtures; these are considered in Chapter 16.

Quantifying Objectives

A program of thinnings is essentially a series of temporary reductions made in stand density (measured in terms of basal area or some similar parameter) to maximize the net value of products removed during the whole rotation. Among the factors determining this net value are the quantity, quality, utility, and size of the products as well as the costs of harvesting and manufacture. Production in terms of quantity or volume of wood is usually considered to be the factor of greatest importance. In this connection, clear distinctions should be drawn between all the different ways in which the growth and yield of stands and trees are measured. Thinning practice cannot be adequately understood without a thorough comprehension of the relationships between the very different mensurational units in which growth can be evaluated.

It must be kept in mind that thinning is primarily intended to optimize economic returns for timber production. This means that the mensurational units used to assess various treatment alternatives, including no thinning, are meaningful only to the degree that such units of measure indicate the true net value of the material grown and removed. Unfortunately, conventional mensurational parameters are, at best, only preliminary approximations of this value, mainly because they do not automatically reflect the variations in value associated with tree diameter and wood quality. It should not be inferred in any situation that each unit of cubic volume, weight, or board measure is as valuable as all others.

Even the board-measure log rules do not perfectly indicate true net value. All that they are intended to show is that the sawing waste that comes from converting round logs into square-edged boards is less in large logs than in small ones. They do not otherwise reflect the higher handling costs of small trees, except perhaps to the extent that inventions such as the Doyle log rule, which discounts small logs, may happen to do so accidentally. Even after wood is converted into flat boards, it is necessary to recognize that a board foot of a certain grade is usually worth more in a wide board than in a narrow one.

It is very important to make a clear distinction between the yield of forests and their production. **Yield** is the amount that is actually harvested or could be harvested. **Production**, which is more difficult to determine, is the amount deposited by growth whether or not it is harvestable.

Foresters reliably observed several centuries ago that thinning could increase the yield of usable wood from stands. However, the total production is ordinarily decreased. Some of the data about production after thinning are subject to large variations because of the inherent variability in productivity within tree species and in the plots of ground on which their growth is measured. It is also difficult to measure production precisely. Much confusion has arisen from attempts to measure production in terms of merchantable or quasi-merchantable cubic volume. These parameters approximate yield better than they do production, and the diameter limits used in their definition are seldom standardized enough to compare results effectively.

If total production is measured in terms of tonnage of dry matter, there are extremes of stand density at which production would suffer. If the trees being measured did not fully occupy the growing space, their production would clearly be less than if they did. Some other vegetation that was not counted would tend to fill any unused space; its production, if counted, might partially offset the production deficiency of the trees.

If the stand density is very great, there are some instances in which total production is diminished. This happens when a large proportion of the fixed amount of stand foliage goes to support the respiration of such a large amount of living and consuming tissue that the surplus available to form new tissues, including wood, is reduced.

There are also cases, in climates conducive to heavy accumulation of undecomposed organic matter, in which thinning can speed decomposition and nutrient cycling enough to cause real increases in production. Some trees, even those of a given species, are inherently more productive than others; thus stand production could theoretically be increased if such trees were favored in thinnings.

Otherwise, it appears that the thinning of ordinary stands decreases rather than increases the kind of gross, total production that counts all organic materials in the stand, including the trees that died during stand development. If yield and production were indeed equivalent and everything that the stand produced could be harvested, it would be logical to confine thinning to the salvage of suppressed trees just before they died. The vacancies in the growing space left by heavier thinning, regardless of how small or temporary they may be, usually seem to cause uncompensated decreases in true total production.

It is economically impossible to turn all production into yield. Therefore, it is practical to consider how variations in stand density induced by thinning affect production in terms of total cubic volume of stemwood or of potentially merchantable wood. This departure from scientific measurement units such as total dry matter opens a Pandora's box full of measurement units that differ in so many large or subtle ways that they have caused much confusion.

Almost any mode of measuring cubic volume of wood involves some restrictive choice of the smallest stem diameter or of the kind of stem material that will be counted. Total cubic volume is customarily taken to be all the wood in the central main stem from tip to ground level; this parameter is one for which bark is specifically included or excluded. It is reasonably well defined for most conifers and some hardwood species that have excurrent branching and single central stems. However, it involves a concept that is difficult to apply to species with deliquescent branching and single stems that fork far below the tip of the crown. Even in excurrently branched trees, so-called total volume does not include the wood of the branches or roots.

Merchantable cubic volume can be defined only in terms of some specification of the smallest diameter or other characteristic of stem components regarded as merchantable or, often, as potentially merchantable. Usually, the restriction is some minimum diameter, but sometimes it is the lowest point on the main stem at which forks or the presence of large branches prevent utilization. Bark may or may not be included. Most European observations involve a diameter limit of 7 centimeters, including bark and branches, but American specifications vary greatly. Usually, whole trees that are below a certain D.B.H. are excluded.

The practical reasons for setting such restrictions are obvious, but they are probably also part of the source for the conflicting evidence about the relationship between stand density and production. The greater the density of a stand, the slower the diameter growth of all stem components, including branches. Therefore, the denser a stand, the smaller the proportion of the total amount of wood that is included in merchantable or even ''total'' cubic volume. When production is assessed in such terms, it often appears to be higher at moderate levels of stand density, such as those left by thinning, than at higher stand densities. The larger the restrictive minimum diameter set on measured cubic volume, the more accentuated is the effect; it becomes even more pronounced if production is measured in the American board-foot unit. Although such practical mensurational adjustments have clouded the scientific study of controlling stand production, they have advantages in considering the technology of thinning, especially if the nature of their effects is recognized.

The Effect of Thinning on Stand Production

Figure 4.5 shows three alternative interpretations that have been proposed about the relationship between stand production in gross merchantable cubic volume after thinning and density to which stands were reduced in thinning. Stand density is expressed in basal area after thinning, and cubic volume includes stemwood to minimum diameters of 4 inches (10 cm). Production is expressed as gross periodic annual increment during periods of 5–15 years between thinnings. Gross production includes wood laid down on trees that died during the period; net production would ordinarily deduct the entire volume of such trees and will be discussed later. In Fig. 4.5, all the differentially thinned stands being compared are of the same single species, the same age, and on the same site. In other words, stand density after thinning is the only independent variable.

The first interpretation (A) is that production increases right up to the highest level of stand density that can be maintained in nature. Any vacancy in the growing space is counted as reducing total volume production.

It may be well to digress here to point out that there is a long-standing procedure under which it is presumed that Alternative A is represented by a straight-line relationship. It has been common in North America to predict yield from adjustments of so-called normal yield tables. These tables are based on stand densities equal to or only slightly less than the highest ones attainable, but such densities are so uncommon as to be abnormal in the ordinary sense of the word. Predictions of yield are often made by the simple assumption that a stand that has 80 percent of the ''normal'' stand density will give 80 percent of the ''normal'' production. This assumption would be correct if 20 percent of the growing space of the stand were vacant and remained so throughout the rotation. Although most stands have some permanent vacancies, these tend to refill to some extent; therefore, the assumption of a straight-line relationship between stand density and production would generally lead to underestimates. Thinnings are supposed to be light enough that all vacancies soon fill up again.

Alternative A would probably fit the relationship between total production of dry matter and stand density, except that production might decline at very high densities in grossly overcrowded stands. However, it is clear that when production is assessed in merchantable cubic volume, Alternative A fits only part of the cases.

The second alternative (B) is that production remains constant and optimum over a wide range of stand density from some lowest level at which there is full occupancy of

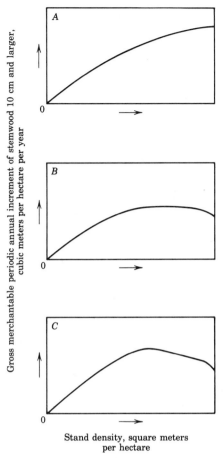

Gross merchantable periodic annual increment of stemwood 10 cm and larger, cubic meters per hectare per year

Stand density, square meters per hectare

Figure 4.5　Graphs depicting three hypotheses about the effect of changes in stand density, induced by thinning, on the production of merchantable stemwood in pure, even-aged stands, all of the same species, site quality, and age. See text for discussion.

growing space up to those levels at which excessive competition is postulated to restrict production. This interpretation received its greatest impetus from studies by Mar:Møller (1947, 1954) and others in Scandinavia. Alternatives B and C rest on the assumption that full occupancy of the growing space can exist at comparatively low levels of stand density.

The data graphically presented in Fig. 4.6 indicate an effect of differences in site quality that may partly clarify the situation. The data are from observations of many thinned stands of loblolly pine in the southeastern United States (Nelson et al., 1961). The data were fitted by statistical techniques that were not obviously biased in favor of any of the three alternatives. On the better sites, production increased to nearly the highest levels of basal area, much as Alternative A would suggest. However, on the poorer sites, Alternative B best fits the results.

It is quite possible that, on the poorer sites, seasonal moisture deficiencies or other soil factors limit production more at high stand densities than at intermediate levels consistent with full or nearly full occupancy. On the better sites, production increases almost

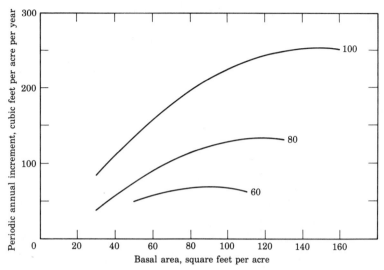

Figure 4.6   The 5-year periodic annual increment of merchantable cubic volume of stands of loblolly pine, averaging 40 years old, thinned to various levels of basal area. The stands were on sites of different quality in the southeastern United States and these curves represent regression equations fitted to the experimental data by Nelson et al. (1961). These relationships suggest that production in thinned stands increases with increasing stand density on the best sites where water is not limiting but it is optimum and nearly the same over a wide range of stand density on more restrictive soils.

up to the limit set by the amount of light because soil factors do not set much restriction on the amount or efficiency of foliage. It is quite possible that if below-ground production could be measured and were added to the total, the curves would have more similar shapes.

There is indeed evidence indicating that on excellent, nonrestrictive soils, the best cubic-volume production comes with light thinning or none (Alternative A). However, on the more ordinary and common soils, which are restrictive, optimum basal area gradually increases with age according to Alternative B or C in youth and middle age, but Alternative A best fits the final stages of a rotation.

The foregoing analysis applies to production of merchantable cubic volume with quite small diameter limits. If the diameter limits were even smaller, Alternative A would probably better approximate the truth in more cases. However, if the minimum diameter limits were increased, the logic of Alternatives B and C and of thinning to ranges or points of moderate basal area would be strengthened. For example, if one used the American board-foot unit, which weights production in large stemwood more than that in small, in many cases Alternative C and rather heavy thinning would presumably become most logical.

Part of the difference in interpretation between Alternatives A, B, and C, results merely from the bewildering variability that would become obvious in the scatter of points representing the actual observations on which graphs such as those of Fig. 4.5 are based. Some of this variance is due to factors such as genetic differences, variation in site quality or stand density within experimental thinning plots, and observational errors, especially those involving the effect of taper on stem volume. In fact, the hypothesis of B may rest on nothing more than the fact that often there is no universally demonstrable upward or downward trend in total periodic annual increment with respect to moderate artificial decreases in stand density.

Even if Alternative B were incorrect, its tacit acceptance would have the thought-provoking effect of focusing attention on the thinning considerations that can be important regardless of what its effect on production of total cubic volume might be. If decades of study have produced so much contradictory evidence, we are entitled to conclude that, as long as stands remain nearly closed, the effect of stand density on production is not large enough to represent an exclusive consideration in determining how to thin.

The Effect of Thinning on the Economic Yield of Stands

Practical understanding of thinning procedures depends on knowing how they can be applied to increase the economic yield, even though they do not substantially increase, and may even slightly decrease, gross production. The general approach is to allocate the production to some optimum number of trees of highest potentiality to increase in value; the other trees are systematically removed in such a sequence as to obtain maximum economic advantage from them. The various advantages that can be gained are as follows:

1. Salvage of anticipated losses of merchantable volume.
2. Increase in value from improved diameter growth.
3. Yield of income and control of investment in growing stock during rotation.
4. Improvement of product quality.
5. Opportunity to change stand composition to prepare for the establishment of new crops.
6. Reduced risk of damage or destruction by insects, disease, fire, or wind.

Salvage of Anticipated Losses

The gross production of wood by a stand should not be confused with the actual yield in terms of usable volume removed in cuttings. Not all the cubic feet of wood produced by the growing stand remain stored on the stump until the end of the rotation. In fact, a high proportion of the total production will be lost from death and decay of the large numbers of trees that do not survive the struggle for existence. From the economic standpoint, any part of this perishable volume that can be salvaged by removing doomed trees in thinnings represents an increase in the quantitative yield of the stand. However, at least in the early years, much of this material is not harvested because harvesting and transportation costs exceed the value of the trees.

If a typical stand is grown on a rotation long enough to produce trees of conventional sawtimber size without thinning, as much as one-third or even half of the potentially merchantable cubic volume may be lost to suppression. This kind of loss can be reduced by ending rotations when such mortality becomes serious, but this approach makes it hard to get trees of the diameters often desired. Diameter growth can be increased by widening the initial spacing of the stand, but this maneuver prevents suppression loss partly by leaving growing space vacant and thus merely preventing the production that would become loss. Thinning is the best solution to this dilemma and is therefore practiced whenever other considerations do not preclude it.

Thinnings designed to anticipate and salvage losses from natural suppression are the only proven means, other than site improvements, of increasing the yield of total cubic volume or tonnage from a stand. Only part of the prospective losses can be recovered unless stands are annually gleaned for dead and dying materials no matter how small. Unless the site is ameliorated such as through fertilization, the practicalities of timber harvesting dictate that the harvested yield can be increased only by decreasing the gross production of the stand.

Not all of the material removed in anticipation of loss need come from small trees of the suppressed and intermediate crown classes. By using crown and selection thinning it is possible to take out part of this volume in larger trees, thus forestalling the death and stimulating the growth of their subordinates. Selection thinning must be used carefully in this way because frequently the lower crown class trees will not respond well to release and the overall growth of the stand will be reduced.

Increase in Value from Improved Diameter Growth

Up to this point, the discussion of the effect of thinning on the quantitative yield of a stand has dealt mainly with production of cubic volume. However, not all units of cubic volume of roundwood are equally valuable; those that come from large trees are generally of greater value than those from small trees. One of the most important objectives of most kinds of thinning is to reduce the stocking of a stand so that it eventually has fewer trees, but of larger average diameter, than it would without thinning.

The most general reason for the greater unit value of trees of large diameter is that the cost, per unit value of ultimate product, is less for processing them than for small ones (Fig. 4.7). This relationship applies to the cost of handling each tree or log individually during harvesting and subsequent processes. The monetary return for growing a tree, stumpage, is the difference between the value of the ultimate unit of product and the cost of harvesting, transporting, and processing it. The forest owner who grows a hundred units of cubic volume in fat trees saves handling costs for the buyer and is thus entitled to a higher price than one who grows a hundred units in a greater number of slender trees. Of all the economic reasons for thinning, this one is usually the most important.

A second reason for trying to improve diameter growth is that wood quality usually increases with tree diameter. This is mainly because the outer rind of wood has fewer knots and grain irregularities than the central core; ordinarily, it also has superior anatomical and mechanical properties. The outer wood is usually stronger, more easily worked, longer-fibered, less subject to warping, and in most ways much easier to use than the inner wood. With some species, the inner wood improves when it turns to heartwood, but this change also depends on diameter growth. The more outer wood that can be added by increasing diameter growth, the greater is the inherent value of the wood.

One of the most common and valuable ways of using trees is as lumber or other solid-wood products. There is probably no other way of producing structural materials that is more economical of the world's energy supply. Unfortunately, it is not correspondingly economical either of the wood itself or of the time required to grow trees of suitable sizes. Both of these problems can be mitigated by thinnings designed to grow trees of larger diameter in shorter time.

The waste of wood in lumber production is in the sawdust, slabs, and edgings generated when round stems are sawn into square-edged boards. Given the ordinary requirement that boards be of certain minimum thicknesses, the proportion of wood that is wasted is greater in trees of small diameter than in large ones. In fact, there is usually some minimum diameter of stemwood below which the material is simply left in the woods; the proportion thus abandoned in severed tops is also inversely related to diameter. One exception is the production of small-dimension products which often results in utilization to

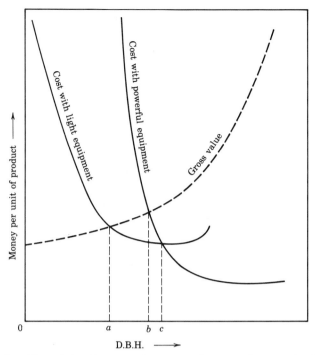

Figure 4.7 The effect of the diameter of a tree on the cost of processing it and the value of the product, per unit of product, when the harvesting and processing are done with two different sets of equipment. The set of light equipment is low in capacity and cost; the powerful equipment is high in initial investment and hourly operating cost. Points a and b are the smallest diameters at which it becomes economical to use each set of equipment; for trees larger than diameter c it would be logical to use the more powerful set. Net value at any diameter is the vertical distance between the appropriate curve of cost and that of gross value. Note that certain costs, such as those of roads or of handling trees collectively in bundles, are not affected by tree size. The curve of gross value would be nearly horizontal for such products as fuelwood or that chipped for pulp or other reconstituted products. Gross value can also decline with increasing tree size if the incidence of rot or similar defect increases with age or size. Note that the ideal thinning program would be one in which trees are caused to move as swiftly as possible into diameter classes that can be harvested most profitably and also with equipment appropriate to those sizes.

a very small minimum diameter. In this case, a large number of small trees will have less total branchwood than a smaller number of large-diameter trees.

The board-measure log rules that are used in North America are, in effect, computational devices designed to estimate the volume of lumber that will remain after round logs are sawn into square boards. A cubic foot of wood could be converted into 12 board feet only if a tree were a square timber throughout its length and could be sliced into boards without sawdust or other waste. Cubic volume is, in other words, round and not square. The ratio of board feet to cubic feet ranges from roughly 5 in trees about 8 inches

($\approx$20 cm) D.B.H. to a nearly stable value of 7 or slightly more in trees that are more than 20 inches ($\approx$30 cm) D.B.H. If trees are grown for lumber and the cubic volume growth is fixed, this difference in proportion of waste is another reason to seek improved diameter growth. Even if the waste is used for reconstituted wood products, loss of potential value still takes place.

Not all units of board measure are equally valuable. Those that are in wide boards are worth considerably more than those in narrow ones. This is yet another advantage that accrues to improved diameter growth.

Merely because the calculations are easier, the results of thinning trials are often interpreted in terms of common mensurational units, especially cubic volume, without any consideration of the distribution of diameters in the stands. In fact, many thinning trials purposefully have tried to duplicate the distribution of diameters before and after thinning for statistical convenience in determining volume-growth changes. The purpose of most regimes of stand management is to maximize the production of net value and not just volume.

Rough comparisons can be made between possible treatments by observing the distribution of numbers of trees or of volume by diameter classes. Another simple test parameter is the average diameter of a fixed number of the largest trees per unit area, such as the 100 largest per acre. It must be borne in mind that the average diameter of a whole stand is automatically increased by removing some of the smaller trees.

To reap the rewards of cutting large trees late in the rotation, it is necessary to cope with the high cost per unit volume of harvesting and utilizing small trees in earlier thinnings. This difficulty can be mitigated partly by using different equipment and logging methods for the small trees than are later used for the large. The kind of equipment that is most efficient for handling final-crop trees is usually not the best for the thinnings; such equipment is ordinarily heavier, more cumbersome, more costly, and more expensive to operate for a given length of time.

The problem can also be evaded by delaying the thinnings until the trees to be removed grow large enough to handle economically. Another evasion is to confine the first removals to any large or medium-sized trees of poor potentiality and to ignore the small ones until they become larger. Trees that are destined never to be worth harvesting can be either left to die of suppression or killed by the cheapest means available.

The ideal solution is to manipulate the equipment and sequence of thinnings so that the equipment is matched with the sizes of trees to be removed and as many trees as possible are cut when they have grown into the range of diameter that can be harvested efficiently. The objective should be to maximize the total value from harvesting the entire crop of one rotation rather than to minimize the cost of each separate operation in the whole sequence of cuttings.

Increase in stem diameter is so advantageous that it is generally best to thin stands down to the lowest density that will not cause poor tree form or other unacceptable side effects. Deliberate sacrifice of production of gross tonnage or total cubic volume is often desirable; sacrifice of production of merchantable volume may or may not be desirable, depending on the relative merits of growing large trees and securing a large total yield of wood of any sizes. Some countries are sufficiently concerned about maintaining high wood production that they prohibit thinning so heavily that yield of merchantable cubic volume is sacrificed. The stand structures that result from heavy thinnings sometimes have other advantages such as foraging areas for the red-cockaded woodpecker in the Southeast and the maintenance of hardwood browse for large herbivores in many areas.

Yield of Income and Control of Investment in Growing Stock During Rotation

An unthinned crop of trees is an asset that increases in value throughout the rotation but remains frozen under passive management. No cash income is realized, yet carrying charges that must be paid for by the crop accumulate. After some trees have become merchantable, thinning can become a form of continual asset management.

The early returns from thinning will lead to longer rotations if the overriding goal of management is timber. Money from the early part of the rotation is worth much more than income received later in the rotation because of the compounding effects of discount rates (Chapter 16). The periodic removal of the poor earners and encouragement of the good is a means of decelerating the speed at which the growth rate of the overall stand value inevitably declines. The rate declines between thinnings and can be elevated again by each thinning. When it becomes impossible to restore the rate to the desired level, according to this mode of analysis it is logical to replace the stand with a new one and thus end the rotation.

In applying this idea, it is best to allocate growing space to the trees that earn the most. This allocation requires estimates of the earning capacity of trees of different classes of diameter, crown class, and stem quality. Once a few calculations are made and the basic relationship between tree characteristics and earning capacity is formulated, it is then necessary to use intuitive judgment in determining which trees to cut or leave because the detailed computations are too complicated to apply tree by tree. It is especially difficult to take differences in tree quality into account. The easy assumption that all trees of the same diameter in a stand are of the same value can impair application of the principle.

Thinning can be applied in such a manner as to reduce the investment in growing stock and increase the value of gross growth by removing trees of low earning potentiality and increasing the growth of those of high potentiality. Because growing stock can be reduced without greatly reducing volume growth, this becomes a way of eating one's cake and having it too.

In whole forests that are young and immature because of earlier heavy cutting in a locality, thinnings and other intermediate cuttings may provide the only source of income for long periods.

Thinning helps meet two financial tests that are often applied to long-term investments in timber production. Both tests involve compound interest and affect decisions about rotation length. Consequently, the two different tests often get confused with each other.

The first test requires an adequate rate of compound-interest return on all out-of-pocket costs of establishing the crop and carrying it to maturity. The second test requires that the rate of interest earned on the stumpage value of the trees or stands not fall below some desired rate. The rate demanded under the first test, which involves whole rotations and hard cash, may be higher than that of the second, which operates over shorter periods and merely involves income temporarily forgone. These are separate tests, and the choice must be made as to which analysis should be used for thinning decisions.

In the first test, the cost of establishing a stand and all subsequent costs are carried at a compound interest rate until the end of the rotation. Thinning helps meet this test by providing income during the rotation which is also compounded to the end of the rotation. By the reckoning involved, the money received at age 20 may be worth twice that at age 35. The compounded costs of stand establishment early in the rotation can be offset by early thinning returns rather than letting them mushroom until the end. If the thinning enables trees to grow fatter and more valuable or increases the total yield, the net income against which the charges can be made is increased. If these improvements make it financially prudent to have longer rotations, the average annual expense for regeneration in the whole forest is reduced, and so the overall result can be still further improved.

The second test involves the crucial task of improving the return on the large investment represented by the liquidation value of merchantable growing stock in managed forests. If an owner requires a rate of return of 8 percent, each dollar left in growing stock at the beginning of the decade must grow to $2.16 in value during the decade to justify having left the tree or stand. In this test, unlike the first, any money invested in growing the tree does not count; only that money for which it could have been sold at the beginning of the period is considered.

The second test embodies one of the most useful ways of determining which trees to cut and which to leave in thinning. Those trees that cannot be made to increase enough in value to yield an acceptable rate of return on their own value are removed. Those that can are not only left but they are also granted enough additional growing space that the compound interest that they earn is increased. The trees most likely to continue yielding the desired rate of return for the longest time are those of highest quality and rate of growth. In a pure application of this approach, the first trees to be cut would be those of low quality or slowest growth, although they would theoretically not be cut until they had acquired a positive value.

Regardless of how rotation length is determined, thinning provides ways of extending rotations and also of lending some flexibility to their control. The rotation necessary to grow trees of a given size can be shortened; larger and more valuable trees can be grown on rotations of fixed duration. Short rotations do not necessarily have the financial values that are so often ascribed to them, especially if all of the financial factors involved are considered. If thinnings have produced enough revenue to amortize establishment costs, are yielding an adequate income, and are maintaining good rates of growth, there may be little need to rush into the risk and expense of starting a new crop that will be some years in regaining full occupancy of the site. The forest manager who refrains from thinning may be condemned to short rotations, frequent regeneration problems, and limited opportunity to maneuver to meet all sorts of emergencies and changes in markets or management objectives.

Improvement of Product Quality

The value of a fixed total production can be enhanced simply by favoring the trees of best potential quality and discriminating against the poor ones. This effect of thinning on wood quality is vastly more important than any other. The superior diameter growth induced by thinning usually improves wood quality because the large trees tend to be of better quality than small ones. The general effect of thinning on wood quality is not harmful, in spite of some opinion to the contrary.

Control of Stand Composition and Effects on Regeneration

The opportunity to intervene in the development of the stand during the rotation produces a number of other economic benefits. It enables continuing control of undesirable species that were not eliminated during the period of regeneration. If mixed stands of dissimilar species are being deliberately maintained, thinning may represent the only means of taking adequate advantage of those species that tend to mature earlier than the major components.

The long period of control over stand composition enables the forester to prepare for natural regeneration by reducing the seed source of undesirable species and fostering that of the good. Thinning also builds up the physiological vigor, mechanical strength, and seed production of the individual trees of the final crop, so that there is wider choice among methods of regeneration cutting. Unthinned stands are likely to have to be replaced during an abbreviated period of establishment, even when more gradual replacement might enable greater use of the financial potentialities of the crop.

Reducing Risk

Thinning can reduce the risk of the stand being destroyed or reduced in value by a variety of causes. One of the main effects of thinning is to increase the vigor of the trees by giving each tree more growing space. Increased vigor increases the tree's ability to respond to most insect attacks. Production of secondary compounds helps protect the tree from many types of herbivory, and the additional vigor allows those trees capable of a second flush of foliage to recover from defoliation. More vigorous trees are also able to withstand reductions in photosynthesis without succumbing to the overriding needs of respiration.

Thinning also increases protection from abiotic agents. Salvaging potential mortality reduces the fuel load that leads to catastrophic fires. Thinning leads to reduced height/ diameter ratios and greater wind firmness.

An appropriate thinning schedule will address all issues of growth and yield in a way that meets the owner's goals and needs. The actual selection of trees to be cut is just as important as the timing and intensity of the thinning and will be discussed in the next chapter.

ADDITIONAL IDEAS ON GROWTH AND PRODUCTIVITY

Interspecific Variations in Production

Further reference in this chapter to values of dry-matter production will, unless otherwise specified, be in terms of net M.A.I. at or near culmination age, for above-ground materials, including leaves of the current year, in metric tons per hectare per year. A metric ton is 1000 kg, and one metric ton per hectare is equal to 100 g per m^2, or 0.446 English short tons of 2000 pounds each per acre.

Clearly, some species produce more than others even when the site factors are the same. Evergreen species are basically more efficient than deciduous species because they use their investment in a year's production of foliage for more than one year. Their foliage can also start photosynthesis whenever light, water, and temperature are favorable without any delay while new leaves form. In fact, in some regions such as the coastal Pacific Northwest, conifers may conduct a major portion of their annual photosynthesis during times when the hardwood trees are without leaves. In these areas, photosynthesis during the summer months is often restricted by drought.

Although conifers are often thought of as more productive than angiosperms, they are not inherently more efficient. Broad-leaved evergreens, such as eucalypts, can be just as productive as evergreen conifers. However, many coniferous species have lighter wood and longer lengths of straight central stems than angiosperms. These attributes may make a conifer more productive than some hardwoods in terms of volume of usable stemwood, even if its biomass production is no greater.

The arrangements of branches within crowns and of leaves on twigs appear to cause some of the interspecific differences of photosynthetic production (Hallé, Oldemann, and Tomlinson, 1978). For example, one highly efficient form of crown structure is that akin to a candelabrum in which leaves are displayed in a single flat plane by wide-spreading branches. Although this form, typified by some understory palms, has high photosynthetic efficiency, it is so weak structurally that the trees must remain small. Many trees growing in overtopped, low-light conditions adopt a single plane type of foliar display. This crown shape has been called *monolayer* and is most efficient if light levels are below 50 percent of full light (Horn, 1971). Some early successional species, such as poplars, exhibit rapid growth in youth because they have branch arrangements that allow them to develop closed sheaths of foliage in balloon-shaped crowns that cover the ground quickly. These species are called *multilayer* and are most efficient at high-light levels. However, it is postulated that certain structural inefficiencies make it difficult for these species to sustain the early high rates of production.

Another highly efficient crown form is that in which substantial gaps exist between foliated branches. However, this structure is not conducive to swift occupancy of growing space. Stands of trees with this crown structure attain high, long-sustained values of C.A.I. only after the early stages of development.

The mode of arrangement of leaves on twigs that is photosynthetically most efficient is the two-ranked one in which the leaves do not shade one another. However, this arrangement is the least efficient in preventing the leaves from losing water and becoming overheated. Because there is no way that the open stomata of land plants can admit carbon dioxide for photosynthesis without letting water escape, high rates of photosynthesis have to be paid for with transpirational water loss.

Some species are adapted to drought periods because they close their stomata and halt photosynthesis earlier during episodes of moisture stress. These species, called **drought avoiders**, avoid internal moisture stress that could cause the foliage to be permanently damaged, but do so at the expense of productivity because photosynthesis cannot continue if the stomata are closed. This adaptation to drought is one of several that reduce production but increase survival. On the other hand, **drought endurers** leave their stomata open and continue photosynthesis at the risk of suffering permanent wilt damage if the period of moisture stress is not short-lived.

Shade-tolerant species, whether evergreen or deciduous, are usually more productive than the intolerants. Actually, it is more accurate to think of shade tolerance as the result rather than the cause of this difference in photosynthetic efficiency. The organs that confer efficiency are leaves that can manufacture much sugar and also live at low-light intensity. Species that have such leaves have more of them and also deeper crowns than the less efficient species. Perhaps because they have more leaves in the lower and cooler parts of the crown canopy, shade-tolerant species do not use such a high proportion of their gross photosynthate in respiration (Assmann, 1970).

Species with dark-green leaves denoting high chlorophyll content tend to be relatively efficient. Another useful criterion is the leaf-area index, which is the number of layers of leaf surface above each point on the ground. Values of this index vary from 3 to 12 in closed stands but are ordinarily about 5 (Waring, Newman, and Bell, 1981).

Interspecific variation among species is well illustrated by estimates of gross M.A.I. (exclusive of roots but including current leaf production and material previously removed in thinnings) for stands of 40–47 years, at or near peak M.A.I., on comparable soils in the favorable climate of England (Ovington, 1957; Ovington and Pearsall, 1956). *Castanea sativa* and *Quercus robur*, deciduous species of moderate shade tolerance, had M.A.I.'s of 3.8 and 3.9 tons per hectare per year, respectively, and the M.A.I. of *Fagus sylvatica*, a very tolerant hardwood, was 4.9 tons. The evergreen conifers were more productive:

Pinus sylvestris, 8.0; *Pinus nigra*, 9.7; *Pseudotsuga menziesii*, 9.9; and *Picea abies*, 9.9 tons per hectare per year. *Larix decidua*, a deciduous conifer, had a rate of 6.7 tons, intermediate between the hardwoods and evergreen conifers. Data from other parts of Britain indicated that even higher rates than those of spruce and Douglas-fir are achieved by very tolerant conifers such as redwood, the hemlocks, and true firs. Intolerant pioneer hardwoods such as *Betula alba* produced at lower rates than those of chestnut and oak. The yield of moderately intensive mixed agriculture in the same locality was 4.5 tons per hectare annually.

A combination of intuition and some data suggest that the highest production rates are in the evergreen tropical rain forest, where site factors are unfavorable to photosynthesis only at night. Annual production rates are at least 8 and possibly as much as 20 or 30 tons per hectare. The high temperatures also induce such profligate respiration that the production rates are not clearly higher than those of the most favorable parts of the temperate zone. Studies show that the range of productivities across soil types is larger than that across temperature regimes, making this comparison very difficult.

The highest proven rates of production are for evergreen species in limited areas of wet, mild, strongly maritime climates at middle latitudes. One of the highest known values of C.A.I. is 30.7 tons per hectare in a stand of western hemlock and Sitka spruce at an effective age of 21 years on the Oregon coast (Fujimori, 1971). The M.A.I. was 9.2 tons per hectare and might culminate at about 20 some decades hence. Such favored midlatitude areas include the coastal fringe of western North America, parts of Western Europe, southern Japan, and certain coastal areas in the strongly oceanic climate of the Southern Hemisphere. If the rainfall is sufficient and the soil favorable, the equable maritime climates of these areas provide very long growing seasons without the high temperatures that stimulate rapid respiration. Production rates approximating 15 tons per hectare have been observed with shade-tolerant conifers, evergreen oaks of southern Japan, and *Pinus radiata* in New Zealand (Cannell, 1982).

Most of the world's forests grow under less favored conditions. In widely varying degrees, production is reduced by seasonal moisture restriction (which in some regions can exist at any latitude), periods when low temperature immobilizes water, deficiencies of soil oxygen induced by excess water, low light intensity at high latitudes, or excessive respiration from high temperatures. However, the production rates of the poorest vegetation that might plausibly be called forest are not less than 1 ton per hectare per year.

Observed production rates vary so much within species and genera because of site variations that it is perilous to assign numerical values to any of them. The total production of a given species on a poor site is commonly only a half or a third of that on the best. In general, the greater the tolerance for shade, the higher the total production, although the trees may be slow to achieve such rates. The true firs are among the most productive of the conifers, yet the peak rates of above-ground M.A.I. can range from 17 tons per hectare in Britain down to a tenth of that on the wettest sites where balsam fir will grow in eastern North America.

The two- and three-needled or hard pines, which are so often grown for timber, are generally not as productive of dry matter as more shade-tolerant conifers. Many observations suggest maximum values of M.A.I. around 6 tons per hectare. However, these species are most commonly grown on sites where there are significant seasonal moisture deficiencies. On better sites in parts of the Southern Hemisphere, production rates of Monterey and other pines can often be 12 tons per hectare or as high as 17 in the case of New Zealand. Thus site factors can be as important as foliar efficiency.

It has sometimes seemed plausible that very fast growing pioneer species or stands of coppice shoots grown on short rotations might sustain remarkably high values of M.A.I. in terms of dry tonnage. The culmination of C.A.I. for these species does come very early. However, such values of peak M.A.I. as can be found are generally less than those of species with slower rates of juvenile growth in height on the same sites. The high respiration rates of these species may offset the rapid gross photosynthesis associated with quick juvenile growth.

The culmination of M.A.I. in terms of above-ground dry-matter production, for a given species and situation, does not appear to come at ages much less than those of the peak values of M.A.I. for total cubic volume of stemwood. This should not be surprising because stemwood accounts for a high proportion of the above-ground dry matter that accumulates in stands of trees.

Studies of total production tell us a great deal about the fundamental biological factors controlling forest production and the tremendous role it plays in matters such as world budgets of energy and carbon. They also show the tantalizing possibilities of any technology that might enable closer utilization of the massive production of organic matter by forests. However, most use of wood is confined to those parts of tree stems that are large enough to be handled economically. One key problem of timber-production silviculture is manipulating this production so that as much of it as possible is channeled into usable stem sizes commonly defined in terms of merchantable cubic and board-foot volume.

There is ample evidence showing that levels of optimum basal area tend to be greater for shade-tolerant species than for intolerant ones and greater for evergreens than for deciduous species. The more efficient the foliage of a species, the greater not only the production but also the amount of growing meristematic tissue that can be nourished efficiently. The approximate ranges of appropriate residual basal area, after thinning, for middle-aged stands of these four categories of species may be tentatively stated as follows:

	ft^2/acre	m^{2}hectare
Tolerant evergreens	130–230	30–53
Intolerant evergreens	80–130	18–30
Tolerant deciduous	70–160	16–37
Intolerant deciduous	50–80	12–18

It should not be inferred that the limits stated here are precise. The basis for the implication that the logical ranges are greater for tolerant than for intolerant species is weak, although stands of shade-tolerant species do seem to be more flexible in their response to thinning than those of intolerant species.

Where differences in site quality are substantial, the logical basal area for stands of a given species would be less on a poor site than on a good one because the foliage per m^2 is so much less. If the site is capable of supporting a stand with a closed canopy, the difference may be smaller than might be assumed because development progresses so much more slowly. The basal area left after thinning would be much less at comparable ages, but if residual basal area is allowed to increase with age, the differences at comparable heights would be small. However, in arid forest regions, such as those of parts of the interior ponderosa pine region, full closure of the root systems is possible with as little as 10 percent canopy cover. Ideas about levels of basal area appropriate to more humid forest regions would scarcely be applicable there.

BIBLIOGRAPHY

Assmann, E. 1970. *The principles of forest yield study.* Translated by S. H. Gardiner and P. W. Davis. Pergamon Press, Oxford. 506 pp.

Bailey, R. L., and K. D. Ware. 1983. Compatible basal-area growth and yield model for thinned and unthinned stands. *Can. J. For. Res.* 13:563–571.

Bennett, F. A. 1980. Growth and yield in natural stands of slash pine and suggested management alternatives. *USFS Res. Paper* SE–211. 8 pp.

Berg, A. B. 1976. *Managing young forests in the Douglas-fir region, 1973 symposium proceedings.* Vol. 5. Oreg. State Univ., School of Forestry, Corvallis. 147 pp.

Burger, H. 1919–1953. Holz, Blattmenge und Zuwachs. I-XIII. Mitt d. Schweizer. Anstalt f. d. forstliche Versuchswesen 15:243–292, 19:21–72, 20:101–114, 21:307–348, 24:7–103, 25:211–279, 435–493, 26:419–468, 27:247–286, 28:109–156, 29:38–130.

Burschel, P., E. Kürsten, B. C. Larson, and M. Weber. 1993. Present role of German forests and forestry in the national carbon budget and options to its increase. *Water, Air, and Soil Poll.* 70:325–340.

Cannell, M.G.R. (ed.). 1982. *World forest biomass and primary production data.* Academic, New York. 400 pp.

Cannell, M.G.R., and F. T. Last. (eds.). 1976. *Tree physiology and yield improvement.* Academic, London. 567 pp.

Clutter, J. L., and E. P. Jones, Jr. 1980. Prediction of growth after thinning in old-field slash pine plantations. *USFS Res. Paper* SE–217. 14 pp.

Clutter, J. L., J. C. Fortson, L. V. Pienaar, G. H. Brister, and R. L, Bailey. 1983. *Timber management: a quantitative approach.* Wiley, New York. 333 pp.

Drew, T. J., and J. W. Flewelling. 1977. Some Japanese theories of yield-density relationships and their application to Monterey pine plantations. *For. Sci.* 23:517–534.

Duvingneaud, P. 1971. Productivity of world ecosystems. *UNESCO Ecology and Conservation Series* No. 4. 707 pp.

Fries, J., H. E. Burkhardt, and T. A. Max (eds.). 1978. *Growth models for long-term forecasting of timber yields, IUFRO Proc.* School of Forestry and Wildlife Resources, Va. Polytechnic Inst. and State Univ., Blacksburg. 249 pp.

Fujimori, T. 1971. Primary production of a young Tsuga heterophylla and some speculation about biomass of forest communities on the Oregon Coast. *USFS Res. Paper* PNW–123. 11 pp.

Gingrich, S. F. 1967. Measuring and evaluating stocking and stand density in upland hardwood forests in the Central States. *For. Sci.* 13:38–53.

Gingrich, S. F. 1971. Management of young and intermediate stands of upland hardwoods. *USFS Res. Paper* NE–195. 26 pp.

Gower, S. T., K. A. Vogt, and C. C. Grier. 1992. Carbon dynamics of Rocky Mountain Douglas-fir; influence of water and nutrient availability. *Ecol. Mono.* 62(1):43–65.

Hallé, F., R. A. A. Oldemann, and P. B. Tomlinson. 1978. *Tropical trees and forests, an architectural analysis.* Springer, New York. 411 pp.

Horn, H. S. 1971. *The adaptive geometry of trees.* Princeton University Press, Princeton, N.J. 144 pp.

Keyes, M. R., and C. C. Grier. 1981. Above- and below-ground net production in 40-year-old Douglas-fir stands on low and high productivity sites. *Can. J. For. Res.* 11:599–605.

Kimmins, J. P. 1987. *Forest ecology.* Macmillan, New York. 531 pp.

Kira, T., and T. Shidei. 1967. Primary production and turnover of organic matter in different forest ecosystems of the western Pacific. *J. Jap. Ecol.* 17:70–87.

Long, J. N., and F. W. Smith. 1984. Relation between size and density in developing stands: a description and possible mechanisms. *For. Ecol. and Mgmt.* 7:191–206.

Mar:Møller, C. M. 1947. The effect of thinning, age, and site on foliage, increment, and loss of dry matter. *J. For.* 45:393–404.

Mar:Møller, C., D. Müller, and J. Nielson. 1954. Graphic presentation of dry matter production in European beech. *Forstl. Forsøgsvaesen i Danmark 21*:327–335.

Mitchell, K. J. 1975. Dynamics and simulated yield of Douglas-fir. *For. Sci. Monogr.* 17. 39 pp.

Mohren, G.M.J., H. H. Bartelink, and J. J. Jansen. 1994. Contrasts between biologically-based process models and management-oriented growth and yield models. *FE&M*, 69:1–350.

Nelson, T. C., et al. 1961. Merchantable cubic-foot volume growth in natural loblolly pine stands. *Southeastern For. Exp. Sta., Sta. Paper* 127. 12 pp.

Ovington, J. D. 1957. Dry-matter production by *Pinus sylvestris* L. *Annals of Botany* 21 (n.s.):287–314.

Ovington, J. D., and W. H. Pearsall. 1956. Production ecology. II. Estimates of average production by trees. *Oikos* 7:202–205.

Reinecke, L. H. 1933. Perfecting a stand-density index for even-aged forests. *J. Agr. Res.* 46:627–638.

Reukema, D. L., and D. Bruce. 1977. Effects of thinning on yield of Douglas-fir: concepts and some estimates obtained by simulation. *USFS Gen. Tech. Rept.* PNW–58. 36 pp.

Satoo, T., and H.A.I. Madgwick. 1982. *Forest biomass*. Nijhoff/Junk, The Hague. 152 pp.

Shugart, H. H., and D. C. West. 1980. Forest succession models. *Bioscience* 30:308–313.

Stage, A. R. 1973. Prognosis model for stand development. *USDA For. Ser. Res. Pap.* INT–137. 32 pp.

Waring, R. H., H. L. Golz, C. C. Grier, and M. L. Plummer. 1977. Evaluating stem conducting tissue as an estimator of leaf area in four woody angiosperms. *Can. J. Bot.* 55:1474–1477.

Waring, R. H., K. Newman, and J. Bell. 1981. Efficiency of tree crowns and stemwood production at different canopy leaf densities. *Forestry* 54:129–137.

West, P. W. 1983. Comparison of stand density measures in even-aged regrowth eucalyptus forest of southern Tasmania. *Can. J. For. Res.* 13:22–31.

Westoby, M. 1984. The self-thinning rule. *Adv. in Ecol. Res.* 14:167–225.

Whittaker, R. H., and G. M. Woodwell. 1969. Structure, production, and diversity of the pine-oak forest at Brookhaven, New York. *J. Ecol.* 55:155–174.

Woodwell, G. M., et al. 1978. The biota and the world ecosystem. *Science* (Washington) 199:141–146.

CHAPTER 5

METHODS AND APPLICATION OF THINNING

METHODS OF THINNING

Four distinct methods are used to determine which trees to favor and which to remove in thinnings. These are: (1) **low**, (2) **crown**, (3) **selection**, and (4) **geometric** or **mechanical**. In the first three methods, the choices are based on the developmental status of the trees and their position within the crown canopy structure. In the fourth, spacing or arrangement of trees in the stand is the first consideration. It is also useful to recognize a fifth method, **free thinning**, which can be any combination of the other four simultaneously applied in a single operation, usually in stands of irregular structure.

Each of these methods will be considered in turn. Vertical profiles through stands are used to illustrate the basic patterns of choices made in the first three methods. Such sketches lack the depth of three-dimensional representation with the result that they make the thinnings seem heavier than they actually are. The stands depicted are also single-canopied, that is, with all trees competing in a single crown stratum. Each profile shows a kind of stand for which the thinning method illustrated is especially appropriate.

Low Thinning

This method, the oldest, is sometimes called *thinning from below* as well as the *ordinary* or *German* method. In low thinning, trees are removed from the lower crown classes. This mimics at an accelerated rate the natural mortality of self-thinning during normal development of single-canopied stands.

Low thinning can be applied throughout a range of severity. This gradient is shown in Fig. 5.1 and is divided into four grades ranging from A for the very light to D for the heaviest. They are differentiated on the basis of the crown classes removed. In A-grade

99

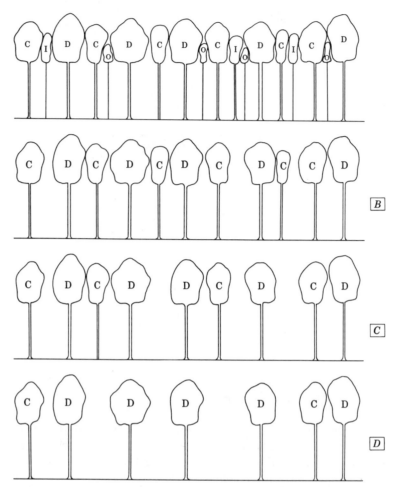

Figure 5.1   The B, C, and D grades of low thinning as they might look if each were simultaneously applied in the same middle-aged stand (top) of loblolly pine that had been rendered uniform by earlier thinning. The letters on the crowns denote crown classes and those at the side denote the grades of low thinning. Note that previous treatments have promoted the crop trees to the dominant class. The nearly useless A grade is omitted.

low thinning, removals are confined to overtopped trees, or merely those dead or nearly dead of suppression. These are little more than salvage operations and the canopy remains unbroken as shown in Fig. 5.2. In the successively heavier grades, additional and higher crown classes are removed. In B-grade low thinnings, the intermediate crown class is also eliminated. The main virtue of A- and B-grade thinnings is that they are practically the only kind of thinning that can be done without any risk of reducing the gross production of wood by the stand. The trees of the overtopped and intermediate crown classes in pure, even-aged stands have done most of their growing already and either die soon or merely exist without much further growth.

Because the object of most thinnings is to stimulate the growth of some remaining trees and not merely salvage, the heavier C and D grades of low thinning are now much

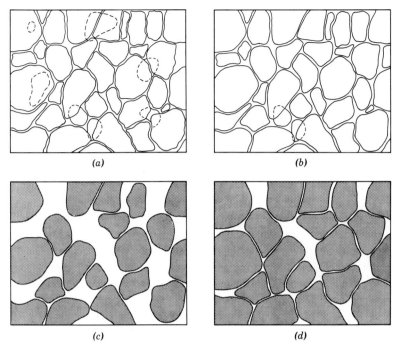

Figure 5.2 The outlines of the crowns of a previously unthinned stand as viewed from above. Overtopped portions of the crowns of trees of the lower crown classes are indicated by broken lines: (*a*) before thinning; (*b*) after an A-grade thinning in which overtopped trees were removed leaving the canopy unbroken; (*c*) immediately after a D-grade thinning in which overtopped, intermediate, and many codominant trees were removed leaving each remaining tree exposed on one or more sides; (*d*) the same stand 10 years after the D-grade thinning showing the canopy so nearly completely closed that another thinning is needed.

more commonly applied than the light A and B grades. Both of the heavier grades involve deliberate creation of temporary canopy gaps to accelerate crown expansion of the remaining trees (Fig. 5.2). In the C grade some codominants are cut along with the lower crown classes; in the D grade, many but not all codominants are cut. The concept of low thinning is based on the premise that all trees smaller or shorter than a given standard are cut. Low thinning does not imply that no main canopy trees are cut, only that if these trees are cut all smaller, intermediate and overtopped trees are also removed. An even heavier thinning, in which only dominants were left, would ordinarily leave so much growing space for new regeneration that it would be better thought of as shelterwood regeneration cutting.

Low thinning has a simple, close, and logical relationship to the natural course of stand development. It is easy to pick the trees to remove; those that remain are, at least in pure stands, sure to include those likely to continue the superior growth that has already put them in the upper crown classes. However, the removals are concentrated in the small trees least likely to be marketable. Since the best growing trees are still left, stand growth is not reduced. However, low thinnings have to be very heavy, or else be frequent and started early, to make the important final-crop trees grow faster than they would without thinning. In other words, so much attention is devoted to the negative step of eliminating losers that positive actions to make the leaders grow still better are somewhat indirect.

The general result is that low thinning is most applicable to stands in which nearly all trees are merchantable. It is no accident that the method developed in times and places in which fuelwood was in demand. Where there is little use for small trees, low thinning becomes most applicable in the later stages of the rotation after all trees have become merchantable. The trees of the lower crown classes have by then lost so much foliar surface that their growth dwindles irreversibly and they become financially mature. Intolerant species are so likely to suffer reductions of live crown ratio that there is less latitude for applying methods other than low thinning to them than is the case with tolerant species that can carry foliage at lower light intensity.

Theoretically, low thinning should be more appropriate than other methods for sites where moisture or other soil factors are seriously limiting. Similarly, but on any kind of site, it would seem comparatively favorable to the development of accessory understory vegetation, which may be desirable or undesirable from the management standpoint. However, the competition for soil growth factors is probably controlled more by the amount of foliage temporarily removed than by the particular thinning method employed.

Crown Thinning

To overcome the limitations encountered in applying low thinnings, a method was developed in which trees are removed from the middle and upper portion of the range of crown and diameter classes rather than from the lower end. This technique is best referred to as **crown thinning**. It is also known as the *French* method of thinning because of its origin and early use in France. The terms *thinning from above*, *high thinning,* and *thinning in the dominant* have also been used for crown thinning, but these are too easily confused with selection thinning.

In crown thinning, trees are removed from the upper crown classes in order to open up the canopy and favor the development of the most promising trees of these same classes (Fig. 5.3). Most of the trees that are cut come from the codominant class, but any intermediate or dominant trees interfering with the development of potential crop trees are also removed. The trees to be favored are chosen, if possible, from the dominants and, where necessary, from the codominants. Crown thinning favors nearly the same kinds of trees as low thinning but more by removing a few strong competitors than by wholesale elimination of the weak. Low thinning and crown thinning differ radically from selection thinning in which dominant trees are cut to favor the lesser crown classes.

Crown thinning is different from low thinning in two respects. First, no matter how lightly applied, the principal cutting is made in the upper crown classes. Second, the bulk of the intermediate class and the healthier portion of the overtopped class remain after each thinning.

The question of whether individual dominant or codominant trees are favored is settled according to the relative potentialities of adjacent trees. If the choice lies between a promising codominant and mediocre dominant, the codominant is favored. A situation of this kind occurs most often where the codominant has a straighter, smoother bole and smaller branches than the dominant. Where all trees are of good health, form, and species, codominants interfering with the growth of dominants are removed, on the premise that position in the crown canopy is the best index of past and future vigor.

Theoretically, overtopped trees and intermediates that do not interfere with crop trees

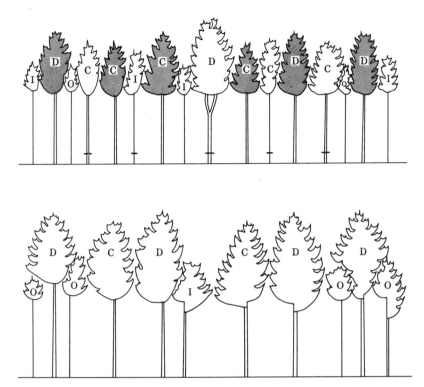

Figure 5.3   The upper sketch shows a coniferous stand immediately before a single crown thinning. The trees to be cut are denoted by horizontal lines on the lower boles and those with shaded crowns are the crop trees. The lower sketch shows the same stand about 20 years after the crown thinning and reclosed to the point where a low thinning would be desirable.

are not cut in crown thinning. A number of factors determine whether this is carried out in practice. The lower canopy trees serve the useful function of restricting epicormic branching on the crop trees as well as the development of unwanted understory vegetation. Their presence also creates a more continuous vertical distribution of foliage, thereby creating a more diverse feeding and nesting habitat for animals. The decision to leave these trees may also depend on aesthetic considerations; such stands have a very different appearance than the more open, uniform conditions following a low thinning. Either kind of stand might be preferable under different circumstances. However, if these considerations are not important, there is little reason to leave such trees if they can be harvested profitably and their continued presence will add value neither to themselves nor to the stand as a whole. If left, many may suffer damage as loggers try to avoid damage to the crop trees. If most of the smaller trees are removed, the operation might be better thought of as an incomplete low thinning .

The immediate cash return from crown thinnings is greater than that from low thinnings of equal severity because the material removed is larger and of greater utility. The smaller trees of the subordinate classes, which would be removed in low thinnings, can be left to grow to larger size, and a few trees that are unmerchantable may grow enough to be harvested at a later thinning. However, foresters often overestimate the release potential of overtopped trees after a crown thinning. In most situations, if such a tree is merchantable at the time of thinning, it might as well be removed.

One result of crown thinning is the division of the residual stand into two categories of trees. The first consists of the favored dominants and codominants, which are destined for removal either in reproduction cuttings or in later thinnings. The second category is made up of the subordinate trees, which are favored only by indirect action and are gradually removed as they grow up and interfere with the trees of the first group. After a series of crown thinnings it is possible that a stand of two distinct stories, but only one age class, may be created (Fig. 5.4). This development is not likely to be persistent unless the sub-

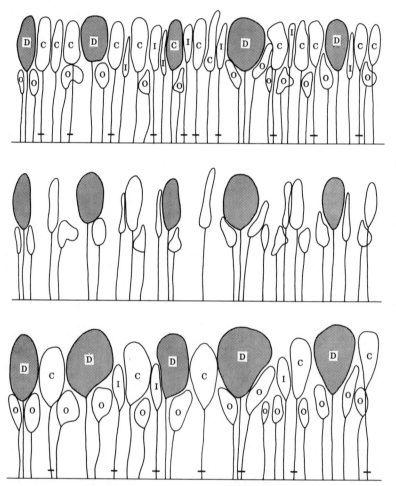

Figure 5.4 Part of a series of crown thinnings in a hardwood stand. The upper sketch shows the previously untreated stand immediately before the first crown thinning. The trees to be cut are denoted by horizontal lines on the lower boles and those with shaded crowns are the final crop trees. The middle sketch shows the stand immediately after the first thinning. The lower sketch shows it advanced to readiness for the second crown thinning. Note that a lower story is being developed by these thinnings. Presumably a third crown thinning could be conducted in this stand.

ordinate trees are of tolerant species. Actually, it is common to have the lower story of trees vanish as a result of the combined attrition of natural suppression and a series of crown thinnings. Under such conditions, a program of crown thinnings pursued to its logical conclusion will be replaced almost automatically by low thinning. With some very intolerant species, the first crown thinning may prove to be the last because of the failure of the subordinate trees to survive after release. A result of this kind represents poor judgment only if the trees that die could have been utilized profitably at the time of thinning.

One important advantage of crown thinning lies in the opportunity to stimulate the growth of selected crop trees without sacrificing quantity production. The favored trees grow to a given size in a shorter time, or to larger size in a given time, than they would with a regime of low thinnings, owing to the freedom for expansion of the crowns. Meanwhile, any growing space not taken up by the selected trees is given over to the subordinates. Most evidence indicates that the total yield, in terms of cubic volume, is no greater from crown thinnings than from a comparable series of low thinnings. With crown thinnings, however, this volume is harvested in fewer trees of greater average diameter than with low thinnings.

Crown thinning is a more flexible method than low thinning, and it demands greater skill on the part of the forester. Because it is not feasible to distinguish different grades of crown thinning, the severity of cutting must be regulated in terms of basal area or some other index of stand density. Even these types of measures are of doubtful applicability when applied to mixed-species stands (see Chapter 16).

Applications of Crown Thinning

This thinning method can be applied uniformly throughout a stand or concentrated very specifically on the release of a limited number of chosen crop trees. The uniform application is most feasible if it is desirable to remove a large volume of wood in a single operation. This approach, sometimes called **crop tree management**, is useful where the object is to stimulate the crop trees as much as possible with the least amount of cutting.

In any crown thinning, it is most logical to identify the potential final-crop trees and then be sure to eliminate the most vigorous competitor of each (Fig. 5.5). Although other trees may also be removed, the leading competitors are the most important. When a stand is young and crown expansion rapid, it may be possible and desirable to free the crop tree crowns on all sides. In older stands canopy gaps do not close so quickly, and it becomes impossible to release the crop trees completely without seriously reducing the volume growth of the stand.

Crown thinning often represents one of the most expeditious means of restricting the investment represented by the value of growing stock without reducing the growth in value that represents the income on this investment. The codominants that are typically removed in crown thinning subtract more from the investment in growing stock than the smaller trees that would be removed in low thinning. At the same time, the potential for increase in value remains essentially undiminished because the dominant trees are favored. Selection thinning, in which numerous dominants are cut, may reduce the investment more, but it usually reduces the growth in value as well. If all of the trees in a stand are of good quality and value, differing mainly as to rate of growth, low thinning is likely to represent the logical way of manipulating the investment represented by the stand because the dominants will increase in value most rapidly.

The principles of crown thinning can be followed in almost any situation where economic conditions are compatible with the application of thinning. This method provides

Figure 5.5 A 32-year-old plantation of red pine in Connecticut just after completion of a third crown thinning. While the crop trees have been thoroughly released, the vigor of the subordinate trees is so poor that they will be removed in a subsequent low thinning. *(Photograph by Yale University School of Forestry & Environmental Studies.)*

such direct and consistent means of fostering and regulating the development of the chosen trees of the final crop that it has become very common. It is also sufficiently versatile in application that the details of procedure are subject to a wide variety of modifications; many of these modifications are recognizable as crown thinning only if the method is interpreted rather broadly.

Crown thinning works best, or at least can be repeated most often, in mixed stands or in pure stands of tolerant species; both kinds of stands are likely to have trees capable of forming a subordinate stratum. It can be applied at appropriate stages in handling pure stands of intolerant species, but must soon be succeeded by low thinnings if any use is to be made of the trees of the subordinate crown classes. One of the shortcomings of crown thinning is that it makes no provision for the ultimate disposition of the subordinates. This drawback is readily correctable by subsequent use of low thinning or other techniques. Often this stratum is left until regeneration cuttings begin. If left, there is a danger that the subordinate stratum will be thought of as a younger age class. This misconception has led to many poor second-generation stands.

Selection Thinning or Thinning of Dominants

This method of thinning differs radically in principle from the two methods already discussed. In selection thinning, dominant trees are removed in order to stimulate the growth of the trees of the lower crown classes (Fig. 5.6). The same kind of vigorous trees that are favored in crown and low thinning are the very ones that are likely to be cut in selection thinning. It is obvious then that selection thinning is suitable only for rather limited purposes and, if not carefully used, readily degenerates into the practice of harvesting the best trees and leaving the poorest. A common American term for this destructive practice is **high-grading**, that is, harvesting only material of high grade.

The term *selection thinning* comes from the very superficial resemblance of this kind of cutting to the selection method of regeneration, which will be discussed in Chapter 15. Both are characterized by the removal of the largest trees in a stand. However, selection thinning involves the cutting of scattered trees from even-aged stands or aggregations of trees without the intent or result of regenerating new age classes. The selection method of regeneration cutting, on the other hand, is applied to uneven-aged stands to make vacancies large enough to allow new age classes to develop; this is likely to require removal of groups of trees rather than individuals. The similarity of the terms causes semantic confusion and, worse yet, confusion about the purpose of some kinds of partial cutting. "Thinning of dominants" is a less ambiguous term that might replace "selection thinning."

There are a number of different kinds of selection thinning, and the method can be understood only in light of these various approaches and the situations in which they are applicable. These approaches differ with respect to the objectives sought and the number of times this kind of thinning is repeated.

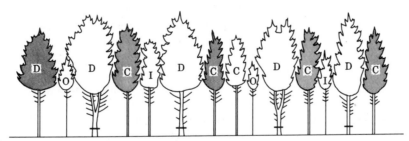

Figure 5.6 A coniferous stand marked for a selection thinning aimed primarily at the elimination of rough, poorly formed dominants and the release of less vigorous trees of better form. The next thinning would not come until the large holes in the canopy had nearly closed and it could not logically be another selection thinning because the crop trees (shaded crowns) would then be the dominants.

Different Forms of Selection Thinning

In the first and most common approach, poorly formed dominants are eliminated in favor of satisfactory crop trees chosen from the highest possible level in the lower crown classes (Fig. 5.6). There is no point in procrastinating about the removal of coarse dominants that become poorer and cause more harm to better trees the longer they are allowed to remain. The removal of such trees takes advantage of the fact that codominant and intermediate trees as well as the smaller dominants often have smoother, straighter boles and smaller branches than the most vigorous dominants.

Such action may be useful for species being grown for clear wood, such as hardwoods used for furniture. Merchantable height, determined by the height of the lowest major fork in hardwoods, is often shortest in the most dominant trees if they are slightly older than the rest of the stand. Codominants with a higher first fork have a higher potential value and should be favored in the thinning. However, with some species, this may stimulate epicurmic branching on previously clear boles.

The trees selected for retention to the end of the rotation should have live crowns sufficiently deep to enable them to respond to release rapidly and then develop into thrifty, fast-growing individuals. Trees with live crown ratios less than 30 percent are rarely suitable. In general, the more tolerant the species, the greater is the possibility that appropriate trees can be found in the lower crown classes.

Selection thinnings designed to improve the quality of the crop trees are best carried out as early as possible in the life of the stand and replaced by other thinning methods as soon as the crop trees approach the dominant position. Occasion arises for such treatment only where the irregularity and sparsity of initial stocking has caused rough dominants to develop (Fig. 5.7). An outstanding illustration is to be found in stands of eastern white pine that have been attacked by the white pine weevil. Because the dominant trees are the ones most likely to be deformed by this insect, it is advantageous to remove them as soon as the straight, potential crop trees from the lower crown classes have attained a height equal to the desired log length.

This kind of treatment is best regarded as a necessary evil dictated by imperfect initial stand structure, especially with intolerant species. The importance of this point should not be minimized because so many valuable species are relatively intolerant. With such species, it is rare that any additional advantage can be secured from selection thinnings after rough dominant trees have been eliminated in favor of better codominants.

In the second approach (Fig. 5.8), selection thinnings are continued until the point is reached when further removals from the main canopy would open holes too large to be filled by the expansion of the crowns of the remaining trees. At that point attention is turned either to low thinning or, more often, to whatever measures are necessary for regeneration. The objective of this kind of treatment is not to develop large trees but to grow as many trees as possible to medium size for pulpwood, small saw logs, posts, poles, or piling. The procedure differs from diameter-limit cutting only to whatever extent discretion is exercised in choosing the trees to cut.

The stands with the greatest capacity to endure repeated selection thinning are of species such as spruces and true firs, which are both shade-tolerant and maintain a single leader when growing in low light (strong epinastic control). The subordinate crown classes of such species are likely to retain high live crown ratios, to respond promptly to release, and to have remained straight. Stands of intolerant species cannot withstand more than one or two selection thinnings before most of the trees capable of useful response are gone. Positively phototropic species, such as many hardwoods, which have stems that bend toward the light as they grow, generally become deformed if they have grown underneath

Figure 5.7 An old-field stand of loblolly pine in which a selection thinning has just been completed. Most of the trees removed were large, rough dominants that produced the kind of material typified by the butt log appearing in the foreground. The men are standing in front of a gap in the canopy created by the removal of trees of this kind. *(A.F.I. Photograph by South Carolina State Commission of Forestry.)*

more vigorous trees. If trees such as these do respond to the increase in growing space, they often develop multiple leaders, which causes major forking in the stem.

Even if repeated selection thinning does not significantly reduce production of total cubic volume or tonnage of wood, it greatly diminishes production measured in units such as merchantable cubic volume and especially board-foot volume for a given length of

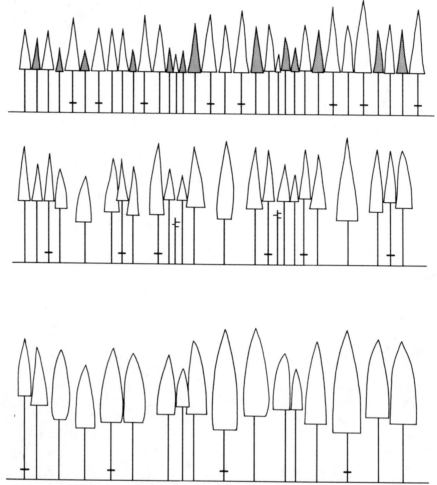

Figure 5.8 A series of three selection thinnings in a stand of balsam fir and red spruce being grown for pulpwood on a good site. Each sketch shows the stand immediately before a thinning. The trees with the shaded crowns in the first sketch are those that persist until the end of the rotation. The choice of trees to be removed in the second and third thinnings is somewhat modified by the necessity of avoiding enlargement of gaps caused by the earlier removal of dominants. The advanced reproduction that would undoubtedly become prominent by the time of the third thinning is not shown.

rotation. If the repeated thinning uses a large minimum cutting or "target" diameter, the result is that the production over the rotation will be in trees of very similar dimensions. The total value removed from the stand over the rotation may be greater, but the rotation will be longer as growth is consistently shifted from more to less vigorous trees.

A program of repeated selection thinnings can have the virtue of holding the investment represented by merchantable growing stock at a low value. The value of the growth obtained then tends to return a high rate of interest on this part of the investment in the forestry enterprise. This can, of course, be an empty virtue if the annual cash return per acre is so low as to constitute a low rate of interest on the capital represented by the whole enterprise. The rotation will be significantly prolonged, but if the overall returns can be kept high throughout the rotation, this can actually be advantageous.

There is a third approach to selection thinning in which it is combined with simultaneous low thinning in order to enhance the development of remaining codominants. This technique is of some advantage in counteracting the baneful effects of selection thinning, especially in previously unthinned stands that have become overdense.

Problems with Selection Thinning

Almost all methods of selection thinning have a tendency to increase the possibility of losses to physical and biotic agencies. The trees that are released are likely to be of less than optimum vigor and to have rather weak and slender stems. Their reservation is, therefore, a calculated risk until they develop thriftier crowns and stronger stems. The risk of major losses to wind can be slightly reduced by keeping the stand edge as strong as possible; this necessitates *leaving* the dominants in the stand edges.

Selection thinning tends to aggravate problems with wind damage because it removes the trees that have the most well-developed butt-swells and the more tapered boles which are the strongest. If the weaker trees do not have strong trees protecting them from the force of high wind and snow, they are apt to blow over and create more of the small gaps in the crown canopy in which funneling effects can accelerate wind speed and cause even more blow-down.

If preoccupation with logging costs leads to addiction to selection thinning that causes wind damage to residual stands, there is a psychological tendency to get desperate and cut even more of the larger trees and leave even weaker stands. This syndrome is one in which the first mistake seems to justify making worse ones.

The better solutions would be to thin more lightly, or by a different method, or simply not to thin at all. Sometimes this problem arises because of the mistaken belief that trees with large crowns are the most threatened by wind because they have the largest ''sail area.'' The susceptibility of a tree is a combination of the horizontal force (of wind acting on the sail area) and the strength of the stem and root structure (indicated by the taper of the tree). These go hand in hand; the more foliage (sail area), the more strength (taper). These trees are often the least likely to be blown down in a normal windstorm.

In theory, the choices made in favor of slower-growing trees would be undesirable from the genetic standpoint, while the discrimination against poorly formed ones would be desirable. It is debatable whether the degree of selectivity and of heritability are great enough to cause major change in the average characteristics of any ultimate progeny, but the rare, best genetic combinations are very likely to be eliminated.

One insidious advantage of selection thinnings is that they are more likely to return an immediate profit than any other kind of thinning. The trees that are cut are ordinarily the most salable to be found in the stand at the time of cutting. It is excessive attention to this short-term return that leads to high-grading. This advantage is gained at the expense of some others normally associated with thinning; for example, the time required to grow crop trees to a particular diameter is increased.

The proportion of the cubic volume removed in any one selection thinning should generally be as low as possible. There is a tendency to make low thinnings and crown thinnings too light; the reverse is generally true of selection thinnings. The intervals between successive selection thinnings are longer than for thinnings of other kinds because the holes created by cutting dominants close so slowly.

There is a desperate sort of cutting, sometimes euphemistically described as one of the kinds of "selective cutting," that is best regarded, at least in its early stages, as selection thinning. Such cuttings involve attempts to extract the fullest potential for growth from the subordinate crown classes by successive cutting of trees that have been brought to the dominant position. This often results from despair over the problem of replacing existing stands with desirable reproduction. If the problem is evaded by very light periodic cuttings of the largest trees, the residual stand grows poorly and an undesirable understory may burgeon. One way to break this vicious cycle is to destroy the understory and open the overstory enough to allow reproduction of the desirable intolerant species.

This kind of cutting can also result from single-minded concern about avoiding the high costs of harvesting small trees. Even more short-sighted is the policy of regarding the stand merely as a magic warehouse into which one ventures sporadically, attempting to find trees that will meet the specifications for current orders. Both procedures ultimately leave stands of poor trees that cannot be harvested economically by the most ingenious logging or the most astute salesmanship.

Geometric Thinning

In the fourth general method of thinning, the trees to be cut or retained are chosen on the basis of some predetermined spacing or other geometric pattern, with little or no regard for their position in the crown canopy. The older, ambiguous designation of such thinning as "mechanical" refers to the mechanistic mode of choices and not to any use of machinery.

Geometric thinning can be advantageous in treating young stands that are densely crowded and previously unthinned. Rather arbitrary choices of trees can be justified if the number of present dominants greatly exceeds the ultimate number, or if the stands are so uniform that there has not yet been much differentiation into crown classes. Geometric thinning is also common as a means of enabling the use of large harvesting equipment in thinning dense stands. The relatively small size of the trees results in high cost and low value per unit of product removed. These problems often make it virtually impossible to initiate thinnings in the very stands that need them most without the use of large, high-volume machines that gather tree stems in bundles.

Ordinarily, geometric thinning is employed only in the first thinning of a stand. If properly done, it usually sets the stage for exercising a much higher degree of selectivity in any subsequent thinnings, especially where large machines can be maneuvered more effectively in the less congested stands.

Two general patterns may be followed in mechanical thinning. In the first, referred to as **spacing thinning**, trees at fixed intervals of distance are chosen for retention and all others are cut. In the second, which is usually called **row thinning**, the trees are cut out in lines or narrow strips at fixed intervals throughout the stand. The rigidity of these specifications may be modified as occasion demands. In general, the justification for adherence to an arbitrary pattern decreases as the variation in size and quality of the trees increases.

Spacing thinnings are most commonly applied in overcrowded stands that have developed from dense natural regeneration or artificial broadcast sowing of seed. The simplest mode of application involves adherence to a rigid spacing. This approach can be modified by leaving the best tree in each square defined by the desired spacing. The amount of work can often be reduced by removing only the most active competitors of trees designated for release. The removal of the adjacent tree with the largest crown causes the greatest improvement in growth of the released tree and would thus be the best one to remove. However, the thinning effect of such limited kinds of removals does not long endure.

Thinning to a predetermined spacing is generally employed only in young stands. The idea of thinning to a certain average spacing is often followed in older stands, but as a numerical guide to the severity of the thinning and not as a way of selecting trees to be reserved.

The overriding concern for spacing that geometric thinning patterns best address leads to their frequent application in **precommercial thinnings**. These are thinnings made purely as investments in the future growth of stands so young that none of the felled trees are extracted and utilized. **Commercial thinnings** are most simply defined as those in which all or part of the felled trees are extracted for useful products, regardless of whether their value is great enough to defray the cost of operation.

Row and Strip Thinning

The removal of trees in rows, lines, or strips is done for a variety of reasons. The fact that this kind of thinning usually represents a compromise between doing a perfect job or doing nothing is not a reason to disdain it.

This kind of thinning plays a very important role in the precommercial thinning of excessively dense stands. A variety of machines can be used to destroy trees in narrow strips so that the ones remaining along the edges of intact strips can grow faster (Fig. 5.9). Among these are rolling brush cutters, bulldozers, brush hogs, and some of the other kinds of machines also used in mechanical site preparation (see Chapter 8). It is also possible to kill narrow strips of trees by applying herbicides by helicopter. If devices are used that cut or kill trees one by one, such as brush cutters or chain saws, it is usually best to employ methods of selecting the trees that are more sophisticated than eliminating lines or strips.

Row or strip thinning has been successfully used, with or without modification, to facilitate the harvesting or products from dense stands that have not been thinned previ-

Figure 5.9   Strip thinning in a dense, stagnating stand of ponderosa pine, nearly 50 years old, in eastern Oregon. The trees in the cleared lanes (right) were crushed with a bulldozer, leaving the narrow strips of trees visible in the aerial photograph (left). Such treatment is crude but effective; the only serious danger is that Ips beetles will breed in the slash and attack the standing trees. *(Photographs by American Forest Institute.)*

ously. The first thinning in a stand is the most difficult. Congestion in the crown canopy and the light weight of the trees make felling difficult. The close spacing of the trees, both cut and left, impedes skidding or forwarding.

With row or strip thinning, however, it is possible to proceed so that trees are successively felled into the vacancies already created by the cutting of adjacent trees. The products can then be easily removed along the lanes that have been cut through the stand. If the trees are limbed in the woods, the logging debris can either be stacked along the edges out of the way or else used as a kind of roadbed to protect the soil from damage by the log-moving machinery. This pattern of initial thinning overcomes many of the biological objections that are raised against thinning because of logging problems. A second cycle of this or any other kind of geometric thinning is rarely necessary or advisable because of the improved accessibility resulting from the first thinning.

Row thinnings are most readily applied in plantations that are already laid out in straight rows. In its purest form, this involves removal of every third row (Fig. 5.10). Every residual tree is freed on one side, and the removal of one-third of the stand is a good approximation of the normal severity of most thinnings in young stands. The removal of every other row would ordinarily be too severe. It is sometimes found that spacing of the trees can be adjusted more satisfactorily in subsequent thinnings if the residual strips alternate in width between those that are two rows wide and those that are three rows wide.

The strips can also be shifted to go around especially desirable trees. In closely spaced stands, the lanes might be made two rows wide in order to allow the passage of skidders or forwarders. The severity of the thinning can be reduced by spacing the cut strips at wider intervals. This course would be desirable where the objective was to stimulate the growth of trees destined to be removed in the final crop with a smaller reduction of gross yield or the production of an excessive amount of small material for which there was a limited demand.

Thinning in strips and rows can also be used effectively in combination with other methods of thinning within the leave strips. The cut strips become closely spaced skid trails. The lanes that are cut across a stand do not have to be straight and at some arbitrary spacing. The transportation lanes should be as narrow as possible and the spaces between them should be kept as wide as possible while being consistent with extracting trees. It often helps to fell the trees in a herringbone pattern, with their butts (or sometimes the tops) pointing in the direction along which they will be pulled out into the lanes and in the general direction along which they will be pulled to the landing.

This general approach helps overcome the logging problems encountered in the first thinning of almost any kind of stand. If avenues are cut through the stand at appropriate intervals, it is then often simultaneously possible to cut and leave trees in much more appropriate patterns between the avenues. It is sometimes possible to thin whole stands with a high degree of selectivity simply by proceeding in straight or gently curving lines from one tree that ought to be cut to another, marking them as one goes along. Sharp curves cause difficulties with timber extraction, especially if whole tree lengths are removed. If such special measures are taken in the first thinning, the avenues remain useful for subsequent thinnings.

Even when most timber harvesting was done with hand tools and animals, it was difficult to thin on any highly selective basis without giving some consideration to the problems of felling each tree and moving the products out of the woods. When logging equipment becomes more mechanized and less maneuverable, it is even more necessary to adjust the patterns of tree removal to the logging equipment. Development of mecha-

Figure 5.10 A row thinning being started in a dense, 25-year-old plantation of red pine in New York. Every third row is removed arbitrarily. *(Photograph by U.S. Forest Service.)*

nized fellers in which the cutting head is mounted on a long hydraulic arm may increase the degree of selectivity possible in dense stands of small trees.

Potential Problems

Row and strip thinning can lead to the development of lopsided crowns that can, if they remain that way for long periods, lead to the development of stems that are elliptical at the base. Except to the extent that unequal crown loading may predispose trees to later damage by wind or frozen precipitation, this effect is seldom of much consequence. However, it can help to try to develop more symmetrical crowns in later thinnings.

The main disadvantages of arbitrary selection of trees in various kinds of geometric thinning patterns lie in the cases where there is substantial variation between individuals. The removal of too many of the dominant trees can have the same undesirable results as selection thinning. Among other things, it usually causes a more definite and longer-lasting depression in the production of a stand than do the more uniformly distributed vacancies in growing space caused by low or crown thinning. Therefore, geometric thinning is best carried out when stands are young and plastic, before the deep-crowned trees become too few and the crowns of the trees so broad that wide gaps result.

Free Thinning

Cuttings designed to release crop trees without regard for their position in the crown canopy are free thinnings, in the sense of being unrestricted by adherence to any one of the other methods of thinning. However, in any free thinning, the mode of treatment accorded any one of the trees cut or released can be identified as representing one of the other four methods. The greatest need for combining several methods in a single thinning is encountered in stands that are somewhat irregular in age, density, or composition. Pure plantations are usually sufficiently uniform that free thinnings are unnecessary in them.

Technical terminology should convey as much meaning as possible. Therefore, vague terms like ''free thinning'' ought to be avoided if there is any reasonable possibility that one of the other terms might apply to a particular operation. It is more informative to use terms like ''modified crown thinning'' or ''crown and selection thinning'' than to throw all variants into a miscellaneous category. It should be noted that all of these terms refer to single operations and not sequences thereof. A program of treatment is not referred to as free thinning simply because crown thinnings are, for example, followed by low thinnings.

Conditions requiring free thinnings are most likely to exist at the time of the first thinning in previously untreated natural stands. This approach is distinctly advantageous for bringing a stand into shape for efficient production. If the objectives of the thinning program are clearly defined and consistently pursued, the irregularities begin to fade and the likelihood increases that a logical thinning operation will conform to a single method. From the standpoint of efficiency of logging and administration, it is desirable to achieve regularity of treatment as soon as uniformity can be imposed without undue sacrifice of the growth potential of the stand.

A typical free thinning operation in an unevenly stocked but even-aged stand might simultaneously include (1) selection thinning to eliminate scattered undesirable dominants, (2) crown thinning to release crop trees drawn mainly from good dominants and secondarily from thrifty codominants in the more sparsely stocked portions, and (3) low thinning to salvage all merchantable overtopped trees throughout the stand and to thin the well-stocked portions to the severity of the D grade. If trees are unmerchantable, a free thinning approach can be applied as a precommercial operation.

Quantitative Definition of Thinning Methods

The ratio d/D, where d is the average D.B.H. of trees removed in thinning and D is the average before thinning, is sometimes used to characterize methods of thinning quantitatively (Bailey and Ware, 1983). The ratio is more than 1.00 for selection thinning, 1.00 or less for other methods, and lowest for light low thinning. It is exactly 1.00 for perfect row thinning in plantations. The diameter used is sometimes the arithmetic mean, but is often the **quadratic mean** (diameter of the tree with average basal area). The ratio is best used to define the thinning rather than as a guide for tree markers.

THINNING SCHEDULES

A schedule of thinning should be a systematic plan for a whole rotation based on deliberate decisions about the kind of vegetation, benefits, and products desired. One should reason from these chosen goals backward to the schedule of treatments designed to achieve them. The ultimate objective is some concept of the stand that will exist near the end of the rotation. The details of a thinning program should remain flexible enough to be settled in the light of conditions prevailing at the time of each treatment.

Although thinning can certainly be used for other objectives of managing forest vegetation, timber production is the chief purpose and also the one that requires the most precise adjustments. Prospective markets for wood products set most of the long-term objectives. Markets are, in a sense, also the tools that are used, often quite opportunistically, to achieve ultimate goals, whether they are timber production or other objectives. Fortunately, the most rapid product innovations tend to apply to the small, poor trees of the kind that are best removed in thinnings.

Changes in utilization sometimes seem to cast doubt on the need to grow the kind of good, moderately large trees usually envisioned as the final goal of thinning. However, the prices paid for the products of such trees have shown more tendency to soar than to remain stable. The most important choices usually lie in determining how much of the total production to channel into trees large enough for sawn or sliced products and how much into bulk commodities such as pulpwood, chips for reconstituted products, and fuelwood. Small trees are almost as good as large ones for the latter uses.

Choosing Methods of Thinning

Three interacting considerations enter into the formulation of thinning programs, which may, more than incidentally, include deliberate programs of not thinning. One set of choices has to do with timing, which includes rotation length, time of first thinning, and intervals between subsequent thinnings, if any; another is over the method or methods of thinning employed at each stage. It is only under the least complicated circumstances that a single method can be followed through an entire thinning program. The most difficult set of choices is about the regulation of stand density or the amount of growing stock to be left after each thinning.

If a stand is thinned only once and is not very uniform, it is likely that the necessity of doing so much with such limited opportunity will dictate simultaneous use of two or more methods in a free thinning operation.

The circumstances under which repeated application of a single method might be found best for a long series of thinnings are rather limited. The stand would have to be very uniform so that the crop trees would be alike and treatment would not have to be varied in different parts of the stand. Then, if markets were excellent and if the crop trees were of good quality and always in a clearly dominant position, one might conceiv-

Figure 5.11   A plantation of eastern white pine at Biltmore, North Carolina, at various stages during a series of five light, low thinnings. The plantation was established at very close spacing to check serious erosion and was first thinned at the twentieth year (Wahlenberg, 1955). Subsequent thinnings were made at intervals of about 6 years, reducing the basal area to approximately 100 square feet per acre each time. (*a*) 20 years: numbered trees being left in first thinning; (*b*) 26 years: freshly numbered trees being left in second thinning; (*c*) 32 years: just before third thinning; (*d*) 45 years: just after fifth thinning. *(Photographs by U.S. Forest Service.)*

ably employ a simple series of low thinnings like those shown later in Fig. 5.11. A series of selection thinnings might be applicable to growing stands of very tolerant conifers for pulpwood, although there might be some problems of securing reproduction and disposing of unmarketable residual trees at the end of the rotation. With species of sufficient tolerance, it might be possible to apply a series of what could be called crown thinnings if the ultimate disposal of any lower layer of trees were not regarded as a deviation from the method.

The crown classes that should be removed to favor the crop trees and those that can be harvested profitably will normally change as a stand grows older. This calls for changes in the method of thinning even in the absence of differences in condition of various parts of the stand.

The high labor costs and low product values associated with small trees often make it necessary to change sequentially from one thinning method to another to evade or postpone the need for cutting small trees. Figure 5.12 illustrates the point that the four well-defined methods involve harvesting trees from different segments of the distribution of diameter classes in a pure, even-aged stand. From this it can be seen that if selection, crown, and low thinnings followed in succession as a stand grew older, there would be some possibility of avoiding cutting trees when they were small. If removals from the dominant crown class are undesirable, the same can be accomplished by crown thinning that is followed with low thinning. Such measures are not perfect solutions to the problem because it is so difficult to make the trees of the lower crown classes increase rapidly in diameter, and often they might as well be left until the end of the rotation. Geometric methods such as row thinning reduce logging costs mainly by facilitating extraction and not by postponing the handling of small trees; the trees removed are an essentially random

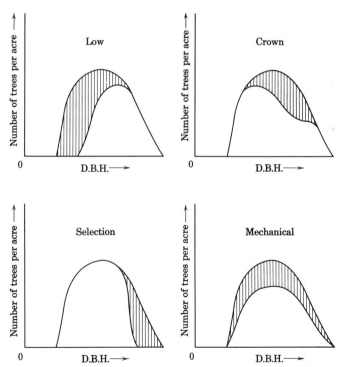

Figure 5.12 Diameter distributions for the same pure, even-aged stand showing, by cross-hatching, the parts that would be removed in four different methods of thinning. In each case about one-third of the basal area is represented as having been removed. It is assumed that no overtopped trees are salvaged in the crown and selection thinnings, that the stands have not been treated previously, and that D.B.H. is closely correlated with crown class.

sample of the existing diameter classes. Nevertheless, the role of large machinery for timber harvesting increases the desirability of using some sort of row or strip thinning initially. The dilemma faced in the need to thin out small trees to capture the subsequent advantage of large trees (and the lower cost of harvesting them) can sometimes be overcome by using light, low-cost machinery in the early thinnings and switching to larger machinery in the later ones.

The changes in thinning methods appropriate to developing good crop trees often follow the same sequence by which the costly logging of small trees is evaded. It is logical that any coarse dominants be removed in early selection thinnings. The enlargement of the crowns of crop trees is, in the early stages, most readily expedited by crown thinnings. Once the crop trees have been successfully promoted to clearly dominant positions, their competitors can be removed in low thinnings. Such a sequence might start with a geometric thinning.

Such changes generally follow the principle of shifting the level from which trees are removed downward through the crown canopy. The compulsion to low thinning late in the rotation comes purely from the fact that the trees of the subordinate crown classes have usually reached the end of their usefulness and only the dominants or vigorous codominants retain much capacity for further increase in value.

Even though a stand is uniform, it is not necessarily desirable that each thinning operation follow the same pattern throughout the stand. For example, if the rotation is long and the thinnings start early, at 30 years it may be useful to accord different treatment to trees destined for removal at 60 years than to those to be held for 100 years (Macdonald, 1961). The ''short-term'' trees may require complete release so that they will lose no more live crown surface and will grow rapidly during the remainder of their allotted time. Meanwhile, the ''long-term'' trees may be left somewhat crowded so that the attrition of their lower live branches will continue and their boles will not develop large branches or excessive taper. In general, when thinnings start early in the rotation, it is necessary to plan proper treatment not only for the final-crop trees but also for the rather numerous ones that can develop substantial value even if they are destined for somewhat earlier removal.

Timing of the First Thinning

In theory, the first thinning can be made as soon as the crowns or the root systems of individual trees grow together and start to interfere with one another. Competition of this sort commences early in the sapling stage in any but the most poorly stocked stands. The best single criterion for determining when to apply the first thinning is the live crown ratio of the potential crop trees. As long as a satisfactory ratio is maintained, a stand need not be thinned until the thinning will pay an immediate profit. When the ratio approaches an amount at which undesirable losses of diameter growth are in prospect, it is time to consider the first thinning. Some planned dwindling of the live crown ratio is usually desirable to restrict branch size and the degree of stem taper as well as for natural pruning. The practice of thinning and regulation of stand density usually involves compromises of this sort between diameter growth and various ''stem-training'' effects.

The various factors that affect the timing of first thinnings are considered mostly in terms of their economic consequences. If there are ample funds for long-term silvicultural investments, the first thinning can perhaps best be done when the value of ultimate benefits, discounted to the present at compound interest, equals the cost of the operation. If such investments cannot be made, the first thinning is delayed until it will give an immediate profit. In some cases the appropriate time never arrives.

Precommercial Thinning

If thinning is done as an investment, it is desirable to proceed in such a manner as to achieve the greatest possible increase in diameter growth with a small amount of work. This line of attack differs from commercial thinning in which it is usually advantageous to harvest the highest possible proportion of those trees that will not contribute to the future increase in value of the stand.

One useful approach to precommercial thinning is that of ''crop tree thinning'' in which only the trees likely to form the final crop are released. Usually, most of the potential benefits are gained merely by eliminating the one or two most serious competitors of each crop tree. It has been shown that cutting of additional competitors caused little further increase in diameter growth of leave trees and that dominant trees responded better than those of lower crown classes (Reukema, 1961). However, the period of enhanced diameter growth is longer the more drastic the reduction of competition. If an investment is required to carry out the first thinning, it is desirable to make it heavy enough to ensure that no further treatment will be required before the stand reaches the stage where a profitable cutting can be made.

The policy of postponing the first thinning until it returns an immediate profit can be defended on the ground that many fine stands have developed in nature without treatment of any kind. In most stands, failure to thin should be regarded as a lost opportunity rather than as an invitation to disaster. However, in many of these same stands, the net return from the whole crop could be increased if an outright investment were made in thinning at the appropriate time. In fact, economic analyses have repeatedly shown that precommercial thinning often is the most rewarding long-term investment that can be made in silvicultural treatment. This is because the investments are not made at the beginning of the rotation and it is not long before the treatment pays off through increased production of merchantable volume.

Precommercial thinning can also reduce the cost of subsequent harvesting by getting rid of small, unmerchantable trees that impede logging. If stands are made more open, there is less reason for skidding equipment to damage the remaining trees because there is more room to go between them.

A strong case can be made for investment in early thinning with stands that would deteriorate without treatment. The most common cases of this kind involve very dense, pure stands, usually from regeneration by natural or artificial seeding, and especially those of species with low genetic variability in factors that affect height growth. The hard pines, especially red, jack, and lodgepole, are notorious for these tendencies, but they can exist in some degree with any species. Some species, on the other hand, such as eastern white pine, appear to have sufficient genetic variability to express dominance even in dense stands.

The likelihood that trees will stagnate or stall in diameter growth or, worse yet, stop growth in height is greatest on poor sites. Sometimes fertilization will favor strong trees and hasten suppression of the weak and thus produce the same effect as thinning. There are even cases of extreme deficiency of water or nutrients in which thinning alone will not produce any effect without fertilization, irrigation, or drainage. Even if the poor growing stands do not stagnate, the trees might become so slender that even a typical wind or snowstorm may cause large patches to topple.

Effect of Stand Structure

The trees of planted, single species stands are exceptionally uniform in age and spacing, so they tend to be evenly matched in size and vigor. Therefore, it is necessary to anticipate that their crowns or root systems will close together rather suddenly and that the time of this event may be the ideal time for the first thinning. Generally, it is wise to make the initial spacing of the planted trees wide enough so that all of the trees will have grown to minimum merchantable size at this time; however, need for competitive training effects may make this impossible. This general kind of problem can also be mitigated by deliberate avoidance of extreme genetic uniformity in the stand. The genetic uniformity that is sought in agricultural annuals is usually undesirable in woody perennials. The dwindling of numbers of trees is a process that must be anticipated; it is also a desirable one that is put to planned use in thinning.

Stands can also be so irregular that some trees race ahead to become branchy, malformed superdominants. If their removal in a first thinning is too long delayed, they are likely to suppress their better-formed subordinate neighbors.

Even if an untreated stand does not stagnate or develop into a collection of malformed dominants, the stage is ultimately reached at which it may be dangerous or fruitless to initiate thinning or any other form of partial cutting. Dominant trees can eventually become weak, slow-growing, and slow to respond to release. If thinning is delayed too long, the residual trees will be slow to reclaim the vacant growing space, and the production rate of the stand will suffer a major, though not necessarily permanent, reduction.

In other words, stands can become so crowded that it may be questionable whether it is wise to thin them even when such treatment would be immediately profitable. Investment in early thinning may provide the only way in which the important financial benefits of later thinnings can be captured.

Timing of Subsequent Thinnings

The effects of a single thinning are not maintained indefinitely. After a few years, the gaps in the canopy close together, and before long the same crowded condition that existed before thinning redevelops. As the crowns expand, the number of trees that can occupy an acre to best advantage decrease and the surplus volume available for removal in the next thinning accumulates.

The rate of growth of the crop trees is the best single criterion for determining when thinnings should be repeated. It is logical to set some realistic rate of diameter growth as a goal and thin when the growth of the crop trees falls below it.

It is also likely to be necessary to wait until the volume that can be removed has accumulated enough to support a profitable operation. If the stand closes too tightly and diameter growth slows down too much before sufficient volume accumulates, it is a sign that the previous thinning was either too light or was not done in such a manner as to give the crop trees enough room to grow. In this respect, close attention must be paid to manipulation of the diameter distribution of trees destined to be removed in thinnings. If too many of the larger ones are cut at once, the next thinning may have to be postponed until long after the small trees have choked the crown canopy in the process of growing to merchantable size.

The proper balance between severity, method, and timing of thinning is not an easy one to strike. This kind of adjustment is still highly intuitive because there are so many considerations to take into account and all are susceptible to unpredictable factors such as weather, disturbances, and prices. Diameter growth of the crop trees is by far the best and simplest guide; unfortunately, it indicates the status of only one component of the stand. The benefit of this measure is that the crop trees represent the largest economic component of the stand.

The heavier the thinnings, the longer is the interval between them. Heavy, infrequent thinnings tend to reduce the total yield of a stand because of the long periods during which parts of the growing space remain unoccupied.

It is seldom possible to thin lightly and frequently enough to avoid losses in production of gross, total cubic volume or tonnage of wood. Ordinarily, it is logical to thin in ways that will increase diameter growth enough to increase the actual yield of merchantable volume or value of wood even if this does involve deliberate sacrifice of the gross production of wood. Almost any kind of commercial thinning augments actual yield through harvest of material that would otherwise be lost to natural suppression.

The ratio of the amount of wood that is harvested over the rotation to the gross amount of wood produced is called the **harvest index.** Figure 5.13 shows how amounts of gross wood and harvestable wood vary over time. Thinnings usually reduce the gross production because there are periods when growing space is left vacant, but increase the harvest index because more of the wood is in larger pieces. Although the biological production of the stand is lower, there is more utilizable production.

The practice of repeating thinnings at equal intervals simplifies administration but has no other virtues. Actually, stands should be thinned more frequently when young than when old because they close more rapidly.

The height of dominant trees of a stand is an integrated expression of age and site quality. Therefore, it is an excellent parameter to use in developing thinning schedules. Provided that there are no large differences in site quality, a stand of a given species will go through the same processes while it is growing from 40 to 50 feet ($\approx$12 to 15 m) of height regardless of whether this takes 3 years or 10. Dominant height becomes the index of the stage of stand development, while increase in height can be used to measure the

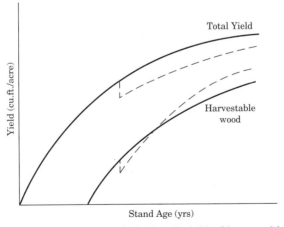

Figure 5.13 Relationship between total yield and yield of harvestable wood as a stand ages. The effect of a typical thinning is shown by the dotted lines. Note that total yield (of standing trees) decreases and remains below that of the unthinned condition. The yield of harvestable wood decreases at first and then increases to a level above that of the unthinned stand.

rate of development. Because height growth and crown expansion both depend on shoot elongation, the relationship between the two is generally close (Mitchell, 1975).

If one planned to thin at equal intervals of attained dominant height, the time intervals between thinning would automatically and logically increase with stand age and be longer on poor sites than on good. Knowledge developed on one site would be applicable to others, but one should anticipate that trees on good sites might culminate in rate of height growth at greater height than those on poor. Height can be translated into age and height growth into time by use of site-index curves of height over age, but this kind of arithmetic can be postponed until some late stage in the development of thinning schedules.

Regulation of Stand Density and Thinning Intensity

Thinning programs should be governed by reasonably definite schedules indicating the density of stands at all stages of development. Such schedules help ensure final crops that consist of trees of the qualities and sizes desired. They are also necessary for purposes of forest regulation to predict yields from intermediate and final cuttings and other purposes such as determination of relative amounts of different wildlife habitats.

Without thinning schedules there are psychological tendencies to thin too lightly, at least if the trees to cut are the ones that are marked. A thinning usually looks much more

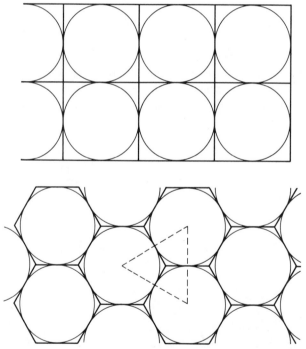

Figure 5.14   Schematic diagram contrasting the arrangement of circular tree crowns with trees standing at the corners of squares (top) and equilateral triangles (bottom), with the same distance between trees and the same crown diameters in each case. Note that the arrangement of nested hexagons with the triangular spacing allows the crowns to cover more of the ground area such that there are also more trees of the same size per unit of area.

severe after the trees are marked than it does after they are cut; within a few years, the rotting remains of the cut trees are the only readily apparent evidence of any cutting. Conscientious foresters are chronically fearful of running out of trees and often fail to anticipate how rapidly the stands will close up again. There are tendencies to make thinnings too severe if the trees to be left are the ones that are marked.

Thinning schedules based on appropriate studies of growth and development after thinning are the best source of assurance. It is desirable for those who mark stands for thinning to check themselves occasionally with the use of prisms or other angle-gauges under the Bitterlich point-sampling scheme to determine residual basal area and average tree size. No one schedule is optimum for all stands, even within a given forest type, because economic as well as natural factors must be considered.

Consideration of the details of any thinning schedule requires recognition of the fact that, at the time of each thinning, the trees of the stand are segregated into three categories. The first and most important group consists of the trees destined to form the final crop. The second comprises those trees that will be removed in subsequent thinnings but should maintained in the meantime to utilize the growing space that will eventually be occupied by the final crop. These two categories, which do not always have to be clearly distinguished, make up the growing stock left after each thinning. The final category consists of the surplus trees to be eliminated in the thinning.

In the kind of analysis that leads to development of schedules, it is generally desirable to approach the matter from two directions. The first, and often the most important, is consideration of the development of the final-crop trees—that is, the amount of growing space they will need at each stage to meet whatever goal is set for their growth. The second approach guards against failing to see the forest (or stand) for the trees. It involves determining how much growing stock should be left after each thinning to give an optimum yield, over the whole rotation, in terms of amount of product per hectare or acre.

One approach cannot be followed to the exclusion of the other. It would be easy to get so concerned about the crop trees that one might end the rotation with a few fat trees and sacrifice too much total yield. Conversely, preoccupation with yield of cubic volume per unit area almost always leads to sub-optimal diameter growth. If both approaches are used, desirable compromises often become obvious.

Many of the basic principles underlying these approaches were discussed in Chapter 4. What follows is a discussion of the ways that are, or have been, used to create thinning schedules based on these principles.

Tree-Spacing Guide

The stand-density parameters discussed in Chapter 4 can be used to describe almost any thinning schedule that has been decided upon, but they do not necessarily define any. The same is true of various ways of using the spacing between trees. Spacing itself is really the same as numbers of trees per acre. However, this can be modified to produce some "rules of thumb" regarding the proper spacing of the trees left after thinning.

According to the useful $D + x$ rule, the average square spacing in feet should equal the average stand diameter D in inches plus a constant x. The values of x that are suggested for various situations range from 1 to 8, the most common being 6. Rules of this form provide steady increases in basal area. Rules of the form Dx provide for constant basal area; the values of the factor x for basal areas of 90 and 120 square feet of basal area per acre are 1.61 and 1.4, respectively.

Another way of regulating the spacing of trees left after thinning is to set the average spacing equal to a constant fraction of the height of the dominant trees. Wilson (1955)

Table 5.1 Numbers of Trees per Unit of Area with Square and Equilateral Spacing in English and S.I. Systems[a]

Trees per Acre		Spacing	Trees per Hectare	
Square	Equilateral	(ft, m)	Square	Equilateral
4840	5589	3	1111	1283
1742	2012	5	400	461
889	1027	7	204	236
681	786	8	156	180
538	621	9	123	143
436	503	10	100	115
302	349	12	69	80
151	174	17	35	40
99	114	21	23	26
48	56	30	11	13

[a]The stand densities for the English half of the table are not equivalent to those on the same lines in the S.I. or metric half.

suggested that the spacing for tolerant species such as spruce and fir be one-sixth of the height and that for intolerants such as red and jack pine be one-fourth of the height. Dominant height is an excellent indicator of the stage of development of a stand and, unlike D.B.H., is independent of the effects of thinning.

In the foregoing and in most cases, it is assumed, purely for convenience of calculation, that trees stand at the corners of squares. This means that if the crowns closed perfectly they too would be square in horizontal cross section. Because they are closer to being round it may be recognized that, were the trees uniform and evenly spaced, they would stand at the corners of equilateral triangles and the crowns would nest together like hexagons (Fig. 5.14). This means that, with crowns of uniform width and circular cross section, the more efficient equilateral spacing would give 15.47 percent more trees per unit of ground area than square spacing.

This phenomenon should be taken into account if one has determined the crown widths of trees of some desired D.B.H. and wants to know how many might stand on an ideally stocked acre or hectare. Table 5.1 gives numbers of trees per unit area for the range of tree spacings that might exist in stands being thinned and in final-crop stands. For closed stands of equilateral arrangement, spacing and crown width are synonymous. However, a good, closed stand will, in practice, have more trees per unit area than that with square spacing but less than that with equilateral spacing (Assmann, 1970). There are always gaps; the crowns will not be perfectly round or of exactly the same size.

Taper Guide

One useful parameter for the guidance of thinning practice, developed by Abetz (1982) for Norway spruce plantations in Germany, is the ratio of total height to D.B.H., with both measured in centimeters as described in Chapter 3. While this does not define stand density, it does indicate when to thin and also quantifies two very important goals. First, it focuses attention on the rate of diameter growth. Because diameter growth is a lower priority use of the carbohydrates produced by photosynthesis than height growth, the ratio steadily climbs if trees become crowded and competition is intensified. Diameter is also the most important economic parameter. Second, the ratio very simply approximates the degree of butt-swell and taper that govern the resistance of trees against the mechanical loads of wind and frozen precipitation. If the ratio becomes too high, trees exposed to higher mechanical loads after thinning will be unstable and vulnerable to windthrow. The goal in the case of spruces in southwestern Germany is to keep the h/d ratio below 70, and stands with an average ratio above 100 are only thinned lightly and with great care.

Thinning Guides Based on Crown Expansion

The basic reason for using tree spacing to guide thinning is that it defines the area, along a horizontal plane, that can be occupied by the crown. The size to which the foliar sugar factory can grow is thus defined. There are high correlations between crown diameter and stem diameter and crown volume and stem volume, as well as between crown expansion and diameter growth.

The key to using these relationships to predict the growth of trees is determination of the rates of crown expansion (Mitchell, 1975). It is possible to ''grow'' tree crowns by computer simulation and present the results in the form of crown maps that change with time. Crowns can be grown in three dimensions and for sufficiently large portions of simulated stands. Stem growth is calculated from the amount of foliage each tree has based on the three-dimensional crown structure. This simulation is a great tool for quickly doing thinning experiments that would take decades to follow in the forest. This approach has the virtue of predicting the growth of wood on the basis of the amount of foliage, which is the real source of the wood. Any predictive method that is based on correlating the periodic growth of a tree with its initial D.B.H. or basal area is founded on the fallacy that wood grows wood; the results amount only to extrapolation of past growth.

With this model (Tree and Stand Simulator: Mitchell, 1975) it becomes possible to do thinning ''experiments'' in minutes that would take decades in the woods (Larson and Cameron, 1986). Different thinning regimes can be tested, and the results can be considered not only in terms of the volume yield but also with respect to important considerations such as the prospective assortment of tree diameters, and even the effect of pruning on wood quality. The best data to construct and calibrate the models come, however, from long-term observations of the growth of trees and their crowns in thinning experiments.

Formulation of Thinning Schedules

The general result of these considerations is that thinning schedules are based partly on the best biological, economic, and mathematical analysis available and partly on intuitive art designed to integrate these considerations. They are not etched in stone, and it is fortunate that stands that are still vigorous enough to thin are flexible enough to respond to changes in plans.

An approximation of a program of intensive, frequent thinnings is depicted in Fig. 5.15. It shows how the interval between thinnings would lengthen with advancing age. The most important line on this graph is that defining the residual basal area left after each thinning; in this case, it is shown as gradually increasing. If the thinnings were less frequent, they would be heavier and stand basal area would fluctuate within a wider range.

The amount of standing tree vegetation, basically the amount of foliage, which should be reserved in thinning varies widely, depending on species, site, and management objectives. The rates of diameter growth that are sought will greatly influence the choices.

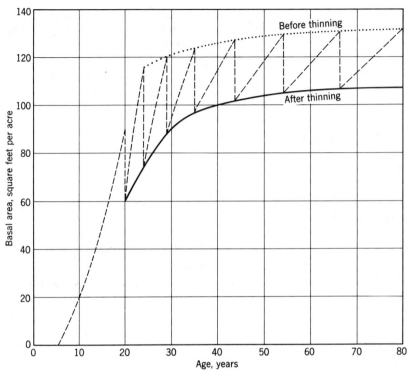

Figure 5.15 A hypothetical example showing how basal area should vary during a series of thinnings. The solid line indicates the lower guiding limit of basal area reserved after thinning. The dashed line shows the fluctuations in actual basal area. The dotted line represents an approximation of the level to which basal area is allowed to increase before thinnings are repeated. The interval of time elapsing between thinnings lengthens with age because the rate of growth in basal area declines as the trees grow larger. The arbitrary assumption is made that, in this case, the basal area left after each thinning should increase gradually from the middle of the rotation onward.

Appropriate levels of residual basal area depend heavily on the total amount of foliage that can be supported by a given species on a given site and may vary from 40 to 175 square feet per acre (≈9 to 40 square meters per hectare). This calls for higher residual stand density with evergreens than with deciduous species. The more shade tolerant the species, the greater is the logical amount. Good sites can carry more foliage than poor, so the amount left on them should be higher. On sites that are so poor that the root systems close but the crowns cannot, the stand density may have to be extraordinarily low.

In formulating a schedule it is first necessary to plan for the degree of initial crowding that will be required to control the size of branches and achieve other training effects in the early stages. Ideally, there should be some definition of the stage at which further dying of branches should be slowed or halted. It is at this stage that it is usually desirable to thin as heavily as possible to maintain good diameter growth of the trees that were previously "trained" by deliberate crowding. When all of the trees have been brought to merchantable size, thinnings usually become lighter because the creation of excessively large vacancies in the stands more likely to waste usable production.

RELATION TO FOREST USES OTHER THAN TIMBER PRODUCTION

Thinning plays a role in the achievement of most management purposes because it represents the primary means by which forest stands can be controlled and altered during the course of their development. It increases the water yield of forested watersheds by temporarily reducing the amount of foliage that transpires and intercepts water. It is seldom possible, however, to increase water yield even temporarily by more than about 10 percent by thinning. Regeneration cuttings cause larger and less temporary increases, but they cannot be repeated as often.

Thinnings are also useful in enhancing the development and controlling the composition of understory vegetation that may provide for age, browse, and seeds for herbivorous animals, both wild and domestic. Tightly closed crown canopies usually keep this food supply above the reach of all animals but birds. The vegetation that is valuable from the standpoint of wildlife management and the grazing of domestic animals is sometimes the same as that which causes problems in the regeneration of species for timber production. Therefore, the understory vegetation must be manipulated with due regard for all the effects that it has on the economic use and overall ecological balance of the forest community. The lesser vegetation is an integral part of the stand and should be regarded as something that can be usefully manipulated for either increase of benefits or reduction of harmful effects. No progress is made in this direction if it is looked upon only as an annoying impediment composed of plants unworthy of identification.

Thinning, like all forms of cutting, poses problems for intensive recreational use because slash is a far greater hazard and eyesore to the layperson than it is to the forester. However, it is better to dispose of the slash than to refrain from thinning in areas of heavy recreational use. Trees must be vigorous if they are to endure the trampling of the soil by their admirers or the more active attacks of vandals. Reduction of the numbers of trees is inevitable and is better used as a means of stimulating the residual trees than as a last resort for protecting the public from injury by falling dead trees and branches.

CONCLUSION

Thinning represents the primary means by which the yield of stands can be increased beyond the best that might be achieved under purely natural conditions. As a consequence, it is one technique that distinguishes intensive silvicultural practice from extensive.

It is possible that the essential generalities about thinning have been buried in the flood of qualifications that must inevitably enter into any discussion. The most important fact to be noted is that the total cubic volume of wood that can be produced by a given stand in a given length of time may be reduced but rarely increased by thinning. However, by skillful control of the growing stock the forester can marshal this production so as to capture the highest net return possible under the economic circumstances. The trees that are cut represent a salvageable surplus that is not needed to ensure full utilization of the growing space; they are, therefore, available for harvest. In the act of removing them it becomes possible to build more crown surface on selected individuals so that they will attain larger diameter, become more valuable, and be less vulnerable to damage.

The positive action of choosing the kind and number of trees to remain for future growth should be emphasized much more than the selection of trees to cut for immediate use.

BIBLIOGRAPHY

Abetz, P. 1982. Zuwachsreaktion von Z-Bäumen and Durchforstungsansätze bei Auslesedurchfor-stung in Fichtenbeständen. (Growth of crop trees and application of thinning treatments using the selective thinning method in Norway spruce stands.) *Allgemeine Forstzeitchrift* 47:1444–1446, 1448–1450. (Engl. sum.)

Bailey, R. L., and K. D. Ware. 1983. Compatible basal-area growth and yield model for thinned and unthinned stands. *CJFR* 13:563–571.

Larson, B. C., and I. R. Cameron. 1986. Guidelines for thinning Douglas-fir: uses and limitations. In: Douglas-fir: stand management for the future. C. D. Oliver, D. Hanley, and J. Johnson. (eds.) College of Forest Resources, University of Washington, Seattle, Contribution No. 55, pp. 310–316.

Macdonald, J.A.B. 1961. The simple rules of the "Scottish electric" thinning method. *Oxford Univ. For. Soc. J.* 5(9):6–11.

Mar:Møller, C. M., et al. 1954. Thinning problems and practices in Denmark. *State Univ. N. Y., Coll. For., Tech. Publ.*, 76.

Mitchell, K. J. 1975. Dynamics and simulated yield of Douglas-fir. *For. Sci. Monogr.* 17. 39 pp.

Oliver, C. D., and M. D. Murray. 1983. Stand structure, thinning prescriptions, and density indexes in a Douglas-fir thinning study, Western Washington, U.S.A. *CJFR.* 13:126–136.

Reukema, D. L. 1961. Response of individual Douglas-firs to release. *PNWFRES Res. Note*, 208. 5 pp.

Sassaman, R. E., Barrett, J. W., and A. D. Twombly. 1977. Financial precommercial thinning guides for northwest ponderosa pine. USFS Res. Paper. PNW–226. 27 pp.

Schlesinger, R. C., and D. T. Funk. 1977. Manager's handbook for black walnut. USFS Gen. Tech. Rept. NC–38. 22 pp.

Wahlenberg, W. G. 1955. Six thinnings in a 56-year-old pure white pine plantation at Biltmore. *J. For.*, 54:331–339.

Wiley, K. N. 1968. Thinning of western hemlock: a literature review. *Weyerhaeuser Forestry Paper* 12. Centralia, Washington. 12 pp.

Wilson, F. G. 1955. Evaluation of three thinnings at Star Lake. *For. Sci.*, 1:227–231.

CHAPTER 6

RELEASE OPERATIONS AND HERBICIDES

Release operations free young stands of desirable trees, not past the sapling stage, from the competition of undesirable trees that threaten to suppress them. A distinction may be made between two kinds of release operations, **cleaning** or **liberation**, which differ mainly as to the ages and sizes of trees eliminated.

In the release of most young stands, the operation is, in effect, an uncovering of them through elimination of overtopping trees. The basic objective is to give the trees that are released enough light and growing space to grow adequately and develop into trees of the main canopy. Sometimes whole layers of overtopping trees are removed; sometimes they are merely interrupted over especially desirable small trees. The degree of release sought depends on the method and cost of release, the minimum amount of vacant growing space that must be created, and other considerations.

Release cuttings are most readily visualized in terms of freeing the crowns of existing desirable trees, but there are several additional considerations. The growing space in the soil may be fully as important as that in the crown, particularly on dry sites. Even when the crown of a tree is fully released, its growth may still be hampered by competition for moisture and nutrients with the root systems of adjacent trees. Some techniques used for releasing stands are very effective in killing the tops of competing trees but do not prevent sprouting if the roots remain alive. Entirely new vegetation of aggressive competitors may claim new vacancies faster than the released trees can expand to fill them.

As with any silvicultural treatment, release cuttings should be conducted with a clear understanding of the way in which all the vegetation of the site will develop after treatment. This is especially true of the release of young stands. There the desirable trees command such a small fraction of the growing space that they must expand considerably to preclude recapture of the site by other species. If the work is conducted with little foresight and much wishful thinking, discouraging developments are likely.

131

Before we consider the techniques of release cutting, let us first examine the ways in which plants can be killed and some characteristics of the chemicals that provide one of the ways.

KILLING VEGETATION IN PLACE

Because the harvest of useful products is the cheapest way of killing trees to guide the development of forest vegetation, it is chosen whenever possible. Some silvicultural treatments, especially those employed to mold stands when they are young and plastic, require killing trees and other plants that are not worth harvesting. The rest of this chapter is devoted to techniques of dealing with such vegetation, except that prescribed burning and mechanical methods employed in conjunction with preparing sites for regeneration are discussed in Chapter 9.

The techniques of killing unwanted vegetation depend on (1) whether the undesirable plants can reproduce by vegetative sprouting, (2) the degree and pattern of selectivity desired, and (3) the size and other characteristics of the stems of the unwanted plants. Virtually all of the ways of killing woody plants depend on killing the roots directly or indirectly. This was first learned by fur-clad scientists who girdled trees with stone axes to make way for primitive agriculture. Sprouting plants cannot be killed without killing the roots. Many angiosperms sprout and most conifers do not.

The surest and most direct way to kill a plant is to uproot it. This is still a dependable way of dealing with perennial grasses and other small plants that sprout profusely from the roots, although there are chemical techniques of killing such species. Sizable trees can be ripped out with some kinds of site preparation machinery (see Chapter 8), but this method is costly and may harm other qualities of the site.

Ordinary cutting merely decapitates the plant. It is effective against those plants that do not sprout but sometimes worse than useless against those that do. The worst cases are those species that are able either to sprout from the roots or to layer because each cut tree is replaced by large clumps of new trees. However, sprouting is somewhat reduced if cutting is done early in the growing season just after rapid formation of new tissues has depleted the carbohydrate root reserves. Sometimes browsing or grazing animals can be used as patient, persistent substitutes for cutting. Care must be used to keep these animals under control and prevent indiscriminant consumption of desired plants.

Girdling is the removal or killing of a ring of bark around the tree stem so that the flow of carbohydrates from crown to roots is blocked. Ideally, this causes the roots to die and the whole tree is killed. Unfortunately, girdling is almost as effective as cutting in stimulating sprouting during the time before the roots die.

Fire kills plants by girdling and also causes sprouting. In some kinds of forests the vulnerability of different species and sizes of plants varies enough that prescribed burning can be used as a selective technique (see Chapter 8).

In the ways in which they are used to control woody perennials, herbicides often operate by killing the stem cambium and phloem, thus producing a girdling effect somewhere between foliage and roots. As will be described later, herbicides kill woody plants mostly by killing the roots directly or by starving them.

Tactics

Two or more methods of tree killing, sequentially applied, are frequently more effective than one alone. For example, one highly effective sequence involves burning or some sort of cutting, soon followed by herbicide spraying; the first treatment stimulates sprouting and the second kills the sprouts, preferably while they are still succulent and have yet to form new dormant buds. Another effective sequence starts with some treatment that covers an area rather completely and is followed by a different one concentrated on those localized or resistant plants that escaped the first treatment.

Emphasis on preventing sprouting does not mean that the goal is always complete killing of an undesirable plant. It may be if there is some reason to eliminate root competition quickly. Sometimes it suffices merely to cripple the undesirable vegetation so that the desirable can gain ascendancy in competition. The objective is to control the living system of the forest; any killing is a means and not an end. It should also be noted that it is not feasible to achieve total eradication of any undesirable species from a stand or site. Eradication would require effort pursued beyond the point of diminishing returns, to say nothing of possible harm to the environment.

Cutting Small Trees and Shrubs

Using hand tools to cut small woody plants is so laborious that they are almost never used except where labor is cheap or the amount of work is so small that it is not expedient to bring in chemicals or power saws, or they are not available. The best tools for cutting very small stems are those that can be held in one hand, are well balanced, and have straight or concave cutting edges. The ax, with its convex edge, is a better tool for stems thicker than 2 inches ($\approx$5 cm). Where muscle power is used, it ordinarily takes less time, effort, and risk of injury to inject or spray a chemical than to sever a stem even "with one quick stroke of an ax."

One can also use tractor-mounted devices with heavy, horizontally mounted, rotating blades or chains, similar to rotary lawnmowers, to thin young stands by cutting narrow swaths through them. The same can be done by crushing trees with rolling brush-cutting drums or by shearing them off with the kinds of bladed equipment described in Chapter 8.

The main objection to cutting is that it causes much resprouting of the very sizes and species of woody plants that one is likely to be trying to kill. The most important exceptions exist where nonsprouters, such as most conifers, are to be eliminated. The selectivity of herbicides has not been improved to the point where broadcast foliage spraying can be used to favor one conifer over another. Therefore, either cutting or some sort of chemical treatment of single stems is necessary for release operations in young conifer stands.

HERBICIDES

Toxicity of Herbicides and Other Pesticides

The pesticides used for killing the woody weeds of the forest are called **herbicides** because of their wider use against the herbaceous weeds of agriculture. Since they are pesticides, about which there is always concern, the forester must know as much as possible about the nature, use, and toxicity of the compounds. There are many more comprehensive accounts of these matters such as those of Walstad and Dost (1984), U.S. Forest Service (1984), Klingman, Ashton, and Noordhoff (1982), Weed Science Society of America (1983), and Newton and Knight (1981).

Most problems with pesticides arise from two categories of ignorance about their chemistry and biological effects. One form is hysteria on the part of the public, and the other is heedlessness of stupidly complacent users. Education is the best solution to both problems. The public is not required to educate itself, but users are. Most foresters who are going to use pesticides, other than those safe enough for anyone to purchase, must pass examinations to demonstrate enough knowledge to get an applicator's license. The manuals that licensing agencies prepare to help users qualify are often excellent sources of information about the necessary precautions.

Registration of Herbicides

In most countries, no pesticide can be marketed until a governmental regulatory agency has "registered" or approved it for some specified kind of use. To get a compound registered, the developer must show that many different kinds of tests have been reliably performed to demonstrate its efficacy and patterns of toxicity as well as whether it is safe enough to use. The label on the container is a document with the force of law that must be read and followed by the user. Among other things it gives the chemical name and concentration of the compound, necessary precautions, allowable uses, and detailed instructions about matters such as dilution rates and modes of application. Sample labels are commonly distributed by manufacturers and are convenient, reliable sources of information.

The greatest hazards with legal pesticides are to the people who handle and apply them. If they protect themselves properly and also confine the pesticides to the places where they are supposed to be applied, the risks of other damage are reduced.

There is also an unplanned safeguard that reduces the general hazard from pesticides used in forestry. The tests of new compounds now required are very costly. The agricultural market is generally regarded as the only one large enough to allow recovery of the large investments involved. Therefore, the chemical industry tends to avoid compounds so toxic that they have no prospect of use on food plants. Although this greatly reduces the potential for problems with forestry pesticides, it has the ironic effect of paralyzing the development of pesticides that are narrowly and ideally specific for any particular forest pest.

Quantitative Toxicology

Regardless of whether one is concerned about target or nontarget organisms, it is necessary to know some elementary toxicology. Much of the popular fear of "chemicals" ignores the simple fact that all substances are chemicals. Furthermore, all chemicals, natural or synthetic, can be poisonous to any organism if administered in sufficient amount by some route that reaches a vulnerable tissue. For example, air-breathing mammals are killed by high liquid dosages of the most common oxide of hydrogen if it is administered to the lungs; that is, they drown. No substance is absolutely safe or nontoxic, and none is absolutely poisonous, except possibly for certain radioactive and carcinogenic substances about which there is both legal and scientific debate.

The best way to interpret matters involving the toxicity of registered pesticides is in terms of the kind of curve (Fig. 6.1) that depicts the basic relationship between dosage of a compound and biological response over the whole range of possible dosages. In the case shown in the figure, dosage is that of some pesticide, and response is the percentage of mortality in a series of populations of one species exposed to the various dosages. Below a certain "threshold" or "no-response" level, there is no mortality at all. In fact, it is

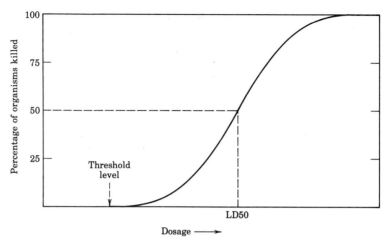

Figure 6.1   The typical form of the sigmoid curve by which the dosage of a biologically active substance is related to the response of a population of organisms.

even possible that a compound might have a beneficial effect at some dosage much below the threshold level.

The most important thing to note is that the dosage-response curve is asymptotic at both ends. This is mainly because there is variation within the tested populations such that some individuals are much more resistant than others, although other factors contribute to this kind of statistical variance.

The asymptotic nature of both ends of the curve gives two warnings. First, it is impractical, expensive, and environmentally unwise to seek 100 percent or even 99.8 percent mortality of target organisms by increasing the dosage. It is much better to be content with something like a 96 percent kill and either tolerate the survivors or attack them later by some other means. Second, it can be presumed that a curve of the same shape applies to humans and similar organisms, so its asymptotic nature near the no-response level becomes important. Biological variance, risks of accidents, and the possibility of hazards (such as long-term cumulative and synergistic effects of multiple chemicals) still unknown are factors so important that the maximum allowable dosage for these organisms should be much less than the no-response level. For example, in the extreme case of human food, the legal "tolerance" or maximum allowable amount of a given pesticide is 1 percent of the highest dosage at which no harm is observed in laboratory test animals.

It is also necessary to distinguish between **acute toxicity**, which involves single exposures and short-term effects, and **chronic toxicity**, which is associated with repeated exposures and long-term effects. Pesticide registration requires determinations of both kinds. Recognition of chronic toxicity serves not only to guard against danger but also as a possible way of using pesticides more efficiently. Two or more applications at low dosages may kill more target organisms than a single high-dosage application. This is mostly because not all organisms are simultaneously at the same stage of development and vulnerability. However, a nontarget organism that is not damaged by one exposure may suffer from several. If the dosage-response curve of Fig. 6.1 is for acute toxicity, one for chronic toxicity with repeated exposures would lie far to the left and perhaps be steeper.

The sigmoid curve form of Fig. 6.1 is useful for revealing the nature of the dosage-response relationship but could not be plotted without a very large amount of data for low and high dosages. Such curves can usually be converted to straight lines if the dosages are expressed as logarithms and the responses as probits. Probits are statistics that convert cumulative normal-distribution curves to straight lines. Because straight lines can be extrapolated dependably, this kind of statistical transformation enables the dosages of the threshold and 99.9 percent response levels to be predicted with reliability from moderate amounts of data covering the range of responses.

The abbreviation LD50 denotes the dosage that kills 50 percent of the test organisms; it is the most common expression of the relative toxicity of different chemicals. It must, of course, refer to the organism used in the tests; most tests rest on such inferences as the assumption that trout react like laboratory animals such as minnows; alders, like beans; and people, like rats. The expression LC50 means "lethal concentration, 50%" and might be used to denote the concentration of some substance in water that killed 50 percent of the minnows in it.

Fate of Herbicides

Another important toxicological consideration is the degree of persistence of a pesticide in the environment. The ideal pesticide would be one that decomposed into such substances as water, carbon dioxide, and chloride ions after it had done its work. Most modern pesticides break down from the actions of sunlight, water, and soil microorganisms. The best of them are, in other words, biodegradable. There are, for example, bacteria that can decompose 2,4-D by using it as an energy source for respiration. However, the breakdown within actual soil conditions is not well understood, and the formulation of byproducts other than the major breakdown chemicals is likely.

Because life processes require water, pesticides should have enough affinity for water that living organisms can react with them chemically. DDT and other chlorinated hydrocarbon insecticides were banned basically because their insolubility in water caused them to be very persistent. Their solubility in oils allowed them to accumulate in potentially active form in the fats of animals and to be concentrated as they were passed along the food chain to eagles and other carnivorous birds. The threatened extinction of these organisms through the thinning of egg shells served as an early-warning symptom of harm that might result from long-term accumulation in people and other animals.

Although risks to nontarget plants must be considered, herbicides (and fungicides) are generally less dangerous to animals, including humans, than insecticides and rodenticides. This is because herbicides have been selected for toxicity to green plants, which have physiological processes quite different from those of insects and other animals. Potential dangers are likely to lie as much with additives, dilutents, and contaminants (such as dioxin in 2,4,5-T) as in the herbicides themselves. It is prudent to regard any substance, whether it be a pesticide, tree sap, or printing ink, as being possibly more hazardous than it is known to be.

Occasionally, certain tactics can be used to improve the effectiveness of low dosages of herbicides. Two low-dosage treatments done in succession can be more effective than applying the same amount all at once. The application of two different herbicides may be more effective than a large total dose of one alone; such effects are synergistic in that the effect of one reinforces that of the other. The combination of different measures, such as herbicide treatment and fire, can be synergistic. However, it should not be presumed that combinations are always more effective than single treatments.

Modes of Entry and Movement of Herbicides in Woody Plants

Some herbicides can be effective if sprayed on intact leaves and bark, but others must be injected through the bark. Still others can be applied to the soil and absorbed by the roots. Many herbicides used to control herbaceous weeds in agriculture kill only the plants with which they come in contact or render the soil surface lethal to the roots of germinating seedlings. These agricultural herbicides, of which there are very many, are used in forest tree nurseries and sometimes to prepare spots for tree planting but are otherwise not useful in silviculture. The killing of woody plants depends on herbicides that can move within them.

In practical application, the main concern is with the physical properties that determine how herbicides enter plants and move after entry (Ashton and Crafts, 1981). Most herbicides will kill the appropriate vital tissues once they get to them; the problem is causing them to get there. Also crucial are those properties that may govern the degree of differential killing or selectivity among species that can be achieved with carefully developed techniques of broadcast spraying.

Penetration

The surfaces of land plants are almost always covered with waxes or other lipids that serve to reduce transpiration. The most important of these materials are the cutin of leaf surfaces and the suberin that impregnates bark. The structure and thickness of the cutinous coverings of leaves vary tremendously between species and in different parts of the leaf. Electron microscopy reveals that cutin consists of woolly strands of wax and is not a solid coating. Molecules that are small enough can pass through the gaps between the strands. The stomata or guard-cell coverings are also potential entry points, but it is not easy to get herbicides to move into leaves. The variability of their entry is indeed a possible way of getting them to kill some species and not others.

After it has suberized, the bark of woody plants is practically impermeable to water or herbicides dissolved in it. However, the absorbing tissues of the roots offer an open pathway.

Water-soluble herbicides can be introduced into plants by injection through cuts in the bark or by root uptake from the soil, but only some of them can enter through the leaves and then under special circumstances. The waxy coverings of plants not only resist passage of water but also tend to prevent aqueous solutions from sticking to them; the droplets may instead ball up into little spheres that merely rest on the surfaces or roll off. If herbicidal materials are to stick to such surfaces and penetrate them, they must be lipophilic or oily, which usually means that they are insoluble in water and do not ionize.

Translocation

After herbicidal materials have penetrated into plants, they cannot move to vulnerable tissues unless they are water-soluble. Those that are not work only because they have been formulated so that they hydrolyze to become water-soluble. Many move readily throughout plants; these tend to be so lethal that it is difficult to use them for selective effects with broadcast foliar spraying.

The compound must move in the living phloem from the leaves to the roots and arrive there at times when the roots can be killed. The movement of the active substance is from areas of high concentration ("source") to areas of low concentration ("sink"). The chemical compound must reach the part of the plant where it is most effective. Some, by blocking photosynthesis, can "starve" the roots. Others, such as 2,4-D, kill actively dividing meristematic tissues and must move to the elongating roots to be most effective. The actual mechanism by which some chemicals work is not well understood.

Formulation of Herbicides

The physical chemistry of herbicides varies substantially in ways that affect methods of application, movement within plants, and selectivity among species.

Highly water-soluble herbicides can be made to enter plants through the roots, gaps in the waxy armor of leaves, or cuts made in the bark. They are quite mobile within the plants. However, it is not easy to achieve selective effects with them unless they are applied to plants individually. Selective effects with foliar application are easier to achieve with herbicides that are less mobile and less soluble in water. Newer chemicals are specific to certain amino acids used in basic plant metabolism. Certain species are tolerant of these herbicides because these amino acids are not present.

Oil-soluble materials can stick to the waxy surfaces of bark and leaves and penetrate such surfaces better than water-soluble materials. However, they must change to water-soluble form after penetration if they are to move to vulnerable tissues, especially if they are to move from the leaves to the roots. Paradoxically, this difficulty of movement is part of the basis of the selectivity among species that can sometimes be achieved by applying these compounds in broadcast foliage spraying.

Some herbicides (such as 2,4-D) are usually made oil-soluble by converting them to esters. This process takes place by combining them with alcohols of sufficiently high molecular weight that their volatility is low. This reduces the risk that phytotoxic vapors will drift out of the target areas. (Water-soluble herbicides may drift as droplets but not as gases.)

Sometimes such materials are simply dissolved in oil, but the esters can also be modified to provide enough affinity for water to make them emulsifiable. In ordinary emulsions, droplets of oil (in which the herbicide is dissolved) are dispersed within a matrix of water. Emulsions are used mainly because water is much cheaper than oil and is often available close to the site. If the active oil-soluble ingredient can be spread over the target surface to an adequate extent, one might as well use the cheapest carrier.

Some herbicides of very low solubility have some specialized uses in silviculture. Most of them are applied to soil surfaces, as in nurseries or around Christmas trees, to kill herbaceous weeds as they germinate. Some of them kill the foliage of plants on which they are applied. Such "contact-killers" can be useful in cases, as in killing annual weeds, where it is sufficient to kill the foliage and the herbicide need not be translocated to the roots.

Herbicides of low solubility can be formulated as "wettable powders" or "flowable liquids," both of which are devised to be sprayed not as solutions but as suspensions in water. Some water-soluble herbicides are available in pellet form and are spread over the ground.

The efficacy of these materials can also be modified with various additives or adjuvants normally made by the manufacturers. These may include wetting agents that cause the sprays to spread over surfaces and stick to them better and penetrants that help break down the waxy cuticle on the leaf surface and increases penetration of buds and bark. Other additives may confer emulsifiability, improve solubility, or stabilize the spray mixtures.

Water-soluble herbicides may have to be formulated in ways that neutralize acid reactions enough to reduce tendencies to corrode metal spraying devices, although it may help to use plastic devices for corrosive materials. The physical chemistry of the various additives is complex enough that applicators should follow the instructions on the container labels and not presume that they know more about the materials than do the manufacturer's chemists. Impromptu mixtures are not only illegal but can also leave intractable precipitates in spray tanks.

Methods of Applying Herbicides

Stem Injection

If stems are more than about 2 inches (5 cm) in diameter, it is convenient to kill them individually by injecting water-soluble herbicides into their xylem. If the herbicides are highly mobile and phytotoxic, they act as systemic poisons and kill many parts of the plant. However, the problem with this is that most of their movement is upward with the sap stream in the xylem and not downward to the roots. If the herbicide could be depended upon to move throughout the plant, it would be necessary to inject it at only one point.

Because this effect is either not attainable or undependable, the herbicide is usually injected at several points around each stem or even in a complete frill-girdle. The purpose is to achieve a kind of chemical girdling in which the cambium and phloem are killed in a ring around the stem; this killing may be from a combination of the effects of both cutting and poisoning of tissues. If the injections are made in bark incisions that have spaces between them, the sideways movement of herbicides through pits in the sides of the conductive elements of the xylem may cause complete girdling at some place higher on the stem than the point of incision. However, with any sort of girdling effect, mechanical or chemical, it must be noted that it will not be effective unless it takes place below all of the foliage.

If spaced incisions are used, they are usually made at intervals of 3–6 inches ($\approx$7–15 cm) around the circumference of the trees, being farther apart with the larger trees. However, it is sometimes stated that the spaces between the cut ends of incisions should not be more than 2 inches ($\approx$5 cm) wide. Injection into continuous incisions around the stem may be necessary during the dormant season or in diffuse-porous angiosperms.

The simplest way of doing the work is to make the cuts with a hatchet held in one hand and to put about a milliliter of solution into each cut with a squirt oilcan held in the other hand. There are faster devices that make the cut and apply the solution simultaneously; these are the ''hypo-hatchets'' or ''tree-injectors'' which cut and inject the chemical near the base of the tree. Although they are more expensive and difficult to maintain, they can still be the most economical equipment to use when many trees need injection treatment.

Under some circumstances, downward movement can occur in the xylem after stem injection. If the herbicide can be applied almost simultaneously with the cutting, the breakage of the sap columns, which are normally under tension, can suddenly pull some of the herbicide downward. Solutions can also diffuse very slowly downward in the xylem at night or at other times when transpiration is not actively pulling water upward. In fact, problems can arise if trees are connected by intraspecific root grafts and one tree pulls the herbicide up from the roots of another. This phenomenon, called flashback, is most common when the herbicide has been injected into some small, overtopped tree and pulled into another by its large, actively transpiring crown. The best way of avoiding this effect is to make the injections higher on the stems and at times when transpiration is active. It is also a reason not to poison small, feeble trees in precommercial thinning; this is usually a time-consuming waste of effort anyhow.

It is sometimes observed that basal sprouting is less common when herbicides are applied in spaced incisions than in complete frill-girdles, but it is not clear why.

The stem-injection technique has the highest degree of selectivity of any. It can be used at any time of year when it is warm enough to allow the solutions to flow, although it is most effective during the growing season. It cannot be used at times when watery sap flows out of the cuts because that flushes the herbicide away. Although it is a common means of precommercial thinning and release cutting, it is too costly for killing undesirable plants that are small and numerous. One must also be able to walk under a stand to use this method of killing trees.

Stump-Surface Treatment

A closely related method involves application of water-soluble herbicides to the surfaces of freshly cut stumps. In such cases, the work must be done within a few minutes after the cuts are made because downward movement will be blocked by air bubbles or the swift formation of wound gums and resins. Various attachments and additions to chain saws have been developed for this purpose, but an oilcan seems to work more reliably. This type of application is generally used when a small number of sprouting hardwoods are growing in an otherwise pure conifer stand.

The herbicides are best applied to rings around the cut surface that are as close to the cambium as possible; there are no tissues to be killed in the inner parts of the stump. All effects depend on the slow downward diffusion of the herbicide. The only purpose of stump treatment is to prevent sprouting; unless the bark is too thick, this can also be done by basal spraying as described next.

Basal Bark Treatment

Woody plants can be girdled by applying oil solutions of certain herbicides in continuous rings around the basal bark. The purpose is to kill not only the plant but also any basal buds from which sprouts might arise. Except for killing such buds, the solution might just as well be applied at some higher and more convenient height where the bark is thinner. Because of the difficulty of getting satisfactory penetration of thick bark, basal treatment is usually limited to control of shrubs and trees too small for stem injection.

Directions usually specify that the oil solutions be applied so that the entire circumference of the lower 6–12 inches ($\approx$15–30 cm) of stem surface are wetted to the point of runoff. This is to ensure that enough runs down over the buried root-collar surfaces to kill the buds beneath them; it is not actually necessary to kill such a wide ring of cambium. There is no fundamental reason to spray the solution; cheap devices, often with split nozzles, which cause it to flow under gravity in fine streams that encircle stems can suffice. If there is no concern about sprouting, the encircling solutions can be applied at a convenient height with paintbrushes.

Sometimes effects similar to basal spraying result from deliberate spraying of oil-soluble herbicides on the upper stem surfaces of woody plants. This is associated with and will be considered with foliage spraying.

This treatment can also be used to prevent stumps from sprouting, if the bark is thin enough. In this case, the oil solutions are applied to the sides of the stumps; there is no reason to put them on the cut surfaces as there would be with water-soluble herbicides.

The main advantages of basal bark treatments are that they can be done with high selectivity at any time of year and are not likely to cause leaves to die and turn to an unsightly brown color. This is because the girdling effect starves the roots slowly and scarcely any herbicide moves to the leaves. The main drawback is the high cost of materials and labor. As will be described in the next section, the localized application of newer herbicides to the soil at the bases of sprouting shrubs and trees is often a superior technique, but it does not prevent "brownouts" of the foliage.

Soil Application

Some herbicides are sufficiently water-soluble, phytotoxic, and mobile within plants that they can be applied to the soil and taken up by the roots of the unwanted vegetation. They are distinctly different from some relatively immobile, "preemergent" herbicides of agriculture; those kill the emerging rootlets of germinating herbaceous weeds by sterilizing the soil surface.

If soil herbicides are to reach deeper roots and be absorbed by them in lethal quantities, the powerful effect of the soil in diluting and immobilizing them must be overcome. The first 6 inches of an acre of forest soil weighs about 1.6 million pounds, so 1 pound per acre of herbicide mixed into it would be diluted to 0.625 ppm. Furthermore, a major portion of that might be adsorbed on the tremendous surface area of mineral and organic colloids in the soil. For this reason, only highly mobile, very phytotoxic compounds can be used. Suitable herbicides must be very low in toxicity to animals and not excessively persistent or mobile in soils.

Soil herbicides are applied either as water solutions or as pellets in which they are mixed with inert ingredients. Either form can be applied directly and selectively at the bases of the plants to be killed. Convenient squirt guns are available to spray metered amounts of solution on the bases of small trees or clumps of woody stems from distances of several meters.

The pelletized forms can also be broadcast over whole areas for certain special purposes. No spraying equipment or liquid is necessary because rain is depended upon to dissolve the herbicide. This technique is suitable for achieving the nearly total kill of vegetation sometimes sought in preparing sites for regeneration, but aerial foliage spraying is usually much cheaper.

Under the right circumstances, selective effects can sometimes be achieved by broadcasting pelletized soil herbicides. One situation of this kind is where there is a good stocking of small seedlings that must be released from the cover of much larger undesirable trees with extensive root systems. If the desirable seedlings have narrow root systems and the pellets are scattered sparsely, the chance is great that the larger trees will be killed but adequate numbers of seedlings will escape.

Foliage Application

Herbicides can also be applied to the leaves in ways that enable them to be translocated to and kill vital tissues in the roots and stems of woody plants. The work is usually done with helicopters because that is faster, cheaper, and requires far less herbicide than spraying with mist-blowers from the ground. It is very important to do the spraying quickly because many chemicals must be closely synchronized with the development of the vegetation and the proper periods are short.

If spraying is done from above, the foliage is covered much more uniformly than when it is done from beside or beneath. Aerial application generally requires lower volumes, higher concentrations, and smaller amounts of total active ingredients than ground application. However, there is less risk of drift beyond the target area with ground application. Because of this problem, fixed-wing aircraft are seldom used and are actually illegal for some chemicals.

Foliage spraying is relatively uncomplicated if the purpose is to prepare sites for regeneration by killing most of the existing vegetation without need for discrimination. Several modern water-soluble herbicides are suitable but so phytotoxic and mobile within plants that species selectivity is not easy to achieve with broadcast application of them. Also uncomplicated are the herbicides that are not translocated and kill only the foliage to which they are applied. These "contact" herbicides are used mainly to kill grass and other herbaceous plants on spots where trees are to be planted.

The difficulties arise when the purpose is to kill a sufficient amount of undesirable vegetation without causing unacceptable harm to the desirable. There are various ways of obtaining such selectivity; none works in all cases, and it takes much experimentation to determine which one, if any, will produce the desired results. Selective foliage spraying sometimes requires use of herbicides in oil-soluble form if they are to stick to and penetrate leaves; more often, special surfactants are mixed with the herbicide. Older chemicals are selective because of differences in mobility in the phloem. Newer compounds that interfere with specific amino acids are much less toxic to conifers than hardwoods and can be applied in low dosages to young conifer stands.

Differential wetting has been the most common source of selectivity in foliage spraying and probably is more important than it is given credit for. Most important is the simple fact that spray droplets are much more likely to fall off narrow leaves of conifers and grass than off the broad leaves of angiosperms. This is why foliage spraying is so commonly used to release conifers from hardwood overstories. Any physical properties of the spray materials or of the surfaces of leaves that alter the retention and penetration of herbicides may also change the pattern of selectivity among species. The wooliness or waxiness of leaves can cause important differences; not enough advantage is taken of possibilities of modifying the physical chemistry of spray materials to change their ability to stick to and penetrate leaf surfaces.

Morphological selectivity results from interspecific differences in the location of vital meristems. Part of the resistance of grasses results from the fact that their intercalary meristems are thoroughly ensheathed. Woody plants do not differ much with respect to this kind of selectivity, but some possibilities exist.

Biochemical selectivity begins to operate after a herbicide gets inside a plant. Some species have enzymatic systems that can inactivate a herbicide, but others may convert an inactive form to an active one. Such phenomena are most common and useful with herbicides that are sufficiently similar to naturally occurring growth regulators that they operate as counterfeits and fatally derange normal processes. Members of the genera *Fraxinus* and *Acer* seem to inactivate several common herbicides and are thus resistant to them. On the other hand, some species can activate a herbicide and kill themselves, whereas those that do not remain undamaged. Detection of useful patterns of biochemical selectivity is difficult, but successes have led to development of some herbicides that can be applied at much lower dosages than less selective chemicals.

Phenological selectivity is possible if there are times when desirable species are significantly less susceptible to damage than the undesirable. The ordinary cases are those in which undesirable plants are susceptible because they have vital tissues that are developing actively, whereas those of the desirable species are not. For example, if soil moisture conditions are still good, some angiosperms remain vulnerable late in the growing season after conifers have become resistant because their shoots have "hardened" and formed buds. It takes close observation to disclose useful patterns. In applying the knowledge, one must be guided more by the phenological stages of plant development than by the calendar. The appropriate stage and time for treatment are not the same with all herbicides, species, or sites, so it is necessary to apply all information and experience available. Careful planning can lead to an adaptable spray schedule that moves from stand to stand as the spring progresses. This is an especially useful type of selectivity in forests that have major elevation ranges.

Finally, in small specific situations, there is some possibility of **selective placement** of herbicides even with foliage spraying. If spraying can be done from the ground, it can be aimed to a limited extent. Large, light containers also make easily movable temporary protective covers for seedlings. If spraying is done from the air, there is the possibility of using an undesirable overstory to shield the desirable understory being released. Decisions should be made about whether helicopters should or should not be operated so as to drive the sprays down through crown canopies. Generally, reliance on this type of selectivity is expensive and requires much planning.

Techniques of Foliage Application Aerial spraying can be done only by highly skilled and pesticide applicator licensed aviators operating under detailed contracts. The work is usually performed under nearly calm conditions early in the morning. The size of spray droplets is critical. If they are too fine, they are apt to drift; if too coarse, the foliage may not be covered with adequate uniformity.

The boundaries of areas to be treated must be clearly visible from the air; maps alone are insufficient. Prominent features such as roads, ridges, and edges of clearcut areas should be used as much as possible, but it is usually necessary to have other markers as well. Flags or other markers placed on trees must project above them to be visible. Balloons are often used because smoke drifts too much. It may also help to mark some edge trees in advance by injecting them with some highly mobile herbicide that will kill their foliage. Dyes can be used in the chemical tank to make swaths more apparent, and global positioning systems are used to assist the pilots.

It is so costly to move in aircraft for aerial spraying projects that operations are feasible only if at least 500 acres ($\approx$200 ha) are to be treated at once. One helicopter does about 60 acres (25 ha) an hour. The individual treatment areas must be at least 5 acres ($\approx$2 ha) but are generally larger. It takes about the same size of ground crew to control the operation and refill the helicopter, no matter how large the spray area.

If smaller areas must be treated, it is best to do the work from the ground. For very small areas there are mist-blowers that can be carried on the sprayer's back; they can cover foliage to heights of only about 12 feet ($\approx$4 m). More powerful equipment can be mounted on skidders. Spraying from the ground is especially effective as a means of controlling understory vegetation underneath stands of trees that might be damaged by foliage spraying, or where the helicopter would have to fly too high such as in shelterwood and some seed-tree situations. This kind of vegetation control is done to free-growing space either for regeneration or, where soil moisture is limiting, to increase the growth of the overstory trees.

Small spraying devices similar to those used in gardens can be used for small-scale operations such as applications of contact herbicides to grass in spots where trees are about to be planted. With this or any other kind of equipment, it is necessary to guard against contamination that may cause problems when it is used for other pesticides. Mixtures of herbicides and insecticides may kill crop plants as well as their insect pests. Physically incompatible mixtures may also clog spray equipment. Sometimes the equipment is so hard to clean completely that it can be used for only one kind of pesticide.

Herbicides can also be applied to low vegetation with wick or moplike devices that are used to wipe the materials over the leaves. Most devices are handheld and are used to release individual seedlings. This kind of equipment, mounted on tractors, can be used to kill strips of grass or similar vegetation prior to planting or between rows of planted trees. Placement of the herbicide can be very accurate, and there is no risk of drift, but the method cannot be used on tall plants.

Stem or Stem-Foliage Spraying   In foliage spraying it is almost inevitable that the surfaces of small branches as well as those of leaves will be coated. If the herbicide is oil-soluble, this can add significantly to the phytotoxic effect. In fact, it is sometimes advantageous to treat the stems of small deciduous trees when they are leafless. This kind of ''dormant'' spraying can, for example, be used in spring and fall to release Douglas-fir, hemlock, and spruce from red alder and other angiosperms in the Pacific Northwest (Newton and Knight, 1981). The effectiveness of the treatment appears to depend on what amounts to a large-scale girdling of virtually all conductive tissues at the top of the plant as well as wholesale killing of dormant buds.

The great advantage of foliage spraying in general and aerial spraying in particular is that they enable quick, large-scale release operations. One hour of aerial spraying can accomplish as much as months of hand brush-cutting and prevent resprouting as well. As with all techniques, it can be done unwisely and unsafely. Among other precautions, it must be borne in mind that the purpose of foliage spraying is to kill the roots; killing leaves is incidental, and the work may fail if they fall off before the herbicide moves out of them.

Herbicide Compounds

As has been the case with all pesticides, improvements in herbicides have led to the development of compounds that are not only more effective but also safer. Developments continue, and it should not be presumed that understanding of the compounds or their use is easily obtained. This chapter merely introduces the subject. Better understanding can be obtained from Garner and Harvey (1984), Buchel (1983), Ashton and Crafts (1981), Newton and Knight (1981), and other sources such as the *Herbicide Handbook* issued by the Weed Science Society of America (1984).

The active ingredients of herbicide compounds have such complicated chemical names and structures that easier generic names have been devised for the convenience of users. For example, N-(phosphonomethyl)glycine is called **glyphosate**, a word that is handy but should not be construed as telling anything about chemical structure or physiological activity. The trademark name Roundup® has been applied to this herbicide and registered by the Monsanto Agricultural Products Company. Trademark names must always be capitalized and are owned by the registrants. The generic names are best for exchange of information and so will be used for the rest of the chapter. Table 6.1 presents the generic names of active ingredients of some of the most common herbicides used today.

Chlorophenoxy Herbicides
The first modern herbicides were developed in the late 1940s when the pseudo-auxin **2,4-D** (for 2,4-dichlorophenoxyacetic acid) came into agricultural use (Bovey and Young, 1980). The oil-soluble esters of the chlorophenoxy herbicides proved to be very useful for selective killing of broad-leaved woody plants in releasing conifers by broadcast foliage

Table 6.1 Names and manufacturers of some commonly used herbicides

Herbicide	Manufacturer	Generic Name
Accord®	Monsanto	gyphosphate
Arsenal®	Cyanamid	imazapyr
Chopper®	"	"
Garlon®	Dow	triclopyr
Oust®	DuPont	sulfometuron methyl
Pathway®	Dow	picloram
Pronone®	Proserve	hexazinone
Roundup®	Monsanto	glyphosphate
Rodeo®	"	"
Princep®	Ciba-Geigy	simazine
Stomp®	Cyanamid	pendimethalin
Velpar®	DuPont	hexazinone
Weedone®	Union Carbide	2,4-D
Weedar®	"	"

spraying. This type of release depends on herbicides entering the leaves without killing them, converting to water-soluble form, and then moving to the roots with the sugars at times when the roots are vulnerable. This works only if the spraying is done when the soil is moist enough that the roots are growing but the time of year has come when the shoots are not expanding and drawing sugars from the leaves. It is also often important to delay treatment until some time late in the growing season when the desirable conifers have formed dormant buds but root growth is still continuing.

A slightly more complicated compound, **2,4,5-T**, was soon found to be more effective on woody plants because it was slightly less toxic and less likely to kill leaves or phloem and thus block its own movement to the roots. The esters of 2,4,5-T were widely used to release conifers from hardwood brush until it was determined that a highly toxic contaminant, a dioxin compound, which could develop during the manufacturing process, made the compound unsafe to use.

The esters of 2,4-D were formerly used as broadcast, selective foliage sprays but more modern herbicides have generally replaced them. Amine salts of 2,4-D are soluble in water and can be applied by stem injection, although some newer, but more expensive, water-soluble herbicides seem superior.

Chlorophenoxy herbicides act by deranging respiration, growth of tissues, and many other biochemical processes in plants. The animal toxicity of the compounds themselves is low, and they are readily broken down by soil microorganisms.

Other Auxin Herbicides

Triclopyr has many of the same properties as 2,4-D except that it is so much more phytotoxic and mobile within plants that selectivity is not easily achieved with foliage applications. It has a water-soluble amine form that, unlike that of 2,4-D, can be sprayed on the foliage; the timing of foliage spraying is about the same as that of the chlorophenoxy compounds. The amine can also be used for stem injection. An oil-soluble ester is available for basal spraying. Grasses are resistant and so are conifers in limited degree if dosages are low. Animal toxicity and persistence are low. Triclopyr is not as effective as broad-spectrum plant killers in foliar application for site preparation, and its primary use is for brush control on rights-of-way.

The most outstanding of auxin herbicides is **picloram**, which has amine salts that can be taken up by leaves or roots and esters that can be used for dormant sprays on stems. The water-soluble amines are very effective in stem injection. Picloram is easily translocated within plants, which are so readily killed that no selective effects are sought from broadcast applications. Picloram is very low in animal toxicity but can persist for about a year, which is longer than the other herbicides now in silvicultural use. Its high water solubility and long persistence have led to an EPA advisory regarding groundwater contamination; it is very important to follow all label instructions exactly.

Triazine Herbicides

Triazine herbicides disrupt photosynthesis and nitrogen metabolism. They vary greatly in their water solubility and have the useful property of being toxic to grasses but not highly toxic to conifers.

Simazine is very low in solubility, whereas **atrazine** is only slightly more soluble; they are applied as wettable powders suspended in water. These herbicides are rather immobile and kill only those plant tissues with which they come into contact. For this reason, they are used as preemergence herbicides in nurseries; the herbicides stay at the surface and kill the emerging roots of germinating weed seeds. They can also be used as contact killers of grass and herbaceous weeds in spots being prepared for planting. Because they do not move readily into the soil, trees can be planted on the spots soon or immediately after the treatment.

A much more soluble triazine herbicide is **hexazinone**. It can be sprayed on the foliage at low dosages to release conifers from broad-leaved species, although the main purpose of such application is in site preparation. Hexazinone is taken up by roots effectively enough that solutions of it can be sprayed at the bases of trees and shrubs to kill them; squirt guns are used for this kind of sharpshooting. It can be applied in pellet form, as was described under ''Soil Application.''

Phosphonalkyl Compounds

Phosphonalkyl compounds include **glyphosate**, which is water-soluble and highly phytotoxic. Glyphosate can be taken up as a water solution by foliage and even by buds and twigs, but it is inactivated in the soil so that it does not reach the roots by that pathway. However, it is so readily translocated to all parts of most plants that it can move from leaves to roots and thus kill them. Glyphosate is effective in the unselective kinds of foliage spraying often used to prepare sites for planting; selective effects can be achieved with directed foliage spraying. Carefully prescribed broadcast foliage spraying with glyphosate can sometimes produce selective killing of broad-leaved species to release conifers if they are sprayed shortly before the conifer buds burst which is after many deciduous plants. This compound is very low in animal toxicity and swiftly decomposes.

Imidazolinone

A more recent addition to the herbicides used in forestry has been those belonging to the chemical family imidazolinone. These compounds inhibit an enzyme that produces three amino acids used in meristematic regions. Because the amino acids are slowly depleted, death is slow and the root kill is high. These herbicides are absorbed through both the roots and the foliage and translocated to the roots. There is high soil residual activity, so newly germinating weeds are killed for months after the application. Because conifers are quite tolerant of these chemicals, timing is not important. Imidazolinones are frequently combined with glyphosate, however, so the timing of the combined mixture is more important. The synergistic characteristics of the combination make extremely low application rates possible.

The development of new and better herbicides should be anticipated. Furthermore, the information presented here about pesticides should not be construed as providing directions for the use of any of them. Additional sources should be consulted, and it is illegal not to read the container labels. In some places, use of some listed compounds is prohibited. Foresters who direct the application of many of the compounds must be specially licensed.

CLEANING

A **cleaning** is a cutting made in a stand, not past the sapling stage, to free the best trees from undesirable individuals of the same age that overtop them or are likely to do so. The need for cleaning arises in situations where the methods of cutting and site preparation employed during the regeneration period have created conditions favorable to undesirable as well as desirable plants.

The principal purpose is to regulate the composition of mixed stands for the advantage of the better species. Often the composition of the stand is radically changed as a result of these cuttings (Fig. 6.2). Cleanings are rarely essential in pure stands of reproduction, although they are sometimes carried out to remove trees of poorer form than those that are overtopped.

The term **weeding** is conventionally reserved for the more thorough removal of all plants competing with the crop species, regardless of whether their crowns are above, beside, or below those of the desirable trees.

Types of Vegetation Removed

The only plants that need to be eliminated are those that are going to suppress, endanger, or hamper the growth of trees needed to constitute the new stand. The cost of the operation should not be allowed to exceed the benefits. The reduction of undesirable vegetation should not be so great that it causes even less desirable plants to invade growing space. If those trees that are truly wanted have been released in sufficient numbers, it is possible that the remaining subordinate plants may have a beneficial effect in excluding very undesirable ones. It is a guiding principle of silviculture that one does not eliminate any vegetation without having a good idea about what will fill the vacated growing space.

The plants eliminated in cleanings may include not only overtopping trees of undesirable species but also poorly formed specimens of desirable species that are likely to compete with potential crop trees. It can be important to control climbing vines, overtopping shrubs, and sometimes rank herbaceous growth (Putz and Mooney, 1991).

Sometimes young stands have scattered but isolated trees that are good but taller than those that are to form the main stand. If these are going to race so far ahead of the main crop that they will develop large branches and wide-spreading crowns, the temptation to leave them should be rejected. If they represent a potential source of seed of some species that will be undesirable in the next rotation, it may be especially beneficial to get rid of them when they are small. However, if they are of good species and the trees of the main stand will catch up with them and restrict their crown development, such trees can be left

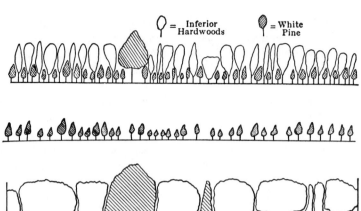

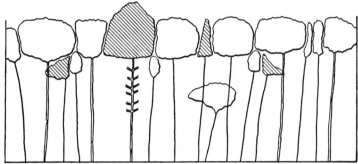

Figure 6.2   A stand of eastern white pine before (top) and after (middle) the removal of overtopping hardwoods in a cleaning. All of the overtopping stratum must be removed in this instance because the young hardwoods grow much faster than the pines. This treatment would best be done by complete foliage spraying, although the large pine would have to be cut or killed by other means. If left it would develop into a limby wolf tree. The lower sketch shows the stand as it would have looked 40 years later if there had been no treatment.

to form a useful component of the new stand. This situation is most likely to arise where the scattered trees are older than those of the main crop and of species that grow more slowly.

Cleanings are sometimes aimed at eliminating the alternate hosts or the favored food plants of dangerous pests of the main crop, such as species of *Ribes* adjacent to five-needled pines. Diseased or insect-infested trees can be removed if such action will really reduce the risk that desirable trees will be attacked.

Timing of Cleanings

If cleanings are needed, they are best done even earlier than precommercial thinning, although there is no fundamental reason why the two operations cannot be combined.

In humid tropical climates, it is often necessary to do the cleanings during the first year, perhaps even several times in the first year as well as afterward. In climates with less luxuriant growth, cleanings can be delayed for as much as a decade or two, especially if the desirable species are shade-tolerant.

It may be desirable to delay release operations while certain developments take place. Sometimes the taller vegetation that may later hamper the development of desirable trees may act initially as a nurse crop by providing crucial protection against sun, frost, insects, or other damaging agencies. Occasionally, nurse crops may choke out even more undesirable plants or, if belonging to leguminous species or *Alnus*, improve the supply of soil nitrogen.

It is also important to recognize that some kinds of pioneer vegetation may seem rampant and overwhelming in the early stages but then simply die off without really causing significant harm to the desirable seedlings beneath them. This is especially true of herbaceous annuals, which may flourish during the first year after some soil-baring disturbance and then vanish; such plants generally retire to the status of dormant seeds that can remain stored in the soil until another disturbance causes them to germinate.

Proper timing of release operations should be based on past observations of how the initial vegetation develops and how the constituent species behave. In fact, it may help to allow it to develop until the competitive status of different individuals and species begins to be asserted. This is especially true if the cleaning is to involve treatment of individual plants. If such treatment is involved, it can also be crucial to schedule the treatments before or after the time when the new stand may be difficult to penetrate on foot. The best time may come when the undesirable vegetation is slightly more than waist high. Very small stems can be awkward to treat, but the difficulty and expense of treatment increase rapidly after they attain 2 inches ($\approx$5 cm) D.B.H. The cost may decrease again when it is possible to walk under the stand, but the desirable plants may have been lost by then.

Occasionally, it is advantageous to delay cleaning until the desirable trees have become tall enough to remain free of any regrowth of competing vegetation after treatment. It may also be necessary to delay until the trees being released have developed roots and conductive systems capable of supporting rapid height growth.

Cleaning by aerial foliage spraying can be done at almost any stage of stand development. If any protective effects of nurse crops are at an end, there is seldom any reason to delay treatment because access to the stand is not a problem.

Provided that their roots extend more than several inches below the surface and their tops require no more shading, seedlings are likely to accelerate in growth after release. It is true that shaded trees may have shade-leaves that suffer some injury and turn yellow when suddenly exposed to the sun. However, such injury is usually only temporary, and when buds next produce new leaves they will be sun-leaves adapted to the new conditions. In other words, an alarming yellow color does not necessarily denote permanent injury. However, large saplings that have developed very small live crown ratios may succumb after release, presumably because they are unable to supply enough sugars for the accelerated respiration that results from increased solar radiation and temperature.

Any benefits from protective cover by overtopping vegetation are soon outweighed by reductions in growth. Continued suppression may cause the death of the better trees. These effects are the result of either direct competition or the mechanical effect of the overtopping trees in whipping buds from the twigs of the desirable trees.

Ordinarily, it is most expedient to try to do all the cleaning necessary in one operation by releasing the desirable trees enough to allow them to grow satisfactorily until the time of the first thinning or even to the end of the rotation. There is no subtle advantage in gradual release if all protective effects have ceased. Repeated cleanings can have a high cost of administration and of moving workers and equipment to the stand.

Repeated cleaning is likely to be needed if the undesirable vegetation resprouts and resumes excessive competition with the desirable crop. One advantage of herbicides over cutting is the greater opportunity to prevent resprouting. As has been the case in agriculture, herbicides have obviated much of the menial labor that formerly had to be expended on weed control. Society is now seldom willing to pay the price that cleaning with hand cutting tools requires.

Cleaning can be used to favor shrubs and other lesser vegetation over trees. Shrub cover may be much more desirable than trees in areas such as roadsides, wild gardens, near electric transmission lines, and patches managed for wildlife food or cover. In fact, it may be desirable to develop dense cover of shrubs in such places in order to prevent the invasion of tree species. In some cases, shrubs may be better ground cover than grass because they exclude conifers as well as hardwoods. This mode of site control is especially important in areas that have to be mechanically cleaned without herbicides, such as electric transmission lines in protected watersheds.

Extent of Release in Cleaning

With broadcast foliage spraying, the extent to which the desirable vegetation is released is controlled only by the selectivity of the chemical applied. Some control is possible with hand-directed foliage spraying, but it is usually neither possible nor necessary that it be very precise.

Where it is necessary or desirable to resort to comparatively expensive treatment of individual stems, there is rarely any justification for bringing through more trees than can reach merchantable size. Ordinarily, it is sufficient to release no more than 150 trees/acre (360 trees/ha), which can be achieved by releasing single trees at intervals of about 17 feet ($\approx$5 m). This number will usually be ample to provide trees for both the final crop and one thinning, as well as a margin for losses. The release of single trees at an average interval of 17 feet will give 150 trees per acre. The less desirable trees left in the process serve to fill out the stand and prevent the favored trees from developing into wolf trees.

In determining how much to remove around each favored tree, one must know the relative rates at which both desirable and undesirable trees grow in width of crown and height. The most serious competition will come from the undesirable trees adjacent to the individual to be released. Trees that are farther away do not offer an immediate threat to the dominance of the favored tree but may eventually overtop it by the lateral expansion of their crowns. Therefore, it is usually best to think in terms of removing the trees that project upward into an inverted cone around the top of the tree to be favored (Fig. 6.3).

The magnitude of the angle represented by the cone should vary directly with the difference in rate of height growth between the released tree and its competitors as well as the extent to which the tree has been suppressed. If the desirable tree will grow faster than the weed trees, the imaginary cone can be narrow, such that only the trees currently overtopping it are removed. More often, the disparity in growth that caused the unfavorable condition to develop is likely to continue; in such cases, the angle must be fairly broad.

The kind of partial reduction of competing vegetation that has been described does not necessarily cause the released trees to grow at rates anywhere near as great as they might if all competition were eliminated. This is especially true if the site is dry or infertile and root competition remains serious. The main object of such treatment is merely to assist the released trees in surmounting their competitors by subsequent growth. The only justifications for not releasing trees more thoroughly are the higher cost, any beneficial effect of moderate competition on the ultimate form of the crop trees, and the possibility that the otherwise unwanted vegetation will resist the invasion of even more undesirable plants.

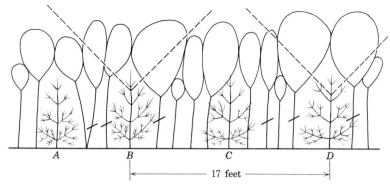

Figure 6.3 An example of how the intensity of cleaning might be regulated in freeing a stand of conifers from overtopping hardwoods. The objective chosen here is to remove all hardwoods that project upward into an inverted cone with an apex that is 2 feet below the tops of the crop trees and has an angle of 90°. The trees to be removed are marked by dashes. Trees *B* and *D*, which are spaced to give a stand of about 150 conifers per acre, are the only trees deliberately released. Trees *A* and *C* are ignored. In a heavier cleaning the angle would be increased or the apex of the imaginary cone lowered. The extent of release would depend on the relative growth rates of the different species present and the amount of time that would be allowed to elapse before another cleaning.

LIBERATION OPERATIONS

Liberation, in the silvicultural sense, refers to the release of young stands, not past the sapling stage, from the competition of distinctly older, overtopping trees. The trees that are cut or killed in place are those left standing when the previous stand was harvested or, for other reasons, were present long before the natural or artificial establishment of the young trees (Fig. 6.4). Liberation cuttings are somewhat like the removal cutting of the seed-tree and shelterwood methods of regeneration in which overstory trees are removed after the new crop is established. The trees removed in liberation are not, however, ones that have been left intentionally to provide seed, shelter, or additional growth.

Isolated trees have so much growing space that they often become stocky wolf trees with little clear length and wide, spreading crowns. The shade that they cast soon ceases to be beneficial and may cause the younger trees beneath them to die or become deformed (Fig. 6.5). Liberation operations can be conducted with most benefit and least damage to the new crop when the seedlings are not more than knee high.

Methods of Liberation

The undesirable trees can be eliminated by cutting, girdling, or chemical treatment. Cutting is, of course, the most advantageous method if the trees can be utilized at a profit or at a loss no greater than the cost of killing them. After an operation of this kind, it may be necessary to cut away any parts of tops that interfere with the young growth. In considering treatments that do not involve utilization of the trees, it is important to decide whether

Figure 6.4   Unreleased (left) and released (right) parts of a 15-year-old plantation of red pine that had been established under slow-growing oaks on a deep sandy soil in Michigan, five years after herbicidal treatment of the plot on the right. *(Photographs by U.S. Forest Service.)*

markets are likely to improve sufficiently to make them merchantable in the near future. Destruction of sound trees of sawtimber size and quality, merely because they are of currently unwanted species, has often proven wasteful.

Cutting

The felling of large, undesirable trees without utilizing them is the most expensive form of liberation. It may also result in far more damage to the young crop than would occur if the trees were killed and allowed to disintegrate in place. Therefore, such trees are felled, limbed, and left on the ground only where absolutely essential. This course of action is desirable where standing dead trees constitute a serious fire hazard. Cutting is also necessary along roads and in other situations where falling limbs and snags are likely to be a dangerous nuisance.

Girdling

The traditional method of killing trees without felling them is girdling. This involves severing the bark, cambium, and, sometimes, the sapwood in a ring extending entirely around the trunk of the tree. The materials already stored in the roots can still be carried up to the crown in the xylem, unless the sapwood has been interrupted. The roots die when the carbohydrates in them have been exhausted, a process that may take several years. If the sapwood is severed, the death of the crown is rendered more certain by the reduced flow of water, stored substances, and inorganic nutrients to it; the vigor of sprouting is, however, increased because the roots are not exhausted.

If any vertical strips of cambium are left intact, bridges of callus tissues are almost certain to form across the girdled ring. These bridges render the entire treatment ineffectual

Figure 6.5 The old post oak (left of center) was present when this 50-year-old stand of loblolly pine became established on an old field in North Carolina. The pine tree in the foreground and the trees in the right background have grown without interference from the wolf tree. An improvement cutting would be highly desirable in this stand, although it should have been conducted as a liberation cutting at least 40 years ago. *(Photograph by U.S. Forest Service.)*

unless they are subsequently invaded by fungi. Even if the cambium is completely severed with a series of single incisions, callus tissue will often bridge over the gap. Therefore, it is desirable to cut out chips of bark and sapwood so that the cambium is actually removed in a visible ring rather than merely severed in a single cut. Trees that have deeply infolded bark around frost cracks or other deformities are almost impossible to kill by girdling. Mortality may also be reduced if the girdled trees are supported by substances transferred to them through root grafts by uninjured trees of the same species. For these reasons, girdling should not be expected to result in complete or immediate killing of all trees that are so treated.

Girdling can be done with axes or by cutting parallel, horizontal grooves through the bark several centimeters apart with chain saws or other cutting devices. Ordinary power saws are quite effective on large trees, especially because the blades can sometimes be poked into the stems deeply enough to sever overgrown inclusions of bark and cambium. Sometimes devices consisting merely of saw-chains with handles on both ends can be used on smaller trees.

The most common and effective method of girdling with axes is **double-hacking.** In this method, a horizontal line of chips is removed by striking two downward blows; the second is made about 2 inches ($\approx$5 cm) above the other so that the chip may be pried entirely out of the cut with a twist of the ax handle. With this technique it is easy to verify that no strands of inner bark and cambium remain.

In **single-hacking** or **frilling**, a single line of overlapping ax cuts is made through the bark. This method is seldom effective by itself because the cuts usually heal. However, if herbicides are applied to the cuts, this becomes one of the best alternatives to purely mechanical girdling; the cutting can be done swiftly, and the chemical generally kills any unsevered tissues.

Notching involves cutting notched rings around the tree through the bark and a half-inch (1 cm) or more into the wood. This usually causes the quickest kill and the most basal sprouting because it is the method that most nearly halts the upward movement of water and nutrients. Unless the functional sapwood is deep, it has almost the same physiological effect as cutting the tree down entirely. It is cheap when done with power saws. However, it is the most expensive, laborious, and dangerous way of ax-girdling because there are so many upward strokes; it can be used to sever ingrown folds of bark with axes but has little else to recommend it.

In **peeling**, strips of bark at least 8 inches (20 cm) wide are stripped off after continuous cuts are made at the top and bottom of the strips. Unless the bark is too thick or tough, this can be done quite effectively at those times when the cambium is highly active and the bark is ''loose,'' as in spring and early summer in temperate climates. If the bark is ''tight,'' however, it is not feasible. Peeling is the best technique from the standpoint of interrupting the phloem without halting the upward flow of stored substances from the roots. The reason for removing such wide strips is that the cambium sometimes can survive and regrow unless there is thorough exposure to desiccation.

The time and cost of girdling, whether mechanical or chemical, depend mainly on the amount of stem circumference treated, so it can be predicted as some function of the sum of the diameters of the trees to be killed.

Girdling is much less effective than herbicides in preventing sprouting. This may be an advantage if the purpose is to increase the supply of food for browsing herbivores. If sprouting is not wanted, then girdling is best reserved for nonsprouting species or stems too large to sprout. For trees of sawlog size, girdling can be as effective as injecting herbicides. It can be done effectively on trees as small as 4 inches (10 cm) D.B.H., although herbicides are better for medium-sized trees. Below that diameter, cutting is easier, cheaper, and safer, but herbicidal treatments are usually superior to either (Wagner and Rogozynski, 1994).

APPLICATION OF RELEASE OPERATIONS

Cleaning is usually associated with comparatively intensive silvicultural practice, but the liberation of young stands from the remnants of old ones is a practice most characteristic of the early stages of new programs of long-term silviculture. Previous high-grading or use of land for grazing or agriculture can cause the development of stands cluttered with wolf trees or other undesirable specimens that encumber smaller and better trees. If the overtopped trees can respond vigorously and speedily to form stands of better quality, liberation is the quickest and cheapest silviculture method of making a silk purse out of a sow's ear (Fig. 6.6). The kind of trees eliminated in liberation operations have large crowns, so that many small trees are released by killing a few big ones.

Liberation operations command high priority in the early stages of intensifying silvicultural programs. They are commonly regarded as means of releasing conifers from

Figure 6.6 Stand of loblolly and shortleaf pine in Louisiana, approximately 5 years after liberation by the girdling of an overstory of oak. Most of the young pines were small seedlings at the time of release. *(Photograph by U.S. Forest Service.)*

hardwoods, but they can play a very important role in refurbishing hardwood stands after their long and typical histories of high-grading.

The time between liberation and the reaping of benefits is shortest if the trees that are released are rather tall, although it is entirely possible for young, vigorous seedlings that have been released to overtake much older saplings that have suffered major reductions in live crown ratio. In liberation operations, care should be taken not to replace high shade with more detrimental crown competition in the crown stratum occupied by the desirable seedlings or saplings. This can happen if the trees that are eliminated sprout profusely or if understory weed species profit more from the release than their desirable associates.

When cleaning could be done only by cutting, it was characteristic only of highly intensive silviculture. Aerial spraying with selective herbicides has made it feasible to contemplate release operations even for natural stands that may receive no other silvicultural treatment until the final harvest cuttings. However, the release of individual plants still belongs only with intensive practice.

Herbicides have increased the power to control species composition on many of those inherently good sites where so many species can grow that the better ones are mostly overwhelmed by the poor. In fact, the tendency is to regard rather dry, mediocre sites as ''best'' simply because certain conifers or other desirable drought-resistant species can be maintained on them without excessive hindrance from many undesirable ones. Now the sites that are "best" in this sense may be found higher on the scale of true biological productivity. This has the important effect of extending the area on which highly productive and useful softwoods can be grown. However, selective placement of herbicides can also play an important role in developing good hardwood stands on good soils, especially by early elimination of malformed trees.

Although improvements can be made in the composition of stands at almost any stage of a rotation, it is usually best to try to make them with early cleanings. When a stand is young, it is much more plastic than after the sapling stage. The longer that undesirable trees dominate, the more will be the loss suffered by suppressed, desirable ones. As time passes the more likely it is that any released stand will be irregular and full of gaps. Planted stands, in particular, represent such a large investment that it is folly not to conduct cleanings in them if necessary.

It is generally easier, though less necessary, to conduct the kind of release that accelerates plant succession than that which retards or reverses it. For example, it is much harder to free eastern white pine from red maple and oaks than from pioneers such as gray birch. The pioneers can usually be counted upon to die after several decades if one is ready to wait that long; they often cast such light shade that they do not entirely arrest the growth of more shade-tolerant species beneath them.

Unless they can be done by selective broadcast aerial spraying, cleanings are so costly that it is important to reduce the need for them as much as possible before stands are established. Efforts should be made throughout the rotation to forestall problems before they arise. Thinning should not, for example, unwittingly be made so heavy that they allow establishment of advance growth of undesirable species. The simplest means of eliminating an undesirable species is to get rid of the seed source; cleaning is the hardest way.

The silvicultural operations that cause new stands to become established naturally or artificially can seldom be conducted so perfectly that composition can be governed entirely by site preparation or skillful treatment of the previous stand. Consequently, it must be anticipated that there will be many cases in which releasing operations will be desirable and will set the stage for subsequent rewarding efforts.

BIBLIOGRAPHY

Ashton, F. M., and A. S. Crafts. 1981. *Modes of action of herbicides.* 2nd ed. Wiley, New York. 525 pp.

Ashton, F. M., and T. J. Monaco. 1991. *Weed science.* 3rd ed. Wiley, New York. 466 pp.

Auburn Univ. 1993. Proceedings of the international conference on forest vegetation management. *CJFR*, 23:1954–2327.

Balmer, W. E., Utz, K. A., and O. G. Langdon. 1978. Financial returns from cultural work in natural loblolly pine stands. *SJAF*, 2:111–117.

Bovey, R. W., and A. L. Young. 1980. *The science of 2,4,5-T and associated phenoxy herbicides.* Wiley, New York. 462 pp.

Burschel, K. H. 1983. *Chemistry of pesticides.* Wiley, New York. 518 pp.

Garner, W. Y., and J. Harvey, Jr. (eds.). 1984. *Chemical and biological controls in forestry.* Amer. Chem. Soc. Symp. Series 238. 406 pp.

Gratkowski, H. 1975. Silvicultural use of herbicides in Pacific Northwest forests. USFS Gen. Tech. Rept. PNW–37. 44 pp.

Gratkowski, H. 1978. Herbicides for shrub and weed control in western Oregon. USFS Gen. Tech. Rept. PNW–77. 48 pp.

Holt, H. A., and B. F. Fischer (eds.). 1981. *Weed control in forest management.* Purdue Univ., Dept. of For. and Nat. Resources. 305 pp.

Klingman, G. C., Ashton, F. M., and L. J. Noordhoff. 1982. *Weed science, principles and practices.* 2nd ed. Wiley, New York. 449 pp.

Newton, M., and F. B. Knight. 1981. *Handbook of weed and insect control chemicals for forest resource managers.* Timber Press, Beaverton, Oregon. 213 pp.

Putz, F. E., and H. A. Mooney (eds.). 1991. *The biology of vines.* Cambridge University Press, Cambridge. 526 pp.

Stewart, R. E., Gross, L. L., and B. H. Honkala. 1984. Effects of competing vegetation on forest trees: a bibliography with abstracts. USFS Gen. Tech. Rept. WO-43. 1 vol.

U.S. Forest Service. 1981. Forest management chemicals. *USDA Agr. Hbk.* 585. 512 pp.

U.S. Forest Service. 1984. Pesticide background statements. Vol. 1. Herbicides. *USDA, Agr. Hbk.* 633. 850 pp.

Wagner, R. G., and M. W. Rogozynski. 1994. Controlling sprout clumps of bigleaf maple with herbicides and manual cutting. *WJAF* 9:118–124.

Walstad, J. D., and F. N. Dost. 1984. *Health risks of herbicides in forestry: a review of the scientific record.* Oregon State Univ. For. Res. Lab., Spl. Pub. 10. 60 pp.

Weed Science Society of America. 1983. *Herbicide handbook.* 5th ed. 515 pp.

PART 3

REGENERATION

CHAPTER *7*

ECOLOGY OF REGENERATION

Even trees do not live forever; the time comes when they must be, or naturally are, replaced by new ones. The world also has a chronic oversupply of treeless areas where reforestation is needed to remedy the effects of previous misuse or natural accidents. Young vegetation is so plastic that what is done during the period of regeneration or establishment determines most of the future development of trees and stands. The events or treatments of the first few weeks or months can govern the future characteristics more than even the most strenuous subsequent tending. Many of the successes or failures of silvicultural treatment are preordained during stand establishment. Physicians can bury their worst mistakes, but those of foresters can occupy the landscape in public view for decades. The ultimate act of regenerating a stand is so crucial that it should be kept in view throughout the whole rotation.

Ecological Role of Natural Disturbance

Introduction

At all stages of stand development, but especially in the earliest ones, it must be kept in mind that one is regulating all of the vegetation, and, for that matter, the animals as well, not just the trees. True forests grow only where the climate is more conducive than average to the development of land vegetation. New vegetation generally appears only after some lethal disturbance has eliminated all or some of the preexisting plants. This is because woody perennials, once established, usually command growing space so tenaciously that only the death of some of them can make vacancies large enough for new ones to start and make rapid growth.

Trees evolved many millions of years before humans. Each species represents some sort of adaptation to a kind of lethal disturbance that occurred in nature. The natural vegetation of any locality is a repertory of species, each adapted to colonize some micro-environment or ecological niche that might commonly have become vacant in nature and is at least barely favorable to plant growth (Grubb, 1977). Evolution has produced lichens that can grow on bare rocks and epiphytes adapted to crevices high on the boles of standing trees. It is through this collective versatility that vegetation is able to approach full occupancy of the growing space. It is seldom a question of whether there will or will not be vegetation but only of what kind.

These rules are not repealed for introduced, exotic vegetation. No exotic can be successful unless its new habitat is at least as favorable as its native one. For example, the annual plants of most agriculture are herbaceous pioneers that colonize drastically disturbed sites around the world. During the growing season the microenvironment of a wheat field in Manitoba differs little from that of one in Tunisia. Exotic trees differ mainly in that they do not thrive on such extreme disturbance and must be adapted to year-round climatic conditions over many decades.

If one wants to create a certain kind of forest the best point of departure is consideration of the natural disturbances that bring it into being in nature. Once this is known, it is possible to conjure up ways, sometimes quite artificial, that simulate the appropriate disturbance. In determining what disturbances to simulate and in assessing the success of the technique devised, it is necessary to remember that large trees all start small. One can tell remarkably little about regeneration simply by observing large trees and old stands, even though it can be very useful to deduce as much as one can from them. The literature of silviculture and forest ecology may be helpful in particular cases.

Often the quickest and best way to get new answers or test the validity of old ones is to crawl around on hands and knees examining the seemingly chaotic vegetation that develops during the first few years after regenerative disturbances. If one sees an abundance of desirable seedlings from a road, it often means that there are too many.

Kinds of Natural Regenerative Disturbances

Natural vegetation ultimately arises after the occurrence of one of the many kinds of **lethal disturbances**. No species is adapted to all kinds, so it is necessary to know which appears after what event.

In most parts of the world the most common natural disturbance is fire (Agee, 1993; Abrams and Nowaki, 1992; Mooney et al., 1981), which was kindled by lightning and volcanic eruptions long before people put it to use and abuse. Fire, by itself, generally kills small trees more effectively than large ones; in other words, it tends to kill old stands from the bottom upward. The species that it favors are generally of two different categories. One of these is perennials that sprout from roots or the bases of fire-girdled stems. These sprouting species include many angiosperm trees and shrubs as well as a few conifers such as pitch pine and coast redwood (Boe, 1975). Frequent repetition of fires in forest climates usually perpetuates stands of sprouting shrubs or perennial grasses.

The other category of species favored by fire comprises those that germinate from the usually small seeds of wind-dispersed species adapted to germinate on mineral-soil surfaces bared by fires. This category of post-fire seeders includes most of the pines and many other important commercial species. The importance of fire-followers in silviculture merely testifies to the fact that most forests are associated with climates that are seasonally conducive to fire.

Severe Disturbances

Erosional geological events constitute the most severe kind of natural disturbance. In ecological terms, the only true primary succession starts with landslides, the melting of glaciers, or the formation of new land by water, wind, or vulcanism. Human-caused erosion or earth-moving can also expose soil parent materials that are free of organic matter and deficient in nitrogen or other nutrients.

The true **pioneer** vegetation adapted to colonize such vacancies is more likely to be herbaceous than woody plants, but there are some tree species, such as the true poplars, alders, and certain river sandbar species, that can do so without some initial herbaceous stage. Ordinarily, trees do not start on such barren surfaces until some other vegetation has begun to build up the organic matter or provide enough shade for tree seedlings to endure microclimatic extremes.

In silviculture it is seldom wise to simulate such drastic disturbances deliberately. However, foresters are often called upon to afforest parent material that has been exposed by erosion, road building, strip-mining, or similar events. This very difficult form of silviculture works best if one has a very clear idea of how the natural pioneer vegetation develops on similar sites. Natural regeneration sources must be seeds or spores spread by wind, water, or animals.

The next most severe natural disturbance is the very hot fire that burns large amounts of dead fuel created by blow-downs or pest outbreaks. Lesser kinds of forest fires, even crown fires in living stands, are fueled mainly by the unincorporated organic matter of the forest floor but usually do not consume all of that. However, if there are large quantities of dead, combustible material on the ground, fires can burn so hot that they can be almost as lethal as a landslide. The main difference is that most of the incorporated organic matter remains intact.

Fires of this severity usually eliminate most sprouting species, thus leaving the site nearly vacant for establishment of new vegetation from seed. The subsequent ecological successions are usually thought of as secondary ones because the remaining organic matter enables some species to start earlier than they would otherwise.

The significance of severe fires for silviculture is their similarity to much more artificial disturbances. Cultivation for agriculture and certain kinds of silvicultural site preparation also leave only bare soil with incorporated organic matter and without woody perennials capable of sprouting. It is noteworthy that many of the illustrations of discrete, orderly, sequential stages of plant succession used in North America come from the natural revegetation of agricultural lands abandoned during a century-long epoch now coming to a close. The pure conifer stands that represent a stage in such old-field succession are so productive that they have been perpetuated in some regions as a silvicultural legacy of that epoch. The silviculture that simulates these cases depends on either natural seeding or planting of species adapted to colonize vacant areas.

Much of the foregoing paints a fiery picture of what can be called ''scorched-earth silviculture.'' This term is not intended to be pejorative because so many forests were and are regenerated in nature by fire. The tree species involved usually grow rapidly as individuals probably because early attainment of seed-bearing status is crucial to their perpetuation and the fires are apt to be frequent. There are important economic advantages in having trees that grow rapidly in diameter and height, regardless of how well they produce in terms of the separate consideration of annual growth per hectare. Consequently, these species are very important in silviculture, especially that in which large initial investments are made in planting.

Releasing Disturbances

The other general category of regenerative disturbance is that in which such agencies as wind or pests of large trees kill forests from the top downward but spare most plants of the lower strata. The species that are adapted to such circumstances are those with foliage that is constructed and displayed in ways that make it tolerant of shade. Their seedlings are usually not adapted to exposed microclimates, and their juvenile growth is slow.

These species can endure for many years as advance growth beneath old stands and retain the capacity to initiate rapid height growth whenever some lethal event releases them (Ashton, 1992a; Gordon, 1973). Some endure as stunted seedlings and saplings; others, as perennial rootstocks that survive while their tops grow up and are repeatedly killed back. Once released, most of these species maintain height growth for long periods. This characteristic, combined with the ability to maintain deep canopies of shade-tolerant foliage, can lead to long-sustained periods of high per-hectare production. However, such species tend to be limited to sites and regions that are continuously moist enough to reduce exposure to fire or drought stress. At risk of overgeneralization, they can be thought of as ''advance-growth-dependent'' species best regenerated naturally under some sort of protective cover.

Patterns in Floristics in Relation to Disturbance

Natural disturbances are not uniform in severity, type, or extent. Even a single hectare seemingly laid bare by a hot fire is actually a myriad of tiny spots, each a regeneration microenvironment different from adjacent ones. The chief differences are degree of exposure of mineral soil, extent of freedom from root competition, and intensity of solar radiation. These factors are the main determinants of the kind of vegetation that appears in a given vacancy of a few square centimeters in area.

In almost all forest regions, natural succession and development of new stands proceeds either from already established survivors, or from the germination of seeds (Glenn-Lawson, Peet and Veblen, 1992; Connell and Slayter, 1977). This occurs almost immediately after a disturbance has liberated or created new growing space. However, the complexity and variety of ecological species groups (Swaine and Whitmore, 1988; Bazzaz, 1979), and their inherent differences in reproductive mechanisms (Grubb, 1977) will vary because climates and the disturbance regimes they create are different across forest regions (Oliver and Larson, 1990). For example in many moist forest regions, the kind and variety of regeneration that appears after a disturbance that kills only some of the forest canopy can be very complex and diverse in nature (Fig. 7.1). This is largely because many forms of advance regeneration and vegetative sprouts already present in the understory before the occurrence of a disturbance are allowed to survive. On the other hand, more severe disturbances can promote a more uniform and simpler species composition that comprise a higher proportion of pioneers, with ecological species groups that are more reliant on strategies that occupy newly available growing space. It is therefore important to recognize that the origin in the stand of both structural and dynamic complexity is established at the regeneration stage.

The Role of Fire as a Disturbance

Late-successional communities are results of long series of small, light disturbances, whereas pioneer stages are the product of catastrophe (Connell and Slayter, 1977). Fire is one of the most common kinds of natural disturbance, and many species, including some of the most valuable of the North American forest, represent natural adaptations of the

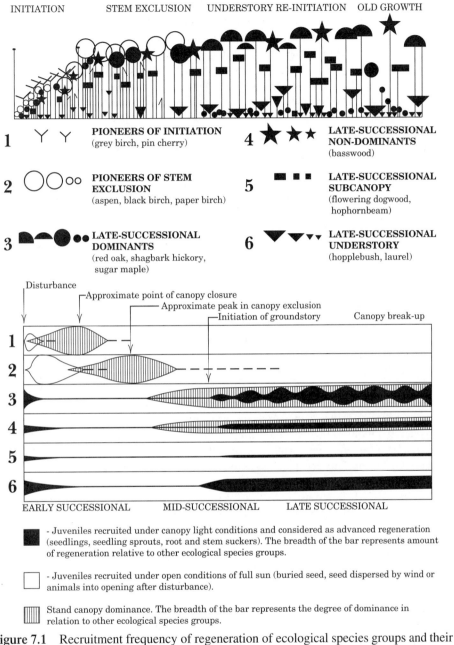

Figure 7.1 Recruitment frequency of regeneration of ecological species groups and their canopy dominance at different stages of stand development. Each of the numbered horizontal graphs shows the time of appearance and the changing abundance of each of the different groups of species that are listed with their respective numbers and crown symbols below the stand diagram. *(Modified after Ashton, 1992a, and reprinted by permission of Kluwer Academic Publishers.)*

vegetation to fire (Kuskela, Kuusipalo and Sirikul, 1995; Agee, 1993; Abrams and Nowaki, 1992; Pyne, 1984; Wright and Bailey, 1982; Kozlowski and Ahlgren, 1974).

Some species are adapted to reproduce after severe fires with effects akin to those of broadcast burning and others to much gentler disturbances like those of prescribed burning (see Chapter 8). Except for landslides, volcanic eruptions, and similar geologic events that set off true primary successions, the most severe natural disturbance of forests is the sequence of blow-down or insect kill followed by fire. Such fires have tremendous amounts of dead, dry fuel. The regeneration of some species appears to require their simulation. One of these is *Eucalyptus regans* of southern Australia, the species that grows almost as tall as coast redwood and does so faster; a lower stratum of noneucalypt understory trees has to be felled to produce enough fuel to get fires of required intensity (Stoneman, 1994). The view has been advanced that some species have evolved to produce enough fuel to provide for their own pyrogenic renewal.

Fire-fostered species may be divided into several ecological species groups depending on the nature of their adaptation to fire. The closed-cone pines constitute the first group. They are best exemplified by jack pine and include lodgepole, Monterey, bishop, and sand pine. In these species, regeneration occurs naturally after a crown fire has killed the old crop, exposed the mineral soil, and opened the cones. The second group is composed of species like the cherries and a number of undesirable shrubs, including certain species of *Ribes*, which have hard-coated seeds capable of surviving for long periods in the forest floor and springing up after fires. The third group consists of the large number of species that can reproduce from sprouts; this group is composed mainly of broad-leaved trees and shrubs but also includes a few conifers. The fourth category consists of light-seeded species that thrive on seedbeds of bare mineral soil exposed by fire but are not outstandingly resistant to fire.

The fourth group includes many valuable species such as the birches, Douglas-fir, eastern and western white pine, the spruces, sweet-gum, and yellow-poplar. Regeneration of these species after fires comes from seeds already present on the old trees or from those subsequently produced by trees that happened to survive because of their size or location. Those that are true pioneers form all or part of the main canopy almost from the time of their establishment. Those that are rather tolerant of shade are likely to become established simultaneously with faster-growing pioneer species and remain underneath until the death of the pioneers, which are usually short-lived. In other words, this fourth group is a heterogeneous one including a wide range of species that do not all respond to fire in exactly the same way.

In all four categories mentioned, the species involved are primarily adapted to regeneration after catastrophic fires that destroy most of the trees of one generation and prepare the way for another. It would be unwise to duplicate these conflagrations because of waste and public danger. The broadcast burning of slash is sometimes used to simulate these effects, but it is commonly necessary to resort to measures that do not necessarily include any burning at all. Finally, there is a fifth category of species that have sufficiently fire-resistant bark to withstand burning at intervals throughout most of a single generation. This group consists of certain hard pines, notably those of the South. The most outstanding is longleaf pine, which thrives only as a result of periodic fires. The others are loblolly, pitch, shortleaf, slash, and ponderosa pine; there is a strong possibility that red pine also belongs in this group.

Occasional surface fires were beneficial to the maintenance of these species in the original forest because they arrested natural succession, exposed favorable seedbeds, and prevented more destructive fires. Up to the present time prescribed burning has been extensively applied only to this group of fire-resistant trees. It has been practiced on a large scale mainly on the Atlantic Coastal Plain from New Jersey southward. Many species have several kinds of adaptations to fire. Pitch pine, a species that grows on sites where it is constantly bedeviled by fire, is an outstanding example of such a species (Little and Somes, 1961). It develops the thick, fire-resistant bark characteristic of trees of the fifth group. Its cones have a tendency to be serotinous, like pines of the first category. Seedlings and saplings killed by fire sprout from the base, like species of the third group; older trees defoliated by fire usually sprout new crowns. Like all species of the fifth group, it can be regarded as a particularly well-adapted member of the fourth.

Another way of viewing differing degrees of adaptation to fire is to note that a given locality may have species with adaptation to different frequencies of fire. In forested climates, sprouting perennial grasses often go with annual burning and sprouting shrubs with less frequent burning. Sprouting tree species are next in line if fires are too frequent for seed production. Serotinous-coned conifers or species with seeds that remain stored in the forest floor often come next. Coequal with these may be the pioneers with long-distance seed dispersal or the fire-resistant species that regenerate after light fires have burned beneath them. Next are usually those that regenerate after fires at long intervals, often under the protective cover of some simultaneously regenerating pioneer. The final members of the series are those that are simply so defeated by fire that they can return only after some intervening stages of succession. Not all members of such a series exist in a locality, but it is useful to know, for each locality, what species would be fostered by a given frequency of fire.

Disturbance and the Environment of the Microsite

New plants germinate and live through the most dangerous weeks of life within tiny worlds not more than several centimeters in any dimension (Fig. 7.2). The chief things that matter initially are the environmental conditions of these small microsites. If these conditions are favorable, it makes no difference whether the spots are in the middle of a huge clearcut area or are the result of thinning. Seedlings respond to water, light, carbon dioxide, chemical nutrients, and biotic influences that make up their particular environment regardless of the name of the silviculture treatment.

As far as initial establishment is concerned, a surprisingly small number of suitable microsites may suffice to regenerate a large area. The establishment of 2000 desirable seedlings per hectare theoretically requires only that 2000 suitable spots are distributed over the hectare. If each microsite occupied 10 square centimeters, this would be only 0.005 percent of the total area. Of course, this presumes that one germinable seed landed on each spot and was overlooked by mice and birds. It is worth noting that the ultimate outcome would also depend on the question of what other vegetation might possibly appear on the other 99.995 percent of the area. Ordinarily the early mortality of seedlings is so very large and the distribution so uneven that it takes more seedlings to achieve satisfactory natural regeneration. On the other hand, it is possible to have conditions so favorable that tens or hundreds of thousands of seedlings survive on a hectare and grow into badly stagnated thicket stands.

The crucial environmental characteristics of seedling microsites are very different from those that will govern the development of the tree after its top and roots have extended a few centimeters above and below the soil—air interface.

Figure 7.2 A newly germinated pine seedling with the seed still encasing the ends of the cotyledons. *(Photograph by U.S. Forest Service.)*

Water Loss

Within the uppermost layers of the soil and organic material water is lost to the atmosphere by direct evaporation. Such loss can proceed swiftly if the surfaces are exposed to large amounts of direct solar radiation. The loose materials of forest floor litter generally lose water so fast that they are seldom hospitable to the roots of plants. The uppermost layers of mineral soil or finely divided organic matter are somewhat more favorable but are also subject to water loss by direct evaporation. This comes about from the effects of surface tension in the very slender water columns within these materials. As water is lost to evaporation from the top ends of these columns, more is pulled upward to be lost in its turn. The finer the materials, the deeper is this capillary fringe. This kind of water loss differs from the transpiration of water by plants, which take up water from the soil through the roots. The important point about direct evaporation is that it explains the loss of water from the uppermost levels of the soil even when these layers contain no roots.

Fortunately, the strata from which water can be lost by direct evaporation are seldom more than several centimeters thick in total. If the roots of a new seedling do not quickly penetrate below this layer or are not repeatedly rescued by rain, they die (Kozlowski, 1949). Shade and mulching effects can greatly slow down the direct evaporational losses but cannot add any new water.

Surface Temperature and Energy Transfers

The same general kind of situation prevails in the first few centimeters of air above the forest floor (Geiger, Aron, and Todhunter, 1995). Here frictional effects greatly impede the turbulent transfer by which most vertical movement of heated air, water vapor, carbon dioxide, and other airborne substances takes place. The sluggish movement thus induced greatly restricts the upward transport of heated air by day and the downward movement thereof that offsets radiational cooling losses of heat by night.

In this connection, it is well to note that much of the energy from the sun comes through the atmosphere almost as if there were nothing there to block it. The ozone of the ionosphere does absorb most of the life-threatening, shortwave, ultraviolet radiation. Otherwise clouds, water vapor, and leaves are the only significant things that can shield the soil surface from solar radiation. The atmosphere is heated from below by heated land and water surfaces just as air is heated when it passes over a hot stove. However, for plants the temperature of their tissues is much more critical than that of the air around them.

The survival of new unshaded seedlings depends heavily on the interaction of the various physical processes by which this stupendous load of solar energy is dissipated. If heat accumulates, the surface temperatures soar; if it is lost too fast by night, the seedlings may freeze. The processes that take heat energy away from the absorbing surfaces are reflection, convection, conduction, evaporation, and outgoing re-radiation. Although the important concern is usually the prevention of extremes of temperature, it is sometimes necessary to provide for temperatures high enough for germination.

Shading, that is, the reflection or absorption of solar energy by leaves or other objects, is the most important physical process subject to silvicultural manipulation. Opaque objects obviously divert the radiation, but leaves reflect, absorb, and transmit. The chlorophyll of leaves not only absorbs visible blue and red-orange light used in photosynthesis but also reflects substantial amounts of green and, more importantly, invisible infrared light (Smith, 1982). A high proportion of the solar heating effect comes from the longwave infrared radiation.

Large amounts of radiation are reflected, without ever being absorbed, by the surface materials themselves. Most soil and dead organic materials do not vary much in their reflectivity or albedo; therefore, not much can be done to foster seedlings by trying to modify surface reflectivity. Charcoal from forest fires, in spite of its blackness, is only slightly less reflective than unburned organic matter or exposed mineral soil almost without regard for their color. The wetting of surface materials slightly reduces their reflectivity.

For practical purposes, the only way in which one can use reflectivity to control forest regeneration is to take advantage of the reflectivity of chlorophyll and absolutely opaque rocks and wood. This does not always have to be by shade; moss and other green vegetation growing around new seedlings can cool them by adding to the reflectivity of the green seedling leaves themselves.

The most important way by which heat moves vertically through air is turbulent transfer, which, with respect to heat, is called convection. This process involves the chaotic movement of large clumps of air molecules driven by the downward transfer of wind energy from the atmosphere hundreds of meters above. Conduction, the movement of heat by collision of single molecules, transfers heat through the air very slowly because there are so few molecules to collide. In fact, if the movement of heat upward from the surface depended only on air conduction, the temperature about 5 meters above would be warmest in midwinter and the midsummer surface temperatures would approach the boiling point of water.

The low heat conductivity of air is the cause of a phenomenon that can make the stratum immediately above the surface as dangerous for seedling tops as that for the roots below. The air in the first few millimeters is so tightly held by friction that convection has scant effect on it. The air molecules right next to solid materials can only vibrate and thus conduct heat; they cannot swirl and join in convection. Although this effect diminishes very rapidly with height, it makes the first few centimeters a bottleneck for heat transfer.

If the surface temperatures rise above 50°C, as they readily can on some kinds of surface, the succulent stems of small seedlings can be girdled by heat injury.

Aside from providing shade, the best way to forestall severe surface heating is to stimulate downward conduction of heat by the soil. The denser the surface material, the more will be the amount of heat conducted and less extreme will be the surface temperature. Leaf litter, with its very large amount of included air, is nearly as low in conductivity as any naturally occurring substance. Some of the highest temperatures naturally induced by the sun on the face of the earth may be those of flat litter surfaces composed of small conifer needles; these can be as high as 75°C. It is because of this and several other factors that the burning, physical displacement, or other modification of the unincorporated organic matter plays such a critical but manageable role in regeneration.

The part of the organic matter that has finely divided humus does have good heat conductivity. Except for coarse sands, most bare mineral soil surfaces conduct enough heat downward to prevent fatally high surface temperatures. The heat conductivity of soil actually depends mostly on the relative amounts of air and water in it. Water is a comparatively good conductor, so the more water and the less air the greater the conductivity.

Water has other important effects in stabilizing the microclimatic extremes of surfaces. Because of its remarkably high specific heat, it can absorb tremendous quantities of heat and yet its own temperature increases slowly. Its latent heat of evaporation is so great that large quantities of heat are absorbed, without change of temperature, when it changes from the liquid to the vapor phase. If the water vapor is then swept aloft, it carries large amounts of heat energy with it. Moreover, the latent heat of condensation is so low that large amounts of heat must be removed to convert it from a liquid at the freezing point to ice at the same temperature. In other words, its benefits to plants can be physical as well as physiological. However, the soaking of surface litter by rain postpones the extreme heating of such material by only about a half hour once the direct rays of the noonday sun hit it.

The remaining heat-transfer process important for plant life is radiation. This goes on all the time and would do so even if there were no air. Much of the heat absorbed by the earth's surface materials is lost to outer space as infrared radiation of wavelengths even longer than that which comes from the sun. The warmer the material, the more it radiates. It does reduce surface heating effects by day, but not much can be done to alter what happens then.

The same surface materials that absorb most of the solar heat by day also lose the most by radiation at night and thus can become much colder than the air a short distance above. This creates risk of frost damage to plants, especially succulent seedlings. The sluggish movement of air in the boundary layer aggravates the problem even more than with heat injury during the day. During daytime heating, the heated clumps of air wrested from the surface layer are less dense than the cooler air around them and so they are convected aloft rapidly. At night any warmer air that is pushed downward to replace very cold air at the surface must penetrate a denser medium. If the air is humid enough to have a dew point higher than 0°C, the condensation of water vapor into dew usually slows the cooling of plant tissues enough to prevent frost damage.

The most practical silvicultural method for preventing excessive radiational cooling is interposition of overhead cover. If the quanta of heat energy that radiate from a surface hit something above, there is a strong statistical probability that some of them will bounce back to the original radiating surface and reheat it. There is a subtle difference between this kind of protective effect and that of daytime shading from direct solar radiation. The sun's rays come in at slanting angles, except at noontime in the tropics. The most effective route by which outgoing radiation can escape is straight up toward the zenith. This means that the side shade that helps in daytime may not help as much at night. In general, the wider the angle of the inverted cone through which radiation can escape unimpeded from any point, the greater will be the radiational cooling of that point.

Light

Light energy, photosynthesis and their effects on seedling establishment and survival deserve particular attention. Radiation that comprises the visible parts (0.4–0.7 μm) and the infrared (0.7–10.0 μm) play an important role in seedling survival. Light regimes beneath forest canopies can be dramatically altered by changes in the quality, intensity, and proportion of direct versus diffuse light (Smith, 1982). Total daily light radiation received at the groundstory beneath a closed canopy of a late-successional evergreen tropical rain forest is normally less than 0.5 percent of that in the full open condition (Ashton, 1992b; Torquebiau, 1986; Chazdon and Fetcher, 1984). The majority of this radiation in evergreen rain forest (60–80 percent of the amount received) can occur over a 10-minute period of a sun-fleck. Studies have reported amounts of light radiation to vary from 1–5 percent for deciduous forests of similar successional stage (Lee, 1989; Canham et al., 1990; Ashton and Larson, 1996). Studies have also shown that light extinction was closely correlated with the shade tolerance and successional status of the canopy (Canham et al., 1994). However, plantations of conifers, such as Sitka spruce in the stem exclusion stage of stand development, can receive levels of daily light radiation that are as low as 0.1 percent. Tree seedlings that survive under these circumstances are usually very shade-tolerant and are remarkably responsive to short periods of direct sunlight (Chazdon and Pearcy, 1986, 1991).

Moisture and temperature affect the times when deciduous forest canopies have leaves that reduce solar radiation at the forest floor. Seedlings of many species thrive by being out of synchrony. Some take advantage of this by early emergence and rapid photosynthesis before canopy leaf out in the early spring or onset of the rains; others, by delay of leaf shedding in the fall or at the beginning of the dry season. In any event, amounts of light radiation at the ground level and beneath a closed canopy are usually considerably higher in seasonally deciduous forests than in evergreen forests. Studies have shown light levels at the ground level of a deciduous forest to be 1–5 percent of light received in the full open with leaves, but 50–80 percent without (Canham et al., 1990; Lee, 1989).

The greatest differences in light radiation quality beneath a forest occur across canopy openings. Many plants grow very well in side-shade where slanting direct rays of the sun are blocked by crowns of adjacent trees in the forest or opening edge. This condition is sometimes called ''blue shade'' because the light that is received is rich in the blue wavelengths of scattered diffuse light from the blue sky. ''Green shade'' exists where much of the light must come through several canopy layers of leaves in full shade. Because green light is photosynthetically useless, light rich in green supports the most shade-tolerant of species. The ratio of red to far red radiation at ground level can vary from 0.27 beneath a fully closed canopy; to 0.97 at the center position of a 200 square meter canopy opening; and to 1.27 in the full open (Lee, 1987; Chazdon and Fetcher, 1984). This ratio signifies the amount of photosynthetically useful red light that is received at the ground level compared to the amount of photosynthetically useless far red radiation. The lower the ratio, the poorer light is in photosynthetically useful red radiation.

Apart from changes in tree phenology and spatial heterogeneity of canopy structure, other factors that affect groundstory radiation regimes are related to physiography (slope, aspect) and latitudinal position. Under these circumstances, latitudes outside the tropics are strongly influenced by the angle of solar radiation during the growing season. Effects of aspect and slope are obviously of greater significance at higher latitudes than those that are more equatorial (Canham, 1988).

The Regeneration Process

The chief purpose of the preceding account of surface microclimate is to show how the success of seedling regeneration and the kind of species obtained can be regulated to some extent by deliberate manipulation of overhead cover and the characteristics of the surface materials.

However, successful natural reproduction depends on the completion of a long sequence of events (Wenger and Trousdell, 1957); failure of any single link in the chain is fatal. As far as sources of regeneration are concerned, there is the basic difference between sexual regeneration from seed and asexual from various modes of vegetative sprouting. All tree species can regenerate from seed, but some are so dependent on vegetative sprouting that the capacity of sexual reproduction seems to be retained only as a nearly vestigial capacity for further evolutionary adaptation.

With natural reproduction, species can be assembled into autecological groups. These categorizations have been based on the nature of the regenerative mechanism and stage of regeneration event (i.e. flowering, seed supply, seed dispersal, storage, germination, succulent stage, growth and establishment). For example, during seed dispersal differences can be observed in the dispersal mechanism between (1) new seed carried in from outside the stand being regenerated, (2) that which is "stored" on site on the trees (as in unopened serotinous cones), or (3) in the forest floor as ungerminated seeds. Lastly, various kinds of advanced growth comprise another kind of regeneration that may be "stored" on site as (4) seedlings or seedling sprouts, or as (5) vegetative sprouts that can arise from (a) basal portions of stems ("stump-sprouts," Fig. 7.3), (b) root systems ("root-suckers"), or (c) rooted branch ends ("layers"). For further description of vegetative reproduction, see Chapter 13. Artificial regeneration can be from the sowing of seeds on the site or from the planting of seedlings or vegetative cuttings started in nursery or greenhouse.

Plants can be dispersed and established in a large number of ways. If one attempts to set up detailed classifications of the sets of ecological adaptations of species, it is soon found that each species of a given locality seems to have its own separate pigeonhole. Similar sets of adaptations for ecologically similar groups of species, which can be called **guilds**, can usually be found among representatives from geographically separated regional flora. However, one species seldom exhibits exactly the same adaptations throughout its natural range. The only real solution to the problem is to find out as much as possible about the regeneration requirements of each species in each locality. For purposes of simplification we have only distinguished five broad guilds of propagule regeneration (Table 7.1).

Mechanisms of Essential Stages of Natural Regeneration

Flowering

Almost all tree species have the capacity to sexually reproduce through a variety of breeding systems (Table 7.2). The nature of sex representation within a flower is an important component of differentiating breeding systems among species and the plant families to which they belong (Bawa and Hadley, 1990). Species that are **monoecious** have both functional sexes on the same plant but in separate flowers. **Dioecious** species have trees

Figure 7.3 Not all forests grow on dry land, nor do all regenerate only from seeds. It may be anticipated that a good tree will develop from one of these 2-year-old stump sprouts of water tupelo in a very wet part of a flood-plain or bottomland forest in Mississippi. *(Photograph by U.S. Forest Service.)*

with flowers of only one functional sex. Plant species that are **perfect** have both functioning sexes within the same flower. Although plant species differ in how sexes are florally represented, all have adopted these various strategies to create a viable propagule that will contribute toward maintaining and advancing the **genetic vigor** of the future population. To the forester this can provide a dilemma that requires a careful balance between maintaining the viability of a population and selecting individuals with certain genetically desirable traits of economic importance (National Research Council, 1991). Unlike annual agricultural crops, trees must endure environmental variability over a much longer period of time with intergenerational responses to environmental change that are considerably slower.

To maintain genetic vigor pollination that crosses one parent tree with another is the most desirable. This is called **outcrossing** and almost all tree species promote it. Many temperate canopy trees that are wind pollinated such as the pines and oaks are moneocious, but the sexes are displayed on different parts of the crown, or mature at different times to avoid self-pollination.

Canopy trees of the tropics often have perfect flowers. Because canopy trees of the same species in a tropical forest can be widely dispersed, many researchers believed that these species were largely self-pollinated. Recent studies have revealed that specific animal pollination vectors ensure much higher outcrossing than originally imagined (Hamrick et al., 1991). Although most forests do not have a major proportion of tree species that are

Table 7.1 Autecology of tree propagules of forests and woodlands

Regeneration Guild	Mode of Dispersal	Life History Traits	Disturbance Type	Species Examples
Seed dispersed into the stand from outside	Wind, small birds, mammals	Seed that is abundant and small; light-demanding pioneers that can be long-lived but usually short-lived	Severe—landslides, sediments from floods, exposed mineral soils	Birch, poplar, pine, Douglas fir, mahogany
Seed stored on the tree or on the ground surface	Gravity, wind, and explosive dehiscence.	Borne within protective woody cones or pods; small-seeded; light-demanding pioneers that can be long-lived	Severe—promotes release of seed—fire	Lodgepole, jack and pitch pines, some eucalypts
Buried seed stored in mineral soil	None—accumulating seed bank from small mammals and birds	Medium-sized seed, light-demanding pioneers	Severe—promotes germination of seed from changes in soil environment	Pin cherry, raspberries
Vegetative propagation that arises from stems, roots and branches of residual parent plants	None	Fast-growing early to mid-successional tree species that are stress adapted	Releasing—promotes death of above-ground parts of residual trees but allows survival of roots.	Oak, maple, ash, alder, poplar, redwood
Advance growth of seedlings, seedling sprouts, and saplings that originate as cohorts during mast years	Hoarding mammals, hoarding birds, and gravity	Seed successfully germinates sporadically as a cohort. Seedlings accumulate beneath a closed forest canopy as a seedling bank carpeting the groundstory. Large-seeded mid- to late-successional species. Intermediate to very shade tolerant. Clumped distribution.	Releasing—promotes release of groundstory regeneration. Windthrows of all kinds, canopy herbivory, and pathogens	Oak, maple, spruce, fir, hickory, walnut, dipterocarp
Advance growth of seedlings, seedling sprouts, and saplings that originate almost continuously from year to year in small amounts	Bats, primates, and large birds. Seeds require ingestion by animal	Seed successfully germinates seasonally but regularly in small amounts beneath a closed forest canopy. Large-seeded, shade tolerant, and as a seedling susceptible to herbivory. Scattered distribution.	Releasing—as above	Durians, persimmon, virola, black cherry

Table 7.2 Modes of pollination and related characteristics of flowers and fruits[a]

Agents	Active Times	Colors	Odors	Flower Shapes	Nectar	Examples
Insects As Primary Agents of Pollination						
Beetles	Day and night	Usually dull	Fruity or aminoid	Flat or bowl-shaped; radial symmetry	Undistinguished if present	*Myristica* (nutmegs), *Annonia* (soursop) *Polyalthia*
Carrion and dung flies	Day and night	Purple-brown or greenish	Decaying protein	Flat or deep: radial symmetry; often traps	If present, rich amino acids	*Sterculia*, *Theobroma* (cacao)
Hover, flower, and bee flies	Day and night	Variable	Variable	Moderately deep; usually radial symmetry	Hexose-rich	Orchids
Bees	Day and night or diurnal	Variable but not pure red	Usually sweet	Flat to broad tube; bilateral or radial symmetry; may be closed	Sucrose-rich for long-tongued bees, hexose-rich for short-tongued	Basswood, horse-chestnut, willows
Hawkmoths	Evenings or nocturnal	White, pale or green	Sweet	Deep, often with spur; usually radial symmetry	Ample and sucrose-rich	*Capiscum* (peppers) Potato, tomato
Settling moths	Day and night or diurnal	Variable but not pure red	Sweet	Flat or moderately deep; bilateral or radial symmetry	Sucrose-rich	*Dipterocarps*, *Genetum*
Butterflies	Day and night or diurnal	Variable; pink very common	Sweet	Upright; radial symmetry; deep or with spur	Variable, often sucrose-rich	*Elaeocarpus*, *Hedyotis*, citrus, coffee

Table 7.2 *(continued)*

Agents	Active Times	Colors	Odors	Flower Shapes	Nectar	Examples
*Vertebrates As **Primary Agents of Pollination***						
Bats	Night	Drab, pale; often green	Musty	Flat "shaving brush" or deep tube; radial symmetry; much pollen; often upright, hanging outside foliage, or borne on branch or trunk	Ample, and sucrose-rich	*Durio* (durian), *Sonneratia* (mangrove), balsawood, kapok
Birds	Night	Vivid, often red	None	Tubular, sometimes curved; radial or bilateral symmetry; robust petals; often hanging	Ample, and sucrose-rich	Catalpa, *Erythrina*, *Spathodea* (African tulip-tree)
Abiotic Pollination						
Wind	Day or night	Drab, green	None	Small; sepals and petals absent or reduced; large stigmata; much pollen; often catkins	None or vestigial	Oaks, pines, beeches, firs, spruces
Water	Variable	Variable	None	Minute; sepals and petals absent or much reduced; entire male flower may be released	None	Podocarps

Data derived in part from Howe and Westley (1988), Baker and Baker (1983), Proctor and Yeo (1972), Faegri and van der Pijl (1971), Baker and Hurd (1968).

dioecious, both Bawa (1980) and Givnish (1980) have reported that the trait increases when moving toward forests of more equatorial latitudes. This can largely be related to the closer coevolutionary relationships that have developed between plants and animals in these climates.

Many tree species in more unfavorable environments have resorted to maintaining genetic vigor through **vegetative propagation**. Although no further advances in genetic adaptation to environmental change are likely via this route, the ability to hold onto a site without needing to relinquish growing space is an enormous competitive advantage. Vegetative sprouting, layering and suckering are most common in montane krumholtz, forest understories, and frequently disturbed sites.

A few species have evolved to solely promote self-pollination. This has favored the retention and passing of certain genetic traits from parent to offspring, particularly where these traits provide a competitive advantage over other species. Environments that promote this reproductive strategy are largely similar to those that promote vegetative modes of propagation such as alpine meadows and highly disturbed sites. Although these plants carry the traits that promote their continued existence on these sites, because of continuous self-pollination they are genetically very similar. Under these circumstances, these species are much more prone to having propagules that have a **lethal recessive genotype**. Fortunately, such genotypes seldom enter the general population because lethal recessive genes usually cause the seeds to abort.

Seed Supply

The first and most obvious prerequisite is an adequate supply of seed. No tree or group of trees is a dependable source unless it is sufficiently old and vigorous to produce seed. Furthermore, seed bearers should be located so that wind or other agencies will ensure pollination and properly distribute the seeds over the area to be regenerated. Regardless of how carefully the seed bearers are chosen and fostered, it must be remembered that most species do not annually produce the abundant crops of seed necessary for satisfactory regeneration. This characteristic makes it difficult to carry out reproduction cuttings with equal chance of success each year. The limited crops of seed that are produced almost every year give rise to little or no reproduction. The origin of many stands can be traced to unusually good ''seed years'' that were followed by satisfactory conditions for germination and establishment of seedlings.

Good seed years occur at intervals that are better thought of as sporadic rather than predictably periodic. Although there might be real cycles of seed (mast) production, they are complicated and have yet to be verified or explained. The first essential is an ill-defined state of physiological readiness for flowering; the supply of nitrogen compounds for protein building may be one of the crucial factors. Even after flowers or strobili appear, it is common for adverse weather or infestations of insects and fungi to intervene and prevent pollination or maturation of flowers and fruit. The depredations of insects are a common cause of irregularity for seed crops. Sometimes good seed crops follow years of total failure that may have caused the collapse of populations of specialized seed predators. In general, the more favorable the conditions of soil and climate for plant growth, the more frequent are good crops of seed.

Seed Dispersal

The modes of dissemination of tree species include almost every imaginable mechanism (Table 7.3). The distances of effective dispersal vary widely but are usually not greater than a few times the height of the seed bearers. With many wind-disseminated species,

Table 7.3 Modes of dispersal and related characteristics of seeds or fruits[a]

Agents	Colors	Odors	Forms	Rewards	Examples
Usually Self-Dispersed					
Gravity	Various	None	Various	None	Oaks, hickories, dipterocarps
Explosive dehiscence	Various	None	Explosive capsules or pods	None	*Acacia, Croton,* dwarf-mistletoe Rubber
Bristle contraction	Various	None	Hygroscopic bristles in varying humidity	None	
Usually Abiotic Dispersal					
Water	Various, usually green or brown	None	Hairs, slime, small size, or corky tissue resists sinking or imparts low specific gravity	None	Bald-cypress, sycamore
Wind	Various, usually green or brown	None	Minute size, wings, plumes, or balloons impart high surface to volume ratio	None	Birches, ashes, elms, maples, willow, cottonwood
Usually Vertebrate Dispersal					
Hoarding mammals	Brown	Weak or aromatic	Tough thick-walled nuts; indehiscent	Seed itself	Oaks, hickories, walnuts, *Dipteryx*

	Color	Odor	Morphology	Reward	Examples
Hoarding birds	Green or brown	None	Rounded wingless seeds or nuts	Seed itself	Pinyon pine, whitebark pine
Arboreal frugivorous mammals	Brown, green, white, orange, yellow	Aromatic	Often arillate seeds or drupes; often compound; often dehiscent	Aril or pulp rich in protein, sugar, or starch	Mangifera (mango), Nephelium (rambutan)
Bats	Green, white, or pale yellow	Aromatic or musty	Various, often pendant or starch	Pulp rich in fat	Banana, figs
Terrestial frugivorous mammals	Often green or brown	None	Tough, indehiscent, often >50 mm long	Pulp rich in fat	Blueberry, Parkia (petai)
Highly frugivorous birds	Black, blue, red, green, or purple	None	Large arillate seeds or drupes; often dehiscent; seeds >10 mm long	Pulp rich in fat or protein	Virola, Myristica (nutmegs) Durio (durians)
Any frugivorous birds	Black, blue, red, orange, or white	None	Small or medium-sized arillate seeds, berries or drupes; seeds < 10 mm long	Various; often only sugar or starch	Cherries, raspberries
Animal fur or feathers	Various	None	Barbs, hooks, or sticky hairs	None	Grasses and sedges
Usually Insect Dispersal					
Ants	Various	None to humans	Oil-secreting seed coat	Oil or starch body with chemical attractant	Banksia, Protea, Macaranga

[a]Data derived in part from Howe and Westley (1988), Gautier-Hion et al. (1985), Janson (1983), Wheelwright and Orians (1982), and van der Pijl (1971).

particularly the conifers, seeding is not uniform in all directions because dispersal is most effective when dry winds are blowing. With species disseminated by birds, bats, mammals, gravity, or flowing water, it is important that the forester proceed with clear understanding of the factors involved.

Some adaptations are quite bizarre. For example, the seeds of some tropical mangrove species germinate on the trees, and the fallen seedlings can then float upright over miles of ocean to take root on some distant shoreline.

Both before and after the seeds fall from the trees, they are likely to be eaten by insects, bats, birds, and rodents. Although seed predators are held in check by natural controls that may be strengthened by direct or indirect artificial measures, the most practical means of combating the menace is to ensure that the supply of seed is sufficient both to feed the predators and to regenerate the forest. Sometimes the most important predators are also the most effective agents of dissemination; if it were not for squirrels and other rodents, the heavy seeds of oak, hickory, and walnut would not be carried farther than gravity would take them. Edible fruits are often adaptations for seed dispersal by animals, especially birds and tropical fruit-eating bats (Table 7.3).

Storage

Seeds are usually produced during a favorable period but must often survive a dry or cold period and be ready to germinate during the next growing season. To do this, they develop varying degrees of dormancy, a condition in which they do not grow and their physiological processes become very slow. This dormancy can become sufficiently pronounced that the seeds will not germinate until particular special conditions have been met. This phenomenon often prevents seeds from germinating prematurely, as during abnormal warm periods of winter.

The required conditions are often those that prevail in the forest litter or soil during periods unfavorable for germination. For example, many species will not germinate without a period of moist storage at low temperature. Some bird-disseminated species have seeds with hard coats that must be abraded by the sand in bird gizzards before water and oxygen can penetrate the seed and start germination. Where seeds are stored artificially, it may be necessary to carry out treatments that imitate those that break dormancy in nature.

The seeds of many species in aseasonal tropical rain forests do not develop dormancy because conditions are always favorable for growth and there is no value in delaying germination. They must germinate in hours or days and are difficult or impossible to store artificially. At the other extreme, the seeds of some species can survive for many years under natural or artificial conditions in which respiration is kept very slow (but not halted) by dryness, low temperature, or limited aeration. For example, the forms of jack and lodgepole pine that are adapted to regenerate after forest fires have seeds that are stored for decades in sealed cones (known as **serotiny**) on the trees. When hot fires melt the cone scales, the seeds are released onto the bare soil left by fires (Burns and Honkala, 1990). The **buried seed** phenomenon is another example whereby certain plant species can store seeds in the forest soil for long periods of time (in some cases thousands of years) and germinate only after a canopy disturbance alters the ground surface radiation and moisture regimes (Leck, Parker, and Simpson, 1989).

Germination

The start of development of the embryo depends on having adequate amounts of moisture, heat, and oxygen, and sometimes certain wavelengths of red to far red light. However, successful germination depends largely on the rainfall and the nature of the spot where the seeds are deposited. Moisture is the most critical and variable factor. The embryos in excessively dry seeds either die or remain dormant; if the seedbed is too wet, the seeds will rot without germinating or they will suffer from deficiency of oxygen. The kinds of dormancy encountered in artificial seeding control the timing of germination. Processes that break dormancy are normally completed in seedbeds that are otherwise favorable for development of the seedling, and are rarely critical in natural regeneration.

In general, the seeds of wind-disseminated species germinate best on or slightly beneath the surfaces of seedbeds that tend to remain moist; bare mineral soil is usually the best seedbed for such species. Seedbeds of horizontally oriented materials, such as coniferous litter, tend to be unfavorable because they inhibit penetration of seeds into moist substrata; vertically arranged materials, like grass, are less resistant. Delays in germination are dangerous because they lengthen the period of exposure to birds and rodents. Losses from germination failure, which are usually caused initially by desiccation of seedbeds, are frequently more serious than losses of germinated seedlings.

Because of differences in ability of roots to penetrate and for many other reasons, many of the microsite differences that can be used to discriminate between species depend on the nature of surface materials and their alteration. Close examination reveals many such differences, but most are poorly documented. Much depends on how deeply fire or physical disturbance has exposed various soil layers.

In general, the only species that can effectively establish themselves on thick leaf litter are those, such as oaks (Carvell, 1979), that have very large seeds. However, this is not because they can germinate on top of such surfaces but because they were buried beneath them by rodents or more falling leaves and because they have enough stored materials to grow back up through the leaves. Oak acorns and other large seeds usually cannot successfully germinate on bare soil surfaces. Their new, blunt roots simply roll them around over the surface without penetrating; they need to be buried to hold them in place and keep them moist. If smaller-seeded species appear to have established themselves on thick litter, close examination usually shows that it was on some localized spot where wind or something such as a little ant mound made the litter very thin. Most small-seeded species can germinate only on dense media such as exposed mineral soil. This is mainly because close contact must exist between the seed and the moisture-supplying medium. Even then, the germination is seldom good unless the action of rain, frost, or other agencies has caused the seeds to be slightly buried.

With these dense seedbed media, it can make a large difference how deeply they are exposed and what sort of materials form the surface. One of the most ideal seedbeds for some very small seeds, such as those of birches, is the finely divided black humus of the layer above the true mineral soil. This has appropriate density coupled with high concentration of chemical nutrients that the small seeds did not bring with themselves. However, some species (subalpine fir, Englemann spruce) can have poor germination from their own unsterilized litter (Daniel and Schmidt, 1972).

The uppermost strata of mineral soil are favorable for many species with small- or medium-sized seeds. For reasons that are not altogether clear, the number of species that can become established seem to dwindle the more deeply the soil is gouged. Sometimes they become limited to sedges or plants other than trees. It is rare that there is any useful purpose in extending seedbed preparation any deeper than the surface of the mineral soil except to bury the seeds slightly or to alter the drainage characteristics of the soil.

Succulent Stage

New seedlings are most vulnerable during the first few weeks of their existence, while their stems are still green and succulent. Heat injury resulting from extremely high temperatures on surfaces exposed to direct solar radiation takes a heavy toll, particularly among conifers. Tender seedlings in shaded situations are subject to **damping-off** caused by a wide variety of weakly parasitic fungi that are normally saprophytic and incapable of attacking larger seedlings. Cutworms and other insect larvae are particularly active during this period; most of the species involved are omnivorous feeders that will eat anything green and succulent. The better known forest insects and fungi generally attack trees that have passed the succulent stage. Seedlings that germinate late and remain succulent in the fall are commonly killed by frost. The dangerous succulent period ends when the outer cortical tissues of the stem become dry and straw colored, a process referred to as **hardening**. Thereafter the attrition of various damaging agencies continues, but the period of catastrophic losses is generally at an end.

Growth and Establishment

Drought mortality is distinctly different from heat injury and can take place after the succulent stage. It will not occur if seedling roots extend themselves rapidly enough to maintain contact with portions of the soil where water is available. The seedling must manufacture enough carbohydrate to enable the roots to grow downward faster than the deepening of the stratum in which water is unavailable. This is one reason why shaded seedlings are sometimes more likely to die of drought than those growing in the open.

It helps to note that the physical environmental factors that control the small, new seedling may not operate the same after the roots have penetrated below the capillary fringe and the tops have grown up into the more turbulent air strata. Once the seedling is tall enough, and at all subsequent stages, it is not threatened as much by extremes of temperature. The risk of frost damage persists longer than that of true heat injury. There still have to be all kinds of adaptations to restrict excessive water loss, but there is usually enough turbulent convection to take care of other microclimatic problems.

A very crucial race against time which the seedling must make is that of extending its roots downward faster than the loss of water through direct evaporation from the capillary fringe can overtake them. After this penetration is achieved, the roots move permanently into soil strata in which water loss is governed by gravity and transpiration but not by direct evaporation. Only the transpiration of established plants can move water upward from these strata. The less established vegetation there is to transpire, the better will be the supply of water beneath the first few centimeters.

The general result of these phenomena is that the ecological and silvicultural rules that govern the lives of new seedlings are quite different from those that apply later. If one reduces the amount of competing vegetation, as by thinning, the supply of water for established trees is increased. On the other hand, reduction of vegetative cover during the early and most crucial stages of regeneration reduces the water supply for new seedlings; it also threatens them with microclimatic extremes that are of little consequence later. If this distinction is overlooked, one can be very puzzled by seemingly paradoxical results.

It might seem from all this that no tree seedling could ever survive in the forest, even though many obviously have for many millions of years. Forests have regenerated because the trees and other plants involved are collectively capable of colonizing virtually any kind of microsite with seedlings. Each species is adapted to a range of microsites; this range is wider for some than for others, but none is adapted to the full range. If one wants to favor any given species, this can, given enough knowledge, be done by deliberately creating a suitable number of appropriate microsites.

Although there are exceptions, the seedlings of most tree species cannot colonize severely exposed microsites without the protective cover of hardy pioneer herbaceous vegetation. Sometimes this herbaceous vegetation is so ephemeral that it is taken for granted or is overlooked because it soon vanishes, leaving only the small trees that it once sheltered. The seeds and seedlings of many tree species have adaptations to take advantage of such cover.

The cotyledons and juvenile foliage of new tree seedlings are often built or arranged in ways that make them more shade-tolerant or efficient in photosynthesis than those of the adult stages. The seeds of many have enough stored material that they are more dependent on water to mobilize these reserves than on their own initial photosynthesis. The supply of carbon dioxide for photosynthesis is quite large because of close proximity to the respiring organisms that feed on soil organic matter. The supply of light becomes much more crucial later on, but even then the requirements and adaptations of different species vary tremendously (Table 7.4).

Table 7.4 Attributes of trees that change in relation to amount of light, with attributes organized in order of increasing scale from leaf to stand[a]

	Shade Intolerant	Shade Tolerant
Leaf Morphology		
Individual leaf area	Low	High
Leaf orientation	Erect	Horizontal
Leaf thickness	High	Low
Cuticle thickness	High	Low
Palisade mesophyll thickness	High	Low
Palisade/spongy mesophyll ratio	High	Low
Stomatal size	Small	Large
Stomatal density	High	Low
Leaf Physiology		
Light saturation rate	High	Low
Compensation point	High	Low
Stomatal conductivity	High	Low
Water use efficiency[b]	High	Low
Nitrogen use efficiency[c]	High	Low
Crown Morphology		
Leaf area index	High	Low
Branch orientation	Erect	Horizontal
Branching pattern	Whorled	Branching
Whole Plant Morphology		
Allocation to leaves	Low	High
Allocation to roots	High	Low
Bole taper	Low	High
Live crown ratio	Low	High
Reproductive effort	High	Low
Seed size	Small	Large
Self-pruning	High	Low
Stand Dynamics		
Self-thinning	High	Low
Stand density	Low	High

[a]Sources: Ashton and Berlyn 1992, 1994; Parker and Long 1989; Givinish, 1988; Boardmann, 1977; Carpenter and Smith, 1976; Jackson, 1967.

[b]The ratio between the rate of CO_2 assimilation and rate of water loss.

[c]The ratio between the rate of CO_2 assimilation and nitrogen content.

The availability of chemical soil nutrients usually does not limit the growth of new seedlings; the seeds themselves supply some nutrients, and more are usually released by the decay of the plants that were killed to make the regeneration vacancy in the first place.

Great variations exist in the means and rates of penetration of seedling roots downward to the relatively stable moisture supplies available at depth. The supplies themselves have different characteristics. Most species germinate at a certain time of year when the soil is well supplied with water. If the ensuing period is rainless, as in the western United States, or subject to drought from high evapotranspiration, as in parts of the South, everything depends on whether the seedling roots can extend downward more rapidly than the drying front that moves downward from the surface. Most of the species involved are adapted to produce roots that grow 6–20 inches (15–50 cm) deep the first year. If the necessary building material cannot be provided from the seeds, there must be enough light for adequate photosynthesis, or the seedlings simply die of drought in the shade.

If there is plenty of continuing rain, the seedlings are often adapted to survive with much shallower root systems. However, in this connection, it is well to note that moistening of the soil from above can proceed only to the extent that there is enough water to bring the uppermost soil strata to **field capacity** (the amount of water that can be held against gravity). This means that there is a wetting front as well as a drying front and both move downward.

Provisions for Initiating Natural Regeneration

Successful regeneration of any sort, natural or artificial, can occur only if the right amount and kind of vacant growing space becomes available for the establishment and subsequent growth of the desired species (Fig. 7.4). The ideal objective is to create vacancies that are not merely favorable to the desired species but are more favorable to them than to any others. This objective would be achieved if an environment were created in which the desired species would start and then grow faster in height than any competing vegetation. If this perfect goal could always be attained, subsequent releasing operations could be avoided.

Regulation of Regenerative Canopy Openings

In very small openings, the development of extremes of surface temperature is impeded by side-shade; in large openings, the wind causes enough turbulent transfer of heat to restrict the diurnal range of surface temperature. There is evidence that the greatest extremes of temperature occur when the diameter of an opening is $1\frac{1}{2}$ times the height of the surrounding trees (Geiger, Aron, and Todhunter, 1995). Presumably this is the situation in which the combined effect of side-shade and ventilation is the least.

When an opening is more than two to three times the height of the surrounding trees, the environmental conditions at the center are about the same as those that would prevail in any very much larger opening (Minckler and Woerheide, 1965). Some important things that would differ would be the effectiveness of seed dispersal from the adjoining trees and effects that had to do with the travel of animals between the opening and the cover of the taller trees.

Microclimatic differences between spots can also be systematized in terms of the distinction shown in Fig. 7–5 between conditions found (1) in unshaded, open areas, (2) in the side-shade and partial protection of uncut timber along the edges of openings, and (3) under full shade. The edge zone is exposed to the scattered, diffuse light from the sky but not to any harmful effects of direct sunlight. Although the total amount of light is

Figure 7.4   Ponderosa pine regeneration under difficult climatic circumstances in northern Arizona. The two pictures show the same spot just after a group-selection
cutting in 1909 and again 1938. The saplings of the second picture did not
become established until abnormally favorable circumstances took place in
1919. Unusual May rains probably overcame the competitive effect of the
grass cover. *(Photographs by U.S. Forest Service.)*

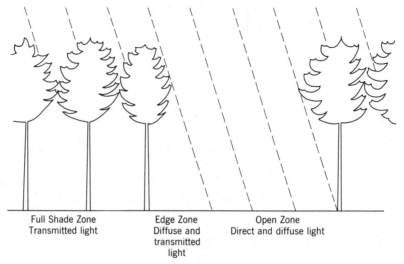

Figure 7.5   The zonation of solar radiation which, in combination with alteration of surface conditions, controls most of the alterable microclimatic factors significant in regulating regeneration from seed.

distinctly reduced, diffuse radiation is relatively rich in the blue wavelengths most effective in photosynthesis. This transitional zone extends, for practical purposes, not only for a fluctuating distance outward from the edge of an uncut stand but also a short distance inward beneath standing trees. The illumination under full, overhead shade consists mainly of whatever direct sunlight penetrates through interstices in the crowns and the filtered light, rich only in the photosynthetically useless green and infrared wavelengths that are transmitted through the foliage. The differences between these zones correspond roughly to differences in degree of root competition, except that sunlit edges of openings are subject to nearly full light and undiminished root competition.

Figure 7.5 does not include depiction of another pattern in which forest cover can be opened to create vacancies for regeneration; all it shows is full crown closure in the adjoining trees. A diffuse cover of scattered trees, such as might be left in what is called uniform shelterwood cutting, would theoretically allow more ventilation; the extent of the effect of increased ventilation in narrowing the extremes of surface temperature is not known. This kind of cutting is the best way of increasing the area of the edge zone. It virtually eliminates the full shade zone but puts varying amounts of surface into the open zone.

Some useful ideas about controlling regeneration from seed can be deduced by observing the patterns of species and their height growth in different parts of forest openings (Ashton, Gunatilleke, and Gunatilleke, 1995; McClure and Lee, 1993; Smith and Ashton, 1993; Canham, 1989; Poulson and Platt, 1989; Denslow, 1980, 1987). These things can be observed and diagnosed most simply in circular openings, as shown in Fig. 7.6; the same things can be observed in openings of any shape but are not as simply analyzed.

Quantities of light radiation received beneath forest cover varies with size of canopy opening and position beneath opening. In a circular opening, those effects associated with the slanting rays of direct solar radiation will be arranged in crescentic patterns. Those that are controlled by root competition of adjoining vegetation, diffuse solar radiation, air

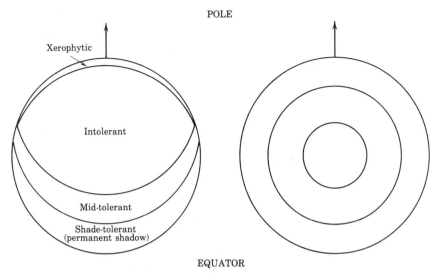

Figure 7.6 Crescentic (left) and concentric (right) patterns of horizontal variation in the effects of microclimatic factors in circular openings at some mid-latitude. The kinds of species favored by different kinds of shading and exposure are indicated in the crescentic pattern. Concentric patterns are more likely to appear during the development of established vegetation than to affect species composition. Recognition of these two kinds of microclimatic variation may help in deducing the factors governing regeneration and determining how to control it.

movements, and outgoing radiation (frost and dew) should have concentric arrangement. Except for the effects of root competition, crescentic patterns, because of the power of direct sunlight, are much more common than concentric patterns. However, these patterns will be altered by the degree and aspect of slope and by latitude. Because there is no environmental factor more readily predictable than the sun, the extent of side-shade can be calculated (Rich et al., 1993; Canham, 1988; Halverson and Smith, 1979; Harrington, 1984; Brown and Merritt, 1970).

On poleward slopes all shade zones are broadened, and the steeper the slope the more they are broadened; on equatorward slopes these effects are the opposite. At high latitudes, the various shade zones become very broad; diffuse solar radiation may even become fully as important as the direct. In the tropics most such patterns become concentric, and side-shade is effective only in very narrow bands.

A case most common at midlatitudes in the Northern Hemisphere is depicted in Fig. 7.6. Usually the only place where new seedlings of species that are tolerant of shade and intolerant of exposure can appear is in a crescent along and under the southern edge. More shade-intolerant and exposure-resistant species find their optima for establishment and height growth in crescents arranged successively northward. At the very northern edge there is often a peculiar zone, almost like a little desert, in which the combination of unimpeded direct sunlight and root competition of taller trees can prevent much of anything from regenerating.

The sizes of the vacancies created in the growing space should therefore be neither so large that the desired seedlings are overwhelmed by new undesirable vegetation nor so small that they are overtopped by preexisting vegetation expanding from the sides. Excessively heavy cutting or overly drastic site preparation can call forth too much unwanted pioneer vegetation.

Openings can also be so small that they create no real vacancies in the soil-growing space. If intraspecific root grafts exist, the root systems of the trees that were cut may simply be taken over by those of adjacent uncut trees of the same species. Even more commonly, the roots and crowns of adjacent trees expand into the vacancies faster than the newly established trees can claim them. This can be deceptive because roots, which do not have to provide for their own mechanical support, often expand into vacancies faster than the more visible crowns. This kind of effect is most spectacular on sites where soil moisture is so limiting that a stand may fully occupy the soil without ever achieving full closure of the crown canopy.

The sizes and shapes of regeneration vacancies may be limited by any requirements for a supply of seed from adjacent trees. It is not absolutely necessary that vacancies be created all at once because they can be enlarged in a series of operations. This sequence is often useful for species that can become established only under some sort of shelter, yet will not commence rapid growth without release. Natural regeneration is easy if any tree species that appears is acceptable and the site is not marginal for natural tree growth. It becomes more difficult when one seeks some chosen species or wants very prompt reestablishment. One of the keys to controlling the process is understanding the adjustable microenvironmental phenomena that govern where given species will appear.

These patterns are greatly modified by differences in seedbed conditions that can vary considerably from spot to spot. Close-by shade of stumps, rocks, and woody debris can also be as effective as that of more distant tall trees. Furthermore, after exposure-tolerant seedlings or saplings are established, seedlings of exposure-intolerant species can start beneath them. However, their height growth is likely to be slow until whatever time the trees above them are eliminated.

The patterns can also be obscured and confused in nature by preexisting advance growth and vegetative regeneration, that is, by plants already present when the opening was created. Observations of these patterns can help greatly in deducing the kinds of microsites that one would aim to create over larger areas to produce stands of the desired species composition. The desired spatial pattern will usually be different from the circular openings that are ideal for the diagnostic purpose.

From these considerations it can be deduced that even small forest openings cannot be regarded as uniform habitats for regeneration. If the regeneration in a quarter-hectare opening is all lumped together and counted collectively without regard for spatial variations, as much useful information will be buried as is revealed.

Regulation of the Groundstory

Unlike the understory which describes all tree strata below the forest canopy, the **groundstory** is the part of the vertical forest profile that includes the ground surface and the plants that are <0.25 meters above. The stratum of the groundstory is therefore the most suitable scale and focus for initiating and observing regeneration. In silvicultural practice, the appropriate kinds of regeneration environment are created partly by regulating the spatial pattern in which the previous stand is removed in the ways just described. Most of the names of methods of regeneration cutting discussed in later chapters refer to these patterns. Differences among these methods are so visible that they tend to dominate thinking about regeneration treatment.

Yet, the pattern of tree removal falls far short of being the only factor that controls regeneration. It is also frequently necessary to regulate the lesser vegetation that can either compete with or favor the desired species. This includes not only the vegetation that is already established but that which may appear in response to the regenerative disturbance (Fig. 7.7). If it is undesirable, it cannot be wished away merely by single-minded concentration on the desired tree species.

Sometimes regeneration, especially that from seed, is regulated by various treatments of the soil and the forest floor. The physical characteristics of the surfaces on which germination takes place exert a powerful control on seedling establishment. Minute differences can cause remarkable variations in the species of plants that appear. (The physical characteristics of the soil in relation to extremes of temperature and water loss were discussed in greater detail in the section "Surface Temperature and Energy Transfer" earlier in this chapter.

The treatments that can be used to regulate accessory vegetation and soil surface conditions include the various forms of release cutting, site preparation, and prescribed burning discussed in other chapters. These treatments are most relevant for regeneration originating from seed because that is the kind of regeneration for which regulation of the microenvironment is most important. Regeneration by planting and from vegetative sprouting evades most of the crucial problems involved in the germination and establishment stages; these forms of regeneration are not immune to microenvironmental effects but are not so tightly regulated by them.

Regeneration Methods as Analogs to Natural Disturbance

Methods that rely on establishing and releasing propagules of both seed and vegetative origin from sources within or adjacent to the stand being regenerated have been termed natural regeneration methods. In these methods, forest canopy manipulations are carried out to favor or discourage certain kinds and amounts of propagules. These manipulations emulate the type and size of a natural disturbance that achieves the same results.

As previously discussed, two major kinds of natural disturbance, severe and releasing, also define the major differences among the natural regeneration methods that are called clearcut, coppice, seed tree, shelterwood, and selection. It is therefore important to consider the name of the method by its predominant origin and guild of regeneration rather than what the cutting pattern looks like.

Clearcut regeneration methods reflect severe disturbance regimes that remove the preexisting vegetation completely. Regeneration depends on dormant propagules that are in the seed bank or that come in on suitable microsites by wind, water, or animal vectors in large amounts from adjacent stands. Examples of natural disturbances that clearcuts emulate are catastrophic fires and landslides. Coppice, seed-tree, and shelterwood methods reflect releasing disturbances that depend on propagules that are established *in situ* as advanced reproduction (seedlings, seedling sprouts, stump, and root sprouts).

Coppice methods rely largely on bud formation and release of vegetative shoots from parts of the residual vegetation that remains alive after a disturbance. As such, this is a very dependable mode of establishment because a new plant never actually has to colonize vacant growing space in the soil. Canopy manipulations for seed tree and shelterwood are tailored to favor regeneration presence and establishment of seedlings and seedling sprouts before complete canopy removal above the microsite. Differences between seed tree and shelterwood, as the names imply, relate to differences in shade tolerance of the species that are being regenerated. Species that are regenerated by seed-tree methods rely on a parent canopy only for adequate dispersal of propagules. These species require the exposed

Figure 7.7   In regenerating forests it is necessary to anticipate the development of all the vegetation and not just that of the desired species. The first picture shows a Douglas-fir seedling barely visible in front of a stake during the year after a cutting in a forest on an excellent site on the Oregon Coast. The second picture shows the same tree 3 years later and almost submerged by red alders and thimbleberry bushes that were not apparent earlier. *(Photographs by U.S. Forest Service.)*

conditions of full sun like that of clearcuts for satisfactory germination and establishment. Shelterwoods cater to species with propagules that require both adequate dispersal and microsites that provide partial shade for germination and establishment.

All four regeneration methods emulate natural disturbances at the scale of the stand. This differs from selection methods that can reflect any of the disturbance regimes and regeneration methods described, but are done at much smaller scales within a stand and are repeated continuously across the whole stand during the course of its development. Site treatments to the ground surface, which are discussed in further detail in the next chapter, also reflect the fundamental differences between severe and releasing disturbances. When the actual forest floor is disturbed, regeneration that is present can be destroyed. Severity of site treatments often match the severity of canopy manipulation for method of regeneration. The most intensive site treatments are therefore usually done in conjunction with clearcutting; the least intensive are usually done with shelterwood cutting. However, there are many exceptions.

With this classification in mind, and with the understanding that as long as the regeneration process is initiated or released, any of these regeneration methods can reflect variations that alter the structure of the young stand (snags, residuals, reserves) or that shape the pattern of the operation (groups, strips). These variations are numerous and serve to maintain a greater diversity of ecological and social values.

In summary, because of the large variety of ways to regenerate stands naturally, it is best to communicate a particular regeneration method for a stand under scrutiny by the focus of propagule origin (clearcut, coppice, seed tree, shelterwood), followed by descriptors that cover site treatment, spatial pattern of operation, and vegetation structures remaining. This might seem wordy, but it does more accurately communicate treatments of stands among foresters and to the public.

BIBLIOGRAPHY

Abbott, H. G. (ed.). 1965. *Direct seeding in the Northeast.* Mass. AES, Amherst. 127 pp.

Abrams, M. D. 1992. Fire and development of oak forests. *Bioscience,* 42:346–353.

Abrams, M. D., and G. J. Nowaki. 1992. Historical variation in fire, oak recruitment and post logging accellerated succession in central Pennsylvania. *Bull. Torrey Bot. Club,* 119:19–28.

Agee, J. K. 1993. *Fire ecology of the Pacific northwest forests.* Island Press, Washington, D.C. 493 pp.

Ashton, P.M.S. 1992a. The establishment and early growth of advance regeneration of canopy trees in moist mixed-species broadleaf forest. In: M. D. Kelty, B. C. Larson, and C. D. Oliver (eds.). *The ecology and silviculture of mixed-species forests.* Kluwer Academic Publ., Dordrecht, The Netherlands. Pp. 101–125.

Ashton, P.M.S. 1992b. Some measurements of the microclimate within a Sri Lankan tropical rainforest. *Agric. For. Met.,* 59:217–235.

Ashton, P.M.S., and G. P. Berlyn. 1992. Leaf adaptations of some *Shorea* species to sun and shade. *New Phytol.,* 121:587–596.

Ashton, P.M.S., and G. P. Berlyn. 1994. A comparison of leaf physiology and anatomy of *Quercus* (section *Erythrobalanus*) species in different light environments. *Am. J. Bot.,* 81:589–597.

Ashton, P.M.S., and B. C. Larson. 1996. Germination and seedling growth of *Quercus* (section *Erythrobalanus*) across openings in a mixed-deciduous forest of southern New England, USA. *FE&M,* 80:81–94.

Ashton, P.M.S., Gunatilleke, C.V.S., and I.A.U.N. Gunatilleke. 1995. Seedling survival and growth of four *Shorea* species in a Sri Lankan rainforest. *J. Trop. Ecol.,* 15:345–354.

Baker, H. G., and I. Baker. 1983. A brief historical review of the chemistry of floral nectar. In: B. Bentley and T. Elias (eds.), *The biology of nectaries*. Columbia, New York. Pp. 126–152

Baker, H. G., and P. Hurd. 1968. Intrafloral ecology. *Ann. Rev. Ent.*, 13:385–414.

Baldwin, J. L. 1973. *Climates of the United States*. U.S. Dept. of Commerce, Nat. Oceanic and Atmos. Admin, Environ Data Serv., Washington, D.C. 113 pp.

Bawa, K. S. 1980. The evolution of dioecy in flowering plants. *Ann. Rev. Ecol. and Syst.*, 11:15–39.

Bawa, K. S., and M. Hadley. 1990. *Reproductive ecology of tropical forest plants*. Parthenon Press, Park Ridge, N.J. 421 pp.

Bazzaz, F. A. 1979. The physiological ecology of plant succession. *Ann. Rev. Ecol. and Syst.*, 10:351–371.

Benzie, J. W. 1977. *Manager's handbook for jack pine in the North Central States*. USFS Gen. Tech. Rept. NC–32. 18 pp.

Boardmann, N. K. 1977. Comparative photosynthesis of sun and shade plants. *Ann. Rev. Plant. Phys.*, 28:315–377.

Boe, K. N. 1975. *Natural seedlings and sprouts after regeneration cuttings in old-growth redwood*. USFS Res. Paper PSW–111. 5 pp.

Brown, K. M., and C. Merritt. 1970. *A shadow pattern simulation model for forest openings*. Purdue Univ. Agr. Res. Bul. 868. 11 pp.

Burns, R. M., and B. H. Honkala (eds.). 1990. *Silvics of North America. Vol. 1, Conifers*. USDA, Agr. Hb. 654. 877 pp.

Canham, C. D. 1988. An index for understory light levels in and around canopy gaps. *Ecol.*, 69:1634–1638.

Canham, C. D. 1989. Different responses to gaps among shade-tolerant tree species. *Ecol.*, 70:548–550.

Canham, C. D., Denslow, J. S., Platt, W. J., Runkle, J. R., Spies, T. A., and P. S. White. 1990. Light regimes beneath closed canopies and tree-fall gaps in temperate and tropical forests. *CJFR*, 20: 620–631.

Canham, C. D., Finz, A. C., Pacala, S. W., and D. H. Burbank. 1994. Causes and consequences of resource heterogeneity in forests: Interspecific variation in light transmission by canopy trees. *CJFR.*, 24: 337–349.

Carpenter, S. B., and N. D. Smith. 1975. A comparative study of leaf thickness among southern Appalachian hardwoods. *Can. J. Bot.*, 59:1393–1396.

Carvell, K. L. 1979. Factors affecting the abundance, vigor, and growth response of understory oak seedlings. In: *Regenerating oaks in upland hardwood forests*. Purdue Univ., W. Lafayette, Ind. Pp. 23–26.

Chazdon, R. L., and N. Fetcher. 1984. Photosynthetic light environments in a lowland tropical forest in Costa Rica. *J. Ecol.*, 72:553–564.

Chazdon, R. L. and R. W. Pearcy. 1986. Photosynthetic responses to light variation in rain forest species. II Carbon gain and photosynthetic efficiency during sunflecks. *Oecologia*, 69:524–531.

Chazdon, R. L. and R. W. Pearcy. 1991. The importance of sunflecks for understory plants. *Bioscience*, 41:760–766.

Clark, F. B., and S. G. Boyce. 1964. Yellow-poplar seed remains viable in the forest litter. *J. For.*, 62:564–567.

Connel, J. H., and R. O. Slayter. 1977. Mechanisms of succession in natural communities and their role in community stability and organization. *Am. Nat.*, 111: 1119–1144.

Daniel, T. W., and S. Schmidt. 1972. Lethal and non-lethal effects of the organic horizons of forested soils on the germination of seeds from several associated conifer species of the Rocky mountains. *CJFR.*, 2:179–184.

Denslow, J. S. 1980. Gap partitioning among tropical rainforest trees. *Biotropica*, 12: 47–59.

Denslow, J. S. 1987. Tropical rainforest gaps and treespecies diversity. *Ann. Rev. Ecol. and Syst.*, 18:431–451.

Derr, H. J., and W. F. Mann, Jr. 1971. *Direct seeding pines in the South*. USDA Agr. Hbk. 391. 68 pp.

Faegri, K., and L. van der Pijl. 1979. *The principles of pollination ecology.* 3rd ed. Pergamon Press, New York. 343 pp.

Finley, J., Cochran, R. S., and J. R. Grace (eds.). 1983. *Regenerating hardwood stands.* Proc., 1983 Pennsylvania State Forestry Issues Conference. 241 pp.

Gautier-Hion, A., et al. 1985. Fruit characters as a basis of fruit choice and seed dispersal in a tropical forest community. *Oecologia,* 65:324–337.

Geiger, R., R. H. Aron, and P. Todhunter. 1995. *The climate near the ground.* 5th ed. Vieweq, Braunschweig, Germany. 528 pp.

Givnish, T. J. 1980. Ecological constraints on the evolution of breeding systems in seed plants: dioecy and dispersal in gymnosperms. *Evolution,* 34:959–972.

Givnish, T. J. 1988. Adaptation to sun and shade. *Aust. J. Plant Physiol.,* 15:63–92.

Glenn-Lawson, D. C., Peet, R. K.,and T. T. Veblen. 1992. *Plant succession: theory and prediction.* Chapman & Hall, London. 352 pp.

Gordon, D. T. 1973. Released advanced growth of white and red fir: growth, damage, mortality. USFS Res. Paper PSW–95. 12 pp.

Grubb, P. J. 1977. The maintenance of species richness in plant communities: The importance of the regeneration niche. *Biological Reviews,* 52:107–145.

Halverson, H. G., and J. L. Smith. 1979. *Solar radiation as a forest management tool.* USFS Gen. Tech. Rept. PSW–33. 13 pp.

Hamrick, J. L., Godt, M.J.W., Murawski, D. A. and M. D. Loveless. 1991. Correlations between species traits and allozyme diversity: implications for conservation biology. In D. A. Falk and K. E. Holsinger (eds.). *Genetics and conservation of rare plants.* Oxford University Press, Oxford. Pp. 75–86.

Harrington, J. R. 1984. Solar radiation in a clear-cut strip—a computer algorithm. *Agric. For. Meteorology,* 33:23–40.

Harris, A. S. 1967. *Natural reforestation on a mile-square clearcut in southeast Alaska.* USFS Res. Paper PNW–52. 16 pp.

Howe, H. F. and L. C. Westley. 1988. *Ecological relationships of plants and animals.* Oxford University Press, New York. 273 pp.

Jackson, L.W.R. 1967. Relation of leaf structure to shade tolerance of dicotyledonous tree species. *For. Sci.,* 13:321–323.

Janson, C. 1983. Adaptation of fruit morphology to dispersal agents in a neotropical forest. *Science,* 219:187–189.

Kozlowski, T. T. 1949. Light and water in relation to growth and competition of Piedmont forest tree species. *Ecol. Monogr.,* 19:207– 231.

Kuskela, J., J. Knusipalo, and W. Sirikul. 1995. Natural regeneration dynamics of *Pinus merkusii* in northern Thailand. *FE&M* 77:169–179.

Leck, M. A., Parker, V. T., and R. L. Simpson. 1989. *Ecology of soil seed banks.* Academic Press, San Diego. 462 pp.

Lee, D. W. 1987. The spectral distribution of radiation in two neotropical rainforests. *Biotropica,* 19:161–166.

Lee, D.W. 1989. Canopy dynamics and light climates in a tropical moist deciduous forest in India. *J. Trop. Ecol.,* 5:65–79.

Little, S., and H. A. Somes. 1961. *Prescribed burning in the pine regions of southern New Jersey and Eastern Shore Maryland—a summary of present knowledge.* Northeastern For. Exp. Sta. Paper 151. 21 pp.

Logan, K. T. 1966. *Growth of tree seedlings as affected by light intensity. II. Red pine, white pine, jack pine, and eastern larch.* Can. Dept. For. Publ. 1160. 19 pp.

Lopushinsky, W. 1969. Stomatal closure in conifer seedlings in response to leaf moisture stress. *Bot. Gaz.,* 130:258–263.

Mann, W. F., Jr. 1970. *Direct-seeding longleaf pine.* USFS Res. Paper SO–57. 26 pp.

Marquis, D. A. 1966. *Germination and growth of paper birch and yellow birch in simulated strip cuttings.* USFS Res. Paper NH–54. 19 pp.

McClure, J. W., and T. D. Lee. 1993. Small-scale disturbance in a northern hardwoods forest: effects on tree species abundance and distribution. *CJFR*, 23:1347–1360.

McDonald, P. M. 1976. *Forest regeneration and seedling growth from five cutting methods in north central California*. USFS Res. Paper PSW–115. 10 pp.

Minckler, L. S., and J. D. Woerheide. 1965. Reproduction of hardwoods 10 years after cutting as affected by site and opening size. *J. For.*, 63:103–107.

Mooney, H. A., et al. (eds.). 1981. *Fire regimes and ecosystem properties*. USFS Gen. Tech. Rept. WO-26. 394 pp.

National Research Council. 1991. *Managing global genetic resources: forest trees*. National Academy Press, Washington, D.C. 228 p.

Noble, D. L., and R. R. Alexander. 1977. Environmental factors affecting regeneration of Engelmann spruce in the central Rocky Mountains. *For. Sci.* 23:420–429.

Oliver, C. D., and B. C. Larson. 1990. *Forest stand dynamics*. McGraw-Hill, New York.

Panel on Aerial Seeding. 1981. *Sowing forests from the air*. National Academy Press, Washington, D.C. 35 pp.

Parker, J. N., and J. N. Long. 1989. Intra- and interspecific tests of some traditional indicators of relative tolerance. *FE&M*, 28:177–189.

Pijl, L. van der. 1972. *Principles of dispersal in higher plants*. 2nd ed. Springer-Verlag. Berlin.

Poulson, T. L., and W. J. Platt. 1989. Gap light regimes influence canopy tree diversity. *Ecol.*, 70:553–555.

Proctor, M. and P. Yeo. 1972. *The pollination of flowers*. Taplinger, New York. 272 pp.

Pyne, S. J. 1984. *Introduction to wildland fire: fire management in the United States*. Wiley, New York. 566 pp.

Rich, P. M., Clark, D. B., Clark, D. A., and S. F. Oberbaur. 1993. Long-term study of solar radiation regimes in a tropical wet forest using quantum sensors and hemisperical photography. *Agric. For. Met.*, 60:107–127.

Ronco, F. 1970. Influence of high light intensity on survival of planted Engelmann spruce. *For. Sci.*, 16:331–339.

Sander, I. L., and F. B. Clark. 1971. *Reproduction of upland hardwood forests in the central states*. USDA, Agr. Hbk. 405. 25 pp.

Schroeder, M. J., and C. C. Buck. 1970. *Fire weather*. USDA Agr. Hbk. 360. 288 p.

Smith, D. M., and P.M.S. Ashton. 1993. Early dominance of pioneer hardwoods after clearcutting and removal of advanced regeneration. *NJAF*, 10:14–19.

Smith, H. 1982. Light quality, photoreception, and plant strategy. *Ann. Rev. Plant Phys.*, 33:481–518.

Stoneman, G. L., 1994. Ecology and establishment of eucalypt seedlings from seed: a review. *Australian Forestry*, 57:11–30.

Swaine, M. D. and T. C. Whitmore. 1988. On the definition of ecological species groups in tropical rain forests. *Vegetatio*, 75: 81–86.

Tappeiner, J. C., and J. A. Helms. 1971. Natural regeneration of Douglas-fir and white fir on exposed sites in the Sierra Nevada of California. *Amer. Midland Nat.*, 86:358–370.

Torquebiau, E. F. 1986. Mosaic patterns in dipterocarp rain forest in Indonesia, and their implications for practical forestry. *J. Trop Ecol.*, 12:561–576.

U.S. Forest Service. 1965. *Silvics of forest trees of the United States*. USDA Agr. Hbk. 271. 762 pp.

U.S. Forest Service. 1983. *Silvicultural systems for the major forest types of the United States*. USDA Agr. Hbk. 445. 191 pp.

Wenger, K. F., and K. B. Trousdell. 1957. *Natural regeneration of loblolly pine in the South Atlantic Coastal Plain*. USDA Production Rept. 13. 78 pp.

Wheelwright, N. T., and G. Orians. 1982. Seed dispersal by animals: contrasts with pollen dispersal, problems in terminology, and constraints on coevolution. *Am. Nat.* 119:402–413.

Willson, M. F. 1983. *Plant reproductive strategy*. Wiley–Interscience, New York. 282 pp.

Wright, H. A., and A. W. Bailey. 1982. *Fire ecology, United States and southern Canada*. Wiley, New York. 501 pp.

CHAPTER *8*

PREPARATION AND TREATMENT OF THE SITE

In considering the regeneration of stands, emphasis is traditionally and logically placed on the pattern by which the preexisting stand canopy is removed. Most of the remaining chapters of this book are in fact devoted to these patterns of cutting and their associated origin of regeneration. Such emphasis should not obscure the fact that regeneration may also depend on accessory measures to dispose of debris, reduce the competition of unharvested vegetation, and prepare the soil. Sometimes it is as logical to prepare the ground for a new forest as for an agricultural field. However, it is necessary to remember that most tree species are of later successional status than most agricultural crop species, which are mainly herbaceous pioneers adapted to colonize severely disturbed soils. The risk of harm to the soil must also be considered.

Acceptable regeneration can often be obtained without deliberate site treatments, but it is important to distinguish between total absence of such preparation and the unintentional kinds resulting from logging and slash disposal or prescribed burning undertaken for fuel reduction. This is not to say that the effects of logging, slash disposal, and prescribed burning always facilitate regeneration. If reliance is being placed on established advance regeneration, it may be logical that the harvesting of the previous stand be accomplished with as little disturbance as possible.

Site treatment, whether deliberate or unintentional, may be more crucial in the establishment of regeneration than how the canopy is cut. For instance, many species can be reproduced by several different intensities and spatial patterns of overstory harvesting but by only one general program of site treatment. The important objective is to prescribe and create environmental conditions conducive to the establishment and growth of the desired species. Efforts can be wasted or damage done if techniques of site preparation are either overdone or automatically applied as ends in themselves. Some of the techniques have overlapping objectives and can also be intended for objectives other than growing trees.

Site preparation is a general term used to describe treatments applied to slash, groundstory vegetation, forest floor, and soil in order to make the site suitable for natural or planted regeneration. Most of the preparation treatments are applied during the period of establishment, but some are started well in advance of harvest cutting or applied occasionally throughout the rotation. Other site treatments that are not necessarily focused on regeneration establishment can be categorized as techniques that are concerned with (1) protecting the site from erosion or sustaining its productivity; and (2) restoring or changing a site's productivity and floristics through the use of techniques that, for example, convert a grassland into a forest or vice versa. The more important site treatments may be imperfectly divided as follows:

1. Disposal of logging slash
2. Treatment of the forest floor and competing vegetation:
    a. Prescribed burning
    b. Mechanical treatments
    c. Herbicide treatments
    d. Flooding
3. Improvement of site
    a. Fertilization
    b. Drainage
    c. Irrigation
    d. Protection

Several regional manuals have been written on site treatments for establishing regeneration in the Southeast (Duryea and Dougherty, 1991) and Oregon and northern California (Hobbs et al., 1992). These manuals provide more detailed reviews of techniques and practices used in establishing and protecting regeneration, particularly on sensitive sites.

DISPOSAL OF LOGGING SLASH

Effects of Slash on the Future Stand

The appearance of **slash**, the debris left by harvesting operations, is so offensive that it is not easy to be entirely objective about determining the appropriate extent of disposal. Slash can be simultaneously harmful and beneficial; its treatment can be very expensive, and the resulting benefits are mostly rather indirect. Consequently, the problems created by slash and other organic materials must be regarded as an integrated whole in terms of their effect on the productivity, utility, and safety of specific stands and site (Cramer, 1974; Martin and Dell, 1978; Kraemer and Hermann, 1979).

Slash in Relation to Forest Fires
Most slash disposal is still applied primarily to reduce the potential fuel for forest fires (Brown and Davis, 1973; Chandler et al., 1983; Pyne, 1984). Slash is a fire hazard because it represents an unusually large volume of fuel; it often dangerously impedes construction of fire lines. Therefore, the debris left after a harvest is potential fuel that would not be

there but for the cutting. The greatest menace exists during the short period in which the foliage and small branchlets remain on the slash; they are readily ignited and burn rapidly. The larger materials are not easily kindled and do not normally burn rapidly, although they give off large quantities of heat if ignited by fires originating in the fine fuels.

When conditions are favorable to very hot fires, the size of the units of fuel is no longer a factor limiting the rate at which a fire will spread. During bad fire weather, fires can burn rapidly in large concentrations of slash and may "blow up" into uncontrollable conflagrations. The main objective of slash disposal for fire control is the prevention of such catastrophes. Fires on cutover areas almost invariably start and spread in the litter of the old forest floor, and only secondarily spread into the logging slash. An area can be rendered temporarily fireproof through the elimination of this blanket of fuel, but the effect lasts only until the first crop of herbaceous vegetation dries out. Therefore, no method of treating the potential fuels of forest fires is a substitute for a good system of fire control.

Policies of slash disposal have been heavily influenced by the popular misconception that the danger of bad fires on cutover lands can only be stopped by destruction and elimination of logging debris. Actually, the menace of slash can be diminished through any measures that break up its continuity, protect it from sun and wind, or decrease the risk of ignition. The prevalence of fires on heavily cutover areas is therefore the result not of the presence of slash but of the desiccation of the exposed forest floor.

Effect of Slash on Reproduction

In addition to being a hazard and impediment in fire control, logging debris often hinders the establishment of reproduction. This harmful influence is caused principally by the heavy shade and injurious mechanical effect of dense concentrations of slash. In such places advance regeneration is buried or crushed, and the establishment of new seedlings is prevented by the shade (Tesch and Korpela, 1993). Slash composed of green branches is more detrimental than that composed of dead branches because of the heavier shade. The magnitude of the harmful effect depends on the proportion of the area covered as well as the thickness and density of the slash.

Thick, dense layers of slash, until broken up by decay, prevent the establishment of reproduction. Sometimes the first plants to appear on the sites of old slash piles are undesirable herbs and shrubs that may usurp the growing space for long periods. Thin, loose layers of slash, on the other hand, may be of real benefit to young seedlings by protecting them from extremes of temperature, desiccation, grazing animals, and the competition of intolerant vegetation. In fact, the complete removal of all potential slash in whole-tree logging can sometimes cause the death of the small crucial advance growth of exposure-intolerant species that have poor sprouting potential, such as the true firs, spruces and hemlocks.

Slash disposal is very often done not to influence the ecology of regeneration, but simply to reduce physical impediments to hand or machine planting.

Management of Slash, Litter, and Soil

Logging debris and forest-floor litter can be viewed as storehouses of carbohydrate energy and inorganic nutrients that can be manipulated in various ways. They can be left to decay naturally, burned, or removed from the site.

Decisions about how to treat such material depend on the extent to which unincorporated organic matter accumulates on the forest floor. Under conditions that do not favor decomposition, organic matter may accumulate and tie up nutrients to such an extent that it may take a hot fire every century or two to rejuvenate the system (Johnson, 1992; Agee,

1993). This situation can exist in the fire-ruled kinds of boreal forests in parts of Canada. At the other extreme is the true tropical rain forest where litter does not accumulate on the forest floor and almost all of the nutrient capital is continually cycling through the living vegetation (Whitmore, 1990). There one should try to conserve every scrap of dead organic matter because the organic part of the system is the only important nutrient reservoir. Most forests lie between these extremes and are places where nutrients are distributed among vegetation, forest floor, and mineral soil. However, the pattern of nutrients and energy distribution varies enough with species and site that one should be aware of the particular situation for a stand at the time of any treatment.

If these organic materials are allowed to decay naturally, most of the nutrients are ultimately returned to the soil and living vegetation. In the meantime, they are apt to remain unavailable to the vegetation. Substantial amounts of nitrogen remain bound away in the body proteins of the microorganisms responsible for the final stages of decay. The effects of immobilization of nutrients can be detrimental on infertile soil or in climates where thick layers of organic matter normally accumulate beneath the forest. If these kinds of dead organic matter decay in place, some of the energy stored in them goes to nourish the large and small soil organisms that churn it and are chiefly responsible for maintaining its good physical properties (Wood, 1989; Killiam, 1994). The resulting incorporation of organic matter into the mineral soil is important in maintaining the capacity of the soil to hold water, oxygen, and nutrients.

A layer of slash covering the soil can have beneficial effects on preventing erosion, but not as much as casual consideration might suggest. The prevention of erosion is achieved mainly by getting the water to infiltrate so that it does not run over the surface, picking up organic and mineral fragments. The root channels of living vegetation and porosity of the forest floor materials and upper mineral soil layers are chiefly responsible for the infiltration capacity of forest soils (Brady, 1990). Slash covering does help by decelerating the rate at which snow melts and by preventing rain-splash erosion. But deposition of slash over actively eroding areas is usually not very effective, except in the cases where it is used as a skid road covering to prevent deep tire ruts from developing.

Effects of Burning on Nutrients and Soil

If dead organic materials are burned, their stored energy goes mostly to heat the air and the stored chemical nutrients are released. Some, but not all, nitrogen compounds are volatilized and lost into the atmosphere. Most of the nutrient elements that are of mineral origin are returned to the soil in more readily available form than before. Sometimes the increased amount of chemical nutrient in the mineral soil stimulates the nonsymbiotic nitrogen-fixers enough that the supply of available nitrogen becomes greater than before. It is possible for some chemical nutrients, especially nitrate nitrogen and potassium, to be made mobile enough by burning to accelerate loss by leaching and surface runoff (Jordan, 1985). In most situations, however, burning either improves the chemical properties of the soil by accelerating nutrient cycling or does them little harm (Walstad, Radosearch, and Sandberg, 1990).

The effects of burning on the physical properties of the soil are either minor or deleterious. They are usually minor because most fires do not burn up all the incorporated organic matter and enough remains to continue most beneficial effects. Severe heating of the mineral soil takes place only where large concentrations of debris or logs burn for an hour or more; this incinerates incorporated organic matter and can cause severe soil damage, but usually only in small spots where the soil is baked red.

Some soils acquire **hydrophobic** properties (that is, resistance to wetting) because the soil particles get coated with waxy substances. These abound on the leathery leaves of xerophyllic vegetation such as that of the California chaparral. Hot fires on steep slopes in the coast range of California are responsible for the debris slides that plague that region, through this process of hydrophobic soil development. These effects are rarely found in other vegetation types.

Fire can damage forest soils but only under unusual circumstances such as combinations of hot fires, steep slopes, and compact soils. The effects are more commonly small or even beneficial. The removal of any organic matter from a site inevitably takes away some nutrients and stored energy. Forest systems can replace the energy easily, so that loss becomes serious only if impoverished animals or people divert almost all of it away from the soil-improving organisms. The nutrient losses are of greatest consequence because they are harder to replace. Aside from the special case of nitrogen, the available inorganic nutrients of the soil are replaced by (1) decomposition of rock minerals in the soil and (2) atmospheric fallout of dust, sea-salts, and the products of other forms of air contamination.

The crucial nitrogen compounds come and go from the huge but remarkably inert reservoir of nitrogen gas in the atmosphere; virtually none are of mineral origin. Most useful nitrogen compounds are captured from the air by nitrogen-fixing microorganisms (Gordon and Wheeler, 1983; Stacey, Burns and Evans, 1992), but they can also be fixed by lightning discharges and high-temperature combustion (Sprent, 1987). Although available nitrogen compounds are almost always in short supply and a cause for concern, they are more easily replaced by nature than the nutrients of truly mineral origin (see Table 8.1).

Effects of Removing Organic Materials

When nutrients are removed from a stand during harvesting, their replenishment comes from the relatively slow processes of rock weathering and atmospheric deposition (and biological fixation in the case of nitrogen) (Perry et al., 1989; Gessel et al., 1990). If nothing but stemwood larger than 4 inches is removed, nutrient depletion is seldom likely to be of consequence. The large stemwood is a major part of the energy storage by forests, but it contains low proportions of nutrients (Weetman and Weber, 1972). Most of the mass of wood consists of molecules of cellulose and lignin, which are composed only of carbon, hydrogen, and oxygen. Usually, the nutrients lost in stemwood harvests are roughly equivalent to the time it takes for the new stand to grow the same amount of wood. Although this balance of nutrient loss and subsequent nutrient input generally holds, there is no inherent reason why it should. It is dangerous to assume it will in cases where the rock weathering cannot act as a source of newly available nutrients, such as peat bogs (Damman, 1978; Paavilainen and Päivänen, 1995), sands composed largely of silicon dioxide (Jordan 1985; Whitmore, 1990), and the strongly leached soils of certain tropical rain forests (Vitousek and Sanford, 1986).

In the most common condition, most of the nutrients are concentrated in the leaves, twigs, rootlets, bark, and especially the litter layers of the forest. Often there is also a large reservoir in the mineral soil, but none of the treatments now under consideration affects that directly. If these materials are left on the ground (or burned in place) at times of harvest, no soil damage is likely. If they are removed for utilization or pushed too far to the side, varying degrees of nutrient depletion can result (Duyea and Dougherty, 1991).

The annual removal of the litter for fuel or agricultural mulches, common where land resources are overstrained, is one of the most damaging things that can be done to forests

Table 8.1   Probable effects of harvest and site preparation treatments on nitrogen cycling[a,b]

Practice	Effects	N-Cycle Consequences	Probable Magnitude
Stem harvest	Removes nutrients, increases soil moisture and temperature, decreases net primary productivity.	Decreases uptake	Moderate
		Increases immobilization	Moderate
		Increases mineralization	Moderate
		Increases leaching losses	Small
		Removal of N in biomass	Large
Whole-tree harvest	Removes nutrients, increases soil moisture and temperature, decreases net primary productivity.	Decreases uptake	Moderate
		Increases immobilization	Small
		Increases mineralization	Moderate
		Increases leaching	Small
		Removal of N in biomass	Large
Chopping	Crushes slash, kills some competing vegetation.	Increases immobilization	Small
		Decreases uptake	Small
Burning	Consumes slash, kills some competing vegetation, blackens soil surface, reduces acidity.	Volatilizes N	Small to large
		Decreases immobilization	Small
		Increases mineralization	Small to moderate
		Increases nitrification	Small
Root raking, blading, windrowing	Concentrate slash in rows, move forest floor and topsoil into rows, expose mineral soil, control competition.	Redistribute nutrients	Large
		Decrease immobilization	Small
		Decrease mineralization	Moderate
		Decrease uptake	Small
		Increase fire loss	Large
Disking	Mixes forest floor and topsoil, reduces compaction, increases aeration, controls competition.	Increases erosion	Moderate
		Increases mineralization	Moderate
		Increases nitrification	Moderate
		Decreases uptake	Small
Bedding	Plows soil into raised rows, creating aerobic zone adjacent to anaerobic zone.	Increases mineralization	Small to moderate
		Increases nitrification	Small to moderate
		Increases denitrification	Small
Herbicide	Inhibits competing vegetation.	Decreases uptake	Small
Thinning	Reduces density, adds slash to the forest floor, increases soil moisture and temperature.	Decreases uptake	Small
		Increases immobilization	Small
		Increases mineralization	Small
		Removes N in biomass	Moderate

[a]A small effect would involve N losses equal to less than the annual N uptake in the ecosystem; a moderate effect would involve losses equal to several times annual N uptake; and a large loss might be 10 times or more greater than annual uptake.

[b]Expanded from Vitousek, Allen, and Matson (1983).

Source: Reprinted from D. Binkley, *Forest Nutrition Management,* © 1986, by permission of John Wiley & Sons, Inc.

and their soil. This activity is particularly prevalent in the oak-pine types of the Himalayas and the central mountains of Mexico (Thadani and Ashton, 1995). It leads not only to nutrient depletion but also to serious erosion because the food supply of the soil-improving organisms is diverted away from them (Brady, 1990). It is possible to develop guidelines for sustainable levels of litter removal in specific forest types, as is being done for collection of slash pine litter (called "pinestraw") in Florida, for use as a landscaping mulch (Duryea and Dougherty, 1991).

Some of the same kinds of problems could result from very close utilization of forest production by whole-tree chipping for fuel, pulp, and animal food (Vitousek and Matson, 1985). These purposes do not require removing the litter, and there is little use in removing green leaves except for their food value. The extent of any potential problems depends on how much of which plant structures are removed and how frequently. Need for replacement of nutrient losses by fertilization may be anticipated, although at rates far less than those common in agriculture.

The problem of nutrient depletion during harvesting can be avoided by trying to leave the leaves, twigs, and small roots where they grew (Vitousek and Matson, 1985; Weetman and Weber, 1972). Their high content of water and poor fiber characteristics make them more valuable for their nutrient content than for fuel or pulp. Their inclusion in chipping processes is often only an attempt to avoid the high cost of delimbing. Of course, it is also theoretically possible to bring some of the nutrients back to the forest in the form of ash from burned wastes.

Timber harvesting is very much less depletive than most forms of agriculture and grazing. Furthermore, fire and, for that matter, herbicides are kinder to the soil than almost any mechanical treatment. These observations are contrary to popular intuition, but they are true so long as forest systems are understood and not pushed beyond their limits (Stone, 1973).

Effects of Burning Forest Fuels on the Air

The most important problems associated with smoke from forest fuels derive from unburned particles that make it a source of dirt and restriction to visibility (Southern Forest Fire Laboratory, 1977). Therefore, air quality is better when fuels are dry and combustion is quick and complete. Furthermore, the conditions conducive to such good combustion are usually those in which smoke columns rise quickly so that the pollutants are soon dispersed thinly in the atmosphere. The more rapidly the air temperature decreases with height, the better is this kind of vertical dilution. If there is a temperature inversion (that is, a situation in which warm air has settled atop cooler air), the smoke will accumulate beneath an otherwise invisible ceiling formed by the warm air.

Unfortunately, the atmospheric conditions that dilute the smoke most rapidly are the same as those most conducive to the dangerously rapid spread of fires. Therefore, compromises have to be made, and these reduce the number of days when silvicultural burning is possible. Greatest care is needed where the spreading of fire would do great harm or where the spreading of smoke would cause dangerous restrictions of visibility along highways or at airports. Sometimes special permits are required for such burning. It is at all times necessary to coordinate the operations with the best available knowledge about how the weather conditions will affect the behavior of fires and the smoke from them. Small amounts of noxious gases, mostly carbon monoxide, are produced by burning forest fuels, but they are so quickly diluted that there is little evidence of harm caused by them. As is the case with the dirt from unburned particles, production of noxious gases arises from poor combustion. In this respect, cool, smouldering fires are more harmful than hot, vigorous ones.

Most of the smoke generated from forest fuels comes from wildfires or land-clearing fires. Many forests grow in seasonally dry climates in which it is not a question of whether the forest fuels will burn but of when and under what conditions. The pollution effects and other kinds of harm will be least if the burning is conducted at deliberately prescribed times rather than determined by accident or human incendiaries.

Both burning and decay of organic substances from the forest and elsewhere change the carbon of those substances into atmospheric carbon dioxide. There is concern that continuing increases in the amount of this gas in the atmosphere will block so much outgoing radiation as to cause significant climatic warming. Unfortunately, it does not appear that the increased carbon dioxide goes entirely toward making the world's vegetation grow faster because the amount in the atmosphere is increasing. There is a net transfer of carbon to the atmosphere when forests are destroyed and not replaced with new ones (Dale, Houghton, and Hall, 1991). However, forests that are actively accumulating wood remove carbon dioxide from the air. Wood put to structural use continues to sequester carbon. Silviculture, especially that aimed at timber production, is not a cause of this problem but is really one of the very best cures available (Burschel et al., 1993; Trexler, 1991; Hendrickson, 1990).

Slash in Relation to Insects and Fungi

The insects and fungi that feed on logging debris are more beneficial than harmful because they are primarily responsible for the disintegration and decay of unburned slash. The vast majority of them are scavengers or saprophytes that do not attack living trees. The few species of bark beetles and heart-rotting fungi that can spread from slash to living trees are found mostly in the cull logs, stumps, and large branches that are rarely eliminated in conventional slash disposal. The injurious fungi that proliferate in large pieces of slash are, for example, those that had already infected the living trees. Some of them may produce fruiting bodies more abundantly after cutting than in living trees. The treatment of slash to control insects and fungi is, therefore, best accomplished by such measures as close utilization, directed application of insecticides, and indirect methods of combating the proliferation and spread of harmful organisms through such techniques as the use of pheromone traps.

Ordinary slash-burning does not necessarily have any effect on the situation and may even aggravate it by killing unmerchantable, standing trees. The most important aspect of the influence of insects and fungi on slash is the rate at which they bring about disintegration. It may take anywhere from several years to several decades for slash to decompose, depending on climate and species. The question of how much time is required has an important bearing on whether or how to treat the slash and also on the planning of subsequent operations.

Hardwood slash, for example, tends to remain moist and decays so rapidly that it is seldom necessary to burn it. Slash generally decays more swiftly on good sites than on those that are extremely wet or dry. Partial shade is conducive to decomposition. For more detail on this topic, refer to Chapter 19 which discusses issues concerning maintenance of forest health.

Slash in Relation to Harvesting Operations

Efficiency in log transportation depends to a large extent on the success with which the slash is concentrated during felling. The interests of both logging and silviculture are usually best reconciled by concentrating the slash in a large number of small compact piles or in long, narrow strips. The consolidation of slash into a few very large piles is as detrimental to silvicultural purposes as the diffuse scattering of debris is to efficient transportation of logs. There are occasions when the slash from one cutting may remain long enough to impede subsequent operations.

The amount of logging residue and the diameters of its components depend on the extent of utilization of the felled trees. Large, untrimmed tree-tops are especially detrimental. They cover much space, and their loosely arranged twigs and foliage allow fires to travel rapidly. Increased intensity of utilization reduces the amount of large debris; moreover, it almost automatically ensures that any fine fuels that remain are left more compact and closer to the ground so that they burn more slowly and rot faster. Debris can also be distributed on trails to reduce compaction when skidding on soils that are wet or fine textured.

Slash in Relation to Aesthetics and Wildlife

Logging debris is so unsightly that it is desirable to dispose of it or refrain from cutting in recreation areas and along public ways (Jones, 1993). Because of this and also to help with fire control, laws often require slash disposal within specified distances from highways, railroads, habitations, and adjoining properties.

In general, slash should not be deposited in waterways because its decomposition may reduce the oxygen content of the water below the level required for many species of fish. However, in some situations it is possible to overdo these precautions. These effects are described in Chapter 20, which covers wildlife habitat.

Methods and Application of Slash Disposal

Tremendous advances in fire control and more complete wood utilization have caused slash disposal to be one American forestry practice that becomes less important with the passage of time. When it is practiced today, it is increasingly for the purpose of facilitating planting or natural regeneration.

Slash disposal for fuel reduction is common practice in western North America, which has long, rainless summers and old-growth forests with many rotten trees. Most of it is done by cheap broadcast burning (Chandler et al., 1983; Walstad, Radoseurch, and Sandberg, 1990). In this region and others, efforts are often limited to eliminating slash in narrow bands along travel routes or at intervals across cutover areas in order to break up concentrations and provide places for fire-line construction in the event of need.

In most other parts of North America, the disposal of slash for fire control has generally been less necessary. In the northern coniferous forests, limited kinds of slash disposal are sometimes desirable. In the southern pine forests, slash rots so quickly that it rarely requires treatment, but it is sometimes used anyway, mainly for the purpose of preparing the seedbed for regeneration. In the eastern hardwood forests and in the Northeast, slash also rots quickly enough that disposal is required only along roads and in similar places.

Broadcast Burning of Slash

In broadcast burning, the slash on clearcut areas is burned as it lies within prepared fire lines. Practically all the remaining vegetation, except for that of sprouting species, is destroyed. This precludes reliance on advance reproduction but eliminates much of any undesirable vegetation that may be present. The extent of exposure of mineral soil is actually rather variable, depending on the moisture content of the forest floor at time of burning. Usually an ample amount of mineral soil is exposed; the areas where fires burn with such sustained heat as to damage the physical properties of the soil are rarely large enough to be of much significance. The sites are left in reasonably good condition for hand planting, direct seeding, or natural seeding from adjacent stands.

Broadcast burning is often associated with the clearcutting of old-growth stands in the West (Fig. 8.1). It can also be applied where scattered seed trees have been reserved, provided that fuel is removed from beneath the trees. Unless some source of wind-dispersed seed is close by, broadcast burning is usually compatible only with artificial regeneration. It is usually important to set the fires quickly under just the right conditions and in patterns that will cause the fires to burn away from the edges. The technique can be expedited by dropping incendiary devices, first developed for use in Australian eucalypt forests, from aircraft.

Spot Burning

If the fires can be depended upon not to spread dangerously, it may suffice to limit broadcast burning to concentrations of slash in spots or patches. Sometimes this is a way of conserving seeds that have already fallen or of reducing any baneful effects of slash burning.

Figure 8.1 Broadcast burning in preparation for planting after the clearcutting of an overmature stand of the western white pine type, Deception Creek Experimental Forest, Idaho. Note the large volume of defective grand fir and western hemlock felled before the burning. If this had been left standing, the resulting dead snags might have become ignited in wildfires and spread burning embers far and wide. *(Photograph by U.S. Forest Service.)*

Burning of Piled Slash

Slash disposal associated with partial cutting usually involves the burning of piled slash. Where much of this sort of work is necessary, it is now ordinarily done by pushing the slash into piles with bulldozers or similar equipment for subsequent burning. Use of such equipment has the important incidental advantage of reducing the competing vegetation and exposing the mineral soil on the treated areas. This effect makes an important contribution to the success of natural regeneration or planting where competition from brush is a serious problem after partial cuttings as in the mixed conifer types of the Sierra Nevada (Fig. 8.2). The older methods of expensive hand piling have little silvicultural effect other than freeing advance regeneration and exposing some mineral soil. As far as the techniques are concerned, distinction is drawn between **progressive burning** in which the slash is laid on piles as the burning progresses and **piling and burning** in which the piles are constructed well in advance of the time when it becomes safe enough to burn them. Piling of slash without burning is often useful for maintaining certain wildlife species that require protection and denning/nesting places under circumstances that otherwise would be open. If such work is done by hand, the cost can be kept in check by planning to move the material over only short distances to many small piles. However, the burning of small

Figure 8.2   Dense, 6-year-old natural regeneration of ponderosa pine resulting from very intensive site preparation at Blacks Mountain Experimental Forest on the eastern slope of the Sierra Nevada. Most of the slash was piled mechanically in windrows along the edge of the opening, which was created in a group-selection cutting. During the next good seed-year the area was disk-plowed and the rodent population was reduced by poisoning. *(Photograph by U.S. Forest Service.)*

piles of slash can be very costly, partly because workers easily become mesmerized watching the fires.

Mechanical piling is well adapted to the rather open forests of the interior ponderosa pine type, provided that the terrain is not too steep or rocky. The opportunity to dispose of large material in which bark beetles might breed represents a highly important advantage as far as management of ponderosa pine is concerned. Mechanical piling has also been used after clearcutting in the lodgepole pine type, and it leads to fully as good fuel reduction and to better reproduction than broadcast burning.

Lopping and Scattering of Slash

Some objectives of slash disposal can be accomplished without burning. The fire hazard can be reduced simply by lopping the tops so that the severed branches lie closer to the ground. Lopping can be advantageous in freeing saplings and seedlings of advance growth that have been bent over by falling tree-tops. A limited amount of this kind of work often obviates the necessity for planting or waiting for new natural reproduction and forestalls subsequent difficulties with malformed trees. Bent trees should be released as promptly as possible; they straighten up much better at the beginning of the growing season than if they are not given an opportunity to do so until growth has been going on for several weeks.

Although lopping is one of the cheaper forms of slash disposal, anything that involves scattering or moving slash can be rather expensive. Lopped slash is sometimes scattered in instances where any sort of burning would destroy too much advance reproduction. The deliberate scattering of slash provides an artificial means of seed dispersal for closed-cone conifers. It is also possible to break up concentrations of slash or eliminate it in strips along travel routes simply by moving it. None of these techniques of redistributing slash reduces the total amount of fuel. However, they are often more compatible with silvicultural objectives than methods involving burning, especially in situations where the risk of wildfire is not very great.

Direct Control of Bark Beetles in Slash

If the hazard of outbreaks of bark beetles is serious and the utilization cannot be close enough to eliminate the felled material in which they breed, it may be most expedient to spray the slash with insecticides. The best time for spraying the bark comes just before the normal time of egg laying. Other techniques of direct control are mentioned in the section on salvage cutting (Chapter 19).

Chipping and Yarding of Slash

Portable chipping devices are sometimes used where slash disposal is essential and equally expensive piling and burning is the only alternative or when burning cannot be done. Such devices are, of course, most useful when the chipped material can be utilized for pulp, fuel, or mulching road and trail surfaces. Slash disposal problems can sometimes be mitigated by transporting large unmerchantable material along with merchantable logs to central landings. The resulting concentrations can be either burned at these points or stored where they will be accessible when market conditions will permit their utilization. This technique of ''yarding unmerchantable material'' is sometimes used on newly acquired forestland on which the owner is focused solely on timber production. Under the circumstances, there may be only one opportunity ever to move large, defective trees out of the woods.

TREATMENT OF THE FOREST FLOOR
AND COMPETING VEGETATION

It has already been stated that the establishment and development of a new forest crop can take place only if sufficient growing space is made available by harvesting or killing all or part of the preceding crop. Similarly, any sort of regeneration from seed is affected by the condition of the seedbed. Basic requirements cannot always be met by judicious adjustment of the pattern of cutting, by intentional or unintentional disturbance in logging, or by treatments of undesirable vegetation that are delayed until after the reproduction is established. Deliberate measures such as prescribed burning, mechanical site preparation, or herbicide applications are sometimes necessary to create the appropriate environmental conditions for natural or artificial reproduction.

Seedbed Preparation

The preparation of seedbeds ordinarily involves treatment of the **forest floor**, which is the layer of unincorporated organic matter that lies on top of the mineral soil and is composed of fallen leaves, twigs, and other plant remains in various stages of decomposition. This material does not make a good seedbed for most small-seeded species. Mineral soil can be exposed by burning or mechanical scarification. Too much exposure can invite excessively dense regeneration or invasion of undesirable pioneer vegetation. In many situations where growing space is not immediately occupied by native vegetation, more opportunistic exotics often colonize and impede succession (Cohen, Singhakumara, and Ashton, 1995; Cheke, 1979). Often the exposure of only a small patch every meter or two is ideal; machines have been developed in Scandinavia for this very purpose.

Competing Vegetation

The harvesting of trees inevitably leaves some vacant growing space, at least temporarily. Even after a very heavy cutting, however, there may be more serious competition, existing or potential, from unwanted vegetation than might be inferred from outward appearances or the severity of the cutting. Unless the markets are good, there is certain to be a residue of trees that were not worth cutting. There may also be low vegetation consisting of grasses, herbaceous plants, shrubs, or advance growth of undesirable tree species. If no vegetation shows above ground, there may still be root-stocks of sprouting species. Finally, if little or no vegetation remains and if the site is reasonably favorable for plants, the way is open for invasion by whole armies of plants. The most serious problems arise when the competing vegetation has existed for many years and is not worth harvesting at all. The low shrubs of brushfields and the grasses or other herbaceous growth of ''open'' lands do not cast as much shade as a closed forest but can cause even more root competition as far as seedlings are concerned (Hill, Canham, and Wood, 1995; Putz and Canham, 1992). Consequently, preexisting vegetation is most likely to have to be controlled when it is so worthless that artificial regeneration is necessary (Stewart, Gross, and Honkola, 1984).

Grass, and other low vegetation, such as ferns, hamper the growth of trees more than might be inferred from their comparatively short stature (Larson and Schubert, 1969; Horsely, 1977). Last year's grass or fern, lying brown and battered on the ground at planting time, can be insidiously deceptive and should be assessed in terms of the height it attains. Tall, dense grass and fern often compete with tree seedlings enough to reduce

survival. Grass competition is almost always detrimental if there are serious seasonal moisture deficiencies (Perry et al., 1994). It is, for example, the chief cause of the "grass stage"
of longleaf pine and of "check" in planted spruce, conditions in which seedlings grow
little or not at all in height until, after some years, they are able to develop large enough
root systems. In certain circumstances where the forest canopy is deciduous or has allowed
more light to the forest floor because of continuous canopy disturbance, woody evergreen
shrubs can dominate the groundstory. Their competitive effects can inhibit establishment
of advance growth of canopy tree species (Ducey, Moser, and Ashton, 1996; Lorimer,
Chapman, and Lamker, 1994; Tesch and Hobbs, 1989; Phillips and Murphy, 1985).

Not all of the effects of grass or other inhibiting vegetation are from competition for
light, water, and nutrients. There is a growing body of evidence about **allelopathic effects**,
which are chemical antagonisms between different species that enable one species to poison the progeny of other species or, sometimes, its own (Daniel and Schmidt, 1972; Ward
and McCormick, 1982; Rice, 1984). The effects of competing vegetation can be useful as
well as harmful. Eastern North America has vast areas of unnaturally pure stands of various
pines and spruces that spontaneously reforested abandoned grassy fields and are testimony
to the ability of grass to exclude many hardwood species (Fig. 8.3) (Larson, Patel, and
Vimmerstedt, 1995). Some of this was the result of selective grazing, but much of it was
probably from the competition and allelopathic effects of grass. These effects of grass can
be great enough that planted stands of hardwoods may have to be cultivated like row-crops
during the first year or two.

In the southeastern United States, the annual and then perennial grasses that normally

Figure 8.3   An old-field stand of shortleaf pine invading abandoned agricultural land in
the Arkansas Ozarks. Because the grass inhibits broadleaved species more
than the conifers, the new pine stand will be unnaturally pure. *(Photograph
by U.S. Forest Service.)*

appear after burning or mechanical site preparation probably serve to inhibit hardwoods. This is on areas being treated to regenerate pine, so the effect is beneficial. This effect has been deliberately used to inhibit reinvasion by the native angiosperm forest in the Jari Valley of Amazonian Brazil after it was clearcut and replaced by planted Caribbean pine (Evans, 1984).

More generally, the establishment of forest trees is often assisted by the temporary protective effects of other vegetation, especially the herbaceous kinds. Such cover may be essential in preventing damage by heat or frost when the tree seedlings are very young and succulent. The theoretically ideal kind of accessory vegetation would be that which protected but did not compete; mosses, which are not capable of pulling water and nutrients from the soil, sometimes come close to this ideal. The next best would be plants that died or became overtopped by young trees promptly after their protective effects were ended. Even this does not always help. The insulating effects of a grass cover can, for example, cause frost damage to hardwoods after they emerge and are exposed to the much colder air above it.

It must be fully anticipated that any lethal disturbance done deliberately or unintentionally in the process of forest regeneration will cause the appearance of some kinds of vegetation other than the species desired. Treatments that expose the mineral soil to sunlight inevitably call forth pioneer vegetation. The forester who carries out such treatments must have full knowledge of what this kind of vegetation will be and be ready to live with the consequences. Any treatments that eliminate all or most of the preexisting vegetation usually fit best with the regeneration of species that are naturally adapted to follow fires. If the object is to regenerate shade-tolerant species not so adapted, such treatments can bring in so much undesirable vegetation as to be worse than useless.

The main point is that one should know the kind of vegetation that will develop after any kind of treatment and conduct the chosen treatments in the light of this knowledge. If the preexisting vegetation is going to hamper the establishment of a new stand of trees, it is usually best to eliminate as much of it as necessary before the regeneration step than to plan to temporize with it afterward. Except for broadcast herbicide spraying, most kinds of selective weeding treatment are costly and cumbersome. In the past, it has often been impractical to provide young trees with anything approaching complete freedom from competition. Herbicides and heavy mechanical equipment have now provided the power to kill the roots of competing vegetation. The spectacular increases in seedling growth that can be attained by such treatment have been observed in a wide variety of species and regions.

Techniques of Treatment

Prescribed burning, mechanical site preparation, and flooding are techniques in which both the forest floor and competing vegetation can be treated simultaneously. Before considering these three methods in detail, attention is called to other means of accomplishing the objectives. Some scarification of the mineral soil and reduction of competing vegetation are accomplished during logging and any disposal of slash that is undertaken. Although the resulting disturbance is often adequate, it seldom exposes any substantial amount of mineral soil or eradicates much sprouting vegetation. Logging does not result in very complete scarification unless there is a deliberate attempt to skid almost every log over a different pathway on snow-free ground. The one kind of slash disposal most likely to achieve complete site preparation is that in which the slash is piled with bulldozers equipped with root rakes or similar equipment. Broadcast burning often eliminates most of the unincorporated organic matter but does not necessarily reduce sprouting vegetation very much. Practically all the other methods of logging or slash disposal have limited and erratic effects.

There is no silvicultural treatment other than those mentioned above that significantly interrupts the continuity of the forest floor, but there are a number of ways of attacking the competing vegetation. Most of these, such as cutting, girdling, and chemical treatment, were considered in Chapter 5. Cutting and girdling are, like fire, entirely adequate for species that do not sprout but are rather frustrating to employ against those that do. Herbicide treatments can be used to eliminate almost all vegetation from most kinds of sites but not always cheaply or with the kinds of species selectivity one might desire. The combination of two or three methods of plant killing in sequence, but with each treatment of limited intensity, is often cheaper and less harmful than reliance on use of one treatment.

Grazing and browsing animals are sometimes selective enough in their feeding habits to cause moderate and temporary reduction of competing vegetation. Most of the grasses and other plants on which they feed are capable of sprouting. Feeding that is heavy enough to reduce the competing vegetation substantially is often associated with effects harmful to tree seedlings. Nevertheless regulated herds of domestic animals can sometimes be used to control undesirable vegetation without suffering from malnutrition.

Prescribed Burning

Fire, like cutting, can be used both constructively and destructively in handling the forest. The practice of using regulated fires to reduce or eliminate the unincorporated organic matter of the forest floor or low, undesirable vegetation is called **prescribed** or **controlled burning**. The burning is conducted under such conditions that the size and intensity of the fires are no greater than necessary to achieve some clearly defined purpose of timber production, reduction of fire hazard, wildlife management, or improvement of grazing (Chandler et al., 1983; Fahnestock, 1973).

These particular terms, the first of which is preferred, have been coined to distinguish the use of fire as a silvicultural tool from its application for purposes bearing little relationship to the maintenance of productive forests. Very similar types of burning have been traditionally employed in many localities, especially in the southeastern United States, to keep the forest open enough for grazing or other uses. For purposes of this discussion, prescribed burning is regarded as involving fires that are set to burn through fuels that naturally occur on the forest floor, usually under existing stands. The burning of slash involves hotter fires and much heavier concentrations of fuel so that it is a kind of treatment easily recognizable as being in a class by itself. However, it could be and often is regarded as a form of prescribed burning.

Purposes and Effects of Prescribed Burning

The fact that a species is adapted to fire does not necessarily mean that fire has a practical, safe, and feasible place in its silviculture. Various cutting practices, herbicidal and mechanical treatments, or other kinds of disturbance can be used to simulate the effects of fire. However, it is the most common of the regenerative disturbances of natural forests, its application is usually cheap, and its silvicultural role needs more use and understanding than it usually gets. When properly done in appropriate situations, burning accomplishes a number of things.

The most common objective of prescribed burning is still **fuel reduction**. In many respects, it is a more satisfactory method than conventional slash disposal because it eliminates most of the readily inflammable fuels rather than just the debris left from logging. It is, however, important to note that prescribed burning does not render any area fireproof, except temporarily, and is, therefore, no substitute for a well-developed system of fire control.

The most important effect of prescribed burning in fuel reduction is the interruption of the horizontal and sometimes the vertical continuity of inflammable materials. The interruption of any vertical curtain of fuel is especially significant in the slash pine type of the Southeast and the oak-pine type of southern New Jersey. Areas of slash pine that have not been burned for a decade or more develop a tall understory of various inflammable shrubs that become draped with fallen pine needles. In fuels of this kind, surface fires can rapidly develop into disastrous crown fires because there is a ready path for the flames to follow from the ground up into the crowns. In the oak-pine type of New Jersey, the presence of a shrubby understory tends to create the same dangerous condition. In each of these regions, successful silviculture depends heavily on the prevention of crown fires and it has been shown that prescribed burning is the only dependable means of forestalling them.

Prescribed burning can be effective in **preparation of seedbeds** for regeneration of wind-disseminated species like the birches, poplars, and pines, which become established most readily on bare mineral soil. This effect is also beneficial in hampering reproduction of heavy-seeded and, therefore, relatively more poorly dispersed species (e.g., maple and beech) that are also prone to stem-girdling from fire after their establishment. These species would ordinarily displace many of the pines in the course of natural succession.

Prescribed burning is also a means of achieving **control of competing vegetation**. This often has the effect of arresting natural succession by killing understories representing stages later than the one desired. However, where the aim is to prevent invasion by hardwoods, as in stands of southern pines, burning must be done fairly often because only the seedlings are likely to be killed by fire; saplings will resprout, and larger trees may only be scarred. If soil moisture is a seriously limiting factor, elimination of the understory may improve growth of the overstory substantially, such as for the Ponderosa pine type (Sutherland, Covington, and Andariese, 1991). Increases of as much as 25 percent have been observed (Zahner, 1955) on fine-textured soils in southern Arkansas, where the rainfall is much less than farther east.

Roots of perennial grasses are killed by prescribed burning only where there are large units of fuel, such as fallen snags or large chunks of wood, that ignite and burn for a long enough period to heat the soil in depth (Weaver, 1951). Both in nature and in practice, surface fires must occur quite frequently if they are to be very effective in keeping brush and other understory vegetation in check. Fire is really a rather cumbersome tool for accomplishing this purpose, and it is fortunate that it can now be supplemented or replaced by herbicides.

The most traditional use of fire in forests is for the **improvement of grazing** for livestock, and for stimulating herbaceous species and sprouts of woody plants for the **improvement of wildlife habitat**. These applications are discussed in Chapter 20.

Burning can be employed in **recreation management** to maintain a parklike appearance in stands that would otherwise develop understory jungles. Large areas of southern pine forests, which now have understory tangles of shrubs, small trees, and briars, were more pleasant places in the era of frequent burning to promote grazing. Openness in the lower strata also facilitates forestry operations.

Prescribed burning has sometimes been successfully used to achieve the effects of **low thinning** in sapling stands of ponderosa and southern pines (Sutherland, Covington, and Andariese, 1991). However, it is very difficult to strike the right balance between fires that kill too much and those that kill nothing.

The various effects of fire can probably be used more than they are for **control of pests**, although it aggravates problems with some. Most of the effects are subtle and indirect. One decisive role is in the direct control of the defoliating brown-spot fungus of longleaf pine, to be considered later. Burning also sometimes reduces problems with annosus root rot.

Potential Damage from Prescribed Burning

Fires started by incendiaries, accidents, or lightning cause so much damage to forests that it is not easy to reconcile prescribed burning with efforts to educate the public about fire prevention. Sometimes the fire-prevention advertising exaggerates the effects of fire and thus plants the seeds of future trouble.

Prescribed burning has its greatest usefulness, and is indeed used the most, in those very regions where difficulties with fire are the greatest, simply because valuable species that are naturally well adapted to fire are most likely to occur there. In many respects, the use of fire in slash disposal and prescribed burning is actually a concession to the inevitable in localities where fires are discouragingly common and fuel reduction almost essential.

As already indicated, only exceptional kinds of soils are harmed by fire. Prescribed burning is useful mainly in forest types where natural fires are common. If burning had any seriously harmful effects on the soils involved, they would probably already be ruined. Successful prescribed burning generally makes fires either less severe or less frequent than they were before the initiation of comprehensive programs of fire control and silvicultural management. The harmful effects of fire are much more likely to take the form of obvious and long-enduring effects on the vegetation than of subtle damage to the soil (Lutz, 1956). However, in some circumstances fire has been used too frequently and soil moisture conditions have been affected by reduced organic matter and changes in soil porosity (Boyer and Miller, 1994).

The effects of prescribed burning on standing timber depend on the size of the trees and the extent to which their stems and crowns are heated by fires. Because the primary source of fuel lies on the ground, the damaging effects ordinarily take the form of complete or partial heat-girdling at the ground line. In hotter fires, the effects extend farther up the trees, especially on the leeward sides of the stems where heated air accumulates. If there is a large amount of fast-burning dry fuel on the ground, there is risk that enough heat will be generated to support burning or overheating of the foliage. No North American species other than longleaf pine can withstand burning until rough, thickened bark has developed on the lower parts of the stems; the main portion of the crown canopy must also be well above the height of the flames.

The extent of injury to any part of the tree depends on whether the living tissues are heated above the lethal threshold of 55°C and how long such a temperature is maintained. Therefore, any factor that hastens the transport of heated air out of the forest reduces the danger of damage. On level terrain it is, for this reason, better to conduct prescribed burning when there is a gentle breeze than in calm weather; however, on pronounced slopes sufficient updrafts develop to allow burning when there is no wind.

The initial temperature of the living tissues is also important in determining the highest temperature that they attain. The risk of injury is lowest when burning is done in the winter because it takes such a large amount of heat to raise the tissues to lethal temperatures. **Head-fires**, which travel with the wind, are often less damaging than **back-fires**, which burn against the wind, because the heat is carried upward more rapidly and high temperatures are not as long sustained close to the ground. However, head-fires are more likely to spread to the crowns and should thus be avoided where the crowns are likely to be damaged.

Prescribed burning is applicable only in stands composed of trees that have reached fire-resistant size. It is not easily used in uneven-aged stands because it damages reproduction. The only way in which controlled burning can be applied in uneven-aged stands is to have intervals between episodes of cutting and burning no shorter than the time required for reproduction to appear and develop resistance to fire.

The surest way to avoid damage to the forest from prescribed burning is to conduct it under the right conditions according to carefully developed plans. Like all silviculture operations, it is not an end in itself but a means of achieving some well-defined goal of management.

Methods of Prescribed Burning

Burning prescription is the term used by fire managers for the plan that defines the appropriate climatic conditions and burning methods desired for a stand. The customary prescription in conducting a prescribed burn is to isolate the area to be treated by means of plowed lines wide enough to contain the fire. Because the construction of these lines is costly, every reasonable advantage should be taken of preexisting barriers such as roads and swamps. The whole operation should be carefully mapped out with special regard for the selected conditions of wind and weather under which the burning is to be conducted (U.S. Forest Service, 1971). The fire-lines should be plowed out in advance according to this plan, but not so early that they are covered over with fallen leaves before the burning is done.

The most common practice is to use back-fires (Fig. 8.4). **Flanking** or **quartering fires**, which are set in lines parallel to the wind, spread more rapidly but are somewhat more likely to scorch the foliage of saplings or taller trees. Head-fires can be used when the forest floor is too moist to be burned by back-fires, thus increasing the number of days when treatment can be conducted. They are in some respects safer than back-fires because they are more likely to go out than to escape from control if the wind shifts unexpectedly. The heat from them is carried away rapidly so that they are less likely to produce fire scars on the stems of trees than are back-fires. If head-fires can be used safely, they are cheaper than back-fires because they can cover an acre much more rapidly.

The area to be treated should be subdivided into units no larger than can be burned in a period of about 10 hours, although several units may be burned simultaneously. It is unwise to let a fire run for a longer period because of the possibility of unforeseen changes in wind and weather.

Weather forecasts and measurements of fire danger should be used to select times for burning that are both feasible and safe. The wind should be steady in direction and not erratic or too strong in speed. The moisture content of both fuel and soil is important not only for safety but also for determining whether the right intensity of burning will be secured. Fires that merely smoulder are costly to apply and cause much air pollution.

The burning is best planned and conducted by the same personnel involved in control of wildfires. Much can be learned about the behavior of wildfires from prescribed burning. There should be enough personnel and fire equipment at hand to deal with any fires that escape.

The cost of prescribed burning depends on the size and shape of the units burned, the

Figure 8.4   Prescribed fire backing into the wind through needle litter and saw palmetto beneath a flatwoods stand of slash pine. *(Photograph by U.S. Forest Service.)*

inflammability of the fuels, the length of fire-line needed, the size of the crew, and the shape of the terrain. If the units are more than 25 acres (10 ha), the costs are usually very low. The cost of fire-lines is the chief variable. High cost may result if there is a heavy accumulation of fuel requiring a series of small fires or if there is so little fuel that the fires have to be relighted often. The costs of burning small tracts, areas with irregular boundaries, or narrow strips are rather high. However, if large areas can be treated in single operations, several burnings can be done for the cost of one broadcast spraying with herbicides. If the undesirable vegetation is small, such a series of burns may be almost as effective in controlling the vegetation and also reduces the amount of fuel very substantially.

Application of Prescribed Burning

The details of prescribed burning vary according to the species and objectives of treatment, the most important variable being the schedule of burning. Longleaf pine is one important species that is ecologically almost totally dependent on frequent surface fires (Grelen, 1978). The seedlings germinate in late fall and thrive best beneath the one year's growth of grass developing after a previous winter burn. Until they form the first dormant terminal bud about nine months later, they can be killed by fire. They stay in the peculiar grass stage, without growing in height, until the large tap-root becomes about 1 inch in diameter; a sudden spurt of height growth ensues with a brief period of moderate vulnerability to fire. During the grass stage, it is necessary to burn about every three years to keep the brown-spot disease in check. The spores that spread this disease can move so far by rain-splash that it is desirable to operate with even-aged stands and regeneration areas of about 250 acres (100 ha). Once the seedlings have become more than 3 feet (1 m) tall, frequent burning remains desirable, but the benefits become the same as with the other southern pines.

The most common purposes of prescribed burning under stands of southern pines are fuel reduction and the control of understory hardwoods that continually threaten to take over the stands (Fig. 8.5). The South has many dry soils on which hardwoods grow fast during the juvenile stages and tend to overwhelm conifers that only later manifest their superior adaptation to such soils.

It may take rather frequent fires to keep hardwoods under control because only the seedlings are killed outright. Saplings usually resprout and can thus be killed back only to the ground. Larger trees have to be killed with herbicides or by girdling or cutting. One

Figure 8.5   The upper photograph shows an untreated 45-year-old stand of loblolly pine with a dense hardwood understory at the Santee experimental forest on the South Carolina coastal plain. The lower photograph was taken on the same spot 6 years later, just after the second of two partial cuttings and after four prescribed burning operations. *(Photograph by U.S. Forest Service.)*

common sequence for eliminating degraded hardwood stands is (1) burning, (2) herbicide spraying of the resulting succulent sprouts, and (3) killing the large survivors by girdling or herbicide injection.

If large amounts of fuel have accumulated from long periods without fire, it may take several light winter fires to skim off enough successive layers to make it safe enough for fires that really favor regeneration. The fires that foster regeneration are often summer fires capable of exposing some mineral soil and killing small hardwoods (Fig. 8.6). Summer fires involve no risk of destroying any seeds that have fallen. Fires can be of many different combinations of timing and intensity; the choice of these can be a fine art.

Prescribed burning plays a very crucial role in prevention of crown fires in the slash pine forests of the Deep South and the pitch pine barrens of southern New Jersey. Although pitch pines can recover from crown fires by sprouting along the charred stems and branches, this impairs the form of subsequent growth of the trees badly. Crown fires in slash pine and most other tree species are lethal to trees and highly dangerous.

Fire has essentially the same role in the ponderosa pine forests of the interior West as it does in the southern pines. On the best sites, fire prevents invasion by the relatively tolerant and less valuable Douglas-fir and true firs. There are also indications that light surface fires improve forage production and reduce the danger of serious fires.

In other kinds of American forests, most silvicultural use of fire, except that for slash disposal, is still in an experimental stage. In the northeastern United States, Native American populations burned beneath hardwood forests to promote forbs and grasses to improve hunting of wildlife (Abrams, 1992; Day, 1953). Then agricultural development involved

Figure 8.6 A stand in the same place and treated similarly to that shown in Figure 8.5, but after an additional 5 years, with dense natural reproduction of pine replacing the original hardwood understory. The treatment here consisted of one winter fire followed by three successive annual summer burnings. *(Photograph by U.S. Forest Service.)*

use of fire for grazing improvement. This may have promoted the development of oak-dominated stands, because of oak's ability to resprout. Though not yet used in most silviculture in this region, fire is being tested mostly to favor oak regeneration by topkilling all shrub and tree regeneration, and then having oak sprout back faster. Results have been variable so far, with successes reported when done once where the forest overstory is absent (Ducey, Moser and Ashton, 1996), or when the groundstory is repeatedly burned at periodic intervals (Nyland, Abrahamson, and Adams, 1982; Niering, Goodwin and Taylor, 1970).

Fires burning beneath natural stands also had an important effect on the maintenance of sugar and ponderosa pine, as well as the big-tree sequoia, in the mixed conifer forests of the Sierra Nevada (Kilgore and Taylor, 1979). These forests are as difficult to regenerate as they are magnificent. The situation is complicated by shrub species that are also favored by fire and by heavy fuel accumulations from decades of fire exclusion.

Many of the better forests of moist, cool, northern climates, especially the birch and coniferous types, owe their origin to the effects of fire (Zasada et al., 1977). However, most of the fires involved were of the catastrophic kind that occurred at long intervals rather than light, frequent ones (Frelich and Lorimer, 1991). Therefore, the desirable effects of fire are normally achieved by broadcast burning of slash rather than by burning under the stands. One characteristic that virtually all Australian eucalypts have in common is adaptation to fire (Hillis and Brown, 1978). In fact, Australian sites that are too moist for fires seldom have eucalypts. Bare mineral soil favors regeneration of this small-seeded genus. Many of the species sprout after crown fires; some can even renew their crowns by sprouting after crown fires. Many of the so-called ash group, usually of the moister sites, are adapted to regenerate after catastrophic fires, best simulated by clearcutting and slash burning. However, in some forests of drier sites, prescribed burning is often conducted beneath the stands for fuel reduction and other purposes described in connection with the southern pines.

There is a significant pattern in the application of prescribed burning in those localities where it is used. When management is started in a forest type that owes its better characteristics to repeated burning, it is most logical to attempt to conduct silviculture by methods that simulate some of the important effects of fire, but do not actually include prescribed burning. If this line of action produces a satisfactory result, there is no reason to use fire, which is at best a treacherous tool.

Nevertheless, in some instances such exclusion of fire created circumstances in which the problems multiplied progressively. The essential conditions for reproduction of the desirable trees were created less frequently, and the understories filled up with tree or shrub species that were adapted to the new conditions and were often of little value. As fuels accumulated, wildfires that formerly caused little damage showed an increasing tendency to develop into conflagrations; often this development proceeded more rapidly than fire control could be improved to meet it. Frequently, the point was reached where undesirable, sprouting vegetation could no longer be eliminated by fire and had to be left or attacked by more laborious methods. For example, overzealous control of fire in certain places of western North America has promoted encroachment of more mesic species prone to root rots and other diseases. This has led to cataclysmic fires that rarely occurred previously on these sites (Sampson and Adams, 1994).

Mechanical Treatments

One mechanical method of site preparation, **reduction of undesirable vegetation**, can be done by uprooting large woody plants, chopping up smaller plants, or plowing under grasses and other herbaceous growth. The objective of this step is to disrupt the roots of the undesirable plants enough to kill them. Of course, mechanical equipment can be used simply to cut or break off the stems of woody plants, especially those that are incapable of sprouting. It is entirely possible that one might attempt to eliminate only the woody vegetation and leave any grasses or other low plants alone, especially if their competition is not likely to be significant and excludes more troublesome plants.

Practically all site preparation involves some **redistribution of dead vegetation**. This may range from concentrating uprooted trees into high windrows to the gentle scarification of thin litter beneath otherwise undisturbed vegetation. With large debris, the objective is partly to remove obstacles to subsequent operations. It may also be desirable to hasten the destruction of debris; this can be done not only with fire but also by pushing it into compact piles or working it into the mineral soil so it will rot more quickly. With the gentler kinds of treatment, there may be no objective other than **exposure of mineral soil**. Any method of mechanical site preparation exposes some of the mineral soil, often in the process of doing something more difficult.

If the site is subject to extremes of moisture conditions, activities occasionally include **reshaping** the soil surface. This usually is done by plowing or scraping to create low ridges in wet places or shallow trenches to collect water in dry areas. Sometimes **smoothing** of the surface with harrows may facilitate machine planting or other future activities.

The equipment used for the work varies widely in kind and size. Most of it requires tractor-powered devices. Simple scarification designed merely to expose mineral soil can be done with light equipment. Single- or double-moldboard plows, including those designed for constructing fire-lines, or special shaping devices, can be used for furrowing or bedding.

Bedding is the mounding up of low ridges or "beds" on flat, poorly drained soils such as those common on the Lower Coastal Plain of the southeastern United States (Fig. 8.7). It is also used widely in Scotland, Ireland, and parts of Scandinavia and northern Canada (Paavilainen and Päivänen, 1995; Rothwell, Woodward, and Rivard, 1993). It has the effect of increasing the volume of soil that is sufficiently well supplied with oxygen and water to lengthen the period of physiological availability of water. This effect speeds the growth of new stands at least until they close, although some slight advantage might theoretically persist longer. Windthrow problems sometimes result from trees developing roots only along the beds with sitka spruce in Scotland (Paavilainen and Päivänen, 1995). Sometimes bedding is extended through force of habit to drier sites where it may do harm without any good.

Figure 8.7   A flat, poorly drained site in North Carolina bedded in preparation for planting loblolly pine. The raised beds or berms on which the trees are to be planted were cast up with a special plow and shaping device to provide ridges of aerated soil in which the roots can start their expansion.

Under extreme cases, the **loosening of compacted soils** may be done by plowing. Ordinarily, the natural soil organisms of the forest keep the soil in a more porous and penetrable state than could be accomplished by mechanical churning. However, mechanical loosening does help if the hooves of herbivores or the weight of heavy machinery have made the soil very compact. The areas used for loading logs can be a particular problem. It has been found in Australia that it may be cheaper to do this with a small explosive charge for each planting hole than to move heavy plows to each landing site. Very deep plowing is sometimes used to rupture subsurface hardpans that have been formed by purely natural processes or as the result of unwise use of certain sensitive soils for grazing or agriculture. This sort of plowing is most common in very humid climates, such as those in parts of Scotland, Wales, and Ireland, where sphagnum bogs and hardpans can form, even on hillsides, because of the leaching associated with the podzolization process (Morgan, Campbell, and Malcolm, 1992). If the accumulation of oxygen-deficient water or merely resistance to root penetration restricts rooting depth, the breaking of impediments to downward movement of water helps growth. The reverse may be true if the hardpans conserve meager supplies of water. As usual, it is well to proceed on the basis of knowledge of which growth factors are deficient in each case.

On very dry sites, it may improve the survival and early growth of planted trees to plow out furrows or to construct contour terraces and depressions in ways that will collect water and uproot or bury competing vegetation. In some arid regions, it is impossible to get trees to start without such artificial means of creating localized concentration of water. Terracing is a common practice in Israel and elsewhere in the Middle East (Goor and Barney, 1968). It has also been successfully used in the northern Rocky Mountains, although there has been some objection to the artificial appearance that it confers on the terrain. The creation of circular depressions for seedlings planted in the center is another similar practice in dry parts of the African Sahel (Evans, 1984).

Most site preparation is aimed chiefly at the destruction of competing vegetation. Plants of sapling size or smaller can be broken up and worked into the soil with heavy disk plows or rolling brush choppers (Fig. 8.8). It is often desirable to induce sprouting with an initial treatment and then follow it with a second treatment to kill the new, succulent sprouts before they form buds. The second treatment can be another mechanical one, or it can be prescribed burning or herbicide spraying.

Devices akin to large, rotary lawnmowers can be used to cut and shred small vegetation and woody debris preparatory to planting, but they do not eliminate sprouting vegetation. Bulldozers or similar devices (Fig. 8.9) can be used against vegetation of any size from small brush to large trees. However, they must be fitted with special toothed blades or "root rakes" (Fig. 8.10) if they are to be effective in uprooting anything without scraping away too much topsoil. Large areas of trees are sometimes uprooted by using powerful tractors to drag battleship anchor chains over the ground. The huge links of these chains pinch around the bases of the trees and remain engaged until the trees are pulled over. Ordinarily, pairs of tractors are used to drag one chain. The uprooted trees must then be pushed into windrows with bulldozers.

Mechanical site preparation of the kinds described is usually a preliminary to planting. If there has been any substantial overturning or upheaval of the mineral soil, it is necessary to delay planting for several months to allow the soil to settle. Otherwise, there is risk that many of the roots of the seedlings will be planted in air pockets and die of drought. The smoothing of areas with heavy harrows is often aimed chiefly at making it possible to plant with machines. Although the competing vegetation is greatly reduced by such drastic treatment, some sprouting root-stocks are usually left in the soil. These techniques are often employed for getting rid of the sparse, small hardwoods in pine stand understories

Figure 8.8   A bulldozer and two rolling brush cutters being used to destroy a cover of poor oak and grass preparatory to planting pine on a dry, sandy site on the Southeastern Coastal Plain. The two water-filled drums are hitched at an angle to one another and to the direction of travel to impart a shearing action to the sharp blades. *(Photograph by Caterpillar Tractor Co.)*

after a pine harvest. They are also advantageous in replacing degraded hardwood stands with pines, because well-established hardwood stands are almost impossible to eliminate by prescribed burning.

Limitations of Mechanical Treatments

In all treatments of this sort, the horizontal movement of organic materials and topsoil should be avoided as much as possible. The churning or overturning of these materials in place is distinctly preferable to any sort of scraping action.

Much of the nutrient capital of the forest is usually tied up in the litter and upper layers of the mineral soil (Binkley, 1986). Most of the forest system's power to resist erosion and protect water quality resides in the protective covering of the forest floor and the infiltrative capacity of the uppermost layer of mineral soil. No silvicultural practice can impair the productive capacity of the forest or damage watersheds more than scraping off the surface materials. This is especially true if the nutrients are moved farther sideways than the future extent of the tree roots that might bring them back. Bulldozers and root rakes should be regarded as instruments with more potentialities for damage than fire, herbicides, or other forestry tools.

The practice of windrowing uprooted material is an especially questionable practice (Binkley, 1986; Morriss, Pritchett, and Swindel, 1983; Ballard, 1978). It is impossible to do this without moving not only nutrients but also some mineral soil. This effect is often made apparent by the fact that planted trees (or weedy vegetation) grow fast in the wind-

Figure 8.9 Clearing of oak on a dry upland site with a ''K-G'' blade. The sharpened blade is set at an angle so that it can be used to shear off the trees at or below the ground line; the upper bar pushes the trees over as they are being severed. The ''stinger'' at one end of the blade is shown splitting the stem of a large tree so that it can be cut off or pushed over in two passes of the machine. *(Photograph by Caterpillar Tractor Co.)*

rows, whereas the other trees may be stunted and chlorotic for a time (Fig. 8.11). The windrows sometimes waste growing space. Closer utilization may reduce the use of the practice, but it is much better for the soil and site productivity to employ rolling brush cutters, plows, or other devices that work material into the soil without moving it sideways. It is ironic that silvicultural efforts to emulate the clean cultivation of agriculture have developed at the same time that efforts have been made to use herbicides for ''minimum-till'' agriculture. Clean-cultivation agriculture is an inherently depletive process; most silviculture is not because it defends the soil against erosion by conserving rather than squandering soil organic matter. Recent studies are elaborating on this work in the southeastern United States (Haywood, 1994), and the Pacific Northwest (Minore and Weatherly, 1990). These studies demonstrate that the best treatments rely on the broadcast distribution of slash without piling it or turning over the soil.

Light scarification and the plowing of shallow furrows are usually cheap, but measures that require powerful machinery can be very costly. Windrowing is one of the most expensive operations. If the investment in equipment is high, the cost may vary considerably depending on how the depreciation of the machines is treated in the accounting. Efforts to reduce costs by clearing narrow lanes through tall, undesirable vegetation do not always prove effective. Mechanical site preparation can be useful, safe, and advantageous, or it can be costly and harmful. Its use should be the result of thoughtful prescription and not

Figure 8.10   A root rake clearing a brush field for planting in the Sierra Nevada of California. *(Photograph by U.S. Forest Service.)*

a matter of unquestioned standard operating procedure. In certain regions where it has been applied under the wrong site circumstances, productivity has declined over successive rotations. For instance, examples have been recorded on heavy clay soils that are sensitive to compaction, or sandy nutrient poor soils of the South (Ruark et al., 1991; Bechtold, Ruark and Lloyd, 1991; Haywood, Tiarks, and Shoulders, 1990), and on certain fragile, nutrient-poor soils of Australia (Squire et al., 1985).

The increase in growth that can be obtained in new stands because of the effects of reducing competing vegetation is limited to the period before the stand itself approaches full occupancy of the growing space. The early gains in growth are not lost, but they should not be extrapolated beyond the time of stand closure.

Herbicide Treatments

As is evident from the previous discussion, herbicides can also be an important part of site preparation. The general use of herbicides has been described in Chapter 5 with regard to release treatments, and much of that discussion applies in this case as well. In site preparation uses, the need for selectivity among species is often absent, so achieving results is not as complicated as it is with release treatments (McCormack, 1991). It is possible to kill all vegetation on a site with aerial or some other form of broadcast spraying, but girdling or cutting of larger trees would be needed. However, herbicide removal of competing vegetation has no effect on forest floor conditions; thus it is not appropriate as the sole treatment for many situations. In addition, the costs of herbicides themselves and the application of labor and equipment are so high that other methods may be preferable. In certain circumstances, the use of alternatives may be more desirable merely to avoid reg-

Figure 8.11 Windrowing of stumps and logging debris and the effect that the associated scraping action can have in moving nutrients sideways. Both pictures show site preparation for planting on the Atlantic Coastal plain in North Carolina. The upper one shows a very thorough job of concentrating the debris in a tight windrow. The lower picture shows tall loblolly pines growing in the nutrient concentration of such a windrow on the right but with stunted and somewhat chlorotic seedlings on the left in the zone from which materials had been pushed with a root-rake. *(Photographs by Yale University School of Forestry and Environmental Studies.)*

ulation or public controversy, especially on public lands. In any case, it is often better to use herbicides in combination with other techniques (McCormack, 1994). Two examples can be cited here: first, when not enough fuel is present to carry a fire, fairly low doses of herbicides can be used to topkill low vegetation and, after drying, will allow burning; second, roller chopping will scarify soil and break up tops of vegetation, causing sprouting, and then applications of small amounts of herbicides to new sprouts will be effective.

Flooding

Flooding is a treatment commonly used to exclude competition and prepare the site for seeding or planting. This is rarely done in North America, but in other regions that have intensively controlled coastal and riparian lands, water, guided by channels and dikes, is directed to seasonally inundate a site to kill off weed growth. After several weeks, these lands are drained and sites are prepared. The obvious example is the practice of paddy cultivation for rice. Another example is the integration of forestry with aquaculture, such as the shrimp ponds and mangrove plantations of some parts of coastal Indonesia.

This might appear novel, but many tree species along rivers, swamps, and coastal margins rely on seasonal flooding as a disturbance regime that promotes their dispersal by water and subsequent colonization on newly opened growing space after the waters recede (Kozlowski, 1984). This in effect is done in bottomland hardwood forests throughout the southeastern United States where fine silts and clays are deposited across extensive flood-plains and backwaters of large river systems, after which certain species of maple, gum, water oak, and hickory establish (Hodges, 1995). In areas where water moves off the land abruptly, with force and volume, the most dramatic disturbances occur along the edges of waterways where new banks are cut and fresh deposits of coarse sands create bars and deltas. Good examples of this occur in the foothills of large mountain ranges such as the Andes of South America and the coast ranges (Cascades, Sierra Nevada) of the Pacific Northwest. Tree species that have been documented to rely on such violent disturbance regimes for their establishment include poplars, sycamores, and sitka spruce of temperate realms (Hodges 1995; Yarie, 1993); and mahogany and Spanish cedar of tropical regions (Salo et al., 1986). However, in most places water levels are often controlled more by hydroelectric and flood-control needs than by plans for silviculture.

IMPROVEMENT OF SITE

Fertilization, drainage, and irrigation are intensive treatments of the soil aimed more at improving or restoring site quality than at preparing for regeneration. The proper conduct of these treatments depends so heavily on knowledge of the complex chemical properties of soils and their moisture relationships that it is logical to refer to works on forest soils and other accounts, such as those of Pritchett (1979), Ballard and Gessel (1983), Bowen and Nambiar (1984) and Binkley (1986) for more complete information.

Fertilization

Few forest soils provide an optimum supply of the nutrient elements essential for the growth of trees. Sometimes marked deficiencies may exist because of improper land management in the past or merely because of inherently low natural fertility of the site. The nutrient elements most likely to be deficient are nitrogen, phosphorus, and potassium (the

NPK elements of most fertilizers), in that order of frequency of deficiency. It is often possible to make forest trees grow better by increasing the supply of nitrogen. The most common cases in which this does not happen are sites where the supply of water or phosphorus is the limiting factor.

The most common nitrogen fertilizers used in forestry are urea compounds that yield positively charged ammonium ions, which can be absorbed in the cation exchange capacity of the soil and are easily taken up by plants. Urea formaldehyde compounds have the advantage of releasing the ammonium slowly. Although nitrate compounds can be used, the negatively charged nitrate ions are easily lost from the soil by leaching. Nitrogen compounds in the forest system are constantly moving to and from the atmosphere or being lost to moving soil water. They can be very mobile and are lost because of leaching. They can also become unavailable to higher plants by being captured by decomposing organisms in the soil or by being locked up in undecomposed organic matter. This often happens where the soil climate is unfavorable to decomposition. It is sometimes possible to remedy nitrogen deficiencies by stimulating the symbiotic and nonsymbiotic nitrogen-fixing organisms (Gordon and Wheeler, 1983). Fertilization with other elements sometimes stimulates the nitrogen-fixers, as does the encouragement of legumes, alders, and other plants that support symbiotic nitrogen-fixing bacteria.

The cost of artificial fixation of nitrogen for fertilizer in terms of both money and energy is high. The effects of a given application are limited to several years, so repeated applications can be necessary. However, nitrogen is so crucial in building the proteinaceous biochemical machinery of life that spectacular effects can result from reducing the chronic deficiencies. Fertilization with nitrogen compounds alone has become moderately common in the humid parts of the Pacific Northwest (Miller and Fight, 1979) and in other forest areas where growing conditions are good and nitrogen may be the most important limitation.

Phosphorous deficiencies remediable by fertilization are common on poorly drained soils. In these and most other cases of phosphorous deficiency, the problem is that the acid condition of the soil or other factors cause too much phosphorous to be tied up in unavailable form in compounds with iron and aluminum (Binkley, 1986). Phosphorous can be applied as ground phosphate rock or in more concentrated form. These do not consume much energy in manufacture. Single applications have long-lasting effects.

The amounts of nitrogen and phosphorous available to plants must be well balanced. What might otherwise be the right amount of one can make a deficiency of the other more acute or even harmful. If a tree has more phosphorous to form the energy-transfer machinery of photosynthesis, for example, it also needs more nitrogen to build the protein components of this machinery.

If there is opportunity for good nutrient recycling, single applications of potassium fertilizer in the forest seem to suffice for very long periods. Potassium is a very soluble and mobile compound. It is not only easily lost by leaching, but it even leaks out of green leaves. Deficiencies have been encountered on easily leached sandy soils previously subjected to highly extractive kinds of agricultural crop removal in New York State (Leaf and Leonard, 1973).

Few cases in forestry show that fertilization with other elements has had actual benefit. It is rather surprising that there is little evidence of improvements from adding calcium, magnesium, or manganese. Deficiencies of trace elements such as zinc, molybdenum, boron, cobalt, and copper are known mostly from a few districts in Australia (Evans, 1984). Australia is an ancient and strongly leached continent where deficiencies of phosphorous and other elements have caused forest fertilization to be much more remarkably successful than in most parts of the world. Forest fertilization is normally done from the air. It is usually desirable to avoid fertilizing the forest waters so as to avoid contributing to eutrophication. The materials to be spread are heavy enough that it helps to minimize flying distances. It is also important to be able to load the aircraft quickly.

In most cases, it is best to restrict forest fertilization to the latter part of the rotation. A given amount of fertilizer seems to produce about the same amount of wood regardless of tree size, and a given cubic volume put on large trees is worth more than the same volume put on small trees. Furthermore, the supply of available nutrients is generally greatest just after the destructive events associated with regeneration (Bormann and Likens, 1979). Deficiencies are most likely to set in after the stands have filled all the growing space and more and more nutrients are getting tied up in living and dead organic materials on the site (Harding and Jokela, 1994). Fertilization is done at the time of planting or regeneration only if the deficiencies are very serious, as on recently drained organic soils (Taylor, 1991). Sometimes fertilization of young stands favors the competing vegetation more than the trees. In some examples, fertilization has changed the composition of the groundstory vegetation (Prescott et al., 1993); in other cases it is not needed, or it induces harmful imbalances between nutrient elements.

Recent research has revealed that use of organic biostimulants as a stress vitamin stimulate plant growth and can provide optimum yields with up to 50 percent reduction in fertilizer use. Organic biostimulant research has shown increased efficiency of nutrient uptake and has demonstrated that increased vigor from supplemental vitamins has enhanced synthesis of defensive polyphenols that make treated plants less susceptible to herbivory (Russo and Berlyn, 1990, 1992).

Drainage

Areas of stagnant, oxygen-deficient water can be almost like arid deserts with respect to plant growth. Paradoxically, it is for the same reason—deficiency of available water (Kozlowski, 1984). The decomposing organisms attacking the organic matter of swamps rob the water of the oxygen that the roots of higher plants must have to function. Therefore, trees growing in bogs can extend their roots downward only a few inches, and the strata below are as impervious to roots as rock would be. This situation prevails only where water is slow moving. Sites along freely flowing streams can seem equally wet but often are biologically the most productive of any of a given locality; in such places, the water is well aerated and freely available.

As is frequently the case, it is important to put intuition aside and recognize that wet sites with ponded water are physiologically dry. At least it can be said that many tree species seem adapted to this phenomenon. Many species of literally dry sites (such as deep sands or thin soils) are also found in physiologically dry swamps. Some of the examples include red maple, black spruce, and some species of pine, although not all species of one habitat also grow in the other. The most common indicators of oxygen-deficient water, the world over, are sphagnum mosses. These mosses not only indicate bogs but, in very humid climates such as in parts of the British Isles, they can even create bogs on hillsides because of their very large water-holding capacity.

Some of the previously described methods of using plows for bedding or puncturing hardpans accomplish some drainage, especially on sloping ground. However, water is viscous enough that such measures do not really move much off flat terrain. This requires ditches or canals as well as enough difference in elevation to provide places to which water can flow (Paavilainen and Päivänen, 1995).

The most common method starts with the use of backhoes or draglines to dig parallel primary canals; the fill placed between them becomes an access road (Fig. 8.12). Secondary ditches are constructed to lead to these usually at right angles and at intervals of about 100 feet (30 m). In the process of site preparation for planting, the sites are often bedded with plows so that the trees can be started on low, well-drained ridges. Unless they have been enriched by nutrients flowing off adjacent uplands, the organic soils of such are as are deficient in nutrients, especially phosphorous and nitrogen. Fertilization is commonly needed, and it usually takes much local experimentation to determine how much of what to add.

It must be anticipated that drainage will cause the newly aerated peat to decompose in such a manner that the surface will sink. Sometimes this proceeds to the point where the ditches have to be deepened. If there is then no longer enough difference in elevation for the water to move, the whole effort is defeated. The worst situation occurs on lands that have a thin layer of light fresh water over dense salty water. The salt water rises as the fresh water moves to the top while the surface subsides. On the other hand, if a treeless area is drained and afforested, the trees can transpire enough water to improve their own growth by removing excess water.

The largest forest drainage projects are those of peat bogs that have formed in former lakes in glaciated lands in Finland (Fig. 8.13), Scandinavia, and Russia. The techniques, which usually have to include fertilization of the sterile peats, are highly developed and based on detailed classifications of the different kinds of wet sites. In these countries, there is growing sentiment that drainage of peat bogs may have gone too far, creating concerns about landscape diversity and plant and animal habitats. Thus far, few attempts have been made to drain similar bogs in Canada and the northern United States. Most North American forest drainage is on the flat Lower Coastal Plain of the southeastern United States. The most successful efforts have been with very flat but somewhat elevated areas distant from major rivers. These are wet mostly because of the flatness of the land and have comparatively thin layers of peat. If the drainage ways are kept open and phosphorous deficiencies are remedied, the drainage of these soils makes them so well supplied with aerated water that they become highly productive (Binkley, 1986; McKee et al., 1984). The drainage of

Figure 8.12 Drainage canal, with water-regulation device, designed to move water off a very flat site in North Carolina. The road is made of soil dug from the ditches on either side of it. The canal is adjacent to the bedded area shown in Fig. 8.7.

Figure 8.13   Scotch pine stands on drained peat land in Finland. The trees were absent or badly stunted before the drainage started 5 decades ago. The surface has subsided about a meter because of decomposition from the aeration of the drained peat. *(Photograph by Yale University School of Forestry and Environmental Studies.)*

depressions in this region is more difficult, and the results vary. Attempts to drain forest areas that are almost at sea level and underlain by salt water have been costly failures (Haywood, Tiarks, and Shoulders, 1990).

Irrigation

Except for seed-orchards, nurseries, and other sites of very intensive tree culture, little use is made of irrigation in forestry for most moist regions where trees grow prolifically. Irrigation water is usually more valuable for agricultural use and for fruit and nut orchards, particularly where rainfall is low but the soils are fertile (i.e., the central valley of California). Care must be taken on these intensively managed sites to avoid salinization. For forestry there are only a few places in the world where hybrid poplars and other species of alluvial flood plains are grown with supplemental irrigation water. The most common ways of increasing the water supply of forest trees are those methods of reshaping the ground surface to concentrate surface runoff water on the roots of planted trees that were mentioned earlier in connection with mechanical site preparation.

The most notable use in forestry is for the purposes of restoration, particularly in relation to reclamation of surface mine spoils (Larson, Patel, and Vimmerstedt, 1995; Messina and Duncan, 1993; Vogel, 1981). The effect of timber cutting and grazing for several millennia have completely degraded sites in arid areas. Irrigation can be used in promoting the establishment of a new forest or woodland. Once the forest ecosystem is functioning, irrigation can taper off. Some of the most advanced mechanical methods of accentuating water capture and also the use of drip irrigation (tubes that supply constant drips of water at the base of each planted tree) have therefore been used to combat these processes of desertification in the Middle East (especially Israel), the African Sahel (FAO, 1977), central China, Pakistan, and central India (Ffolliot et al., 1994).

Protection

Almost all of this chapter has focused on manipulative techniques for preparing the site for establishing various kinds of vegetation. In many other circumstances, rather than preparing a site by exposing mineral soil or by excluding unwanted vegetation, the site must be protected from erosive forces of climate or land use. Sites where protection is required are often prone to continuous or frequent disturbances that prevent vegetation establishment and soil stabilization. Examples can be found in nature along coasts, such as in dune systems, or along the banks of fast-flowing streams. Overuse of land from chronic grazing or continuous cultivation of steep slopes are also examples of sites that can become unstable.

There are many protective techniques that can be broadly defined as **contouring**. All contouring involves constructing physical or vegetative obstructions that are arranged parallel to the contour of the slope or disturbance-prone edge, and serve to stabilize earth movement and soil erosion. Examples of contouring include hedgerows, wind breaks, stone walls, rip-rap, ditches (that collect and channel water away from sensitive sites), and ground covers aligned in rows (deep-rooted grasses). The best examples of this used currently in North America are in the Midwest (Gillespie, Miller, and Johnson, 1995). For most applications in silviculture, contouring is used to temporarily stabilize a site and promote succession. However, where silvicultural techniques are used to arrest successional processes, as in the case of many agroforestry systems, contouring may be a more permanent part of site protection. This is particularly the case where crop cultivation or animal grazing is at least in some form present continuously on the land. On certain sites, the ground surface may be completely protected by either vegetative covers (often nitrogen-fixing legumes), organic mulches (hay, wood chips), or synthetic mats.

BIBLIOGRAPHY

Abrams, M. D. 1992. Fire and the development of oak forests. *Bioscience*, 42:346–353.

Agee, J. K. 1993. *Fire ecology of the Pacific northwest forests*. Island Press, Washington, D.C. 493 pp.

Alexander, M. E., and F. G. Hawksworth. 1975. Wildland fires and dwarf mistletoes: a literature review of ecology and prescribed burning. USFS Gen. Tech. Rept. RM-14. 12 pp.

Ballard, R. 1978. Effect of slash and soil removal on the productivity of second-rotation radiata pine on pumice soil. *New Zealand J. For. Sci.* 8:248–258.

Ballard, R., and S. P. Gessel (eds.). 1983. I.U.F.R.O. Symposium on Forest Site and Continuous Productivity. USFS Gen. Tech. Rept. PNW-163. 406 pp.

Bechtold, W. A., Ruark, G. A., and F. T. Lloyd. 1991. Changing stand structure and regional growth reductions in Georgia's natural pine stands. *For. Sci.*, 37:703–717.

Binkley, D. 1986. *Forest nutrition management*. Wiley, New York. 290 pp.

Bormann, F. H., and G. E. Likens. 1979. *Pattern and process of a forested ecosystem*. Springer-Verlag, New York. 253 pp.

Bowen, G. D., and E.K.S. Nambiar. 1984. *Nutrition of plantation forests*. Academic, Orlando, Fla. 506 pp.

Boyer, W. D., and J. H. Miller. 1994. Effect of burning and brush treatments on nutrient and soil physical properties in young longleaf pine stands. *FE&M*, 70:311–318.

Brady, N. C. 1990. *The nature and properties of soils*. Macmillan, New York. 621 pp.

Brown, A. A., and K. P. Davis. 1973. *Forest fire: control and use*. 2nd ed. McGraw-Hill, New York. 686 pp.

Burschel, P., Kuersten, E., Larson, B. C., and M. Weber. 1993. Present role of German forests in the national carbon budget and options to its increase. *Water, Air and Soil Pollution*, 70:325–340.

Chandler, C., P. Cheyney, P. Thomas, L. Trabaud, and D. Williams. 1983. *Fire in forestry*. 2 vols. Wiley, New York. 789 pp.

Cheke, A. S. 1979. Dormancy and dispersal of a secondary forest species under the canopy of a primary tropical rain forest in northern Thailand. *Biotropica*, 11:88–95.

Cohen, A., Singhakumara, B.M.P., and P.M.S. Ashton. 1995. Releasing rainforest succession: A case study of *Dicranopteris linearis* fernlands of Sri Lanka. *Restoration Ecology*, 3:261–270.

Cramer, O. P. (ed.). 1974. Environmental effects of forest residues management in the Pacific Northwest: a state-of-knowledge compendium. USFS, Pac. Northwest For. and Range Expt. Sta. Publ. D-l-D-23. 537 pp.

Dale, V. H., Houghton, R. A. and C. S. Hall. 1991. Estimates of the effects of land-use change on global CO_2 concentrations. *CJFR*, 21:87–90.

Damman, A.W.H. 1978. Distribution and movement of elements in ombrotrophic peat bogs. *Oikos*, 30:480–495.

Daniel, T. W., and J. Schmidt. 1972. Lethal and non-lethal effects of the organic horizons of forested soils on the germination of seeds from several associated conifer species of the Rocky Mountains. *CJFR*, 2:179–184.

Day, R. K. 1953. The Indian as an ecological factor in the northeastern forest. *Ecol.* 34:329–346.

Ducey, M. J., Moser, W. K., and P.M.S. Ashton, 1996. Effect of fire intensity on understory composition and diversity in a *Kalmia*-dominated oak forest, New England. *Vegetatio* 123:81–90.

Duryea, M. L., and P. M. Dougherty (eds.). 1991. *Forest regeneration manual*. Kluwer Acad. Publ., Dordrecht, The Netherlands. 433 pp.

Evans, J. 1984. *Plantation forestry in the tropics*. Clarendon Press, Oxford. 472 pp.

Fahnestock, G. 1973. Use of fire in managing forest vegetation. *Trans., Amer. Soc. Agr. Eng.*, 16:410–413, 419.

FAO. 1977. *Savanna afforestation in Africa*. FAO, Rome. 200 pp.

Ffolliot, P. F., Brooks, K. N., Gregerson, H. M., and A. L. Lundgren. 1994. *Dryland forestry, planning and management*. Wiley, New York. 632 pp.

Frelich, L. E., and C. G. Lorimer. 1991. Natural disturbance regimes in hemlock-hardwood forest of the upper Great Lakes Region. *Ecol. Monogr.*, 61:145–164.

Gessel, S. P., Lacate, D. S., Weetman, G. F., and R. F. Powers. 1990. *Sustained productivity of forest soils: 75th North American forest soils conference*. Univ. Brit. Col., Faculty of Forestry, Vancouver. 525 pp.

Gillespie, A. R., Miller, B. K., and K. D. Johnson. 1995. Effects of groundcover on tree survival and growth in filter strips of the Cornbelt Region of the midwestern U.S. *Agric. Ecosystems & Env.* 55:263–270.

Goor, A. Y., and C. W. Barney. 1968. *Forest tree planting in arid zones*. Ronald, New York. 504 pp.

Gordon, J. C., and C. T. Wheeler (eds.). 1983. *Biological nitrogen fixation in forest ecosystems: foundations and applications*. Nijhoff/Junk, The Hague. 342 pp.

Grelen, H. E. 1978. May burns stimulate growth of longleaf pine seedlings. USFS Res. Note SO-234. 5 pp.

Harding, R. B., and E. J. Jokela. 1994. Long-term effects of forest fertilization on site organic matter and nutrients. *Soil Sci. Soc. Am. J.*, 58:216–221.

Haywood, J. D. 1994. Early height-growth reductions in 2nd rotation of short-rotation loblolly and slash pine in central Louisiana. *SJAF*, 18:35–39.

Haywood, J. D., Tiarks, A. E., and E. Shoulders. 1990. Loblolly and slash pine height and diameter are related to soil drainage on poorly drained silt loams. *New Forests*, 4:81–95.

Hendrickson, O. Q. 1990. How does forestry influence atmospheric carbon? *For. Chron.*, 41:469–472.

Hill, J. D., Canham, C. D., and D. M. Wood. 1995. Patterns and causes of resistance to tree invasion in rights-of-way. *Ecol. Applic.* 5:459–470.

Hillis, W. E., and A. G. Brown (eds.). 1978. *Eucalypts for wood production*. Australia, Commonwealth Sci. and Industrial Res. Org., Camberra. 434 pp.

Hobbs, S. D., Tesch, S. D., Owston, P. W., Stewart, R. E., Tappeiner, J. C., and G. E. Wells. 1992. *Reforestation practices in southwestern Oregon and northern California*. Forest Research Laboratory, Oregon State University, Corvallis. 465 pp.

Hodges, J. 1995. The southern bottomland hardwood region and brown loam bluffs subregion. J. W. Barrett (ed.). In: *Regional silviculture of the United States*. pp. 227–270. Wiley, New York. 643 pp.

Horsley, S. B. 1977. Allelopathic inhibition of black cherry by fern, grass, goldenrod, and aster. *CJFR*, 7:205–216.

Johnson, E. A. 1992. *Fire and vegetation dynamics studies from the North American boreal forest.* Cambridge Univ. Press, New York. 142 pp.

Jones, G. T. 1993. *A guide to logging aesthetics: practical tips for loggers, foresters, and landowners.* NRAES-60, Ithaca, N.Y. 28 pp.

Jordan, C. F. 1985. *Nutrient cycling in tropical forest ecosystems.* Wiley, New York. 190 pp.

Kilgore B. M., and D. Taylor. 1979. Fire history of a sequoia-mixed conifer forest. *Ecol.*, 60:129–142.

Killiam, K. 1994. *Soil ecology.* Cambridge University Press, Cambridge. 241 pp.

Kozlowski, T. T. (ed.). 1984. *Flooding and plant growth.* Academic, Orlando, Fla. 368 pp.

Kozlowski, T. T., and C. E. Ahlgren (eds.). 1974. *Fire and ecosystems.* Academic, New York. 542 pp.

Kraemer, J. F., and R. K. Hermann. 1979. Broadcast burning: 25-year effects on forest soils in the western flanks of the Cascade Mountains. *For. Sci.* 25:427–439.

Larson, M. M., and G. H. Schubert. 1969. Root competition between ponderosa pine seedlings and grass. USFS Res. Paper RM-54. 12 pp.

Larson, M. M., S. H. Patel, and J. P. Vimmerstedt. 1995. Allelopathic interactions between herbaceous species and trees grown in topsoil and spoil media. *Jour. Sustainable For.* 3:39–52.

Leaf, A. L., and R. E. Leonard (eds.). 1973. Forest fertilization. USFS Gen. Tech. Rept. NE-3. 246 pp.

Little, S., and H. A. Somes. 1961. Prescribed burning in the pine regions of southern New Jersey and Eastern Shore Maryland—a summary of present knowledge. Northeastern For. Exp. Sta., Sta. Paper 151. 21 pp.

Lorimer, C. G., Chapman, J. W., and W. D. Lamber. 1994. Tall understory vegetation as a factor in the poor development of oak seedlings beneath stands. *J. Ecol.*, 82:227–237.

Lutz, H. J. 1956. Ecological effects of forest fires in the interior of Alaska. USDA Tech. Bull. 1133. 121 pp.

Martin, R. E., and J. D. Dell. 1978. Planning for prescribed burning in the Inland Northwest. USFS Gen. Tech. Rept. PNW-76. 67 pp.

McCormack, M. L. 1991. Herbicide technology for securing naturally regenerating stands. In: C. M. Simpson (ed.), *Proc. Symp. natural regeneration management*. Fredrickton, New Brunswick, Canada. 261 pp.

McCormack, M. L. 1994. Reductions in herbicide use for forest vegetation management. *Weed Technology*, 8:344–349.

McKee, W. H., Hook, D. D., DeBell, D. S., and J. L. Askew. 1984. Growth and nutrient status of loblolly pine seedlings in relation to flooding and phosphorus. *Soil Sci. Soc. Am. J.*, 48:1438–1442.

Messina, M. G., and J. E. Duncan. 1993. Establishment of hardwood tree and shrub species in a Texas lignite mine using irrigation, mulch and shade. *Landscape and Urban Planning*, 25:85–93.

Miller, R. E., and R. D. Fight. 1979. Fertilizing Douglas-fir forests. USFS Gen. Tech. Rept. PNW-83. 29 pp.

Minore, D., and H. G. Weatherly. 1990. Effects of site preparation on Douglas-fir seedling growth and survival. *WJAF*, 5:49–51.

Mooney, H. A., et al. (eds.). 1981. Fire regimes and ecosystem properties. USFS Gen. Tech. Rept. WO-26. 394 pp.

Morgan, J. L., Campbell, J. M., and D. C. Malcolm. 1992. Nitrogen relations of mixed-species stands on oligotrophic soils. In: M.G.R. Cannell, D. C. Malcom, and P. A. Robertson (eds.). *The ecology of mixed-species stands of trees.* Blackwell Scientific Publ., London. 312 pp.

Morris, L. A., W. L. Pritchett, and B. F. Swindel. 1983. Displacement of nutrients into windrows during site preparation of a flatwoods forest. *Soil Sci. Soc. Amer. Journal*, 47:591–594.

Niering, W. A., Goodwin, R. H., and S. Taylor. 1970. Prescribed burning in southern New England: Introduction to long-range studies. *Proc. Ann. Tall Timbers Conf.*, 10:267–286.

Nyland, R. D., Abrahamson, L. P. and K. B. Adams. 1982. Use of prescribed fire for regeneration of red and white oak in New York. *Publ. Proc. Soc. Am. For.*, 1982:164–167, Bethesda, Md.

Paavilainen, E., and J. Päivänen. 1995. *Peatland forestry: ecology and principles.* Springer-Verlag, New York. 270 pp.

Perry, M. A., Mitchell, R. J., Zutter, B. R., Glover, G. R., and D. H. Gjerstad. 1994. Seasonal variation in competitive effects on water stress and pine responses. *CJFR* 24:1440–1449.

Phillips, D. L., and W. H. Murphy. 1985. Effects of rhododendron in regeneration of southern Appalachian hardwoods. *For. Sci.*, 31:226–232.

Prescott, C. E., Coward, L. P., Weetman, G. F., and S. P. Gessel. 1993. Effects of repeated nitrogen fertilization on the ericaceous shrub, Salal (*Gaultheria shallon*), in the coastal Douglas-fir forests. *FE&M*, 61:45–60.

Pritchett, W. L. 1979. *Properties and management of forest soils.* Wiley, New York. 500 pp.

Putz, F. E., and C. D. Canham. 1992. Mechanisms of arrested succession in shrublands root and shoot competition between shrubs and tree seedlings. *FE&M*, 49:267–275.

Pyne, S. J. 1984. *Introduction to wildland fire; fire management in the United States.* Wiley, New York. 566 pp.

Reitveld, W. J. 1975. Phytotoxic grass residues reduce germination and initial root growth of ponderosa pine. USFS Res. Paper RM-153. 15 pp.

Rice, E. L. 1984. *Allelopathy.* 2nd ed. Academic, Orlando, Fla. 440 pp.

Rothwell, R. L., Woodward, P. M., and P. G. Rivard. 1993. The effect of peatland drainage and planting position on the growth of white spruce seedlings. *NJAF*, 10:154–160.

Ruark, G. A., Thomas, C. E., Bechtold, W. A., and D. M. May. 1991. Growth reductions in naturally regenerated southern pine forests in Alabama and Georgia. *SJAF*, 15:73–79.

Russo, R. O., and G. P. Berlyn. 1990. The use of organic biostimulants to help sustainable agriculture. *J. Sustainable Agric.*, 1:19–42.

Russo, R. O., and G. P. Berlyn. 1992. Vitamin-humic acid-algal biostimulant increases bean yield. *Hort. Sci.*, 27:847.

Salo, J., Kalliola, R., Häkkinen, I., Mäkinen, Y., Niemalä, P., Puhakka, M., and P. D. Coley. 1986. River dynamics and the diversity of Amazon lowland forest. *Nature*, 322:254–258.

Sampson, R. N., and D. L. Adams. 1994. *Assessing forest ecosystem health in the inland west.* Haworth Press, New York. 461 pp.

Southern Forest Fire Laboratory. 1977. Southern forest smoke management guidelines. USFS Gen. Tech. Rept. SE-10. 140 pp.

Sprent, J. R. 1987. *The ecology of the nitrogen cycle.* Cambridge studies in ecology, Cambridge University Press, Cambridge.

Squire, R. O., P. W. Farrell, D. W. Finn, and B. C. Aeberli. 1985. Productivity of first and second rotation stands of radiata pine on sandy soils. II. *Australian Forestry*, 48:127–137.

Stacey, G., Burns, R., and H. J. Evans. 1992. *Biological nitrogen fixation*. Chapman & Hall, New York.

Stewart, R. E., Gross, L. L., and B. H. Honkala. 1984. Effects of competing vegetation on forest trees: a bibliography with abstracts. USFS Gen. Tech. Rept. WO-43. 1 vol.

Stone, E. 1973. The impact of timber harvest on soil and water. In: *Report of the President's Advisory Panel on Timber and the Environment*. Govt. Printing Off., Washington, D.C. Pp. 427–467.

Stone, E. L. (ed.). 1984. *Forest soils and treatment impacts*. Proc., Sixth No. Amer. Forest Soils Conf. Univ. of Tennessee, Knoxville. 454 pp.

Sutherland, E. K., Covington, W. K., and S. Andariese. 1991. A model of Ponderosa pine response to prescribed burning. *FE&M*, 44:161–174.

Taylor, C.M.A. 1991. *Forest fertilization in Britain*. UK Forest Commission Bull. 45 pp.

Tesch, S. D., and S. D. Hobbs. 1989. Impact of shrub sprout competition on Douglas-fir seedling development. *WAF*, 4:89–92.

Tesch, S. D., and E. J. Korpela. 1993. Douglas-fir and white fir advance regeneration for renewal of mixed-conifer forests. *CJFR*, 23:1427–1437.

Thadani, R., and P.M.S. Ashton. 1995. Regeneration of banj oak (*Quercus leucotricophora* A. Camus) in central Himalaya. *FE&M*, 78:217–224.

Tippin, T. (ed.). 1978. *Symposium on principles of maintaining productivity on prepared sites*. USFS, Southern Area State and Private Forestry, Atlanta. 171 pp.

Trexler, M. 1991. *Minding the carbon store: weighing U.S. forestry strategies to slow global warming*. World Resources Institute. 81 p.

U.S. Forest Service. 1971. *Prescribed burning symposium proceedings*. Southeastern For. Exp. Sta., Asheville, N.C. 160 pp.

Vitousek, P. M., Allen, H. L., and P. A. Matson. 1983. Impacts of management practices on soil nitrogen. In: *Maintaining forest site productivity*. Appalachian Society of American Foresters, Clemson, SC, pp. 25–39.

Vitousek, P. M., and P. A. Matson. 1985. Disturbance, nitrogen availability, and nitrogen losses in an intensively managed loblolly pine plantation. *Ecol.*, 66:1360–1376.

Vitousek, P. M., and R. L. Sanford. 1986. Nutrient cycling in moist tropical forest. *Ann. Rev. Ecol. and Syst.*, 17:137–168.

Vogel, W. G. 1981. A guide for revegetating coal minesoils in the eastern United States. USFS Serv. Gen. Tech. Rep. NE-61, 190 pp.

Walstad, J. D., Radoseurch, S. R., and D. V. Sandberg (eds.). 1990. *Natural and prescribed fire in the Pacific northwest forests*. Oregon State Univ. Press, Corvallis. 317 pp.

Ward, H. A., and L. H. McCormick. 1982. Eastern hemlock allelopathy. *For. Sci.*, 28:681–686.

Weaver, H. 1951. Observed effects of prescribed burning on perennial grasses in the ponderosa pine forests. *J. For.*, 49:267–271.

Weetman, G. F., and B. Weber. 1972. The influence of wood harvesting on the nutrient status of spruce stands. *CJFR*, 2:351–369.

Whitmore, T. C. 1990. *Tropical rain forests*. Oxford University Press, Oxford. 226 pp.

Wood, M. 1989. *Soil biology*. Chapman & Hall, New York. 154 pp.

Wright, H. A., and A. W. Bailey. 1982. *Fire ecology, United States and southern Canada*. Wiley, New York. 501 pp.

Yarie, J. 1993. Effects of selected forest management practices on environmental parameters related to successional development on the Tanana river floodplain, interior Alaska. *CJFR*, 23:1001–1014.

Zahner, R. 1955. Soil water depletion by pine and hardwood stands during a dry season. *For. Sci.*, 1:258–264.

Zasada, J. C., et al. 1977. Forest biology and management in high-latitude North American forests. In: *Proc. Symp. North American Forest Lands North of 60 Degrees*. Univ. Alaska, Fairbanks. Pp. 137–195.

CHAPTER *9*

SITE CLASSIFICATION AND SPECIES SELECTION

The ideal goal of silviculture is to place the right tree in the right place with just the right amount of growing space at each stage of development. Questions about what is right are matters involving much opinion and analytical thought about ecology and human demands. One extreme view is that the genetic material composing a stand should consist of those species and genotypes best adapted to survive and reproduce on the site as a result of many generations of natural selection. However, the attributes that provide for survival of the species are not necessarily those needed to meet human requirements. The natural composition is, furthermore, not static or well defined; it is instead dynamic and subject to continual changes resulting from developmental processes initiated by competition or natural disturbance. This means that even if one adheres to the natural composition, a choice must still be made about which developmental stage to imitate. Another problem with this approach is that human disruptions have made it very difficult to know what natural compositions might have been in many locales.

At the other extreme is the view that silvicultural prowess can make human wishes about species composition come true, as is the case with agriculture. However, problems such as those of getting woody perennials to survive through dormant seasons cause foresters to stop short of full emulation of the agronomy of herbaceous annuals. One may move maize from Central America to Minnesota, but the same cannot be done with Honduran mahogany. In fact, one cannot safely move trembling aspen from a moist site to a dry one within a Minnesota farm woodlot.

Because neither natural factors nor human wants can be ignored, the most logical courses lie between these extremes and also vary with the circumstances. The first step in this course-setting is to determine the limitations imposed by the environmental factors that collectively constitute the site. This restricts the number of species to be considered in the second step, which is the choice of species (plural or singular) that will most nearly meet the humanly ordained objectives of stand management. A third step is consideration of the degree of artificial control that will be exerted over the genetic constitution of the species that are selected. This may range from simply accepting the existing genetic makeup of a species to using intensive breeding methods in order to develop more desirable genotypes.

The basic objective is to use genetic material that will not only survive and thrive on the site but also yield the wood or other benefits at some optimum rate. However, it should be noted that suitable ''genetic material'' is seldom any single genotype. Even if only a single species is to be used, it is best to maintain some degree of genetic variation within each stand. The best choice may also be a combination of many species.

The choices that are made with respect to the matching of species and site affect and interact with decisions about the entire program of silvicultural treatments that will be devised.

METHODS OF IDENTIFYING AND CLASSIFYING SITES

Nature of Site or Habitat

Anything that is done in silviculture should be based on knowledge of the capacities and limitations of the site or habitat in which the trees are to be grown. Although the term **site** is the traditional one denoting the total environment of a place, **habitat** more fully connotes the idea that the place is one in which trees and other living organisms subsist and interact. As far as trees and other green plants are concerned, the site is controlled mainly by the total physiologically available supply of solar energy, water, carbon dioxide, and various chemical nutrients. The primary driving force that controls these supplies is the input of solar radiation and precipitation from above, which defines the **climate** of a region. Total annual levels of radiation and precipitation are important, but growing conditions are also affected by their distribution throughout the year, producing wet and dry seasons and summer and winter.

The sunlight regime is controlled largely by latitude, and so it varies on a regional scale. It sets the absolute limits on the productivity of a site both as the energy input for photosynthesis, and more generally in controlling the temperature range in which organisms must function. Precipitation levels vary in more complex patterns largely associated with the movement of humid air from oceans onto land masses. The balance between temperature and precipitation strongly affects plant survival and productivity. Much greater precipitation is required in Florida than in Ontario to support forest growth, because of the greater potential for evaporation from both soils and leaf surfaces in hotter regions.

For a particular location on the landscape, the effect of climate is modified by the **landform**, which is the nature of the surficial geological material plus the surface shape or topography. Landform characteristics influence the amount of solar radiation that reaches the ground through the steepness and orientation of the slope. In mountainous regions, elevational differences can directly affect the climatic inputs of precipitation, but most landform modifications of water supplies deal with how water moves after it has reached the ground. The depth and porosity of the surficial material govern whether water is drained freely or is retained for some time in the rooting zone; the topographic position on the landscape (e.g., hilltop, lower slope, or flood-plain) further affect the water status of a site. The specific nature of the soil characteristics adds a further modifying force by affecting water retention and the capacity for nutrient exchange, but many of the important attributes of a site can be described by climate and landform.

Sites can be usefully categorized in terms of the limitations imposed on plant growth by shortages of one or more of the basic factors. The supply of water is generally the most important factor differentiating between sites. Its physiological availability is limited by absolute shortages, immobility induced by low temperature, or inability of roots to take up water that does not have enough oxygen to allow root respiration. Carbon dioxide is probably always deficient, but its amount varies so little that it is not a factor that seems to cause differences between sites. Solar energy mostly varies on a regional scale, but slope aspect in steep landscapes can alter temperature regimes and consequently the water status of a site, even if precipitation is not affected. The effect of soil nutrients is exceedingly variable; in general, they affect growth rates more than species composition and tend to be less important than water.

The interaction of all these physical and chemical factors clearly determines what kinds of organisms can survive on a site and also how well they can grow there. However, these organisms themselves become additional constituents of the local environment, especially in how they may affect the particular species that the forester may be trying to grow there. In this sense, browsing animals, mycorrhizal fungi, insects, or other plant species, to name only some of the biotic forces, become part of the site or habitat. Regardless of the semantics, they are part of the environment that must be carefully considered in any silvicultural decisions. Choices of species are commonly dictated by pests and other damaging agencies, as well as by physical conditions of the site.

The Use of Site Analysis in Silviculture

Because it plays so fundamental a role in controlling the factors important to forest growth, the first consideration in site classification is the regional climate. Koeppen's classification of climate (Trewartha, 1968) is especially helpful because it is an attempt to categorize physical climatic data by using the vegetation as the most sensitive measuring device. However, if the silviculture relies entirely on species and genotypes already existing in the region, there is little real necessity of assessing the climate because it can be presumed that the plants are well adapted to it. However, it becomes critical to know about the climate when artificial movements of species or genetic strains are involved.

The usual scale for differentiating among sites for silvicultural purposes deals with smaller units within a landscape (Fig. 9.1). Site quality assessment is commonly incorporated into silviculture in the following way. The first step in managing a forest is to make maps of stands, which are defined principally by the species composition and age of the overstory. When treatment of a stand is being considered, site quality is assessed at one or more points in the stand by making observations or measurements; these usually involve tree heights and ages, but other methods are used as well, as will be described further in this chapter. The goal of such an assessment is either to determine quantitatively the productive potential of the land in order to predict yields, or to identify the stand as belonging to one of several classes of site types. So much emphasis is placed on yield prediction that the utility of site classifications for guiding decisions about silvicultural treatment and species composition is often overlooked. Such classifications provide diagnostic clues about those factors of the particular site that will control such variables as the susceptibility and vulnerability of trees to damaging agencies, the nature of problems with competing vegetation, and responses to various silvicultural treatments.

A problem common to many methods is that they do not provide a clear way of expanding the assessment of site conditions from a point or plot to a spatial scale (Carmean,

Figure 9.1 a–e   A series of photographs showing old-growth forest vegetation characteristic of markedly different sites, each requiring correspondingly different silvicultural treatment, all in northern Idaho. The lowest and driest site with a pure stand of ponderosa pine. *(Series of photographs by U.S. Forest Service.)*

Figure 9.1b   An open stand of ponderosa pine on a south-facing, dry slope at middle elevations. A closed stand of the so-called western white pine type, such as shown in Figure 9.1c, occupies the opposite north-facing slope.

Figure 9.1c A mixed stand of the western white pine type on a mesic north-facing slope. The nearest tree is a western larch; the one to its left is a western white pine; the one with vertically striped bark to the left of that is a western red-cedar; some of the understory saplings are white firs.

Figure 9.1d A 225-year-old stratified mixture of the western white pine type on a mesic valley-bottom site below the stand shown in Figure 9.1b. Among the other species present are western larch, western red-cedar, and western hemlock, with the two latter species in the lower strata.

238

Figure 9.1e   A pure stand of white-bark pine characteristic of very cold sites at high elevations in the same locality as Figures 9.1a–d.

1975; Rowe, 1984). The size of the landscape unit that has a uniform site quality is not necessarily equal to the stand size. Stand characteristics are the result not only of site conditions but also of an enormous variety of human and natural disturbances. Foresters generally must rely on their knowledge of correlations between soils, landforms, and vegetation in a region to make an estimate of the spatial extent of uniform site conditions. Frequently, no systematic technique is available to guide this step in the process.

The problem of the spatial nature of site assessment has been considered more closely for the implementation of intensive management. This interest has developed as much for the diverse objectives of ecosystem management as for timber production. The idea is that there should be two kinds of maps available for foresters to use in managing forest land— one of current vegetation in the form of a stand map, and the other of landscape units defining site or habitat conditions and giving predictions about suitability for different species and growth rates. The development of the second type of map would require a good deal of time and effort, but once available, it would reduce or eliminate the need for point assessments of site when considering treatment alternatives for a stand. For these reasons, there is increasing interest in developing site assessment techniques that lend themselves to mapping.

The use of various methods of site assessment in silvicultural practice will be described in this chapter. More complete reviews of these techniques can be found in Spurr and Barnes (1980), Hagglund (1981), Tesch (1981), and Vanclay (1992).

Use of Tree Growth As an Indicator of Site Quality

The most common methods of assessing site quality depend on using direct measurements of the productivity of the vegetation as the basis of classification. Prediction of the timber yield of a given single species on a site is frequently the specific interest, and the best criterion of this is a recorded history of production on the specific tract itself. However, this is available only where detailed records of careful management have been kept for many decades. In the absence of such records, it has become common to use rates of growth in height of the largest trees of a given species as substitute indicators of stand productivity.

The most common method in use is **site index**, the average height of the dominant and codominant trees of an even-aged aggregation of trees at some index age. This age is 50 years unless otherwise stated; the logical index age is ordinarily somewhat less than that of a normal rotation. Curves of average height over age for dominant and codominant trees have been developed using stem analysis techniques for many species in many regions (Fig. 9.2). In practice, site index is measured by determining the height and age of a sample of trees of a single species on a site; the appropriate standard curves are then used to extrapolate from the measured heights and ages to determine height at the index age. This height is referred to as the site index, and the inference that can be drawn from these calculations is that the trees of that species growing on that site will follow the height-growth pattern of the standard curve for the calculated site index.

The number of trees measured and the method of choosing those trees can affect estimates of site index (Zeide and Zakrzewski, 1993). The criterion of crown class is

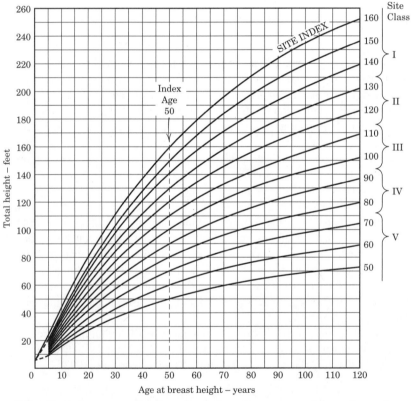

Figure 9.2 Site index curves for Douglas-fir based on stem analyses of trees in western Washington. These curves use age at breast height as the basis for measurement and indicate five site classes that are frequently used in place of a more precise estimate of site index. *Redrawn from King (1966).*

frequently used, with the average height of a sample of dominant and codominant trees being common; the average height of dominant trees alone is also used. The number of trees selected for measurement varies widely. More precise selection methods have also been developed, and two that are frequently used are: **predominant height**, defined as the average height of the tallest 40 trees/acre (100 trees/ha); and **top height**, defined as the average height of the 40 trees/acre with the largest breast-height diameters (Spurr, 1952; Clutter et al., 1983). Other systems use a fixed percentage of trees of a certain class rather than a fixed number. Although sampling techniques vary and are identified by different names, they do not differ from site index in concept.

The site index approach is based on the observation that the rate of height growth of the leading trees is well correlated with the productive potential of a site but is not altered significantly by ordinary variations in stand density (although extremely high or low density may influence the height growth of some species). In the simplest application, it is necessary that the index trees have always been the leading trees and not been suppressed at any time. Periods of very slow diameter growth observed in increment cores taken to determine age are used as an indication that a tree was overtopped for part of its life; such trees are generally eliminated from use as sample trees for site index measurement. Many dominant trees grew slowly in height only during an early establishment period, as occurs when seedlings compete with dense weeds or start as advance regeneration. One way of avoiding this problem is to assess age at breast height and to take that point as the zero height level. This approach has been incorporated in some standard site index curves (Fig. 9.2).

Use of site index has become widespread, partly because it is incorporated into the yield tables developed for many species. In fact, site index is sometimes expressed in terms of the mean annual increment of fully closed stands instead of height, but this is simply a matter of using the value from the associated yield table. The basis for determining site class of a stand is still the measurement of the height and age of a sample of largest trees.

Site index is most useful for conifers because they have a well-defined height that is relatively easily measured. Many hardwood species have rounded, spreading crowns that are more difficult to measure. The range of heights obtained in even-aged hardwood stands, even with careful measurement, is sometimes so wide as to make site index calculations nearly useless.

Determining tree age from increment cores is the most time-consuming part of measurement and often limits the data collection to only a few trees. Site index is easiest to use in plantations or other stands with a narrow range in ages, where only heights need to be measured for most trees. These are the situations in which a sample size of as many as 40 trees/acre (100 trees/ha) is sometimes used. In many other cases, as few as five trees are measured for stands of many acres.

In cases where measurement of a large sample of trees can be used in conjunction with height curves developed for that region, a precise estimate of site index can be determined. However, in most situations, there are enough sources of error in data collection, extrapolation using height-growth curves, and other aspects of ''measuring'' site that it is well to refrain from the spurious precision implied by expressing site index to the nearest foot or meter. One common antidote to this is the practice of assigning sites to not more than five or six rather broad categories of site quality classes, commonly denoted by Roman numerals (Fig. 9.2). This is often a sufficient level of precision for making decisions about silvicultural treatments.

The different uses of site index can be seen in the example of management of eastern white pine in the Northeast. White pine can grow on a wide range of sites, with productivity increasing on landforms and soils with greater moisture-holding capacity. Site index curves and associated yield tables can be used to predict the productivity of pure pine stands. However, because of the difficulty of controlling early hardwood competition on moist sites, pine is generally grown in pure stands only on relatively dry soils. The site index of oak is frequently used as a standard to classify sites based on the competitive ability of hardwoods. Sites have been divided into three categories, with recommendations for converting to pure pine, favoring mixed pine and hardwood, or favoring pure hardwood stands being associated with increasing values of oak site index (Lancaster and Leak, 1978).

Some methods of expanding or simplifying the use of site index have been developed. If the tree species of interest is not present on the site, procedures can be developed to predict the site index of one species from that of another (Doolittle, 1958; Foster, 1959). One can also use the height-intercept method in which the index variable is the number of years required to grow from one stated height level to another. This method works with trees that produce one internode annually, so that height growth can easily be determined for levels close to the ground.

Other methods developed to assess site conditions are not devised to replace site index entirely, but rather to overcome some of its limitations. The underlying rationale is to find methods that (1) lend themselves to dividing a landscape in mappable units; (2) work when no trees are present on a site; (3) work in tropical regions where trees do not form annual rings; or (4) are based on field observations that do not require measurement of tree heights and ages. Many of these involve prediction of site index from other parameters.

Understory Plant Indicators and Habitat Types

Species composition can also be used to assess potential productivity, limitations set by environmental factors, and species suitability. The best-known methods, originally developed in Finland by Cajander (1926), involve use of the lesser vegetation that grows beneath stands. The underlying principle is that some of the small plants are much more sensitive to variations in site factors than large trees. Some such plants have high indicator significance, whereas others, those with broad environmental tolerances, have little. This mode of classification seems to be most successful where climatic conditions are restrictive, as in the boreal forests where it was developed. In those and other forests where large areas are dominated by a single overstory species, distinguishing among several understory species can be used to predict site index without measuring tree heights or ages. There are, for example, large areas covered with Douglas-fir, but much information can be obtained about the site by noting whether the understory has sword ferns or rhododendrons. In most cases, understory plant indicators are no longer used alone but are incorporated into more holistic approaches.

Especially in western North America (Daubenmire and Daubenmire, 1968; Steele et al., 1981), some success has been achieved with using the late-successional plant communities as the basis for site classification. In this approach, a set of site units, referred to as **habitat types**, are identified by the characteristic overstory and understory species that occupy certain elevation and physiographic conditions; the names of the habitat types are taken from dominant species in each layer (e.g., *Pseudotsuga menziesii/Symphoricarpus albus* habitat type). A taxonomic key based on the presence of indicator species is developed and is used as the basis for field identification; field techniques involve the use of **relevés**—a method of plot measurement that quickly assesses the relative abundance of species in each vegetation layer without detailed measurement. Precise measures of species abundances are not necessary because the presence or absence of certain indicator species is often the criterion most indicative of site conditions.

Habitat-type mapping has been used throughout the Rocky Mountain region as the basis for predicting site index, assessing wildlife and livestock forage productivity, estimating water production, and other purposes. This mode of site analysis and mapping works best where species composition is mostly the result of natural processes, as in the series of sites depicted in Fig. 9.1. Even in those cases, it is necessary to use elevation and other landform features to guide mapping where late-successional vegetation is not present. This is most easily done where strong variation in elevation and topography exists.

The habitat-type approach has been used on a more limited basis in areas that have been heavily disturbed by agriculture, such as the Great Lakes region (Kotar, Kovach, and Locey, 1988) and the White Mountains of New Hampshire (Leak, 1980, 1982). Important differences in site types have been determined by studying correlations between physical site characteristics and the composition of relatively undisturbed vegetation in the limited areas where such vegetation could be found. In these regions, landforms and soils have been used as the basis for mapping because of the scarcity of undisturbed vegetation over most of the landscape. The importance of landform in identifying habitat types in the field is evident in the White Mountains classification system, in which the type of glacial deposit is the fundamental basis for mapping land into one of 11 habitat types and is also used as the name for each type, instead of using the names of indicator species (Leak, 1982).

Analysis of Soils and Topography

Considerable interest has long existed in developing ways to predict potential forest productivity directly from soil and topographic variables, without using trees or other vegetation as indicators. Many studies have developed predictive mechanisms in the form of multiple regression equations in which site index is determined from a number of independent site variables (Armson, 1977; Pritchett, 1979). This approach to site classification often requires deep digging to examine the soil structure and to collect samples for subsequent determination of physical and chemical properties (Soil Survey Staff, 1975). Because it has generally proven impractical, there are few examples of such systems being put into practice. These techniques have been most useful where large plantation projects have been established in areas in which forest vegetation was sparse or absent. No alternative methods were available in these situations, and financial resources were sometimes available for detailed soils analysis and mapping.

The greatest value of soil-site studies has generally been as basic research that has elucidated the most important factors controlling forest growth, even if some of these are not easily measured. Most have identified the variables that define the availability of water as being the most critical; these include depth to bedrock or hardpans, which determines the depth of the rooting stratum, and soil texture, which governs the capacity of that stratum to store water (Spurr and Barnes, 1980). Variables that measure nutrient status have generally been less important. To be useful in silvicultural practice, this technique depends on assessing factors that can be measured on one visit to the site. Therefore, the annual regime of soil moisture must somehow be deduced from appropriately selected, semipermanent, observable parameters of soil and site.

The shape of the terrain is often a key, usually because it tells so much about the water relations that are the chief ruling factor. The lower slopes of most hillsides are, for example, concavities that usually receive more water than falls on them from the sky. Water from the convex hilltops seeps down to them, leaving the upper slopes robbed of soil moisture. The boundary between convexity and concavity sometimes defines differences in species composition and productivity. Unless they have sandy soils, very flat areas can be poor sites full of ponded, oxygen-deficient water. If the terrain is steep, the slopes that face the sun can be very dry while the opposite shaded ones are comparatively moist. This difference can be enough to induce grassy brushfields on sunny slopes and closed forest on shaded ones in climates with long dry seasons.

An analytical and predictive understanding of site variables can often be based on good knowledge of geology and especially geomorphology. The factors that have weathered rocks and moved the products of natural erosion around determine the composition and shape of the parent materials from which soils are formed. If it is known how the parent materials got where they are, it becomes possible to learn a good deal about the extent of a particular kind of terrain form and its ability to support forest growth merely by viewing it from a distance or on a topographic map. Knowledge of surficial geology helps not only in determining species composition and predicting yield, but also in building roads, planning logging, and controlling erosion.

It might be expected that the shortcomings of the soil analysis techniques described above would be overcome by using maps of soil types. These are available for many parts of the world and would seem to contain information on the variables that are difficult to determine in field sampling. However, these maps have proven to be of surprisingly little value for silvicultural use. Nearly all soil classifications have been devised with agricultural needs in mind and focus on the characteristics in the top meter of soil. Although soils mappers must have a detailed understanding of the relationship between topography and soils, landform characteristics are generally not an integral part of the soil series that are mapped, yet they control much of the ability of large perennial plants to prosper (Rowe, 1984).

Each kind of landscape—mountain ranges with volcanic deposits, ancient peneplains with heavily weathered soils, rolling terrain with glacial deposits, and many others—has a distinctive set of landforms. Knowledge of these is often more important to foresters than are the details of soil structure. Maps of soils that use landforms as an integral part of the definition of mapping units would be valuable for silvicultural planning (Rowe, 1984).

One example of the effects which different kinds of landforms and soils have in controlling the composition and vigor of forest vegetation is found on the flood-plains of snakelike meandering rivers. These continually destroy and rebuild the ''bottomlands'' through which they flow. Figure 9.3 shows most of the kinds of terrain; differences in elevation of as small as 3 feet can cause major changes in forest composition. Hodges (1995) described these remarkable differences in the bottomland hardwood forests of the southeastern United States. The highest terrain, best soils, and richest species composition are found in the ''fronts'' right on the banks of the rivers where fine sand is deposited

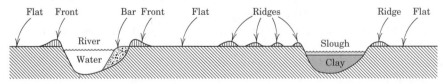

Figure 9.3 A cross-section, in exaggerated vertical dimension, of the different landforms that develop on the flood-plains of meandering rivers, shown at a time when the water level is normal (*i.e.,* not at flood stage). See text for description of landforms and vegetation.

when floodwaters are decelerated as they spill over the banks of the river. "Ridges" are former fronts that have been left when the course of the river shifts sideways or a loop is suddenly cut off. Their luxuriant forest vegetation, which may include such species as cherrybark oak, yellow-poplar, and sycamore, differs little from that of the fronts. The series of ridges shown in Fig. 9.3 is a place where the river course gradually shifted from left to right before it was cut off. The "slough" that formed in the cutoff loop is a place where water stands most of each year and very tight clay formations are deposited. The species that grow slowly there are those such as bald-cypress, tupelo gum, and water hickory that can survive in standing water. "Bars" composed of coarse sand form along the main river where the current is decelerated slightly as it flows around bends. Fast-growing, intolerant, light-seeded species, such as willows and cottonwood poplars, start on the bare soils exposed there. The "flats" are level soils full of clays that are deposited there when water stands for many weeks after typical floods. These wet, poorly aerated soils support such species as red maple, green ash, sweetgum, Nuttall oak, and sugarberry that can withstand physiological dryness. Similarly distinctive guilds of species grow on these kinds of landforms along meandering rivers throughout the world.

Ecological Site Classification

Most of the site classification methods described in this chapter are not used alone, but are combined with others to make them more useful and more efficient to apply. Some multifactor systems explicitly recognize that a combination of the four basic parameters—climate, landform, soil, and vegetation—provide the most complete approach. In many regions, maps already exist for all or most of these factors, and some efforts at multifactor classification are based on using these multiple layers of information to define a set of distinctive site units. However, unless these are based on recognition of the relationships between these four elements, such systems cannot be considered "ecological" (Barnes et al., 1982), and may be difficult to implement.

Ecological site classification is based on a hierarchy with climate as the dominant factor at the regional scale, and landform and soil at the landscape unit scale (Bailey, 1995). These factors generally vary along gradual gradients rather than across sharp boundaries. The importance of vegetation in the development of such systems comes in defining distinctive climates, landforms, and soil characteristics that are correlated to important differences in vegetation composition and productivity. Thus, an ecological approach would seek to define a classification of these ecologically distinctive units before mapping could commence.

This approach is the basis for defining different climate types, as in British Columbia (Pojar, Klinka, and Meidinger, 1987), where large geographic units with similar climates and forest types have been defined and mapped. These "biogeoclimatic" units are useful for such things as identifying fire-climate zones for protection purposes or defining zones for the movement of seeds, but their most fundamental use is for defining the boundaries within which a smaller scale site classification system can be applied.

Site units ranging in size from hundreds to ten or fewer acres are of greatest value in silviculture, and are defined by landform and soil within a climatic zone. These fundamental units are considered "natural" or "ecological" units because they are not based on a derived feature (such as the height-growth rate of a particular tree species). Rather, they are defined by a combination of climate, landform, and soil that controls the energy and material inputs to the ecosystem and are correlated with characteristic vegetation. Mapping these units can be done using a combination of topographic and surficial geology maps, examination of soils, identification of indicator plants, and other techniques for efficiency. Because each unit defines a consistent set of environmental conditions, it can be used to predict many parameters, including but not limited to stand productivity.

The potential of an ecological site classification system is illustrated in Fig. 9.4. The system shown was developed for a glaciated, rolling landscape in the Great Lakes region, where northern hardwoods, red pine, and white pine are the principal species (Barnes et al., 1982). A principal use of the system has been to identify those sites that are suitable for conversion to red pine and those that are best managed for mixed hardwood stands

Figure 9.4 a-c   The major features of an ecological site classification system for the McCormick Experimental Forest in the Upper Peninsula of Michigan.

DRYLAND SITE UNITS

I.   Deep soils (bedrock below 100 cm)

    A.   Level to gently sloping terrain (usually 0–5 percent)

        1.   Excessively drained sand; jack pine/Vaccinium

        2.   Somewhat excessively drained sand and gravel; sugar maple/Maianthemum

        3.   Somewhat poorly drained sand; maple–yellow birch—conifer/Clintonia

    B.   Moderately to steeply sloping terrain (usually >5 to <30 percent)

        4.   Well-drained loamy sand; sugar maple/Gymnocarpium

        5.   Moderately well-drained sandy loam on northerly aspects; sugar maple/Viola

        6.   Excessively drained sand on steep southerly aspects, white pine—hardwoods/Maianthemum

---

Figure 9.4a   A portion of the classification of land units, defined by landform and basic soil characteristics, with indicator species listed; a total of 21 site units were used in mapping.

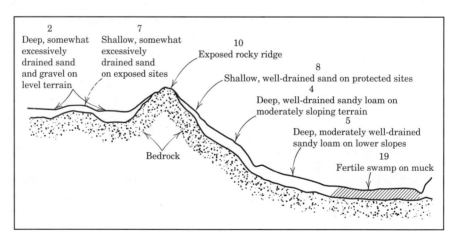

Figure 9.4b   A physiographic cross-section showing the relationship of some of the site units to surficial geology and landforms.

Site unit 2: Flat, outwash sand plain dominated by low-vigor and low-quality sugar maple: the highest sand content, and greatest susceptibility to fire of any sugar maple ecosystem.

- total height of old-growth sugar maple 21 m (low productivity)
- no erosion hazard
- suitable for mechanized equipment
- low recreation and wildlife values
- moderately high fire hazard
- light competition from hardwoods

Recommendation: convert to red pine for greater productivity; fertilize to improve pine yield.

Site unit 5: Lower slope with moist, relatively fertile sandy loam soil; dominated by high-quality, fast-growing sugar maple; significantly less sand, more nitrogen, and higher pH than units 1, 2, and 4.

- total height of old-growth sugar maple 28 m (high productivity)
- moderate erosion hazard
- suitable for mechanized equipment during July through November
- moderate recreation value created by large trees and spring flora
- low fire hazard
- heavy hardwood competition

Recommendation: Manage for high-quality hardwoods.

Figure 9.4c A more complete description of two site units, giving assessments of productivity and other parameters and suggesting appropriate management objectives. Modified from Barnes et al. (1982). Copyright 1982, Society of American Foresters; used with permission.

dominated by sugar maple. Because the site units are fundamental divisions of the landscape, they should maintain their usefulness as new management questions arise.

Choice of species and other silvicultural decisions should be guided by an analysis of site or habitat, regardless of the method of classification or how the categories are named.

SELECTION OF SPECIES AND PROVENANCES

Adaptation to Site

The first consideration in determining which species to grow is the degree of their adaptability to the site. Therefore, the most logical action is to choose from the species that occur naturally in the region—that is, the **native species**, whose natural ranges include the site in question. These have the virtue of a long history of inherited adaptation to their environment. However, this kind of adaptation favors the perpetuation of their own kind, which does not necessarily include high production rates, stem straightness, or other humanly desirable attributes.

It is also possible to bring species from other areas; these are referred to as **exotic species** in the area where they are planted outside of their natural range. Some confusion exists about this term because it is sometimes used to denote movement of species across national boundaries, but that distinction is not very useful. The natural range of a species, controlled largely by climatic conditions, is the most important criterion. By either definition, eucalyptus grown in California is an exotic; Douglas-fir seed from Canada grown a short distance away in the United States is not an exotic in the ecological sense. Sometimes, a species is extended beyond its range but not into an entirely new area; an example of this is the range of red pine being extended south from its natural limit throughout the northeastern United States. Such species are often referred to simply as **non-native**, but they are exotic in terms of ecological and silvicultural management considerations.

Thus, there are three basic alternatives in choosing species: (1) accept the native species adapted to the local climate, (2) move exotic species from distant places with similar climates, or (3) move non-native (exotic) species relatively short distances beyond their natural range. Dramatic movements of species across major geographic barriers have been more successful than seemingly modest extensions of their natural ranges. Intercontinental movements ideally will place the species in a location where they were adapted to grow but had been prevented from doing so only by intervening oceans or inhospitable land surfaces. On the other hand, extensions of the natural ranges of some American conifers by distances of 100 miles or less have sometimes proven to be either marginally successful or highly unsuccessful. This has been true of the southward extension of the range of red pine and the northward movement of loblolly and slash pine. The use of exotics is discussed further in a later section of this chapter.

It is also important that the species to be used on a site are drawn from those that not only survive there, but also grow well, preferably more competitively than others. This consideration should include not only growth rates in average conditions, but also during the less common incidents that are inevitably part of the site. Trees must survive long periods through seasonal changes in temperature and precipitation, while spread over large areas that can receive little artificial protection. Events such as the coldest day of the century, infrequent outbreaks of defoliating insects, rare severe fires, or bad windstorms must be anticipated. For the forester, these should not be surprises but events for which one plans by fostering resistant species or other means. What happened once can and probably will happen again. Such factors are dependent not only on climate, but also on specific microclimatic and soil conditions. This is where the site classification techniques described in the first half of this chapter become important.

Utility of Species for Different Objectives

Although it is often ranked as the first consideration, the utility of a species for wood products or other objectives is best considered somewhat secondary to its adaptability. Difficulties commonly arise if one species is planted up hill and down dale because its timber brings the best price at the time of planting. Some of the worst problems ascribed to "monocultures" derive from this approach. Choices made by nonforester executives in distant offices have a bad history of squandering the monies that these people are so proud of being able to manage.

Certainly, however, there are reasons to favor particular species from among those well-adapted to a site, based on their utility for wood and paper products. There is a long history of choosing conifer species over angiosperms for many timber purposes. One important reason is that evergreens are more productive than deciduous trees in many (but not all) climates and sites. This amounts to a choice between conifers and angiosperms in most temperate and boreal regions, but not in the tropics and subtropics. In addition, conifers have an excurrent growth form, with a greater proportion of the wood produced being contained in a single central stem. This is in contrast to the more decurrent form of many angiosperms, with much of the tree's wood production being contained in the large branches forming the crown. Conifer stems also tend to be straight, whereas the stems of many angiosperms bend toward gaps in the canopy.

The usefulness of different species is also affected by their wood characteristics. The differences in wood anatomy among species gives each a unique set of characteristics for use in wood and paper products, but general patterns in species use can be understood by considering only a few of these characteristics here. More detailed treatments have been written by Panshin and DeZeeuw (1980) and Hoadley (1980). Specialty uses exist for nearly every kind of wood, but the main commodity uses are for solid wood for constructing buildings and furniture, and for paper and cardboard for packaging and printing.

For solid wood products used for structural purposes, wood density is the single most important characteristic because it controls load-bearing strength. It is generally expressed as specific gravity, which is the ratio of wood density to the density of water. At specific gravities below 0.30, wood is not strong enough to be used for most building construction purposes. Wood in the range of 0.40 to 0.55 is ideal for such purposes because it has adequate strength but is not so heavy that it cannot be easily transported and milled, or nailed and cut with hand tools or small power tools. Strength and durability to denting and scratching increase significantly above specific gravity of 0.60, but these woods are no longer as easy to work with. Above levels of 0.80, there are serious limitations to machine milling and sometimes even to felling the trees.

Conifer wood has a transition within each annual ring from low-density earlywood with large thin-walled cells to high-density latewood with thick cell walls. The strength of different species depends primarily on the density of the latewood. Species with the highest density, including Douglas-fir and many species of hard pine, are ideal for framing lumber and plywood. These species have the structural design of an engineered composite, in which sheets of dense material are embedded in low-density matrix, producing a strong material that can still be easily milled, cut, and nailed. Other conifers, such as spruces, firs, and hemlock, have latewood of lower density; this gives them lower overall strength, but they are still useful for many structural purposes. A very mild transition from earlywood to latewood density is found in such species as the soft pines and western red-cedar. These are generally not used for structural purposes, but they are easily machined and are important where load-bearing strength is not the primary consideration. Soft pines are prized for furniture and interior finish, and the rot-resistant properties of western red-cedar make it valuable for uses where it will be exposed to the elements.

Many of the most useful hardwood species produce wood with a density somewhat greater than that of the high-density conifers. Diffuse-porous species and tropical species lacking growth rings also have a more uniform density within each growth ring. Wood of this density, from such genera as oak, maple, beech, teak, and some members of the eucalyptus genus, is used for flooring and furniture, where durability is required. Some tropical species of even higher specific gravity are used only where minimum millwork is required, such as for industrial flooring, railroad ties, and construction where extraordinary strength is needed.

Hardwood trees of small diameter or poor stem form that cannot be used for the markets described above have often been underemployed. They are useful for pallets, barrels, and crating, but these markets are so limited compared to the supply of such trees, that it is often difficult to recover the costs of removing these trees in thinnings. Some hardwood species also have wood that is too low in density for many uses, but this problem appears secondary to that of size and form. For example, yellow-poplar has been used for a variety of purposes over the years, in spite of its relatively low density; its utility is increased by the fact that it has a conifer-like excurrent growth form with large straight stems and rapid growth.

The development of engineered solid wood products, in which chips are reconstituted into panels and framing lumber, has increased the use of small trees, and has opened the possibility of using hardwoods of low density or poor stem form such as aspen poplar and red maple, which for many years had low timber value. The uniform density of the wood of these diffuse-porous species is favored for some of these products because of the uniform chip quality produced.

The production of paper from wood fibers depends on somewhat different properties of wood. (It should be noted that in wood technology, the term *fiber* is used for both the fiber cells in angiosperms and the tracheids of conifers). The strength of paper depends on having long fibers with relatively thin cell walls, which allows the cells to collapse during the paper-making process. In a sheet of paper, wood cells lie in a random arrangement like pine needles on the forest floor. Pulp that is composed of long, flattened cells with most of the lignin and any resins removed produces the strongest cellulose bonds between cells, and thus the strongest paper. Conifers have a distinct advantage in cell structure because conifer tracheids are longer than hardwood fiber cells, and their cell walls are generally much thinner.

Among the conifers, those species with latewood of lower density have cells that collapse more easily. When paper was first made from wood on a large scale, it could only be done from species with long fibers, moderate to low density (thus thin cell walls), and very low resin content. Spruces and true firs fit these criteria, and were the chief species from which paper was made. Paper-making initially was done by mechanical pulping, but the strength of this paper was limited by the lignin content which interrupted the cellulose bonding. Mechanical pulping of these species is still widely used for producing newsprint.

Much paper is now made from the fibers of hard pines, but this was impossible for a long time, because of the high resin content and the high proportion of thick-walled latewood cells that resisted collapse. Only by the development of two processes was it made possible: (1) the Kraft method of chemical pulping, which used alkaline solutions not only to remove resins, but also to convert them into turpentine and other useful chemicals; and (2) mechanically "beating" the pulp, which partially shreds the outer layer of cellulose of the cell walls, thus giving greater surface area for intercell bonding, which compensates for the cells' resistance to flattening. The development of these processes in the middle of the twentieth century had great impact by allowing the use of the long fibers of hard pines and Douglas-fir. This resulted in strong brown paper and corrugated cardboard that could be used for packaging, and an industry developed to turn hard pines into such material. Pulp made by the Kraft process can also be bleached to produce white printing papers. These high-density species are so valuable for solid wood products that paper-making is not necessarily the most logical principal management objective for them, but it is a valuable outlet for thinnings and sawmill residue.

These advanced pulping technologies also made it possible to make paper from the very thick-walled fiber cells of hardwoods. Because these fibers are short, paper produced from hardwood fibers alone is not strong, but a mix of long conifer and short hardwood fibers can produce an adequately strong, fine-surfaced paper, the kind of which this textbook is made. The use of papers for printing has increased so greatly that the world demand for hardwood pulp is approaching that for conifer pulp.

This very simplified discussion of wood technology is included here mainly to demonstrate that certain conifer species have characteristics that have made them highly useful for producing the basic commodities of building materials and paper and cardboard, using relatively simple techniques. This has been one factor that has led to the conversion of large areas to conifers, but another factor is that conifers have better ability than many hardwood species to grow in disturbed sites or with highly competitive grasses. In parts of Europe and Great Britain, the shift from a natural deciduous forest to conifers was the result of deforestation of the native forests for agriculture, followed centuries later by afforestation with conifers. In some cases, forest industries developed where agricultural abandonment was followed by the establishment of conifers from natural seedfall. Examples are the white pine industry of the northeastern United States in the early part of this century, and the southern pine industry in the southeastern part of the nation in the second half. A major use of exotics throughout the world has been to bring conifers to parts of the world lacking species with their fine wood characteristics.

Fortunately, the technology of wood utilization has become continually more versatile and less selective as to species. A shift has occurred from demand for certain wood properties supplied by conifers to a much wider ability to use all kinds of tree species. Much of this technological shift has occurred during a 50-year period, which is approximately the length of a rotation of fast-growing temperate species. This illustrates an important point—that wood utilization tends to change faster than foresters can change the forest composition. Products are continually being made from trees that have been regenerated with a different goal in mind. Consequently, it is perhaps best to grow species that are adapted to a particular site, choosing from among the inherently useful species that will grow there.

There is always the risk that any chosen species may fall completely from favor. However, this has actually happened only when the utility of the wood was limited to specialized purposes. For example, some of the first silviculture in America was based on the fast growth of insect-deformed, branchy eastern white pines on abandoned agricultural lands in New England. The wood was good for making wooden boxes but not much else. Such material became technologically underemployed when it became cheaper to ship goods in corrugated boxes made of brown Kraft pulp. In contrast, the market for good white pine lumber remains excellent. It is unwise to be content with growing trees that are, regardless of species, of such poor quality or small size that their use is limited to some very specialized purpose. It is logical to ponder whether hard pines should be grown in ways that make them suitable only for pulp for making cardboard boxes, or similarly, if potentially high-quality hardwoods should be grown in ways that limit their use to burning for industrial biomass energy production.

It would be wrong to dwell on these large-scale commodity uses to the exclusion of others. For much of the world, the most highly valued wood product is fuelwood for home heating and cooking. The choice of species in these cases leads to hardwoods of high density and low moisture-content, which sprout vigorously from stumps or roots and thus are amenable to the coppice system of management. There is also great interest in choosing **multipurpose tree species** for use in agroforestry. In this case, the ideal species would have wood that is useful for fuel and light construction, foliage that is highly nutritious for livestock, and fruits that are food crops for humans.

Frequently, the choices to be made revolve not entirely around the species themselves but around the stage of natural succession or of natural stand development to be maintained. In general, later stages will be easier to maintain, but some earlier stage may have more productive or valuable species. The choice can also be viewed as that of an assemblage of species, both plant and animal, with pests and parasites included, representing some stage of vegetational development and not just a species for timber. The choice of successional stage or general stand structure is usually adequate for meeting most objectives in watershed and wildlife habitat management, but favoring a particular tree species or genus is sometimes necessary. Such choices are discussed in the chapters devoted to those management objectives.

Selection of Provenances

Many species extend over such wide geographical ranges that, as a result of natural selection, they develop different forms suitable for different environments. The genetic variation within a species may occur as a continuum across the range, or where the range has been broken up by barriers such as mountain ranges, it may occur as distinct ecotypes. In either case, the geographical origin of seed is called its **provenance**. If ecotypes develop which are distinct in their geography, morphology, and genetics, they are recognized as **subspecies** or **varieties** (no clear distinction exists between these two terms), which are formally described and given Latin names. For example, lodgepole pine has three recognized subspecies: *Pinus contorta* ssp. *contorta* on the Pacific Coast, ssp. *murrayana* in the Cascade and Sierra Nevada Mountains, and ssp. *latifolia* in the Rocky Mountains. Where ecotypes are less distinct, they are referred to as **races** and are usually given a name associated with their geographical location (Fig. 9.5). Not all of the variation among provenances is due to environmental adaptations; some appears to be the result of more random processes, but several of these traits may still be very important in meeting human objectives.

As is the case with species, the safest provenances to use are the indigenous ones. If a native species is being used, there is usually the temptation to bring in seed from a provenance of more favorable climate. In general, trees from lower elevations, lower latitudes, or regions of higher precipitation grow faster when moved to higher elevations or latitudes or regions of lower precipitation. However, the more rapid growth of these trees is gained at the expense of increased risk of damage. This is because trees growing in harsher climates have become adapted to those conditions by developing such characteristics as early stomatal closure during periods of water stress or very early bud set and cessation of shoot elongation. These characteristics reduce photosynthesis and growth, but also favor survival. Movements of provenances adapted to moderate conditions may produce more rapid growth for considerable periods, but then when predictable climatic extremes occur, the result will be outright damage from cold or drought, or reduced vigor leading to insect or disease attack. The wisdom of using the faster-growing plants may depend on whether thinning, weed control, fertilization, or some other treatment will reduce these effects. Rules have been developed for some species regarding the safe distance to move seed in elevation or latitude. Sometimes, general guidelines are given, suggesting that seed can be moved safely up to 100 miles (160 km) in latitude or 1000 ft (300 m) in elevation. Actually, these guidelines are useful only for indicating the order of magnitude of such limitations, because they will differ for each species and location (Zobel, and Talbert, 1984). For example, there is evidence that Douglas-fir is specialized in its adaptation to local climate in the Northern Rocky Mountains, whereas western white pine is much more genetically homogeneous across its range in the same region.

Figure 9.5 Foliage characteristics of two distinct races of ponderosa pine. Left: Open, plume-like arrangement of long, slender needles of North Plateau race from eastern Oregon. Right: Compact, brush-like arrangement of short, thick needles typical of race found east of the Continental Divide from Colorado northward. The needles shown on the right are more resistant to frost and winter injury than those on the left. (*Photographs by U.S. Forest Service.*)

Exotic Species

Exotic plant species are used widely in agriculture, horticulture, and landscaping. Although their use is less widespread in silviculture, in some regions of the world they dominate forest management. The principles of using exotic tree species in silviculture have been reviewed by Zobel, van Wyk, and Stahl (1987).

Generally, an introduced species should come from a locality whose climate and soils are closely similar to those of the home region. The climate of a site is controlled mainly by its latitude and position on the continent. Climate can be further defined by the following quantitative parameters, which have proven useful for mapping areas of the world suitable for particular tree species: (1) mean annual temperature; (2) mean maximum temperature of the hottest month; (3) mean minimum temperature of the coldest month; (4) mean annual rainfall; (5) rainfall regime (uniform or seasonal); and (6) length of any dry season. These parameters identify both average conditions and climatic extremes that may cause damage to trees. The matching of soils can also be important, at least to the extent that the movement of species to extremely different soil types from which they naturally grow (from acid to basic soils or from clays to sands) may cause problems.

Exotics have been used largely to supplement the native species in areas that naturally have low species diversity; this has often consisted of adding conifer species to regions where they are few or entirely lacking. In other regions with a large complement of native species (particularly in the tropics), a few well-known exotic species have been used to replace the native forests just to simplify silvicultural techniques. The latter practice is more difficult to defend than the former.

Use of exotics in North America is rather limited because in every region of the continent there are many species from which to choose. Rather, North America has been important as a donor region. Many conifers from the West Coast of North America have been remarkably successful at the same latitudes in Western Europe; Douglas-fir, Sitka spruce, and lodgepole pine are among the most important, and dominate much of the forestry of Great Britain, Ireland, and parts of Sweden. The forest economies of many countries in the warm temperate and subtropical regions of the Southern Hemisphere are heavily dependent on slash, loblolly, jelecote, and other pine species introduced from localities of comparable climate in the southern United States and Mexico.

The single most widely exported tree species in the world is Monterey (radiata) pine. This species is native to a few small areas on the coast of California; it has been limited from expanding into adjacent areas by too low precipitation or too low winter temperatures. It is planted in the Southern Hemisphere in Australia, New Zealand, Chile, and South Africa, but has thrived only in places with the same general climatic pattern as the California coast. Monterey pine is an example of an exotic that can grow better in the new environment because the climate is more favorable than the native one. The territory suitable for Monterey pine as an exotic is much larger and often has more rainfall than the small parts of California to which geologic events have confined it. The vast oceans of the Southern Hemisphere provide a mild climate that is very favorable to this and other tree species from the more harshly variable climates of the Northern Hemisphere.

Some of the widest use of exotics has been in the tropics and subtropics. This has included the planting of tropical pines from Central America and the West Indies, most notably Caribbean pine, throughout the tropical regions of the world. Also used are a large number of exotic hardwoods, especially species of teak, eucalyptus, and gmelina. Among the hundreds of species of eucalyptus are species that are adapted to all kinds of climates in the tropical to warm temperate latitudes.

Exotics sometimes are very successful because they have left their pests behind. This advantage may continue for extended periods, but it is sometimes counterbalanced by the fact that natural pathogens and predators of pests have also been left behind; thus, when outbreaks do come from a new pest becoming adapted to the tree or an old pest being transported inadvertently from the native range, they may have more serious effects than in the native environment. For these reasons, all phytosanitary precautions must be observed in transporting plant materials for introduction; the movement of seedlings or soil is generally prohibited. Disinfected seeds are the most safely transported forms; however, this becomes a problem when attempts are made to move some tropical species whose seeds can survive only a few hours.

Introduced species should be used with caution until their safety and superiority have been tested in trials extending over all or most of a rotation. The success or failure of particular exotic species cannot be presumed entirely on the basis of generalizations. Only long-term tests will tell, and even then there can be no certainty that the most appropriate genotypes were being tested. It is highly important that seed be selected from the appropriate provenance. Many failures have resulted from choosing provenances that were poorly adapted to the new climate, just as would occur in moving seed within a species' natural range. When a species has been introduced, it is wise to develop a **land race**, which is in essence an artificial provenance adapted to the new location. This is done by letting nature take its course in the stands developing from the initial plantings, resulting in the elimination of those trees that are poorly adapted to the new environment. The best of the remaining trees are then used for seed collection for establishing plantations. It is important that the original plantings be derived from at least several hundred parent trees, rather than just a handful, as is sometimes the case with the very first introduction of a species into a new region.

Problems With the Use of Exotic Species

The problems of exotic species of all kinds are becoming widely recognized. Plant and animal species are being moved around the world (both purposely and accidentally) at increasing rates. A small fraction of these introduced species prove to be highly adapted to the new environment and become invasive, competitively excluding native species. Hawaii is an extreme example of a place where introduced plant and animal species have become more prevalent than the native species. The potential for such a problem to exist with an introduced plant species can be predicted to a certain extent by the ''weediness'' of its characteristics. Plants that invade disturbed sites or reproduce clonally are likely to cause the worst problems. Herbaceous species with short generation times will be more problematic than trees; however, the trees usually chosen for plantation silviculture are shade-intolerant species that colonize disturbed sites. In a number of cases throughout the world, trees that were introduced in part for timber production purposes have become the focus of eradication efforts. Such problems appear to be greatest in tropical and subtropical areas. In the United States, some of the most damaging invasive trees are tropical ash and silk-oak in Hawaii and melaleuca in Florida. Hughes (1994) has outlined procedures to reduce the risk of introducing species that may become invasive pests. The most obvious and important of these procedures is to minimize the use of exotics.

Particularly in the tropics, some foresters have succumbed to the temptation to carry their favorite species with them as they move from one region to another. Species trials established in one location in the tropics have tended to look very similar to those established elsewhere, concentrating on the same set of eucalyptus, teak, gmelina, and pine species. Most native species are excluded from use simply because there is little knowledge concerning their growth characteristics. New interest has arisen in establishing native species trials for use in plantation forestry, and perhaps not surprisingly, some native species appear to outperform the favored exotics (Butterfield and Fisher, 1994; Hughes, 1994).

Some of the strongest public opposition to exotics in forestry has involved the planting of tropical eucalypts. This practice has received such a long list of charges against it that a worldwide review of such problems has been conducted (Poore and Fries, 1985). The alleged problems include excessive use of water in semi-arid areas, reducing levels available for other uses; creation of erosion problems; depletion of soil nutrient levels; and production of poor habitat for wildlife. These problems were found to recur in various combinations of the large number of eucalypt species and locations studied. However, many of these problems do not appear to be related to eucalypts in particular or to its status as an exotic. Rather, they result from the use of short-rotation plantation monocultures of a highly productive species with dense roots, which take up much water and nutrients and eliminate understory vegetation through competition. These same conditions may occur if the species involved were a vigorous native species used in this fashion. This does not make the problems any less real or serious where they occur, but it is useful to separate effects that arise from the use of exotics from those that may come in general with intensive plantation silviculture.

GENETIC IMPROVEMENT

Choices of species are choices of genetic material made on a grand scale. Those discussed in the previous section are rather passive because they involve finding and using the best populations already available. It is possible to do better with more active intervention in the processes by which trees pass their characteristics to their offspring. The techniques of forest genetics and tree improvement are described in detail by Wright (1976) and Zobel and Talbert (1984).

Most tree improvement depends on finding trees with desirable characteristics and testing their progeny to determine whether the good characteristics are transmitted by sexual regeneration. Progeny do not always resemble their parents. In fact, so many possible combinations are possible that, except for identical twins from single seeds, no two products of sexual regeneration are exactly the same genetically.

From the genetic standpoint, the best that can be said of an outstandingly good tree found in a wild population is that it is a good **phenotype**; the interaction of its genetic constitution and its environment has produced a tree with desirable anatomical and physiological attributes. To separate these two interacting factors, the sexually or vegetatively reproduced progeny of trees of good phenotypes are grown in uniform environmental conditions; it can then be concluded that those displaying characteristics superior to others are from parents with desirable **genotypes**. The opportunities for genetic improvement of a species depend on the degree of genetic variability in the wild populations. If the variability is large, there are more genotypes from which to pick so that the possibility of genetic gain is large. However, the same high variability is also likely to mean that each individual is strongly heterozygous and may thus transmit mixtures of good and bad characteristics to its progeny. Trees have not been subjected to the centuries of human selection that have tended to produce strains of agricultural crop plants that are genetically very uniform or homozygous.

Tree breeding is quite different from that of agricultural crops, especially of herbaceous annuals. Some widely recognized risks are inherent in creating crops that are genetically very uniform. The most immediate danger is that an insect or a disease may infect agricultural fields in which all the crop plants are equally vulnerable. This problem can be dealt with by having each field contain genetically uniform plants, but mixing genotypes from field to field on the landscape scale; it is also possible to alternate genotypes from year to year. Uniformity within a field is advantageous for timing the many treatments and the final harvest in agriculture. In contrast, trees must live for many years and endure various waves of insects and diseases, pollutants, and climatic extremes; it is wise to maintain genetic variability within each stand to ensure that all trees are not equally susceptible to such effects. Furthermore, stands of trees, unlike those of most crops of agricultural annuals, start with many more individuals than are present in the final crop. A small number of the fastest growing are likely to endure natural or artificial thinnings. If the trees of a stand are too similar, especially in their rate of height growth, most will survive and tend to stagnate in diameter growth. There are, in other words, important reasons besides reduction of risk to try to maintain some degree of genetic variability.

Procedures in Tree Improvement Programs

The first step in most tree improvement efforts is **mass selection** (also referred to as "plus-tree selection") in which promising phenotypes are identified from the wild population. Immediate use of this selection process can be made by harvesting seed from the selected parent trees from the wild stands. These stands can be thinned to leave only those trees with good phenotypes, thus increasing the probability that seed collected in the stand will have arisen from both parents of good genetic stock. These **seed production areas** also have larger crops of seeds because of the thinning. However, the method of relying solely on phenotype as an indicator gives only a modest genetic gain, especially for growth rate; such traits as stem form and disease resistance are likely to show greater improvement. One of the important effects of mass selection is avoiding the degradation of genetic quality of seed used for planting, which occurs when seeds are collected from short, branchy, easily climbed trees. There are grounds to suspect that many Scots pines plantations are full of crooked trees because this method of seed collection was used in the early days of European forestry.

The next step involves testing the offspring of the chosen trees from wild populations; seedlings are grown in uniform environments in the field to assess the degree to which the good characteristics are hereditary. The seeds from each parent tree are kept separate and sown in a replicated experiment called a **progeny test**. The groups of trees arising from each female parent are referred to as a family, so this level of selection is called **family selection**. In the case described above, they are ''open-pollinated'' families in which each member shares the female parent, but the pollen parent is unknown and may be different for different progeny in the family. These are sometimes also referred to as ''half-sib'' families (i.e., half-siblings), but this term is technically correct only if each of the progeny is known to have a different male parent. Controlled pollination methods can also be used to produce seed in which both parents are known. Because this procedure requires collecting pollen and applying it to flowers that have to be covered to prevent wild pollination, it is costly, and not widely used at this stage of selection. ''Full-sib'' families, in which each tree has both parents in common, can also be evaluated in progeny tests. In either case, family selection provides greater gain than mass selection, because choice of desirable parent trees is being made on the basis of genotypes, by observing phenotypes of trees of known parentage grown in uniform conditions.

The results of progeny tests can be used to return to the wild stands or seed production areas and eliminate those trees whose progeny had undesirable traits. However, if the expense of progeny tests has been undertaken, then it is logical to go to the next step of creating a **seed orchard.** Cuttings are taken from the desirable parent trees, grafted onto root stocks, and planted in areas that are isolated from sources of contaminating pollen. Trees of the same genotype are separated so as to enhance cross-pollination. Results of the initial progeny tests can be used to identify families with unsatisfactory genotypes; the parent trees of such families are eliminated (''rogued'') from the orchard. To maintain a large and variable gene pool in the progeny, it is desirable to start with enough trees to have at least a dozen genotypes represented in each orchard unit after roguing has been completed. Tree seed orchards can be treated much like fruit orchards. The trees are widely spaced with large crowns, and are cared for with timely applications of fertilizer, insecticides, and fungicides. The trees will not resemble good forest-grown trees, but this is of no consequence to the genetic constitution of their progeny.

The production seed orchards described above are ''first-generation'' orchards, with the trees in the orchard having come from the original mass selection from the wild trees. Progeny testing usually continues while the orchard is in production. This is the stage when making selections based on full-sib families is useful and more efficiently conducted because controlled pollination can be carried out among trees in the seed orchard. Through full-sib family progeny tests, it is possible to identify parent trees that have particularly good combining abilities.

These procedures are even more complex than they sound. Successful genetic improvement programs are highly specialized research programs. The desirability of making selections from large populations and the high cost of the work have made it expeditious for different workers and entities to share genetic material on an organized basis. Such sharing helps develop the combinations of diversity coupled with good hereditary characteristics needed in the planted progeny. Tree improvement programs are usually continuing operations that switch from one approach to another and back, or use different methods simultaneously. Regulated sexual regeneration aimed at seeking genetic gains may alternate with vegetative regeneration to preserve the gains that have been made. The use of rooted or grafted cuttings often speeds up progress at certain stages of the effort, but these techniques must be perfected for each species. Progress is slowed if cuttings will not root or prove to be incompatible with root-stocks on which they are grafted. As testing progresses, many of the genotypes that were the best in the initial stages may be replaced either by new ones or by superior ones derived from their progeny. Usually, swiftest improvement comes when selection is made for only one or two different characteristics.

Many species throughout the world have undergone mass selection to develop seed production areas. First-generation seed orchards exist for a more limited number of species. For relatively few species used in large-scale plantation silviculture, trees that are the progeny of mass selection have themselves been rigorously selected for desirable traits and combined in second-generation orchards, and the process can continue to even more advanced generations.

Hybrids

Tree improvement programs have, in a few cases, been based on interspecific hybridization. Most tree species are sufficiently far apart in their heredity that they do not easily cross. However, examples of pairs or groups of species that naturally hybridize within a genus are found among pines, spruces, larches, oaks, poplars, willows, and eucalypts. Trees arising from some of these crosses exhibit **hybrid vigor** in which the hybrid has superior growth characteristics compared to either parent; however, this appears to be the exception rather than the rule.

Some hybrids prove to be practical because the crosses can be made merely by planting the different species together. The vigorous Dunkeld hybrid of Japanese and European larch is the best example of the use of this method. It becomes very laborious to produce artificial hybrids of species that only cross as a result of controlled pollination. In many cases, this process must be used with the parent species to produce each hybrid seed because it is not possible to obtain usable seed from the hybrid offspring. Many hybrids are infertile, and even for those that do produce seed, the offspring arising from crossing hybrids is generally too variable in genotype to be used as planting stock. The only artificially cross-pollinated hybrid in mass production is aimed at combining the straighter form of loblolly pine with the winter-hardiness of pitch pine. This hybrid has proven highly successful in plantations in Korea, and controlled pollination is used there for seed production. This hybrid is also being tested for use north of the natural range of loblolly pine.

The problems of producing hybrid planting stock are decreased when superior crosses can be perpetuated through vegetative propagation. This is possible with the vigorous hybrids of poplar species because their cuttings root very easily. Hybrid poplar plantations from cuttings have been established for fiber production in many parts of the world.

Vegetation Multiplication and Tissue Culture

One of the impediments to application of genetic improvement is the cost of the important step of multiplying the desirable progeny. Seeds produced from tree improvement programs are very costly unless they come from species such as birches and eucalypts that produce vast quantities of very small seeds. Thus the seedlings must generally be reared in nurseries and planted simply to achieve high efficiency in use of expensive seed. For some species, rooted cuttings can be used for establishing plantations. This assures the duplication of the parent tree's genotype but has only been developed for willows, poplars, sugi, and some eucalypt species (Zobel and Talbert, 1984).

The development of techniques for propagating whole plants vegetatively from culturing small bits of tissue in nutrient media *in vitro* would be especially useful for silviculture (Bonga and Durzan, 1982). Breeding programs that must rely on the sexual regeneration of tree species, which are characteristically very heterozygous and slow to bear seeds, are maddeningly slow. It can be especially frustrating if some excellent genotype has attributes that seem to be the result of some rare combination of genes that are not likely to be produced again even by the same parents. If such a genotype can be multiplied by mass-production tissue culture, many problems of breeding that arise with cross-pollination can be bypassed. However, this kind of tissue culture is still under development, and seedlings produced by it are not likely to be cheap. It is significant nevertheless that most varieties of woody plants used as ornamentals and for fruit production are vegetatively propagated, commonly from aberrant seedlings detected by observant growers.

Use of Genetic Improvement

If stands are to be established by planting, it is folly not to use the best genetic material available. Conversely, the availability of improved material can be a reason to regenerate stands by planting.

Genetic improvement of forest trees has not produced either the spectacular gains or the worrisome overreliance on intensive cultural practices associated with improved strains of agricultural crops. Most of the genetic manipulations of agriculture have been aimed at diverting more of a fixed amount of plant production into fruits, seeds, roots, or other plant parts that happen to be most useful to humans (Silen, 1982). Where total production has been increased, often immensely, it has been mainly by improving the site through such measures as fertilization or irrigation or by controlling weeds and other pests. However, it has also been accomplished to some extent by breeding varieties that not only can take advantage of intensive cultural measures but are also heavily dependent on them.

Perennial woody plants are different but not completely so. Most of the production of trees is tied up in the stem. This is the very organ that is usually viewed as wasteful in agriculture and is "robbed" to increase some other parts. In a certain sense, this may mean that trees are hard to improve because nature has already done much of the work. It would help timber production if trees had more main stem and fewer branches, but somehow supported the same amount of sugar-producing foliage. It would also be good if seed production could be turned on and off at will by either genetic control or application of appropriate hormones.

Much effort in forest genetics has been aimed at increasing production per acre. Because of the variable effects of site conditions on production, it is very difficult to determine whether this objective has been or can be achieved. The growth rates of individual trees have definitely increased, but these do not expand productivity per unit of land area unless trees of given size occupy less growing space because of their more efficient use of resources. Much confusion and heightened expectations in this regard have come from projecting percentage gains in individual tree height and volume achieved in young trees in small plots to predicted gains for operational plantations. More rigorous testing is needed to assess the realized productivity gains from tree improvement.

It is worth noting that if the individuals grow faster, they may attain optimum sizes on shortened rotations even if the production per unit area is not improved. Since different species clearly differ in production per acre, it seems likely that similar differences must exist within species (Cannell and Last, 1976). Much of the evidence suggests that the faster-growing material is generally that adapted to favorable sites or parts of the natural range. As long as the growing conditions happen to remain favorable or can be kept so by artificial measures, these faster-growing varieties can persist on sites where they might ordinarily suffer damage from drought, cold, or similar problems. In other words, this kind of increase in forest production may come with the same kind of price paid for such improvement in agriculture.

Tree improvement work has also been aimed at important factors other than productivity, and some of the greatest gains have been achieved with these. Among the characteristics that can be improved are stem form, branching characteristics, resistance to insects and diseases, wood density, fiber length, and resin production. There is evidence that success comes most rapidly with artificial selection for characteristics, such as stem quality or wood density, that have so little importance for survival that nature has not selected for them. In such cases, traits that are both good and bad from the human standpoint are very likely to have persisted in the natural populations.

Conservation of Genetic Resources

Maintenance of genetic variation is vital to the long-term survival of any species. It is also the raw material with which tree breeders work to develop more useful genotypes for various objectives. The goal of tree improvement is to narrow the genetic base with respect to the important characteristics being selected for, but at the same time maintain the variability that exists for characteristics that affect general adaptability. This can be done by breeding for the same desirable traits (such as straight stems or high wood density) simultaneously in a series of different populations (Namkoong, Barnes, and Burley, 1980; Zobel and Talbert, 1984). In some cases where intensive tree breeding has produced improved seed that is used on a wide scale, depletion of the genetic base may occur, if these or other efforts at maintaining variability have not been followed. However, this is generally more a risk for the landowner of such plantations than for the species as a whole. There usually are plenty of wild populations still existing. The risk is greatest when using clonal reproduction, such as with poplars. It is important to maintain 10 or more clones or families within each planting area in order to reduce such risk of loss at the stand level.

The dangers to the genetic base of entire species do not come from tree improvement programs, but rather from the various forces that underlie the conversion of large areas of forest to agricultural fields, pastures, or degraded woodlands from firewood harvesting. High-grading or diameter-limit cutting of stands can also be more damaging to genetic resources than it may appear because the best genotypes are removed from a stand.

Most threats to tree genetic resources are risks not of species extinctions, but of loss or reduction of varieties or races, decreasing the genetic base to a fraction of its original level. Natural populations of trees are such a storehouse of hidden genetic resources that it is increasingly obvious that significant samples of truly natural forests must be maintained or allowed to redevelop. In this way, potentially important gene combinations can be maintained as one part of a strategy to maintain biodiversity of all kinds of plants and animals. Such places are also sites where silviculturists can truly learn about the natural developmental processes that they claim to be able to direct. In some cases, it is advantageous to supplement these large nature reserves with small stands scattered across the range of a species, principally for the purpose of tree genetics conservation. In contrast to wildlife conservation areas, such stands need be only a few acres, because a genetically diverse population of trees can be maintained in a much smaller area than can populations of owls or wolves. Seed can be used from these areas to incorporate into breeding programs, but regeneration within the conservation areas is strictly from the mature trees of the stand involving no artificial selection. A series of gene conservation stands has been established for various species across their native ranges and has also been created for exotic pine species in South Africa by planting small stands with no artificial selection in each area where large plantations of improved pines are being established. In the case of a number of tropical pine species in Central America, widespread forest clearing poses a real danger of losing entire species. Cooperative efforts are underway to plant conservation areas with these species.

Other methods, such as grafted clone banks or long-term seed or pollen storage using freeze-drying techniques, may be used in the future; it is likely that these methods will be developed for agricultural crops first. The entire range of alternatives for conserving forest genetic resources have been reviewed by the National Research Council (1991).

Genetic Improvement with Natural Regeneration

The principles of genetic manipulation of trees are usually applied to the production of improved planting stock. However, choices that are made regarding which trees to cut and which to leave during partial cuttings can affect the genetic quality of the progeny arising from natural regeneration. Thinnings, seed-tree cuttings, and shelterwood cuttings are selections made among trees based on multiple phenotypic characteristics, which are similar to those made in mass selection for seed collection (Wilusz and Giertych, 1974; Bassman and Fins, 1988). Leaving the best phenotypes at each stage will increase the probability that seed fall in the stand will come from genetically superior trees. The selection intensity is generally not as great in a series of thinnings and regeneration cuttings as in mass selection for seed collection (i.e., a larger proportion of the population contributes to the seed production with natural regeneration), so genetic gain is not expected to be as great. As in intensive breeding programs, improvement is likely to be greater for stem form and pest resistance than for growth rate.

It is perhaps most important that foresters recognize that the opposite effect on genotypes can occur—that poor judgment in selection of trees during stand tending can lead to genetic degradation in naturally regenerated stands. These effects may not be apparent after one cycle of high-grading; enough high-quality trees will exist in the regeneration to produce a good stand. However, successive rotations of this kind of treatment will seriously diminish the genetic potential of the regeneration (Zobel and Talbert, 1984). Even if genetic gains are not achieved in natural stand management, simply preserving the level of genetic quality and variation of the original stand is a worthwhile goal. Indeed, maintaining genetic quality is second in importance only to maintaining site quality in the task of sustaining the long-term productivity of managed forest stands.

BIBLIOGRAPHY

Armson, K. A. 1977. *Forest soils: properties and processes.* Univ. Toronto Press, Toronto. 390 pp.

Bailey, R. G. 1995. *Ecosystem geography.* Springer-Verlag, New York. 199 pp.

Barnes, B. V., K. S. Pregitzer, T. A. Spies, and V. H. Spooner. 1982. Ecological forest site classification. *J. For.,* 80: 493–498.

Bassman, J. H., and L. Fins. 1988. Genetic considerations for culture of immature stands. In: *Future forests of the mountain west: a stand culture symposium.* USFS Gen. Tech. Rep. INT-243. Pp. 146–152.

Bonga, J. M., and D. J. Durzan. 1982. *Tissue culture in forestry.* Nijhoff/Junk, Hingham, Mass. 420 pp.

Booth, T. H. 1990. Mapping regions climatically suited for particular tree species at the global scale. *FE&M,* 36: 47–60.

Butterfield, R. P., and R. F. Fisher. 1994. Untapped potential: native species for reforestation. *J. For.,* 92: 37–40.

Cajander, A. K. 1926. The theory of forest types. *Acta For. Fenn.,* 29. 108 pp.

Cannell, M.G.R., and F. T. Last (eds.). 1976. *Tree physiology and yield improvement.* Academic, New York. 567 pp.

Carmean, W. H. 1975. Forest site quality evaluation in the United States. *Adv. Agron.,* 27: 209–269.

Clutter, J. L., J. C. Fortson, L. V. Pienaar, G. H. Brister, and R. L. Bailey. 1983. *Timber management: a quantitative approach.* Wiley, New York. 333 pp.

Daubenmire, R., and J. B. Daubenmire. 1968. *Forest vegetation of eastern Washington and northern Idaho.* Wash. Agr. Exp. Sta. Tech. Bul. 60. 104 pp.

Doolittle, W. L. 1958. Site index comparisons for several forest species in the southern Appalachians. *Soil Sci. Soc. Amer. Proc.,* 22:455–458.

Foster, R. W. 1959. Relation between site indexes of eastern white pine and red maple. *For. Sci.,* 5:279–291.

Hagglund, B. 1981. Evaluation of forest site productivity. *For. Abstr.,* 42:515–527.

Hoadley, R. B. 1980. *Understanding wood.* Taunton Press, Newtown, Conn. 256 pp.

Hodges, J. D. 1995. The southern bottomland hardwood region and brown loam bluffs subregion. In: J.W. Barrett (ed.), *Regional silviculture of the United States.* 3rd ed. Wiley, New York. Pp. 227–269.

Holdridge, R. R., et al. 1971. *Forest environments in tropical life zones—a pilot study.* Pergamon, New York. 747 pp.

Hughes, C. E. 1994. Risks of species introductions in tropical forestry. *Comm. For. Rev.,* 74: 243–252.

Jahn, G. (ed.). 1982. *Application of vegetation science to forestry.* Dr. W. Junk Publishers, Boston. 402 pp.

King, J. E. 1966. Site index curves for Douglas-fir in the Pacific Northwest. Weyerhauser Forest. Pap. No. 8.

Kotar, J., J. A. Kovach, and C. T. Locey. 1988. *Field guide to forest habitat types of northern Wisconsin.* Dept. For., Univ. Wisc. Madison, Wisc. 139 pp.

Lancaster, K. F., and W. B. Leak. 1978. *A silvicultural guide for white pine in the Northeast.* USFS Gen. Tech. Rep. NE-41. 13 pp.

Leak, W. B. 1980. The influence of habitat on silvicultural prescriptions in New England. *J. For.,* 78:329–333.

Leak, W. B. 1982. *Habitat mapping and interpretation in New England.* USFS Res. Paper NE-496. 28 pp.

National Research Council. 1991. *Managing global genetic resources: forest trees.* National Academy Press, Washington, DC. 228 pp.

Namkoong, G., R. D. Barnes, and J. Burley. 1980. *A philosophy of breeding strategy for tropical forest trees.* Comm. For. Inst. Tropical For. Pap. No. 16. Oxford, UK.

Panshin, A. J., and C. deZeeuw. 1980. *Textbook of wood technology.* 4th ed. McGraw-Hill, New York. 722 pp.

Pojar, J., K. Klinka, and D. V. Meidenger. 1987. Biogeoclimatic ecosystem classification in British Columbia. *FE&M*, 22: 119–154.

Poore, M.E.D., and C. Fries. 1985. *The ecological effects of eucalyptus.* FAO For. Pap. No. 59. United Nations FAO, Rome. 87 pp.

Pritchett, W. L. 1979. *Properties and management of forest soils.* Wiley, New York. 500 pp.

Rowe, J. S. 1984. Forestland classification: limitations of the use of vegetation. In: J. Bockheim (ed.), *Forest land classification: experiences, problems, perspectives.* Univ. of Wisconsin, Madison, Wis. Pp. 132–147.

Silen, R. R. 1982. Nitrogen, corn, and forest genetics. USFS Gen. Tech. Rept. PNW-137. 20 pp.

Soil Survey Staff. 1975. *Soil taxonomy: a basic system of soil classification for making and interpreting soil surveys. USDA, Agr. Hbk.* 436. 768 pp. (Reprinted by Wiley, New York)

Spurr, S. H. 1952. *Forest inventory.* Ronald Press, New York. 476 pp.

Spurr, S. H., and B. V. Barnes. 1980. *Forest ecology.* 3rd ed. Wiley, New York. 687 pp.

Steele, R. W., et al. 1981. *Forest habitat types of central Idaho.* USFS Gen. Tech. Rept. INT-114. 137 pp.

Tesch, S. D. 1981. The evaluation of forest yield determination and site classification. *FE&M*, 3:169–182.

Trewartha, G. G. 1968. *An introduction to climate.* 4th ed. McGraw-Hill, New York. 282 pp.

Vanclay, J. K. 1992. Assessing site productivity in tropical moist forests: a review. *FE&M*, 54:257–287.

Wilusz, W., and M. Giertych. 1974. Effects of classical silviculture on the genetic quality of the progeny. *Silv. Gen.*, 23:127–130.

Wright, J. W. 1976. *Introduction to forest genetics.* Wiley Interscience, New York. 282 pp.

Zeide, B., and W. T. Zakrzewski. 1993. Selection of site trees: the combined method and its application. *CJFR*, 23:1019–1025.

Zobel, B., and J. Talbert. 1984. *Applied forest tree improvement.* Wiley, New York. 505 pp.

Zobel, B. J., G. van Wyk, and P. Stahl. 1987. *Growing exotic forests.* Wiley, New York. 508 pp.

CHAPTER *10*

ARTIFICIAL REGENERATION

PLANTING

The most expeditious and certain way of establishing trees is to rear them in some protected environment and then plant them where they are to grow. In this way, they are not exposed to the rigors of the site to be restocked until they have successfully passed the most critical early stages. Many of the microenvironmental problems that were discussed in Chapter 7 on the ecology of regeneration are simply bypassed. The planted trees can be **wildlings** of natural origin dug from elsewhere in the forest. It is usually more efficient, however, to raise planting stock in nurseries or greenhouses. The plants that are moved can be rooted or nonrooted cuttings as well as seedlings.

Most forest planting involves conifers because the prospects of successful establishment and high yield are ordinarily greater with them than with hardwoods, or at least with deciduous hardwoods. This chapter deals with conifers and planting stock grown from seed, except in those cases where it is stated otherwise. Among the sources of detailed information on continuing developments in artificial regeneration are the U.S. Forest Service periodical *Tree Planters' Notes* and the journal *New Forests*. There are also comprehensive accounts by Lavender et al. (1990), Shepherd (1986), Nwoboshi (1982), Evans (1982), Cleary, Greaves, and Hermann (1978), Goor and Barney (1968) and Wakeley (1954). Most of these works include material on the production of planting stock which is a specialized kind of plant culture even more intensive than that of most agricultural crops.

Planting is one of the most crucial investments made in silviculture. The success of a whole rotation can depend on decisions made at planting time. The establishment of a plantation commits the forester, and usually the forester's successor, to many subsequent actions and investments. Errors made at the time of planting usually cause future problems

264

that are not always soluble. As is the case with all powerful techniques, planting affords opportunity not only for great success (Fig. 10.1) but also for the biggest blunders in silviculture.

Most of the blunders come from overzealous site preparation or putting species on sites to which they are not adapted. The only unavoidable drawback of planting is the inevitable deformation of the root system. This can reduce windfirmness and may possibly increase root rots. Other difficulties are generally surmountable. One of the greatest of these is the necessity of organizing people and facilities for a major effort and satisfactory performance during each short planting season.

Kinds of Planting Stock

After the all-important choices about species and genetic strains have been made (see Chapter 9), the next choices have to do with the kind of planting stock. The object of planting is to put a viable root system into the soil. The roots must go deep enough and grow fast enough to maintain contact with the soil-moisture supply. As far as the top of the plant is concerned, all that is basically necessary is that it have some shoot, bud, or apical meristem capable of growing into a tree. There must also be enough carbohydrate and other nutritional material stored in the seedling to enable it to resume growth.

The main choice is between bare-rooted and containerized seedlings. **Bare-rooted seedlings** have roots that are separated from the soil in which they were grown before they are taken to the ultimate planting site. Their survival after planting depends on quick reestablishment of functional contact between the roots and the soil. This problem is avoided with **containerized seedlings**, which are moved to the field in the soil in which they were grown and are planted with the soil still attached to the roots. With some species, it is also possible to plant **cuttings** which are cut segments of stems or roots with buds capable of sending out new roots.

Size of Plants

With regard to the size of plants, a dilemma is usually posed because bigger is not necessarily better. Large planting stock can be more costly to grow, harder to plant, and less likely to grow well. Growth is poor when too much root tissue has been left behind in the nursery but the top is intact. The top of the seedling should not be so large that it transpires more water than the roots can provide. Many roots are unavoidably killed or broken off when bare-rooted seedlings are removed from nursery seedbeds. Containers restrict the sizes of root systems unless the seedlings are planted when very small and tender.

A planted tree with a large top is readily visible and looks very impressive, but it is not better because it cost more. Even if it survives, it may be so slow to resume active growth that a smaller planted seedling may actually overtake it in size. The most important virtue of large or tall tops is the greater likelihood that they will not be submerged by competing vegetation. A tall seedling can also project above strata in which damage by animals or other agencies is common. For example, in planting saplings for street trees (Grey and Deneke, 1978), the tops should be taller than the level of people's eyes and the itchy fingers of children. Short seedlings are sometimes too easily buried by fallen leaves or eroding soil. Plants often have to grow to a certain size before they have enough stored material or conductive tissue to resume development after the damage inevitably associated with planting. Deep-rooted seedlings tend to be less susceptible to frost-heaving, and those with thick bark are less prone to suffer heat injury.

Figure 10.1 A crew planting loblolly pine by the bar-slit method on eroded land in southwestern Tennessee (top) with a view of the same spot 6 years later (bottom). *(Photographs by U.S. Forest Service.)*

Many of the crucial characteristics of nursery stock have little or nothing to do with seedling size. Physiological health, capacity for root regeneration, and similar attributes are controlled by the conditions of nursery culture, the care in handling, and the degree to which the timing of lifting the seedlings from the soil is synchronized with the seasonal regime of plant development.

Bare-rooted Seedlings

The outstanding advantage of bare-rooted seedlings is that no heavy, bulky soil is attached. Therefore, it is comparatively easy to transport large numbers of such seedlings to the field. It is also possible to produce and handle very large numbers of them expeditiously and comparatively cheaply in outdoor nurseries. However, the reestablishment of contact between roots and soil is such a crucial and fundamental problem that the planting can be done only during those short seasons during which rapid root growth takes place.

Ordinarily, the seedlings are simply grown in the seedbeds until the root systems are extensive enough and the tops tall enough to be planted. Plantable conifer seedlings should usually be at least 0.08 inch (≈3 mm) in **caliper** (basal diameter), 4–14 inches (≈10–35 cm) in above-ground height, and with a ratio, by weight, of top to root of not more than 4 to 1.

If the roots are too small for the tops or are too sparse and sprawling, the seedlings can be lifted and replanted in rows elsewhere in the nursery. This step is called **transplanting** in silviculture, and the word "planting" is reserved for final placement. Trees produced by this method are called **transplants** to distinguish them from **seedlings** grown in the original seedbeds. Transplanting makes the root systems more compact with a larger number of fine lateral roots than found in seedlings. If the seedbeds were crowded, the operation makes the tops become larger. The regrowth necessary for reforming the root systems takes place if the plants are left to grow in the transplant beds for a year or, less commonly, two years.

Root pruning or **wrenching** is an expeditious substitute for transplanting. If the object is to make the root systems shorter but more compact, this can be done by drawing thin, horizontal blades, mounted behind tractors, under the beds at appropriate depths. Wrenching is a more severe form of root pruning in which the horizontal blade is tilted so that the whole seedbed is lifted slightly in order to rupture the roots of seedlings enough to reduce their growth or induce early dormancy. The root systems can also be trimmed off at the sides with U-shaped blades if the seedlings are in drills. The seedlings are usually left to grow for another year after these treatments.

Sometimes it helps to bring the tops into better balance with the roots by mowing back the tops of the terminal shoots. If this is done when the tops are succulent and actively growing, most conifers will form new buds that create new tops that are nearly normal. This treatment can also be used if the plants must unexpectedly be left in the nursery for another year and might otherwise grow too large. If the roots are longer than the anticipated depth of the planting holes, they should be pruned back enough to facilitate planting. Because the ends of the roots are often killed by the lifting operation, root pruning may not actually sacrifice any useful tissue.

Tropical angiosperms, such as teak or Spanish-cedar, often grow very large before they become dormant at the end of the first growing season. If they are capable of sprouting, they can be trimmed down to plantable size by severe pruning of both top and roots. The stubby **stump plants** or **truncheons** thus created typically have 5 centimeter tops and roots 15 centimeters long.

Classification of Nursery Stock

The primary classification of bare-rooted nursery stock depends on the age and treatment of the plants. These two characteristics are designated by two figures, the first showing the number of years through which the plants grew as seedlings, and the second the number of years they were grown as transplants. For example, "1–0" indicates a 1–year-old seedling that has not been transplanted; "2–1" designates a 3–year -old transplant grown 2 years as a seedling and 1 year as a transplant. Similar letters and numbers are used for plants grown in containers or plants given certain special treatments.

Before nursery stock is shipped into the field, it is usually culled to eliminate trees infested with virulent insects or fungi, those that have been badly damaged in handling, and those with distinctly poor roots. Further segregation must be based more on indicators of the internal physiological qualities that affect survival than on the external appearance of the seedlings (Grossnickle and Folk, 1993; Rose, Campbell, and Landis, 1990; Duryea and Brown, 1984). The viability of whole batches of seedlings can be assessed by various destructive sampling techniques for determining their moisture status, carbohydrate content, or ability to form new roots. If seedlings are to be machine-planted, it may help to sort them into batches of uniform size.

The culling, grading, and counting of seedlings is best done under shelter as part of the packing operation. These operations should be completed swiftly under cool, moist conditions with a minimum of handling of the seedlings.

Transportation and Storage of Bare-root Seedlings

If bare-rooted stock is to remain viable, respiration and transpiration must be held to a minimum during transit. It is for this reason that success usually depends on doing both lifting and planting during periods when the trees are dormant or nearly so. It may be possible to move seedlings at other times if the seedlings can be planted in moist soils quickly, as on the same day. The crucial test is the ability of the roots to renew rapid growth after planting. Bare-rooted stock quickly loses this capacity if water and carbohydrates are squandered in transit because the tops are growing too actively or because the plants become too warm or dry.

Planting stock is usually packed in bundles in ways that keep the roots moist and cool but with provision to allow oxygen to enter the package while carbon dioxide and other gases that might be noxious can escape. The traditional practice is to construct bales consisting of two tight bundles of trees arranged in opposite directions so that the tops are open to the air at each end and the roots of each bundle are in close contact with each other. The roots are then covered with sphagnum moss or similar water-holding material, and the whole bales are wrapped with waterproof paper. Provided that they can be kept cool, the seedlings can simply be sealed in bags of paper impregnated with polyethylene; such bags are appropriately permeable to oxygen and carbon dioxide but impermeable enough to water that it is not necessary to include any moss.

The best way to store packages of nursery stock, whether for a few days or over winter, is under refrigeration at temperatures slightly above the freezing point. It is usually important to avoid the desiccation of bare roots that results from freezing. Although they may survive brief periods of freezing if thawed slowly, it is often best not to waste time planting seedlings that were accidentally frozen after lifting. Leafless hardwoods endure storage better than conifers. Storage of this sort is most commonly done to hold the stock dormant for a few weeks until planting sites in localities colder than the nursery become ready for planting. Seedlings lifted in the fall are stored over winter to plant on sites that become ready for planting in the spring before stock can be lifted in the nursery.

Care of Planting Stock in the Field

If it becomes necessary to store planting stock after it is received from the nursery, it is better to store it in the original bundles in cool places than to store it in the field. As an emergency measure, it can also be heeled in on the planting site and taken out as needed. **Heeling-in** consists of storing planting stock in a trench dug in moist soil with a smooth, vertical or slanting side as deep as the roots of the trees are long. The plants are put in this trench in a relatively thin layer and covered with soil up to the root collars. Protection from sun and wind is essential. Ordinarily, several days' supply of planting stock, if kept shaded and moist while on the planting site, will be damaged less in the original bundles than if heeled in.

The trees can be carried in pails, baskets, bags, or trays by the planting crews. It is absolutely essential that the roots of the planting stock remain moist until planted. Moss, a puddle of wet mud, or other material for keeping the roots moist is often placed in the container.

Containerized Seedlings

Under most circumstances, bare-root planting is cheapest and most expeditious, but it has the fundamental disadvantage that contact between root and soil is broken. This constraint and other problems can be mitigated by growing the seedlings in containers of soil or soil-like medium and planting them without ever breaking contact between root and soil (U.S. Forest Service, 1989–1990; Guldin and Barnett, 1982; Scarratt, Glerum, and Plexman, 1981; Tinus and McDonald, 1979). Containerized planting has often been necessary in arid regions. It is commonly used in the humid tropics, where many species become too large to move several weeks after germination and must be planted when actively growing and succulent. This problem has traditionally been overcome by planting both container and seedling. Pots made of quickly disintegrating bamboo or paper are often used, although polyethylene bags (Fig. 10.2) can be used if they are removed or thoroughly slit before the planting.

Trees more than a meter tall usually have to be dug out of the ground with large amounts of attached soil, which is then wrapped or "balled" in burlap for moving and replanting. The root systems of such trees are so much wider than the tops that many roots are left behind. The chief advantage of this very costly kind of containerized planting is that it gives instant beauty. It is a long time before the trees are able to resume active growth, and their initial survival may depend on frequent irrigation.

The old technique of planting container-grown trees has now been systematized and mechanized for a variety of reasons. The most important is that seedlings individually grown in containers in greenhouses can be cared for much more effectively and grown faster than in the open. Among other things, it takes much less seed, an important consideration given the high cost of seeds produced in genetic improvement programs. Even though containerized seedlings are costly to produce, they can be much cheaper to plant if labor costs are high. Much of the cost of hand planting is that of making a hole in the ground. It takes less labor to punch or bore a hole to install a plant unit of uniform size, sometimes with a special planting tool or machine, than to dig holes for stock of irregular sizes.

Figure 10.2   Mahogany and eucalyptus seedlings growing in polyethylene bags beneath a shade house at a nursery of the Puerto Rico Department of Natural Resources.

Kinds of Containers and Their Characteristics

Containerized seedlings can be grown (1) as **tubelings** in plantable pots or tubes, (2) as cohesive **plugs** pulled out of the containers before planting, and (3) in **blocks** of pressed peat or pulp that are both container and rooting medium. Somewhat similar are "mud-pack" seedlings, which are bare-root stock individually wrapped with soil around the roots in the nursery shed.

The ideal plantable pot would be one that held together through the act of planting and either disintegrated promptly thereafter or did not impede the subsequent growth of roots through the walls. Porous, biodegradable materials such as paper and pressed peat, are more suitable than plastic in this respect. When they are planted, however, the container tops should be covered with soil so that they will not act as evaporational wicks desiccating the soil. The volume of the containers should be uniform and is often about 40 cubic centimeters.

Plastic containers are usually best for producing plugs. These are often grown in cylindrical cavities closely spaced and inset vertically in reusable molded blocks of Styrofoam. Another kind, sometimes called a book planter, is made of two hinged sheets of thin molded plastic that form proper kinds of cavities when folded together. When it is time to remove the plugs, or replace an empty one with one that has a developing seedling, the ''book'' is opened. Sometimes these containers are just individual tubes that can be set into, and taken out of, fixed racks. Figure 10.3 shows some plug seedlings, the containers in which they are grown, and a tool for punching out planting holes of the same dimensions as the seedling plugs.

There is concern about the tendency of elongating roots to be deflected into spiral arrangements when they impinge on the walls of any kind of container. They can even be deformed if they grow against a minute flange on the inside of a plastic casting. Because root systems thus deformed tend to remain so and produce ''pot-bound'' seedlings, ways have been developed to counteract the tendency. One of these is to cast vertical grooves or ridges into the container walls so that the roots are directed downward. Another, called **air-pruning**, involves containers with holes in the bottoms supported so that there is air beneath them. When the roots reach the air, they simply stop growing, but they retain the capacity to resume when planted in contact with soil. The same effect is sometimes produced by treating the container walls with toxic copper salts.

Theoretically, growing pot-bound seedlings can be avoided by getting them out of the containers before the roots touch the walls, but this calls for impossibly good timing. Such problems are reduced by growing seedlings in blocks of peat or pulp, but not if the blocks are confined inside containers. Another way of dealing with this problem is to grow seedlings in plugs and then transplant them in the nursery. These so-called **plug + 1**

Figure 10.3 Plug seedlings of longleaf pine with the tool that punches holes in which to plant them. When each hole is made the plug of soil left from the previous hole is pushed out the top of the cutting tube.

transplants usually develop dense, cylindrical root systems and can be produced much faster than ordinary transplants.

Most methods of planting bare-rooted stock also lead to deformities of root systems that are merely less obvious than those caused by containers. Unless they have been confined within long-enduring plastic tubes, containerized seedlings probably ultimately develop more symmetrical root systems than bare-rooted seedlings that are usually planted with their roots in a single vertical plane like a crack in a rock.

Handling of Containerized Seedlings

If necessary, dormant containerized seedlings can be stored under whatever conditions of low temperature or dryness they are conditioned to endure in nature. Since the roots are not exposed, there is no need to worry about freezing them if they are dormant and the species are adapted to frozen soils. If they are about to be planted, however, they should be nondormant or capable of surviving from the time of planting until the normal, natural ending of dormancy. Containerized stock is shipped to the field whenever the time is suitable for planting. There are no problems with the timing or mechanization of lifting as with bare-rooted seedlings. Sorting and culling have ordinarily been finished during the growing process. If the containers are shipped out with the seedlings, packing consists only of putting the material in crates or on racks designed to prevent crushing the seedlings. If the containers are for growing plug seedlings and are to be reused, there is the problem of returning them.

If plug seedlings are withdrawn from the containers before shipping, it is usually necessary that they still be dormant. Generally, they are packed in polyethylene bags. They can be put in cold storage if desirable. Because such plugs lose their cylindrical shape during all this handling, they cannot be planted in standardized holes. The chief advantage is that the containers do not have to make a round trip to the planting site. Whether or not the containers make the trip, the main shortcoming of containerized planting is the simple but arduous and costly task of moving so much material to the characteristically dispersed spots where trees are planted.

Vegetative Propagation of Planting Stock

Planted trees do not always have to be grown from seed, and important advantages can accrue from bypassing sexual reproduction and germination problems. In the first place, asexual regeneration by vegetative propagation is the only sure way of perpetuating trees with the same genetic qualities as the parent. Second, if cuttings form roots readily, it may be much easier and give quicker results to plant them rather than to grow seedlings. Even if it is difficult to make pieces of old plants form new ones, it can still be advantageous to do so in order to multiply a few individuals with highly desirable genetic characteristics.

With some species, such as the cottonwood poplars, one can merely plant long pieces of thin shoots and depend on most of them to form roots and become established (McKnight, 1970). This usually happens only on moist soils and with some species of those sites. Sometimes, as with sycamore, dense stands can be created by burying branches in shallow trenches and having each branch produce many shoots.

Most tree species are not propagated from cuttings as simply or easily. In fact, for most species it is so difficult that it is done mainly for multiplying valuable ornamentals, fruit trees, and trees for seed orchards. Such work is usually done in greenhouses with the use of rooting hormones such as indoleacetic acid.

Even more sophisticated techniques are used to produce whole new plants from culturing small fragments of tissue in media that supply the complex biochemical substances needed for plant development (Bonga and Brown, 1982). Plantlets produced by such intensive methods are normally transplanted to containers for further growth before planting.

The potential advantages of creating forests from asexually regenerated genetic material are very great. Forest tree species have not been subjected to millennia of genetic selection by people and are therefore very heterozygous. This genetic heterogeneity can make it difficult to replicate individuals with many fine genetic traits through controlled pollination schemes. With vegetative propagation, it is not possible to improve upon nature, but the best individuals found there can be multiplied. However, the very nature of the development of stands of woody plants and the vicissitudes that long-lived perennials must endure are such that it would be undesirable to create highly uniform stands of single genotypes. Genetic resources must be managed both to multiply good genotypes and to preserve diversity.

Season of Planting

Success in bare-root planting depends on doing it when periods of 2–3 weeks of active root elongation can be anticipated. The use of containerized stock gives more latitude in timing than is available with cuttings or bare-rooted material. Roots normally elongate at times of the year when supplies of physiologically available water can be captured, except when rapid shoot or leaf formation commands most of the supply of constructive materials (Jenkinson, 1980).

In climates with periods of cold or dryness that induce dormancy, the period of most rapid root growth occurs just after the breaking of that dormancy. The best time for planting comes just before that not only because of the opportunity for quick reestablishment of contact between plant and soil, but also because the whole growing season is available for further development.

Bare-root planting must, in other words, normally be done in a rather short period at the beginning of the growing season. If the period must be brief, it means that the operations that start with lifting and end with planting are apt to be frantic, costly, and sloppy. If the lifting must be delayed by waiting for the nursery beds to thaw or stop being muddy, the season can be shortened even if the planting sites are ready. Given such problems, it is well to consider ways of extending the planting season.

It may help to lift the stock much earlier and store it so that the lifting operation does not delay the start of planting. If soil moisture and temperature are adequate, there may be another period of active root growth in the autumn so that trees with tops that have hardened and formed buds can then be planted with fair success. However, not all species in the same locality have the same seasonal regime of root growth. Thus it is best not to plant at any time without knowing from experience or direct observation that the roots can be depended upon to grow actively after planting. A major advantage of both containerized planting and direct seeding is that they can be done at times when bare-root planting cannot. It can be advantageous to have planting crews do that kind of work when bare-root planting is not possible.

Bare-root planting should not be attempted when the air temperature is below 0°C or when hot or dry winds are blowing. If trees are planted in the autumn in cold climates, there can be serious problems with frost-heaving during the winter, especially if the trees are not well covered by an insulating blanket of snow. Another drawback of autumn planting is that the seedlings do not get the benefit of a full growing season of carbohydrate production. It is well to remember that trees are generally storing carbohydrates at those times of year when water is available but formation of new tissues has ceased. They may continue to gain in weight late in the autumn, even if their dimensions do not change. In some warm climates, such as that of the southeastern United States, spring and autumn are, in effect, merged so that planting is done during the mild winters. It may, however, have to be suspended during temporary periods of freezing weather.

Site Preparation

Successful planting often depends on measures such as reduction of competing vegetation, removal of physical obstacles to planting, and drainage of water toward or away from the planted trees. Most of these measures are dealt with in Chapters 6 and 8.

Control of competing vegetation is especially important because many planting sites are already crowded with preexisting vegetation. Unless a distinct vacancy is found or created in the growing space, it will be difficult for a planted tree to survive or grow satisfactorily. Site preparation as a means of controlling competing vegetation is usually done as a separate operation. However, planting spots can also be prepared at the same time as the planting by spot applications of herbicides or **scalping**, which is the scraping of competing vegetation from the planting spot with hand tools. Scalping is moderately effective against thin grass or very low shrubs and can be done with the grub hoes sometimes used in planting.

Preplanting site preparation by itself is not necessarily sufficient to eliminate all problems with competing vegetation. In fact, it may aggravate such problems by opening the way for overwhelming invasions of weeds. There will be competing vegetation simply because the planted trees will not promptly fill all the available growing space. Sometimes it helps to limit site preparation to small planting spots and use the other residual plants to exclude more aggressive pioneer competitors.

Methods of Planting

An increasing number of kinds of hand tools and machines (American Society of Agricultural Engineering, 1981) can be used for making planting holes, inserting trees, and packing soil back around their roots.

Successful planting depends on the ability of the roots of the planted trees to regain contact with the soil so that the uptake of water and nutrients can resume. The first essential is that the roots be placed in soil in which water is immediately available. The moisture content must be well above the wilting coefficient, and the soil temperature must be sufficiently far above the freezing point that water can move readily. Moist mineral soil is by far the most dependable water source. Organic materials are not reliable unless they are peats and mucks situated so that they remain close to saturation without being deficient in oxygen. The soil must also be warm enough for roots to grow.

When bare-rooted stock is lifted from the nursery, virtually all of the mycorrhizal mycelia and root hairs that constitute the main absorptive portion of the roots are destroyed. Until these unicellular strands resume proliferation, the plant has no means of absorbing water that is not immediately adjacent to the existing roots. Fortunately, brown suberized roots can absorb water directly (Kozlowski, Kramer, and Pallardy, 1991), thus providing planted trees with an immediate but limited, supply of water. Presumably this is one of the reasons why it is so essential that the soil be solidly packed around the roots of planted trees.

Until the main absorbing network redevelops, the planted tree has only slightly more ability to capture moisture than a cut flower in water. The development of new root hairs depends on the elongation of new adventitious roots and root tips, many of which are broken off during lifting. New root tips must form before any major portion of the permanent and readily visible part of the root system can extend itself in the soil. The root system of a planted tree elongates rapidly only during rather limited periods when the normal physiological rhythm of growth is directed toward this process. Furthermore, it is reasonable to suppose that such elongation of the roots will not become very active until photosynthesis becomes vigorous again. Because this is not likely to occur until the active absorption of water and nutrients is resumed, any complete dependence on elongation of root tips and on absorption by root hairs would seem to jeopardize the survival of planted trees.

Mycorrhizal mycelia are probably much more effective than root hairs in reestablishing early contact with the soil. The fungi that are involved need only a supply of reserve foods from the plant and the proper conditions of moisture and temperature to resume growth; consequently, they do not depend on resumption of full activity by the whole plant. It is significant that most tree species normally enter into mycorrhizal relationships and may require them to survive or thrive after planting. Unfortunately, the soil sterilants that are often applied before seeds are sown in nurseries to kill weed seeds and control other pests also kill mycorrhizal fungi. This problem is overcome by artificial inoculation of the soil with appropriate fungi (Marx, Cordell, and Kormanik, 1989) which are sometimes of strains better than the natural ones. Sometimes the appropriate inoculum is nothing more than litter from beneath trees of the host species growing in their natural habitat. When exotic tree species are introduced to a locality, it may be necessary to bring the appropriate fungus species as well. However, it is prudent to test native fungi first and it is mandatory to adhere to regulations about the introduction of species that can become pests.

The root system's ability to redevelop is diminished more than appearances would suggest because so many bare roots die between lifting and planting. The roots closest to the root collar are more likely to survive than those farther from the stem.

Trees should usually be planted at the same depths that they occupied in the nursery. Sometimes it is advantageous to plant them more deeply but never more shallowly; the greater the depth, the more dependable the water supply.

If the roots are more than 8–10 inches (20–25 cm) long, they may be too long to plant properly, and it may be desirable to prune the ends off before giving the seedlings to the planting crews. Every precaution must be taken to ensure that the roots remain *visibly moist* at every stage of handling. Seedling roots should be left covered until the moment comes to stick them into the planting hole and not waved around in the air. Moist soil must be packed around the roots, and the planting holes should be filled completely, leaving no air spaces. Litter or other organic debris should not be used for filling the holes unless the soil itself is organic. The trees should be planted firmly enough to resist a gentle tug with thumb and forefinger.

Any mode of root placement resulting from planting is unnatural (Van Eerden and Kinghorn, 1978). Trees are ordinarily planted with stems erect and roots spread out in a vertical plane. This pattern is least desirable with species such as the spruces, which have roots that characteristically grow in a horizontal plane; there are special techniques for planting trees with the roots so arranged.

The greatest problems arise from sloppy planting that leaves the roots curved upward like the letters *J*, *L*, or *U*. The initial difficulty that can result from this is that the roots are not placed as deeply as they might be. The deferred problem is that the roots may not develop the symmetrical structural pattern necessary to confer windfirmness so that the trees become unusually vulnerable to uprooting after they grow tall. This problem is greatest with root systems that are wound into balls when stuffed into shallow holes or left to grow too long in containers. If roots are too long, it is definitely better to prune them than to force them into contorted shapes when planting them. Natural growth of new roots does correct some of the abnormalities of root systems of planted trees but seldom perfectly. The stems of most species will straighten up if the stems are not perfectly erect, but some, such as the larches, will not.

Regardless of whether the work is done by hand or with machines, there are two general methods of making holes and planting bare-rooted trees. In the **compression methods**, holes are made for the plants by pushing the soil aside with sharp instruments driven into the ground; after the tree is inserted, the soil is pressed back around its roots. In the **dug-hole methods**, the soil is actually removed and set aside to be repacked around the roots after they are arranged in the hole. All methods of hand planting also involve the use of the heel to repack the soil around the tree roots.

Several types of heavy dibbles or planting bars, such as the two shown in Fig. 10.4, can be used for planting trees in sandy soils. The most common technique for using such tools, known as the **bar-slit method**, and shown in Fig. 10.5, illustrates the basic principle of the compression methods. In this method of planting, it is important to push the seedling down deeply into the hole and then partly withdraw it so that the roots will go as deeply as possible but be straightened out and free of J- or U-bends. One continues to hold the seedling upright while pushing the soil back against its roots. Because one person does all the work in this method, it helps to carry the seedlings in a shoulder bag rather than in a container that is set on the ground; by using a planting bag, one bends over only once for each tree. One can plant an average of 1500 trees daily with this fastest method of hand planting.

On soils that are stony or have so much clay that they become compact under true compression, a modified compression method can be employed. Grub hoes are used in

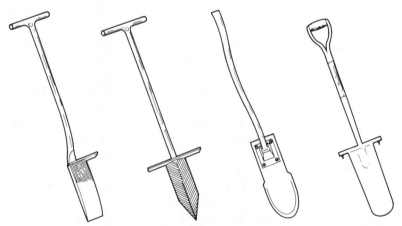

Figure 10.4 Four tree-planting tools (left to right): planting bar; a pointed planting bar useful in stony soils; the Rindt grub-hoe (L-shaped) for making straight-sided planting holes, and a tile spade planting shovel for digging deep holes for large planting stock.

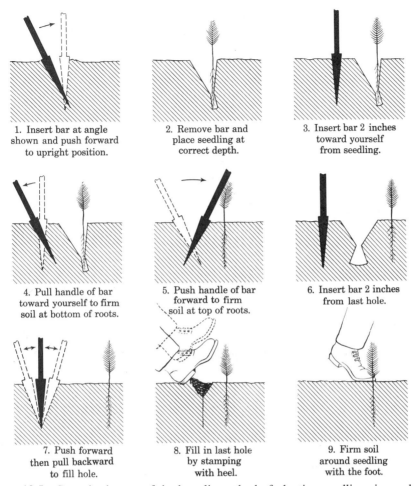

1. Insert bar at angle shown and push forward to upright position.

2. Remove bar and place seedling at correct depth.

3. Insert bar 2 inches toward yourself from seedling.

4. Pull handle of bar toward yourself to firm soil at bottom of roots.

5. Push handle of bar forward to firm soil at top of roots.

6. Insert bar 2 inches from last hole.

7. Push forward then pull backward to fill hole.

8. Fill in last hole by stamping with heel.

9. Firm soil around seedling with the foot.

Figure 10.5   Steps in the use of the bar-slit method of planting seedlings in sandy soil. *(Sketch by U.S. Forest Service.)*

this **grub-hoe-slit method**. The soil is partially lifted from the hole rather than pushed aside. A person can plant about 700 trees daily with this method. One advantage of the grub hoe is that it can be used for scalping, whereas planting bars cannot.

The main drawback of the compression method is the difficulty of being certain that the holes are completely closed or that the roots of the seedlings are free of U-shaped bends. These problems can be partially avoided by refraining from working the blades of the tools back and forth after driving them into the soil. The holes created as a result of this error are shaped like hourglasses in vertical cross section; it is difficult to work the roots past the constriction and impossible to be certain that the lowest part of the hole is closed. Even the grub-hoe modification of the compression method may be difficult to apply in soils that are very stony or cohesive.

The dug-hole methods must be used where the compression methods fail to give good results. Grub hoes (Fig. 10.4) are usually employed in this method, sometimes with one person digging the holes and another doing the planting. The holes are made sufficiently wide and deep to accommodate the roots of the trees. The mineral soil is piled beside the hole separate from leaves, sod, and other debris to facilitate planting and to ensure the use of clean soil next to the roots. The trees are planted entirely by hand immediately after the holes are dug so that the exposed soil will not become desiccated. Several different methods can be used to shape the holes and plant the trees.

The simplest dug-hole method is called the **side-hole method** (Fig. 10.6). One side of the hole is left smooth and vertical; the roots are spread out against this wall and packed firmly in place by hand with a thin layer of fresh, loose soil. The rest of the mineral soil is then scraped into the hole and trodden down with the heel. Usually, the area around the hole is scalped before the hole is dug so that the tree will not be immediately adjacent to competing vegetation. This method can be used under almost any conditions but is not rapid; the production rate is about 600 seedlings per person-day. It leaves the roots in a vertical or slightly curved plane.

The best way to get the roots placed in three dimensions is the **center-hole method**. This involves putting the tree in the middle of the hole and then sifting and packing the loose soil between the various strands of roots, without forcing them into either a horizontal or vertical plane. This is a slow method, but it can be speeded by boring the holes with motorized augers (Fig. 10.7).

The **wedge method** is another way of improving root placement. It involves creating a hole with a ridge in the middle such that a vertical cross section would resemble the letter *W* (Fig. 10.8). Half of the roots of the tree are spread out on each sloping surface and tamped down in that position. Rudolf (1950) described how the holes could be made with several strokes of a grub hoe and stated that 500 trees per person-day could be planted by this method. This procedure is well adapted to the planting of species such as the spruces, which have distinctly horizontal root systems.

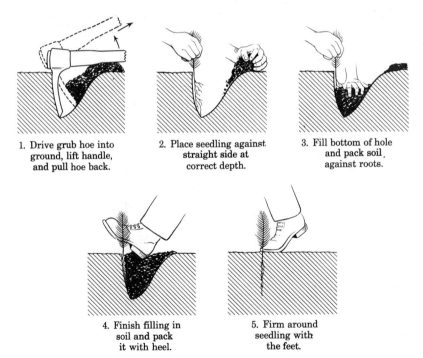

1. Drive grub hoe into ground, lift handle, and pull hoe back.

2. Place seedling against straight side at correct depth.

3. Fill bottom of hole and pack soil against roots.

4. Finish filling in soil and pack it with heel.

5. Firm around seedling with the feet.

Figure 10.6 The side-hole method of planting. *(Sketch adapted from one by U.S. Forest Service.)*

Figure 10.7   Motorized augers being used for planting after timber cutting in an open forest of ponderosa pine and Douglas-fir in Montana. One auger can produce enough holes for a crew of three planters. *(Photograph by U.S. Forest Service.)*

Perfectly horizontal placement of roots can be achieved by techniques such as making T-shaped incisions in turf with spades, turning up the resulting flaps, spreading out the seedling roots, and then tamping the flaps back over them.

Containerized seedlings (Barnett and McGilvary, 1981) can be planted with special devices of varying sophistication, provided that the units to be planted have retained sufficiently uniform size and shape. One of the advantages of using uniform units is the opportunity to poke standardized holes in the ground rather than doing laborious digging. This advantage is of greatest importance with stony soils where digging is difficult. The simplest devices, suitable only for small stock, are nothing more than rods stuck in the ground; these can also be used for planting cuttings of such easily rooting species as

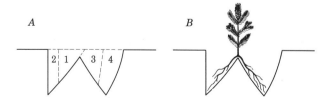

Figure 10.8   The wedge method of planting as described by Rudolf (1950). *A* shows the sequence in which soil is removed by four strokes of a grub hoe to create the hole. *B* shows the tree ready to be tamped in place.

the cottonwood poplars. There are also devices that make holes by compression and either set or shoot the containerized material into the soil; the device shown in Fig. 10.3 punches out standardized holes without compressing the soil around them.

If the container units to be planted are not or have ceased to be of uniform size and shape, they are put in the ground by the same hand or machine methods used for bare-rooted seedlings. This is common with plugs that were extracted from the containers before being shipped to the field.

Planting Machines

Sites that are level or gently sloping and free of stumps, rocks, slash, and tree roots are the places for planting machines. Except where labor is cheap or where the terrain is unsuitable, it is both better and cheaper to plant trees with machines than it is to do the work by hand. A few well-trained people can do a better job than some large crew recruited for a brief flurry of work. It is also more nearly possible to compress bare-root planting into the phenologically appropriate period.

Abandoned agricultural fields are the ideal places for the machines. However, it can be unwise to remove stumps and logging debris simply to enable use of planting machines. Damage to the soil may be too great, and the resulting total costs may exceed those of hand planting. For these reasons, most replanting after logging is done by hand.

Most planting machines operate by the compression method. The typical machine (Fig. 10.9) cuts a narrow trench through the soil with a specially designed plowshare. The trees are placed in a slot just behind this trencher, and the soil is pressed firmly back around the roots by two wheels that are mounted close together in slightly tilted positions at the rear of the machine. The machines are heavy enough that seedlings are usually planted more firmly than is possible with use of the human heel. They can plant from 4,000 to

Figure 10.9   The essential parts of a tree-planting machine. The round cutting disk or coulter at right cuts through roots and sod ahead of the split plowshare or trencher. The trees are put down in the slit made by the trencher between the two wings or guides at its back end. They are then tamped in place by two tilted rubber-tired wheels mounts at the rear of the machine. *(Photograph by Lousiana Forestry Commission.)*

15,000 seedlings per day, depending mainly on site conditions. There are other kinds of planting machines such as those that make discontinuous slots or isolated holes in the soil.

Placement of Seedlings and Reshaping of Soil Surface

Most techniques of treating vegetation and soil in preparation for planting were discussed in Chapter 8. Some involve making furrows or ridges. With these, the basic principle is that, if the soil is too wet, one plants on the tops of the little ridges, but if it is too dry one plants in or near the bottoms of the furrows. If, however, furrows are made to use folded-over sod or other turf to uproot and also cover competing vegetation, it becomes logical to plant trees along the lines where the ridges and troughs intersect. In this way, the trees are kept as far as possible from competition and also have their roots partly in a double thickness of the soil strata in which nutrients tend to be concentrated.

If the soils of planting microsites are too cold, they can be made warmer by removing litter and creating small ridges on which to plant the trees. Sometimes slotted circles of mulch-like paper with holes in the centers are slipped around planted trees to shade out competing vegetation. Localized applications of herbicides can be used to produce the same effect, but care must be taken not to kill the planted seedlings.

In very dry situations, it is desirable or even essential that trees be planted in basins or on terraces designed to concentrate runoff of water (Goor and Barney, 1965). Stones can be put around the seedling stems to shade them, redistribute water, and restrict direct evaporation from the soil. It is well to note that it is possible to store significant quantities of water if infiltration can be concentrated in localized spots. Water sinks into the soil only to the depth that can be wetted to **field capacity** (the amount of water that can be held against gravity). If it is uniformly distributed over the surface, much of it is likely to be lost to direct evaporation or transpiration by shallow-rooted vegetation. However, if it can be concentrated around planted trees, deeper columns of water can be stored in such a way that trees may still have some available even when the soil close by has been depleted. The same general effect can be achieved where snow is caused to accumulate in drifts. The main point is that trees can be started in semiarid conditions if water can be concentrated in the planting spots.

Density of Plantations

One advantage of planting is that it gives close control of the initial density and spatial arrangement of the new stand (Sjolte-Jorgensen, 1967). There is a tendency to think primarily of how many trees will be needed to achieve early occupancy of the growing space. However, it is better to envision the kinds of trees wanted in the stand late in the planned rotation and to work backward from this goal. The question of whether or how the stand may be thinned is a crucial consideration in planning the initial arrangement (Lundgren, 1981).

If the trees are planted close together, their crowns and roots will soon close and full occupancy of the site will be achieved early. The high stand density will induce small branches, slow diameter growth, a low degree of stem taper, and rapid upward retreat of the bases of the live crowns. The trees may seem to be taller than those of less crowded stands, but this is ordinarily an optical illusion, at least as far as the dominant and codominant trees are concerned. The height growth of the leading trees of a stand remains the same over a wide range of stand density but is reduced if stands are extremely crowded or very open.

Total production of wood is maximized at the highest stand density that does not cause diminution of height growth of the leading trees. If the stand density is any lower, there will be some loss of potential total production during the delay in achieving full occupancy. Although the rate of production is generally the same after attainment of full occupancy, there is no way to recover any production loss resulting from such delay. However, there would be purpose in very dense planting to maximize total production only where wood was precious and the labor of gathering it so low that saplings could be harvested profitably.

If timber production is an objective, the purpose is to optimize the yield of economically utilizable wood. Because this always involves some specification of a minimum diameter of tree or piece and because tree value increases rapidly with diameter, total production is deliberately sacrificed to enhance diameter growth. The greater the target diameter, the lower the initial density. An owner seeking only sawlogs from trees more than 10 inches (25 cm) D.B.H. would plant at wider spacing than an owner who could also use pulpwood from trees as small as 4 inches (10 cm).

The ideal number to plant is precisely that which can be grown to the smallest size that is profitably utilizable. However, unless small trees can be utilized, it is commonly necessary to plant more trees than this in order to restrict branch size and prevent the trees from being too tapering. Sometimes it is even desirable not only to plant these extra "trainer" trees but also to kill some of them in later precommercial thinnings.

The question of whether thinnings will be done has bearing on the initial density. If thinnings cannot be made, it is best not to have the stands close any earlier than is essential for the development of acceptable stem quality. Conversely, if early thinnings are certain, it can be desirable to plant more trees to enhance the yield from thinning.

The initial density of stands can also be varied with differences in site quality. In general, the better the site, the greater is the number of trees that can grow at an acceptable rate. The crucial variable ordinarily seems to be the amount of foliage that can be supported by the soil moisture supply. Those parts of alluvial flood plains that have continuously moist but well-aerated soils are good examples of the sites where trees can be close together but still grow well. At the other extreme are dry sites where it may be most logical to plant trees so far apart that their crowns never close even though the roots presumably do. In such cases, artificial pruning is sometimes substituted for the natural pruning associated with closed canopies.

Planting density should also vary with species, but so many kinds of factors are involved that generalizations should not be accepted uncritically. Theoretically, stands of shade-tolerant species can be denser than those of intolerants because they can carry more foliage. However, considerations such as product objectives, crown forms, or cost of planting stock also control the choices. If there is small variability in height growth, as is the case with red or lodgepole pine, there is risk that all of the trees will grow slowly and uniformly in diameter if there are too many of them. For such species one might logically plant fewer trees than with one in which there was enough variability to cause early differentiation into crown and diameter classes.

Spatial Arrangement of Plantations

Consideration should be given not only to the number of trees to plant but also to their geometric arrangement in relation to the future management and development of the stand. The arrangement that normally comes first to mind is the very common square spacing. The chief advantage of this arrangement is that it facilitates control of the work of hand-planting crews. It is the only arrangement compatible with driving such things as grass-mowing machines in straight lines both between and across the rows. Such treatments may be needed in plantations of Christmas trees.

Square spacing does provide for early development of uniformly closed stands. However, two other patterns provide even more uniform closure, whereas others offer alternative advantages coupled with adequate development of uniformity. The ideal arrangement is **hexagonal spacing** (Fig. 5.14), in which the tree crowns fit together in the horizontal dimension like hexagons and stand at the corners of equilateral triangles. As the crowns expand, competition is joined uniformly around them and not unevenly as with square spacing. This kind of arrangement requires costly measurements and is seldom used. The **staggered** arrangement is virtually the same and much easier to establish. This involves planting the trees in parallel rows with each tree opposite the middle of the gaps between trees in adjacent rows.

The important aspect of the hexagons and uniformity of crown coverage is that they are much more crucial considerations in the later stages of stand development than at the beginning. When the stand is planted, the important final-crop trees are mixed with and generally indistinguishable from the more numerous ones that will be thinned out as the stand develops. Any total production lost because some spots are tardily closed over by trees is likely to be in the form of small-diameter wood of debatable economic utility. Only as the stand gets older does it become desirable in thinning to secure uniform coverage and work toward the hexagonal arrangement. The numbers of trees for various spacings and kinds of arrangement are shown in Table 10.1.

Some early mortality will almost always occur, so it is necessary to consider how this might affect the number planted. If it can be anticipated that the losses will be randomly and uniformly distributed, it is then logical to plant enough extra trees to equal the expectable losses. However, if, as is usually true, the losses tend to occur in patches, it is

Table 10.1 Numbers of Trees per Unit of Area for Different Approximately Equivalent Spacings

Spacing		Number of trees	
Feet	Meters	Per Acre	Per Hectare
2 x 2	0.6 x 0.6	10,890	27,778
3 x 3	0.9 x 0.9	4,840	12,346
4 x 4	1.2 x 1.2	2,722	6,944
5 x 5	1.5 x 1.5	1,742	4,444
5 x 10	1.5 x 3.0	871	2,222
6 x 6	1.8 x 1.8	1,210	3,086
6.6 x 6.6	2.0 x 2.0	1,000	2,500
7 x 8	2.1 x 2.4	778	1,984
7 x 10	2.1 x 3.0	622	1,587
8 x 8	2.4 x 2.4	681	1,736
8 x 10	2.4 x 3.0	544	1,389
9 x 9	2.7 x 2.7	538	1,372
9 x 10	2.7 x 3.0	484	1,235
10 x 10	3.0 x 3.0	437	1,111

[a]In English and S.I. (metric) systems, with arithmetic perfectly consistent only within each system.

best not to deviate from the optimum spacing. If one planted more trees to allow for patchy losses, the surviving parts of the stand would be too dense and the vacant patches would still be empty. In such cases, the best course of action is to determine whether to fill the gaps by supplementary planting or simply to accept the vacancies.

Regularity of spacing should not be pursued to the exclusion of other considerations. Deviations from the spacing pattern are fully justified if patches of poor soil or competing vegetation can be avoided or if trees can be placed on the shady sides of obstacles such as stumps or logs. The effect of natural obstacles and mortality among the planted trees always reduces the initial stocking below the figure theoretically indicated by a given interval of spacing and should be taken into account in planning operations. Uneven spacing of trees does cause their crowns to become asymmetrical, and this can occasionally be detrimental. Any associated asymmetry in the stem appears to be most pronounced in the region of the basal butt-swell, whereas the upper part of the stem remains more nearly circular in cross section.

If trees are planted in furrows such as those created by most planting machines, it is cheaper to have them close together within the rows but with the rows far apart. This reduces the number of times the machine needs to be turned around as well as the power used to plant a given number of trees. Furthermore, if the rows are far enough apart, it may be much easier to enter the stand with logging machinery at the time of the first thinning. In determining the number of trees required for this kind of row planting, it can be thought of as a rectangular spacing even though the placement within the rows is not likely to produce a perfectly rectangular arrangement. If increasing the width of all rows would cause the overall density to be too low, it may be better to have only some of the spaces between rows wide enough for later thinning operations or similar activity.

Close initial spacing induces better stem form in trees destined for the final crop. However, if the final crop consists of trees 25 feet apart, it seems wasteful to plant trees 6 feet apart just for the training effects of competition. It could be enough to have clusters of densely spaced trees centered at 25-foot intervals with more widely spaced ones between the clusters. The clusters can be planted first and the more widely spaced trees afterward. It is almost mandatory that the clusters be precommercially thinned when the training effects have been secured; otherwise, the well-formed central trees are likely to be suppressed by the larger-crowned outer trees. However, because of its administrative complexity, this **cluster** arrangement is seldom used.

Mixed Plantations

Although most plantations are pure "monocultures," there is no fundamental reason why they cannot be mixtures of species. It is more complicated to create and manage mixed plantations, but the benefits can outweigh the complexities. As is set forth in more detail in Chapter 16, the management of plantations of mixed species usually depends on recognizing that no two species grow in height at the same rate. The best and simplest way to deal with this disparity is to take advantage of it by mixing a comparatively small number of seedlings of quick-growing intolerant species with more numerous seedlings of tolerant species that start off more slowly. The fast growers form an upper stratum and, if they are sufficiently far from each other, grow as rapidly in diameter as if they had been heavily thinned. The slower starting ones stay in the lower stratum; there they fill the remaining growing space and usually retain the capacity to grow rapidly once the overstory species have ceased rapid growth and have been removed.

In some parts of Europe it is customary to develop mixed plantations in which the different species are kept in the same canopy stratum. This requires careful attention to removing the trees that race ahead and encouraging those that would lag behind in nature. It can also be done by planting different species in different years.

Mixed plantations can also be created by planting small, pure clumps so that interactions between species are reduced in extent. When different species are planted in alternate rows, they usually interact as they would in even more intimate mixtures.

Enrichment Planting and Underplanting

Planting can sometimes usefully be done beneath existing stands or in gaps within them. In mixed stands, especially in the tropics, **enrichment planting** is sometimes done in lines or gaps that are deliberately opened in the forest in order to establish a few trees of highly desirable species. Such trees *must* be released by frequent cleanings, and it helps if the openings are created with the use of herbicides to kill the roots of the trees that are removed.

The deliberate planting of understories of shade-tolerant species beneath taller trees is called **underplanting**. It may be done to establish advance regeneration of desirable but absent species some years before the taller ones are to be harvested. It is also used to introduce understory species to provide such benefits as accelerated nutrient cycling or food and cover for wildlife.

Protection of New Plantations

Plantations are fully as subject to damage from biotic and atmospheric agencies as natural stands in the same stage of development, and often even more so. Losses in artificially regenerated stands represent greater wastage of direct investment than comparable damage in stands that have been reproduced naturally. The large investment in plantations requires a correspondingly heavy outlay for protection that should be regarded as part of the cost of artificial regeneration.

Wild animals are an important cause of damage to plantations. Often it seems that they find fertilized nursery-grown seedlings more palatable than natural seedlings of the same species. Although there are no certain measures of control, it is generally best to attack the problem by indirect means. Rodents tend to prefer dense cover and can sometimes be discouraged by eliminating low vegetation before planting (Barnes, 1973). Temporary reductions of the rodent population may be achieved by distributing poisons on the planting area or applying them directly to the stems of the trees.

Progress in the development of effective repellents for use against larger animals has been slow, but some compounds now on the market are occasionally satisfactory. Another solution is to avoid planting in places or during years with high populations of the most damaging animals. Sometimes valuable planted trees can be protected by enclosing planted areas with fences or planted trees with tall tubular tree shelters. Overpopulations of deer and other large herbivores are generally controllable only by deliberate hunting programs.

Damage by domestic grazing animals may be controlled by proper herding or fencing, depending on the custom of the locality. Sometimes livestock can be lured away from planted areas by putting out salt elsewhere.

One of the most important causes of plantation failure is the competition of woody vegetation. All too often plantations are established and then left to fend for themselves. Even those made in open areas can be overwhelmed by fast-growing vegetation. The costs of releasing plantations can be reduced by avoiding the temptation to plant up brushy areas or spots adjacent to existing trees. The underplanting of trees beneath brush or existing stands of trees usually involves a definite commitment to carry out subsequent release cuttings.

The gaps that develop in new plantations from scattered losses of planted trees can sometimes be corrected by subsequent planting or **refilling** ("beating up" in British terminology). If this is not done soon (and sometimes even if it is), the trees of the second planting get overgrown by the initial survivors. Because most losses occur during the first year, the ideal time for refilling comes a year after the first planting. Even then it is well to plant stock that is at least as tall as the established competitors or is of some intolerant species that will grow faster than the original species.

Refilling is most effective in replacing large patches of dead trees, provided that the causes of the initial failure are identified and either corrected or evaded. If only scattered trees have survived, it is often best to eliminate them and start a whole new plantation. In the common situation in which the losses are well scattered, refilling usually proves to be a costly waste of effort.

The Role of Planting

For many foresters and most people, trees do not grow unless they are planted. Planting is politically popular when it is used to reforest obviously devastated landscapes; it is not when it is used to regenerate clearcut forests. Society would support the practice better if it were employed where it was clearly the optimum regeneration alternative chosen on the basis of professional analysis and not from mindless adherence to some arbitrary standard operating policy. In many cases it is the optimum procedure or even the only feasible one.

The establishment of new forests, or the regeneration of old, by planting is one of the most costly steps in silviculture. These costs, however, vary widely depending on factors such as cost of labor, capital, nurseries, planting stock, equipment, site preparation, and treatment after planting. The benefits can be very high, and there are plenty of cases in which the ratio of benefit to cost is higher with planting than any other technique of stand establishment. The least costly kind of planting is that done where labor costs are low and some devastating event such as fire or agricultural use has made good, level soil free of competitive vegetation. At the other extreme are cases requiring control of aggressively competitive vegetation and hand planting of containerized stock with costly labor.

The financial case for planting can seem very dismal when the investment, necessarily made at the very beginning of a rotation, is carried at compound interest to the end. Society, foresters, and landowners exhibit some ambivalence in dealing with this matter. Sometimes the planting of trees seems so soul-satisfying that society evades the problem with various direct or indirect subsidies. Where the workforce, techniques, equipment, and facilities necessary for large planting programs have been conscientiously developed, there is a compulsion to use them to the exclusion of alternative means of regeneration.

If regeneration after timber harvest is required by law or ownership policy, the cost is often counted simply as a cost of the harvest rather than as an investment in the future. Sometimes this is a case of evading the issue. It is, however, also common that analysis of the situation has shown that a combination of expeditious logging with seemingly expensive planting is the least costly way of harvesting the old stand and starting a new one.

Sometimes, as with reforestation where forests have been eliminated or in afforestation of sites never forested, the only decision is whether to undertake the effort. Direct seeding is an option but often not feasible. Similarly, planting can be virtually the only way of starting stands of newly introduced exotics or plants produced in genetic improvement programs. These programs are, in fact, one of the most common reasons for using planting rather than natural regeneration to replace good stands.

A distinction should be drawn between planting for timber production and that for the forestation of devastated areas to prevent erosion and similar sources of injury to adjacent lands and waters. Where timber is the goal, priority is given to those areas where planting is most rewarding. Reforesting vacant areas commands priority over replacement of degraded stands, and that, in turn, usually outranks planting to regenerate good stands that already have desirable species. If the objective is protection of soil and water, the first areas treated should be those that are the source of greatest damage. Planting for wildlife or aesthetic improvement, just as that for timber, is done first where the prospective benefits are greatest. Planting is the most common means of growing Christmas trees and trees that are grown for wood, fruit, or other purposes, in combination with grass or agricultural crops as described in Chapter 21.

Most planting, especially in the temperate zone, is done with conifers. This is partly because coniferous wood is usually in high demand and short supply. It is also because conifers are often easier to plant successfully than angiosperms, especially where grass competition exists. Sometimes the best way to establish hardwoods on old grassy fields is to grow a rotation of planted conifers first and then depend on adjacent natural seed sources of hardwoods to provide advance growth for the second rotation. However, except for problems with grass, there are no basic natural barriers against the planting of angiosperms. In the warmer parts of the world, such important species as the eucalypts (Jacobs and Métro, 1981), teak, fast-growing species such as *Gmelina arboraea*, and various multi-purpose leguminous trees are commonly planted (Evans, 1982; Nwoboshi, 1982). Just as with conifers, there are cases in which early-successional angiosperms can be planted on exposed sites to create conditions favorable for establishing advance regeneration of species characteristic of later stages of plant succession.

DIRECT SEEDING

The direct sowing of seeds to establish forest stands is a technique of artificial regeneration that ought to work but does so only under special circumstances. Success in natural regeneration usually depends on whether a very small proportion of a very large quantity of seed lands in favorable spots and is overlooked by all the animals that feed on seeds. This random process could ordinarily be duplicated artificially only by broadcast distribution of hundreds of thousands of seeds per acre; that would be much more costly and not often as successful as planting. In establishing stands by direct seeding, the odds against success must be reduced by (1) control of seed-eating animals and (2) distinctly favorable conditions of site and seedbed. Success also usually depends on whether there is sufficient rain after sowing to keep the uppermost layer of soil adequately moistened throughout the period of germination and the succulent stage.

Seed Supply

Direct seeding requires large amounts of seed that should be gathered and handled in the same ways in which it is used to supply tree nurseries. If possible, seeds should be harvested from special seed production areas or seed orchards. The precautions about choices of species, geographical origins, and genetic characteristics discussed in Chapter 9 should be observed.

Most problems arise from the difficulties of gathering cones or fruits from tall standing trees. The work is hard, and the period during which it can be done is usually short. Crops worth harvesting are likely to be produced only during sporadic good seed years.

The best seed producers are ordinarily dominant trees that have attained middle age and are healthy. The likelihood that their seeds will be of acceptable genetic quality is greatest if similarly good trees are close by to provide pollen. If good, younger, and shorter trees produce adequate amounts of fertile seed, it is all right to collect from them. Unfortunately, it is simpler to gather seed from short, easily climbed, deep-crowned trees of poor form than from tall, straight, well-pruned ones. Although poor form in the parent does not necessarily indicate that its progeny will be undesirable, there have been enough bad experiences to suggest that it can increase the possibility.

It is well to examine the seeds of a tree in the field before effort is expended in gathering more of them. This can often be done by cutting them in cross section to see how many are hollow. Operations can be expedited by finding stands with large crops of sound seeds a few weeks before the seeds mature.

Cones or fruits must be collected after the seeds have completed their development but before they have been dispersed. The time during which the work can be done varies greatly. If it is short, it is sometimes possible to collect somewhat prematurely, but only if there is guidance from research or previous experience. With some species, collections can begin when the stored substances in the seeds lose a soft and milky appearance. Most cones and some fruits are mature when desiccation starts; if so, this is the most reliable indicator. Changes in color are associated with maturity and are commonly used even though they are difficult to define and evaluate.

Cones or fruits can be harvested with long poles fitted with hooked blades; climbing is often necessary. The simplest method is to collect from trees felled just when the seeds have matured. Seeds that have been cut down or stored by animals can often be used, but discrimination should be exercised because the animals may start their work prematurely.

The best practices for the collection or any treatment of tree seeds vary widely with species. It is always desirable to consult such sources as manuals of the seeds of woody plants (U.S. Forest Service, 1974; Young and Young, 1992) for the details of treatment and quantitative data about the seed of various species.

The methods of extracting seeds depend on the nature of the fruit. The cones of most softwoods and the dry fruits of some hardwoods will shed their seeds if dried in open air or in kilns. In fact, the seeds of the closed-cone pines can be extracted only in kilns. The seeds of trees with fleshy or pulpy fruits may be removed by macerating or crushing them in special machines. Seeds borne in pods or husks can be extracted by threshing.

The cones of many softwoods can be opened merely by spreading them out in trays in open sheds or in direct sunlight, although this method is less satisfactory than heating them in kilns.

The infrequency of seed crops makes it advantageous to have some means of storing seeds for several years without loss of viability. Fortunately, the seeds of most commercial species can be stored for periods of 3 to 10 years and sometimes longer if held at low temperature and low moisture content in sealed containers. The proper moisture content varies from 4 to 12 percent, depending on the species, and the temperatures should be below 5°C, preferably in the range from $-18°$ to $0°C$. It is important to dry the seeds uniformly and to prevent fluctuations in moisture content during storage. Under these conditions, respiration continues at the low level necessary to keep the embryos alive; only small amounts of the stored carbohydrates are converted into carbon dioxide in the process. Polyethylene bags make good containers because they are impermeable to water but less so to oxygen and carbon dioxide. These attributes prevent changes in moisture content but allow slow exchange of the gases involved in respiration to continue through the container walls.

If seeds can be stored against times of shortage, there is less reason to resort to use of poor or unsuitable seeds. Unfortunately, some species are difficult or impossible to store for long periods. Large nutlike seeds, such as those of oaks or walnuts, are difficult to store past the time when they would normally germinate in the first spring. They must, in any case, be kept in cold, moist media; if kept thus in refrigerated storage, they may last a year or two. Some species, such as the true poplars and many species of moist tropical forests, are so thoroughly adapted for immediate germination after seedfall that their seeds are almost impossible to store or even to transport. The seeds of most species of woody plants deteriorate if stored either moist or dry in heated rooms.

Treatment of Seeds to Hasten Germination

The seeds of species that grow in distinctly seasonal climates are often adapted to remain dormant and not germinate until conditions of moisture or temperature become favorable for germination. The conditions under which they are stored or treated in nature usually provide for the breaking of dormancy at the right times so that the seeds will germinate only when conditions are favorable for survival. However, if seeds are stored under artificial conditions, it may be necessary to treat the seeds in ways that simulate the natural processes that set the stage for germination.

There are two main causes of dormancy. The first is called **internal dormancy**, a condition ordinarily resulting from incomplete digestion of the fats, proteins, and other complex insoluble substances stored in the seed. Before germination can occur, these materials must be broken down into simpler organic substances, such as sugars and amino acids, that can be translocated to the embryo. The essential conditions are created by storing the seed in cool, moist substances just as it would be in the natural forest floor. This process is called **stratification** because it was originally done by putting layers of wet seeds between layers of wet sand. The enzymes that catalyze the breakdown of the complex stored substances are capable of functioning in a cool, moist environment, but those that cause the rapid respiration necessary to release energy for growth of the embryo are not. Consequently, the stored substances eventually used by the embryo are mobilized without being converted into carbon dioxide by useless respiration. The seeds of many conifers develop internal dormancy to some extent.

The stratification of seed in mixtures with wet sand or peat moss is a time-consuming procedure and also has the disadvantage of exposing the seeds to mold. With most species, it is fully as effective and much simpler to store wet seeds in polyethylene bags at 2°–5°C for one or two months.

The second main cause of imperfect germination is **seed-coat dormancy**. This occurs in seeds that have protective coverings so impervious that either oxygen or moisture is excluded from the embryo. This kind of dormancy can be broken by mechanical abrasion or chemical softening or etching of the seed coat. The first kind of treatment is referred to as **scarification** and may be accomplished by grinding the seeds in mixture with coarse, sharp sand or in machines equipped with disks or cylinders of abrasive paper. Chemical treatment is usually done by putting the seeds briefly in concentrated sulfuric acid. Black locust is one species that can be treated effectively by either method. Some species, notably basswood, have both kinds of dormancy and require stratification after treatment of the seed coat. Seed-coat dormancy is probably broken in nature by the grinding of seeds in the gizzards of birds or by the gradual decay of seed coats by fungi. It is most common in hardwoods with hard seed coats and rare in conifers.

Control of Seed-Eating Animals

Almost any site that supports a forest or is capable of doing so also supports a large population of small mammals and is combed over periodically by birds seeking food. These seed predators are rarely obvious to casual inspection, especially during the middle of the day. These populations vary widely in time and space; variations between seasons and between years are large. In general, areas of rather barren soil support lower populations of small mammals than those covered with vegetation or litter; they also offer less shelter to seed-eating birds. In direct seeding, it is well to take advantage of and even to create times and places of low animal population, but dependable results usually require more direct measures to ward off these animals.

Site preparation to reduce the amount of cover over the soil is very important in improving germination and survival of seedlings but is not especially effective in controlling the seed predators. Screens and traps are amply effective but so expensive that they are useful only for investigative purposes. Rodents have a good sense of smell and are rarely outwitted by covering the seeds. Even the wholesale poisoning of rodents on regeneration areas is a rather inefficient and questionable technique, although it can be made to work. If the population of seed predators is reduced, this merely creates an ecological vacuum; breeding or reinvasion can restore the population even during the period that seeds remain edible.

During the 1950s direct seeding became practical because of the development of chemicals that effectively repelled small rodents and birds (Abbott, 1965; Derr and Mann, 1971). An ideal repellent is a compound that makes the seeds distasteful to the seed predators but does not necessarily kill them. If the seed-eaters of a seeded area can be ''educated'' not to feed on the seeds, their presence tends to forestall invasion by their ''uneducated'' brethren. The rodent repellents were compounds that had suitable effects at the low dosages in which they were applied but were prohibited because they were either dangerously toxic at higher dosages or excessively persistent. At the present time there are no adequately persistent substitutes, but efforts are being made to develop some.

Some species, such as birches, spruces, eucalypts, and western red cedar, have seeds small enough that they are not especially subject to predation. Sometimes these can be sown without use of repellents. However, the lack of generally acceptable repellents has greatly curtailed use of direct seeding.

Effect of Site Factors and Their Modification

Success in direct seeding also depends on rendering the microsite as favorable as possible and ensuring prompt germination. The seeds should be placed in contact with mineral soil and, if possible, covered to the greatest depth consistent with successful germination. Moisture must be almost continuously available at or close to the surface of the mineral soil until the seedling roots have penetrated to a stable moisture supply. The amount of water several inches below the surface that might support the damaged root system of a newly planted tree is not necessarily sufficient. Consequently, some climates and soils are too dry for successful direct seeding.

The most favorable soils are those that are well supplied with moisture because of their topographic position, although poor aeration in excessively wet spots can also cause failure. Moist but well-drained soils that are supplied with water by seepage from higher ground are also very favorable. Soils of extremely coarse or extremely fine texture are less favorable than those of loamy texture.

As far as the germination and initial establishment of seedlings is concerned, a light cover of vegetation over bare mineral soil is almost ideal. It allows the seeds to come in contact with the mineral soil and yet it shields the surface from direct sunlight, thus reducing heat injury and direct evaporation from the soil. Although desiccation of most of the soil layer results from removal of water by plants, that of the surface and about the first half-inch or centimeter of soil is caused more by direct evaporation than by transpiration. Therefore, the vegetation that restricts the supply of water to a well-rooted seedling can actually increase the amount of water available during germination and the crucial days thereafter.

Sometimes a thin cover of herbaceous annuals or grass can improve the success of direct seeding by shielding seeds and seedlings from birds, heat, or frost. Such annuals as mustard and rye have been useful when sown in mixture with conifer seeds on bare mineral soil. They do not reappear during the second year and may serve to exclude undesirable plants as well as provide shelter during the first year.

The shade of woody plants is beneficial in the early stages, but its ultimate effect depends on the extent to which it competes with established seedlings for light, water, and nutrients. The best shade is that cast by dead materials, such as stumps, logs, or light slash. However, there is no fundamental reason why mixtures of fast-growing intolerant species of trees cannot be established to form stratified mixtures with simultaneously established tolerant species as described in Chapter 16. In fact, the intolerant species may act as a nurse crop for the tolerant species.

On most sites, the outcome of direct seeding still depends on the vagaries of rainfall. The careful selection and preparation of sites can only mitigate the baneful effect of one or two rainless weeks during the period of germination and initial development. If climatic records indicate that failures ought to be anticipated, for example, in one year out of every four, the true cost of each successful operation should be regarded as increased by 33 percent. In some regions the rainfall is so undependable that direct seeding may have to be restricted to sites where the soils remain moist during drought periods.

The chances of success are significantly increased if the seed is treated so that it will be capable of prompt and vigorous germination. This shortens the period of exposure to predators as well as that during which unfavorable weather can be critical. The same seed of mediocre quality that germinates well in the nursery may fail under the more rigorous conditions of the wild. Therefore, fresh seed is distinctly preferable to that which has been in cold storage for several years. If the seed is dormant and cannot be after-ripened naturally by fall sowing, it must be stratified. Sowing should follow within a day or two because stratified seed is very perishable. Finally, results are greatly improved if the seed is buried at optimum depth on favorable microsites.

Relative Merits of Direct Seeding and Planting

There is much more risk of poor survival with direct seeding than with planting. New seedlings that germinate and grow in the field have scant protection from the numerous lethal agencies that can be controlled in the nursery. Trees established by direct seeding grow no more rapidly than natural seedlings, so they suffer more than planted ones from competing vegetation. Furthermore, there is no opportunity to shorten rotations as there is in planting.

Direct seeding is inherently cheaper than planting because it involves less labor and equipment; investments in nurseries and the overhead charges involved in their operation are avoided. Large areas can be seeded more quickly, on shorter notice, and with fewer organizational problems. The only preliminary step is seed collection, although large quantities are required.

The roots of trees established by seeding develop naturally and are not subject to the deformities that are suspected of making planted trees susceptible to windthrow and root rot. Species that develop taproots or distinctly shallow, lateral root systems grow best if their roots develop naturally.

Direct seeding is possible on soils where planting is not feasible because of stones, stumps, or other obstructions, *but only if the sites are otherwise favorable.*

If dense stocking is desirable, it is more economically secured by direct seeding than by planting. However, the risk of localized understocking and overstocking is far greater. Stands established by direct seeding are likely to require more subsequent treatment, such as refilling, release cutting, and precommercial thinning, than planted stands.

Direct seeding can often be conducted over longer periods than planting and during colder or wetter weather. The main limitations on timing are those imposed by excessively dry weather and any need for minimizing the period of exposure to seed-eating animals and for avoiding unseasonable germination. Because germination often occurs after older seedlings have broken dormancy, the best time for direct seeding may actually come shortly after the regular planting season, thus enabling continuation of activity even where planting is the standard method of artificial regeneration.

Broadcast Seeding

The simplest type of direct seeding consists of scattering seeds uniformly over the area to be restocked. This is likely to be a waste of seed unless the conditions of the seedbed surfaces and competing vegetation have been made appropriately receptive. In the common case in which direct seeding is being used to regenerate some species adapted to follow fire, it is ordinarily essential to expose enough spots of bare mineral soil and to eliminate most preexisting vegetation. Accidental or intentional burning often suffices. Mechanical site preparation for broadcast seeding is justified mainly as a means of removing vegetation that will ultimately compete with the new crop after it is established.

Broadcast sowing is very rapid and is most often used when it is necessary to cover large areas quickly. Its most serious drawback is the lack of any provision for covering the seeds. For this reason it is best done during moist weather.

The seeding rates vary considerably depending on the favorability of site and anticipated weather. Recommendations ordinarily call for 10,000 to 25,000 viable seeds per acre (25,000 to 62,000 per ha), provided that these seeds are treated with appropriate repellents. However, the seeding rate should be carefully adjusted from year to year on the basis of quantitative observations of the results of previous seeding projects.

Most broadcast seeding is done from the air (Panel on Aerial Seeding, 1981). The use of aircraft is usually feasible only if several hundred acres or more can be done in a single project. Uniformity of seeding is possible only if the distributing equipment is closely calibrated and if moving flagmen are stationed on the ground. Fixed-wing aircraft do a faster and cheaper job than helicopters but are best adapted to seeding over gentle terrain where the areas to be seeded are also large and uniform. Helicopters are more suitable if the terrain is rugged or if the areas to be seeded are intermingled with those on which seeding is unnecessary or unlikely to succeed. The cost of aerial seeding itself is small compared with that of the seeds, logistical organization, and any associated site preparation. Broadcast seeding can also be done on the ground at rates up to 20 acres or 8 hectares per day with crank-operated "cyclone" seeders like those used for sowing grain crops. The cost is about the same as that for aerial seeding, but it takes much longer to cover a given tract.

Strip and Spot Seeding

Failure in direct seeding is least likely if the seeds are sown in spots or strips that are specially prepared or selected (Fig. 10.10). The sowing itself is more expensive than broadcast seeding but cheaper than planting. Strip and spot sowing are more economical of seed than broadcast sowing and can be done successfully in a much wider range of times and places.

If the terrain is suitable and there is no need to attempt to eliminate all the vegetation that will ultimately compete with the new crop, the mechanical preparation of the soil is most efficiently accomplished by furrowing or disking in strips. Although the seed is sometimes broadcast on the strips, it is usually much better to take full advantage of the preparatory work and apply the seeds so that they are covered with soil or pressed into it. If the site preparation has left air pockets in the soil, it may be necessary to lightly recompact the soil in the spots or strips to be sown.

Tractor-drawn equipment has been or can be devised to do light soil preparation and to sow and cover seeds individually in one pass of the equipment. Such machines can

Figure 10.10 A stand of longleaf pine in Louisiana established by direct seeding on disked strips. The seedlings are in the middle of the fourth growing season and have been sprayed once with fungicide for control of brown-spot disease. *(Photograph by U.S. Forest Service.)*

cover 25 to 60 acres (10 to 25 ha) per day on gentle terrain at total cost roughly half that of planting. In preparing spots or strips for the seeds, it is important to manipulate soil surfaces in ways that will prevent soil or litter from being moved by water, ice, or wind in ways that undermine or bury germinating seedlings. Shallow furrows can be plowed to anchor blowing leaves, and ridges can be constructed to deal with poor drainage, but it is otherwise better to keep the soil surfaces as level as possible to avoid erosional effects. Most efforts to develop small containers or tiny greenhouses in which seeds can germinate in the field have not been very effective.

Spot seeding is limited to operations that are too small to justify use of tractors or where steep terrain and obstructions prevent their use. The spots are prepared with hand tools or merely by kicking the debris to the side. It is difficult to arrive at the proper relationship between the spacing of seed-spots and the number of seeds applied to each. If one seed is sown on each spot, many spots must be prepared and seeded to allow for all the failures; if enough are sown on each spot to ensure that each has one established seedling, some spots will be choked with seedlings. In the ordinary compromise, several seeds are sown on each spot and these are spaced at intervals closer than would be used in conventional planting. It must then be anticipated that the stocking of the stand will be erratic and that it may become desirable to thin overcrowded spots and patches.

The seeds can be placed by hand, although any repellents that are used may be poisonous enough to dictate use of rubberized gloves. Various special seeding devices resembling corn planters are much more convenient; they require less contact with the seed and enable the sowing of up to 5 acres or 2 hectares per person-day. The only other tools especially designed for spot seeding are the dibbles used for planting acorns and other large nuts. These are bars shaped to punch holes into the soil obliquely so that the seeds can be planted on their sides.

Application of Direct Seeding

Broadcast seeding from the air was more common in the 1960s in North America than it has been subsequently. Planting came into greater favor mostly because it provides more economical use of the costly seed produced in tree improvement programs.

Where direct seeding works, it can enable swift reforestation of various kinds of barren, devastated areas, especially when done from the air. It can also provide a useful supplement to planting programs, which almost always fall behind schedule because of their cumbersome logistics.

Direct seeding is often one of the best ways for regenerating species with very small seeds, provided that there are enough microsites with bare mineral soil suitable for their germination. The supply of seed must also be abundant and cheap. This technique is often used for Australian eucalypts, which have seeds so small (300–800 per gram) that they are coated with clay to facilitate application and ward off insects.

One of these species is the longleaf pine of the South (Mann, 1970) (Fig. 10.10). The stemless, taprooted seedlings are difficult to plant successfully because the buds are so likely to become buried or excessively exposed by the slightest erosion around them. The discouraging tendency of the seedlings to linger in the grass for years without growing in height has also led to the common practice of reforesting extensive areas of old longleaf lands with planted slash and loblolly pine. Direct seeding of longleaf pine can be done by broadcast seeding, preferably with one year's growth of grass after prescribed burning, or by drill sowing on prepared strips. In fact, the reduction of grass competition on prepared strips sometimes causes speedy emergence from the grass stage.

Heavy-seeded hardwoods, such as oak, walnut, and hickory, have large taproots that are easily damaged in planting. The best artificial regeneration of these species is obtained by direct seeding, although the control of rodents is still a serious problem. Screens and other physical barriers are usually necessary.

Direct seeding is potentially useful as a low-cost means of establishing advance growth of shade-tolerant, exposure-intolerant species beneath full or partial cover of older trees. The slow juvenile height growth of most such species so delays return on planting investment that cheap initial establishment is advantageous.

Direct seeding is employed mostly for quick reforestation of large, barren areas, on unusual terrain that is physically difficult to plant but is otherwise favorable, and for rather peculiar species that do not withstand transplanting well. The potentialities of this low-cost technique could be exploited better if suitable repellents were developed. The most serious inherent shortcoming is the need for frequent rain after the sowing of seeds.

BIBLIOGRAPHY

Abbott, H. G. (ed.). 1965. *Direct seeding in the Northeast.* Mass. AES. 127 pp.

Ackzell, L., B. Elfung, and D. Lindgren. 1994. Occurrence of naturally regenerated and planted main crop plants in plantations in boreal Sweden. *FE&M*, 65:105–113.

Adams, D. L., R. T. Graham, D. L. Wenny, and M. Daa. 1991. Effect of fall planting date on survival and growth of three coniferous species of container seedlings in northern Idaho. *Tree Planters' Notes*, 42:52–55.

Allen, M. F. 1991. *Ecology of mycorrhizae.* Cambridge University Press, New York. 184 pp.

American Society of Agricultural Engineers. 1981. *Forest regeneration, proceedings.* Amer. Soc. Agr. Engineers, St. Joseph, Mo. 376 pp.

Bajaj, Y.P.S. (ed.). 1986–1991. *Biotechnology in Agriculture and Forestry: Trees I & II.* 2 vols. Springer-Verlag, New York. 1137 pp.

Barnes, V. G., Jr. 1973. Pocket gophers and reforestation in the Pacific Northwest: a problem analysis. *U.S. Dept. Int., Fish and Wildlife Serv., Spl. Sci. Rept.-Wildlife No.* 155. Denver. 18 pp.

Barnett, J. P., and J. M. McGilvary. 1981. Container planting systems for the South. USFS Res. Paper SO-167. 18 pp.

Baumgartner, D. M., and R. J. Boyd (eds.). 1976. *Tree planting in the inland west.* Washington State University, Cooperative Extension Service, Pullman. 311 pp.

Bonga, J. M., and G. N. Brown (eds.). 1982. *Tissue culture in forestry.* Nijhoff/Junk, Hingham, Mass. 420 pp.

Bullard, S., J. D. Hodges, R. L. Johnson, and T. J. Straka. 1992. Economics of direct seeding and planting for establishing oak stands on old-field sites in the South. *SJAF*, 16:34–40.

Camm, E. L., R. D. Guy, D. S. Kubien, D. C. Goetze, S. N. Silim, and P. J. Burton. 1995. Physiological recovery of freezer-stored whte spruce and Engelmann spruce seedlings planted following different thawing regimes. *New Forests*, 10:55–77.

Chapman, G. W., and T. G. Allen. 1978. *Establishment techniques for forest plantations.* FAO, Rome. 183 pp.

Clark, F. B., and S. G. Boyce. 1964. Yellow-poplar seed remains viable in the forest litter. *J. For.*, 62:564–567.

Cleary B. D., R. D. Greaves, and R. K. Hermann. 1978. *Regenerating Oregon's forests.* Oreg. State Univ., Ext. Serv., Corvallis. 286 pp.

Conrad, L. W., III., T. J. Straka, and W. F. Watson. 1992. Economic evaluation of initial spacing for a 30-year-old unthinned loblolly pine plantation. *SJAF*, 16:89–93.

Derr, H. J., and W. F. Mann, Jr. 1971. Direct seeding pines in the South. *USDA, Agr. Hbk.* 391. 68 pp.

Duryea, M. L., and G. N. Brown (eds.). 1984. *Seedling physiology and reforestation success.* Nijhoff/ Junk, Hingham, Mass. 322 pp.

Duryea, M. L., and P. M. Dougherty. 1991. *Forest regeneration manual.* Kluwer Academic Publ., Hingham, Mass. 440 pp.

Evans, J. 1982. *Plantation forestry in the tropics.* Clarendon, Oxford. 472 pp.

Felker, P. (ed.). 1986. *Tree planting in semi-arid regions.* Elsevier, New York. 444 pp.

Ffolliott, P. F., K. N. Brooks, H. M. Gregerson, and A. L. Lundgren. 1995. *Dryland forestry: planning and management.* Wiley, New York. 453 pp.

Fleming, R. L., and D. S. Moosa. 1995. Direct seeding of black spruce in northwestern Ontario: temporal changes in seedbed coverage and receptivity. *For. Chron.*, 71: 219–227.

Fox, J.E.D. 1984. Rehabilitation of mined lands. *For. Abst.*, 45:565–600.

Goulet, F. 1995. Frost heaving of forest tree seedlings: a review. *New Forests*, 9:67–94.

Grey, G. W., and F. J. Deneke. 1978. *Urban forestry.* Wiley, New York. 279 pp.

Goor, A. Y., and C. W. Barney. 1968. *Forest tree planting in arid zones.* Ronald, New York. 504 pp.

Grossnickle, S. C., and R. S. Folk. 1993. Stock quality assessment: forecasting survival or performance on a reforestation site. *Tree Planters' Notes*, 44:113–121.

Guldin, R. W., and J. P. Barnett (eds.). 1982. Proceedings, Southern Containerized Forest Tree Seedling Conference. USFS Gen. Tech. Rept. SO-37. 156 pp.

Harley, J. L., and S. E. Smith. 1983. *Mycorrhizal symbiosis.* Academic, Orlando, Fla. 496 pp.

Haywood, J. D., and J. P. Barnett. 1994. Comparing methods of artificially regenerating loblolly and slash pine: container planting, bareroot planting, and spot seeding. *Tree Planters Notes*, 45: 62–67

Heidman, L. J. 1976. Frost heaving of tree seedlings: a review of causes and possible control. USFS Gen. Tech. Rept. RM-21. 10 pp.

Heidman, L. J., F. R. Larson, and W. J. Rietveld. 1977. Evaluation of ponderosa pine reforestation techniques in central Arizona USFS Res. Paper RM-190. 10 pp.

Hobbs, S. D., S. D. Tesch, P. W. Owston, R. E. Stewart, J. C. Tappeilner II, and G. E. Wells (eds.). 1992. *Reforestation practices in southwestern Oregon and northern California.* Oregon State University, Forestry Research Laboratory, Corvallis. 465 pp.

Hocking, D. (ed.). 1993. *Trees for drylands.* Oxford and I.B.H. Publ., New Delhi. 370 pp.

Huebschmann, M. M., and R. F. Wittwer. 1992. Comparison of seeding versus planting loblolly pine in rips. *Tree Planters Notes*, 43:114–118.

International Poplar Commission. 1979. Poplars and willows in wood production and land use. *FAO Forestry Series*, 10. 328 pp.

Jacobs, M. R., and A. Métro. 1981. Eucalypts for planting. *FAO Forestry Series*, 11. 677 pp.

Janas, P. S., and D. G. Brand. 1988. Comparative growth and development of planted and natural stands of jack pine. *For. Chron.*, 64:320–328.

Jenkinson, J. L. 1980. Improving plantation establishment by optimizing growth capacity and planting time of western yellow pines. USFS Res. Paper PSW-154. 22 pp.

Johnson, C. M. 1994. Field performance of container systems in British Columbia. *For. Chron*, 70: 137–139.

Kittredge, D. B., M. J. Kelty, and P.M.S. Ashton. 1992. The use of tree shelters with northern red oak natural regeneration in southern New England. *NJAF*, 9:141–145.

Kozlowski, T. T., P. J. Kramer, and S. G. Pallardy. 1991. *The physiological ecology of woody plants.* Academic, San Diego. 567 pp.

Kramer, P. J., and J. S. Boyer. 1995. *Water relations of plants and soils.* Academic, San Diego. 495 pp.

Laurie, M. V. 1974. Tree planting practices in African savannas. *FAO Forestry Development Paper* 19. 185 pp.

Lavender, D. P., R. Parish, D. M. Johnson, G. Montgomery, A. Vyse, R. A. Willis, and D. Winston. (eds.). 1990. *Regenerating British Columbia's forests.* University of British Columbia Press, Vancouver. 372 pp.

Lundgren, A. L. 1981. The effect of initial number of trees per acre and thinning densities on timber yields from red pine plantations in the Lake States. USFS Res. Paper NC-193. 22 pp.

Mann, W. F., Jr. 1970. Direct-seeding longleaf pine. USFS Res. Paper SO-57. 26 pp.

Marquis, D. A. 1966. Germination and growth of paper birch and yellow birch in simulated strip cuttings. USFS Res. Paper N-54. 19 p p.

Marx, D. H., C. E. Cordell, and P. Kormanik. 1989. Mycorrhizae: benefits and practical application in forest tree nurseries. In: C. E. Cordell (ed.), Forest nursery pests. *USDA, Agr. Hbk.* 680. Pp. 18–21.

McClain, K. M., and D. M. Morris. 1994. The effects of initial spacing on growth and crown development for planted northern conifers: 37-year results. *For. Chron.* 70: 174–182.

McDonald, P. M., and O. T. Helgerson. 1991. Mulches aid in regenerating California and Oregon forests: past, present, and future. USFS Gen. Tech. Rept. PSW-123. 19 pp.

McKnight, J. S. 1970. Planting cottonwood cuttings for timber production in the South. USFS Res. Paper SO-60. 17 pp.

Mroz, G. D., and J. F. Berner (eds.). 1982. *Artificial regeneration of conifers in the Upper Lakes Region.* Mich. Tech. Univ., Houghton. 453 pp.

Needham, T., and S. Clements. 1991. Fill planting to achieve yield targets in naturally regenerated stands. In: C. M. Simpson (ed.), *Proceedings of conference on natural regeneration management.* Forestry Canada, Maritimes, Fredericton, NB. Pp. 213–224.

Newton, M., E. C. Cole, and D. E. White. 1993. Tall planting stock for enhanced growth and domination of brush in the Douglas-fir region. *New Forests*, 7:107–121.

Noble, D. L., and R. R. Alexander. 1977. Environmental factors affecting regeneration of Engelmann spruce in the central Rocky Mountains. *For. Sci.*, 23:420–429.

North Central Forest Experiment Station. 1982. Black walnut for the future. USFS Gen. Tech. Rept. NC-74. 151 pp.

Nwoboshi L. C. 1982. *Tropical silviculture, principles and techniques.* Ibadan Univ. Press, Nigeria. 333 pp.

Panel on Aerial Seeding. 1981. *Sowing forests from the air.* National Academy Press, Washington, D.C. 35 pp.

Potter, M. J. 1991. Treeshelters. *U.K. For. Comm. Hbk.* 7. 48 pp.

Putnam, W., and J. Zasada. 1986. Direct seeding techniques to regenerate white spruce in interior Alaska. *CJFR*, 16:660–664.

Puttonen, P. 1989. Criteria for using seedling performance tests. *New Forests*, 3: 67–87.

Read, D. J., D. H. Lewis, A. Fitter, and I. Alexander. 1992. *Mycorrhizas in ecosystems.* CAB International, Tucson. 448 pp.

Roberts, S. D., and J. N. Long. 1991. Effects of storage, planting date, and shelter on Engelmann spruce containerized seedlings in the Central Rockies. *WJAF*, 6:36–38.

Rose, R., S. J. Campbell, and T. T. Landis. 1990. Target Seedling Symposium. Proceedings, Combined Meeting of the Western Forest Nursery Associations. USFS Gen. Tech. Rept. RM-200. 286 pp.

Rudolf, P. O. 1950. Forest plantations in the Lake States. *USDA Tech. Bul.* 1010. 171 pp.

Scarratt, J. B., G. Glerum, and C. A. Plexman (eds.). 1981. Canadian Containerized Tree Seedling Symposium. *Can. For. Service COJFRC Symp. Proc.* O-P-10. 460 pp.

Schneller-McDonald, K., L. S. Tschinger, and G. T. Auble. 1990. *Wetland creation and restoration: description and summary of the literature.* U.S. Dept. of the Interior, Fish and Wildlife Service, Washington. 198 pp.

Shepherd, K. R. 1986. *Plantation silviculture.* Martinus Nijhoff, Boston. 322 pp.

Shiver, B. D., B. E. Borders, H. H. Page, Jr., and S. M. Raper. 1990. Effect of some seedling morphology and planting quality variables on seedling survival in the Georgia Piedmont. *SJAF*, 14:109–114.

Singh, S. P. 1986. *Planting of trees.* B. R. Publ., Delhi. 190 pp.

Sjolte-Jorgensen, J. 1967. Influence of spacing on coniferous plantations. *Intl. Rev. For. Res.* 2:43–94.

Sirvastava, H. C., B. Vatsya, and K.K. G. Menon. 1986. *Plantation crops: opportunities and constraints.* 2 vols. Oxford and IBH Publishing, New Delhi. 776 pp.

Smalley, G. W. 1985. Growth of 20-year-old Virginia pine planted at three spacings in Tennessee. *SJAF* 1(9):32–37.

Tinus, R. W., and S. E. McDonald. 1979. How to grow seedlings in containers in greenhouses. USFS Gen. Tech. Rept. RM-60. 256 pp.

U. S. Forest Service. 1974. Seeds of woody plants in the United States. *USDA, Agr. Hbk.* 450. 883 pp.

U. S. Forest Service. 1989–1990. The container tree nursery manual. *USDA, Agric. Hbk.* 674 1992. 7 vols.

Van Damme, L. 1988. Sowing method and seed treatment effects on jack pine direct seeding. *NJAF*, 5: 237–240.

Van Eerden, E., and J. M. Kinghorn (eds.). 1978. Proceedings, Root Form of Planted Trees Symposium. *Brit. Col. Min. For. and Can. For. Service, Joint Rept.* 8. 357 pp.

Wakeley, P. C. 1954. Planting the southern pines. *USDA Agr. Monog.* 18. 233 pp.

Wiersum, K. F. (ed.). 1984. *Strategies and designs for afforestation, reforestation and tree planting.* Pudoc, Wageningen, Netherlands. 432 pp.

Williams, R. D., and S. H. Hanks. 1976. Hardwood nurseryman's guide. *USDA, Agr. Hbk.* 473. 78 pp.

Young, J. A., and C. G. Young. 1992. *Seeds of woody plants in North America.* Rev. and enlarged ed. Dioscorides Press, Portland, Oreg. 407 pp.

Zobel, B., and J. Talbert. 1984. *Applied tree improvement.* Wiley, New York. 505 pp.

Zsuffa, L. 1976. Vegetative propagation of cottonwood by rooting cuttings. In: *Proceedings, symposium on eastern cottonwood and related species.* Southern Forest Experiment Station, New Orleans. Pp. 99–108.

PART 4

STAND DEVELOPMENT AND STRUCTURE

CHAPTER 11

DEVELOPMENT OF SILVICULTURAL SYSTEMS AND METHODS OF REGENERATION

Systematic programs are required to administer the harvest cuttings and other treatments necessary to create and maintain the kinds of stands that were described in Chapter 2. The most crucial of these treatments are the various kinds of cuttings that can be devised to create vacancies for new trees and provide microenvironmental conditions favorable to regeneration of the desired species. This chapter introduces programs of treatment and discusses how such programs are formulated to fit management objectives and natural circumstances.

A **regeneration or reproduction method** is a procedure by which a stand is established or renewed. Each method consists of the removal of the old stand, the establishment of a new one, and any treatments of vegetation, slash, or soil that are applied to create and maintain conditions favorable to the start and early growth of reproduction. Any procedure, intentional or otherwise, that leads to the development of a new stand of trees can be called a method of reproduction; sometimes it may be a combination of various methods. A **silvicultural system** is a planned program of silvicultural treatment extending throughout the life of a stand; it includes the regeneration treatments and any tending operations, protective treatments, or intermediate cuttings.

Classification of Methods of Regeneration

The simplest classification of reproduction methods is based on (1) the spatial arrangement of cuttings and age classes and (2) the distinction between seedlings and sprouts as sources of regeneration. The following version has six general categories and is the one most widely accepted in North America. As with many classifications, there is a wide range of variation in each category, and some borderline cases may not clearly fit any pigeonhole.

1. High-forest method—production of stands originating mainly from seed.

2. Clearcutting method—removal of the entire stand in one cutting with reproduction obtained artificially or from seeds germinating after the clearing operation.

3. Seed-tree method—removal of the old stand in one cutting, except for a small number of seed trees left singly or in small groups to provide for the establishment of advance regeneration.

4. Shelterwood method—removal of the old stand in a series of cuttings, which extend over a relatively short portion of the rotation, by means of which the establishment of one cohort of advance regeneration under the partial shelter of seed trees is encouraged.

5. Selection method—continual creation or maintenance of uneven-aged or multicohort stands by means of occasional replacement of single trees or small groups of trees with regeneration from any source.

6. Coppice-forest methods—production of stands originating primarily from vegetative regeneration.

    a. Coppice method—any type of cutting in which dependence is placed mainly on vegetative reproduction.

    b. Coppice-with-standards method—the combination, on the same area, of short-rotation coppice growth with scattered trees, which are grown on longer rotations and may be of seedling origin.

Each one of these methods has many modifications, and foresters will devise more; there are also combinations of methods. The more detailed classifications that exist do not differ in basic principle from this simplified classification. Some of the most sophisticated schemas are found in the German literature (Dengler, 1990; Burschel and Huss, 1987; Mayer, 1977). Matthews (1989) describes and classifies many silvicultural systems as applied in many parts of the world.

Paradoxically, the various methods of regeneration do not control the regeneration process as much as the words imply. The actual processes and treatments that cause forest regeneration were those discussed in several previous chapters. The names of the regeneration methods do, however, convey some information about whether the regeneration comes from seedlings, advance regeneration, or sprouts as well as the degree of shading of regeneration.

The main purpose of the terminology of the methods is to indicate the effect that the spatial arrangement and timing of final harvest cuttings has on the sizes, shapes, and arrangements of new stands. These matters are so important for the administration, harvesting, use, and protection of forests that the name of the regeneration method is often applied to the whole silvicultural system by which a stand is managed. The patterns of stand structure exist in terms of acres, but the factors that control the establishment of new plants, as brought out in previous chapters, depend on what happens or is done on square yards or inches. The term *regeneration method* is mostly a device to carry information about silviculture into overall forest management planning.

The terminology of regeneration methods also tells only a little about the details of the silvicultural systems by which the dynamic, rotation-long developmental processes of stands are guided. However, the name of the regeneration method describes enough that it provides a good starting point for further consideration of all the details built into silvicultural systems. A silvicultural system partly guides and is partly driven by one of the dynamic stand development processes described in Chapter 2.

The Basis of Distinction Between Methods of Regeneration

The first distinction is between high-forest methods involving regeneration from seed and the coppice-forest methods, which rely mostly on vegetative regeneration from stump-sprouts, root-suckers, lignotubers, or layered branches. Regeneration by planting goes with high-forest methods, as does that from the release of advance growth and the resprouting of small advance growth.

Coppice-forest methods were once called "low-forest" methods because they were thought of as being limited to growing short trees on short rotations for fuelwood. Because tall, old trees of many species, including mighty oaks and skyscraping coast redwoods, can be grown from sprouts, "low" is no longer intepreted literally. High- and coppice-forest methods are actually differentiated on the basis of the *predominant* sources of regeneration. The coppice-forest methods may include some regeneration from seed, and many broadleaved high forests have some important sprout regeneration.

In this rather traditional classification of methods and systems, no distinction is made between pure and mixed stands. This is probably because it was once assumed that all managed stands were composed of pure, even-aged, single-canopied aggregations of trees and that uneven-aged and mixed stands were composed of pure groups with that same structure. Actually, mixtures of species within stands are almost always more complicated than that. Usually, the different species within a single age-class unit or cohort grow in height at different rates, so that they tend to develop into different horizontal crown strata as shown in Fig. 2.2. Mixed stands composed of more than one cohort or age-class have even more complicated structure.

With regard to arrangement of cuttings in time, the most important distinction is between (1) methods for the maintenance of even-aged or single-cohort stands in which regeneration cuttings are concentrated at the end of each rotation and (2) methods in which reproduction cuttings extend throughout the rotation leading to the creation of uneven-aged or multicohort stands.

The complexity of many classifications results from recognition of the almost infinite variations that can be created in the horizontal, geometric pattern of cutting areas. Each general method can be applied so that openings and uncut timber are left either in uniform distribution or in concentrated strips, groups, wedges, and the like. In the general classification used here, simplicity has been achieved largely by restricting the amount of attention given to differences in spatial arrangement within stands.

The clearcutting methods create highly uniform stands that extend over substantial areas (Fig. 11.1). Reliance on advance regeneration established in reduced light under older stands that are removed in stages is characteristic of the shelterwood methods; the new stands are usually even-aged or of single cohorts (Fig. 11.2). The seed-tree methods produce essentially the same result but with fewer reserved trees. The selection methods produce uneven-aged or multicohort stands (Fig. 11.3) that may have many different spatial patterns. Coppice methods usually produce uniform even-aged or single-cohort stands, although they can be comprised of different age classes. The coppice-with-standards methods usually maintain sparsely stocked older trees of one or more age classes above uniform stands of low trees of sprout origin.

The method of reproduction being employed in a given stand may not be evident to

Figure 11.1 A 25-year-old pure, naturally regenerated stand of black-butt eucalyptus in New South Wales.

the casual observer. The identification and definition of a method of cutting depend fully as much on the results actually obtained, the intent of the treatment, and the nature of subsequent operations as they do on the pattern according to which the trees are removed. A regeneration cutting is a regeneration cutting only to the extent that it leads to the initiation of new trees. If something intended as a thinning results in the establishment of vigorous reproduction, it is best regarded as having been a shelterwood cutting. However, if the residual stand is then allowed to suppress the reproduction to the point of elimination, the initial cutting was indeed a thinning. A partial cutting aimed at starting the development of an uneven-aged stand is selection cutting only if subsequent operations are sufficiently consistent with the first to result in the ultimate creation of such a stand.

Formulation of Silvicultural Systems

A silvicultural system is designed to deal with a whole complex of biological, physical, and economic considerations, including logging problems, administration, manipulation

Figure 11.2   A single-cohort stand of ponderosa pine, western larch, and other conifers in Flathead National Forest, Montana, after a shelterwood cutting that will lead to the establishment of a second cohort composed of the same species. *(Photograph by U. S. Forest Service.)*

of growing stock, wildlife, and protection of stands, soils, and watersheds. It is also a program for initiating the kind of stand that will produce the desired benefits and for guiding its developmental processes through one or more rotations. Formulation of a silvicultural system should start with analysis of the natural and socioeconomic factors of the situation. A solution is then devised to go as far as possible in capitalizing on the opportunities and conquering the difficulties found to exist. If such programs are not formulated and followed, the management of stands may degenerate into rudderless drifting, governed more by current demands and other exigencies of the moment than by any intentions about the future.

For example, a certain application of a shelterwood system in pine stands might be formulated to produce high-quality lumber, habitat for birds and mammals dependent on the pines, and evergreen shade over snowdrifts to retard spring snowmelt. There might be early cleanings in established natural reproduction, followed by pruning and a sequence of free, crown, and low thinnings designed to maintain rapid diameter growth on some chosen crop trees. These thinnings would merge into shelterwood cuttings for the purpose of establishing partially shaded natural reproduction and setting the stage for repeating the sequence in a new stand. The rotation might be prolonged to provide some old trees for cavity-inhabiting animals. There would be shade on the ground continuously if full stocking of new pines was achieved under the old stands. A less intensive application of the

Figure 11.3   An uneven-aged mixed stand along a scenic highway leading to Crater Lake, Oregon. *(Photograph by Yale University School of Forestry and Environmental Studies.)*

same system might involve nothing more than final harvest cutting in two stages. Both silvicultural programs might have the same or different steps designed to provide for fire protection, logging safety, supplemental planting, or any of a myriad of considerations.

A silvicultural system is designed to fit a specific set of circumstances. It should not be presumed that it is something that has already been invented and can simply be selected ready-made from classifications or schematic descriptions of silvicultural systems. The sweeping prescriptions that have emerged from legislative bodies and courts usually solve one problem at the expense of creating many more. This book, for example, and in spite of certain superficial resemblances, is definitely not a catalog from which such choices can be made, nor is it a cookbook for their application. It may help in constructing them, however.

A silvicultural system also evolves over time as circumstances change and as knowledge of them improves. A classic history of the management of red pine in Minnesota, by Eyre and Zehngraff (1948), exemplified this evolutionary process. The point of departure was the seed-tree system, which was found wanting because of the inadequacy of the seed supply. As markets improved, it became obvious that the shelterwood system had important advantages in enabling good regeneration, reducing invasion by shrubs, and allowing retention of the best trees for growth to optimum size and quality. In the years after 1948, more intensive plantation silviculture became more common, but the evolution will probably not end.

Elements of Silvicultural Systems

If silvicultural systems are not to be chosen ready-made from a manual, it is logical to examine the various considerations that enter into their construction and evolutionary development (Florence, 1979). In the first place, a rational silvicultural system for a particular stand should fit logically into the overall management plan for the forest of which the stand is a part. Second, it should represent the best possible amalgam of attempts to satisfy all of the following major objectives, each of which will be discussed here and in subsequent chapters. These basic objectives are as follows:

1. Harmony with goals and characteristics of ownership
2. Provision for regeneration
3. Efficient use of growing space and site productivity
4. Control of damaging agencies
5. Protection of soil and water resources
6. Provision for sustained yield
7. Optimum use of capital and growing stock
8. Concentration and efficient arrangement of operations
9. Maintenance of desired plant and animal populations
10. Execution of policies about landscapes, scenery, and aesthetic considerations

These objectives are as likely to conflict as to harmonize with one another. For this reason, the procedures followed in applying systems with the same name vary widely depending on the relative importance attached to different objectives. Some of the contradictions are partially resolved in the development of management plans for the whole forest. That process is essentially the same as what has recently come to be called ecosystem management. The main purpose of this chapter is to consider the ways in which silvicultural systems are constructed to deal with the various conflicting objectives.

The conflicting considerations can come in ways too numerous to mention except by examples. As will be pointed out later in this chapter, identification of the objectives of ownership usually simplifies the formulation by reducing the number of alternatives that might be considered. Unfortunately, both owners and regulatory entities sometimes have conflicting or excessively demanding objectives.

The requirements for regeneration are discussed at length in many other chapters. These requirements are so crucial that the naming of systems for regeneration methods is a custom designed to remind foresters that some other objectives must be sacrificed to the need for timely regeneration. Continuity of any forestry enterprise, even the maintenance of museum-piece stands with trees thousands of years old, ultimately and absolutely depends on replacing old trees with new ones.

Efficient use of growing space depends on more than securing prompt uniformly distributed regeneration. Consideration must also be given to subsequent measures to deal with deficient or excessive stocking as by thinning and other techniques covered in Chapters 3–6 on the tending of stands. Provisions may have to be anticipated to exclude undesirable plants or control them if they appear. Efficient use of site productivity usually requires planning for the termination of rotations when trees become mature and their net growth declines. The growing space can be depended upon to fill with vegetation. Silvicultural systems are, among other things, programs for determining what fills growing space and for how long.

It is inevitable that stands of trees will suffer some damage during their rotation periods. Some of it can be forestalled by building stands that are low in susceptibility to damage. In many cases it is unwise to grow certain kinds of stands because they may, for example, produce too much forest-fire fuel, invite defoliating insects, or blow down in windstorms. However, there are silvicultural techniques for making otherwise susceptible stands less vulnerable to damage. Any system should also include provisions for emergency actions to be taken when serious damage is underway or has just occurred. The silvicultural control of damaging agencies is considered in Chapter 19.

Whereas silvicultural manipulations of forest vegetation do not greatly affect soil and water, any disturbance of soil associated with them may do so. For example, the spatial patterns of cutting controlled by silvicultural systems may determine the extent and location of the roads and trails that inevitably damage soil and may cause erosion and stream sedimentation. These matters are considered in Chapter 18 and, as they relate to site preparation and prescribed burning, in Chapter 8.

Silvicultural provisions for sustained yield of benefits from forests are usually thought of in connection with wood supply and will be dealt with in more detail in Chapter 15 on uneven-aged stands and in Chapter 17 on timber management. However, the same balanced distribution of age classes needed for sustained yield of timber also makes a major contribution to sustaining the diversity of habitats that is the foundation of biodiversity. Because it is seldom feasible to obtain sustained yield from individual stands, it is sought from combinations of stands. Sustained yield is a goal that can be approached through decades of deliberate effort but is seldom perfectly achieved. Therefore, each silvicultural system should be designed with sober understanding of the degree to which ownership will support efforts at achieving the goal.

Just as the medical profession is committed to curing patients, so is the forestry profession committed to keeping forests healthy and productive forever. However, both professions must deal with patients who will not or cannot invest in preventive and curative measures. In the case of forests and forest owners, the stumbling block is resistance to making long-term capital investments. Even the act of leaving salable trees standing in the woods for additional growth is a kind of investment in growing stock. Planting trees can be one of the most far-seeing investments that people make. Thinning and many other details of silvicultural systems are designed to bring forest owners adequate returns on their investments. The degree to which sustained yield can be achieved often depends on how successful these measures are. Since most monetary returns to owners from silviculture come from timber management these measures are also dealt with in Chapter 17.

In forestry it is necessary to deal with large areas of land and to have the means of keeping them accessible. Having operations simultaneously dispersed lightly over wide areas and frequently repeated yields important advantages. However, these lead to high transportation costs and all the problems that roads make for soil, water, and wildlife. For example, even-aged systems usually provide a more economical concentration of operations than uneven-aged systems. These matters are discussed in more detail in Chapter 17 on timber management and Chapter 18 on watershed management.

Forest ecosystems are not just trees; they always encompass animals and plants in addition to trees. In fact, sometimes the nontree organisms, wild or domestic, are more important to owners and society than the trees. Even if ownership does not want them, they must be considered in forest management either because they can become pests or weeds or because, under most legal systems, the public has a protective and regulatory interest in them. Animals are governed mainly by plants; silviculture is a means of governing plants. Chapter 20 introduces the principles of modifying silvicultural systems to produce the desired kinds of habitats.

Forests and all their plants are part of the landscape and scenery for which the public is asserting a growing interest. An increasingly urbanized population regards the ugly appearance of cuttings areas as clear evidence of forest devastation and is pressing for arbitrary regulations about silvicultural practices. This calls for measures to screen or mitigate ugliness, whether or not they are required. Positive steps can also be taken to make forests beautiful. Silvicultural treatments can be arranged in ways that expose attractive features or distant vistas to public view along roads or trails. In addition, the edges of cutting areas should be manipulated so that they do not look artificial from a distance. It is usually easier to do partial cutting along roadsides than to explain that a clearcutting and planting operation is environmentally benign and will ultimately produce a beautiful stand. These matters are discussed in Chapter 17 on timber management and by Jones (1993), Crowe (1978), and Bacon and Twombly (1980).

All these different considerations represent forces that pull foresters and forest owners in many directions. Analysis will show that single-minded concentration on any one of these objectives can ultimately lead to ridiculous results. The best solutions obviously lie in finding the most appropriate blends of partial fulfillments of all significant objectives. Fortunately, the various considerations do not always conflict, and often the goals are complementary.

Role of Ownership Objectives in Formulating Silvicultural Systems

Decisions about the design of silvicultural systems are greatly simplified by clarifying the objectives of ownership, whether it be public or private. This logical first step automatically eliminates many of the possible alternatives. It also forces recognition of the fact that it would not necessarily be in the interest of two different owners to manage the same stand in the same way. Except to the extent that law may dictate certain actions, there is no justification for a forester to embark in arrogant wisdom on any "standard" procedure for the growing of a particular kind of stand, regardless of whether the technique fits the owner's purposes.

Foresters must often help owners select their own objectives before formulating a silvicultural program. This is partly because owners may not have a clear idea of what objectives are reasonably attainable. The objectives of ownership clearly dictate the relative amounts of attention that is to be paid to management for timber, wildlife, forage, water, recreation, scenery, or other potential benefits of forests. The objectives of ownership are always modified by various public laws and regulations that reflect the objectives of society.

On land managed for multiple use (Burns, 1989), the silviculture logically differs from that on similar land that might be owned by a lumber company primarily for growing sawtimber; this would in turn differ from a paper company's silviculture for pulpwood production. If an owner is most interested in wildlife or in preserving some old-growth timber purely for aesthetic purposes, the forester should modify the silviculture accordingly. In fact, the forester's occupational bias in favor of efficient timber production can sometimes be more of a liability than an asset.

Analysis of the objectives of ownership will normally define the kind of vegetation to be maintained, the kind of trees that are to be grown, and the amount of time, money, and care that can be devoted to the process. The intensity of practice is determined by the amount of money that the owner is willing and able to place in long-term silvicultural investments as well as by the interest return required on such investments. If long-term investments cannot be made, the silvicultural treatment may be limited to those things that can be done in the process of harvesting merchantable timber. Where the future of the enterprise seems limited to the life expectancy of the owner, attention may be restricted to securing maximum benefits during the owner's prospective lifetime; manipulating the existing growing stock would probably take precedence over securing regeneration. If the owner's objectives do not include production of timber or profits, the forester should manage stands accordingly.

Resolution of Conflicting Objectives

There is no inherent harmony among the various major objectives sought in managing forests. Such harmony can be created only by weighing the various objectives individually and inventing silvicultural systems that represent analytical compromises within plans for forest or ecosystem management created by the same kind of procedure (Forest Ecosystem Management Assessment Team, 1993). The disharmony is most evident if one considers all forestry in general. Fortunately, the conflicting objectives need be resolved only for particular forests or stands. Analysis of each situation will usually reveal a few ruling considerations; the necessity of giving first attention to these will simplify and govern the solutions.

Sometimes one is faced with two or more problems, each of which is separately insoluble but which can be neatly combined into a single solution. The process starts with a consideration of the goals of the forest owner. Each of the remaining objectives must then receive some attention; it would almost invariably be a mistake to pursue any single one to the bitter limit. The analytical process generally works downward from the forest to the stand, but not without the formation of some preliminary idea of the range of treatments and results that is silviculturally feasible. Some immutable and absolutely restrictive natural factors are bound to exist, and these must be recognized early in the process. However, the remaining latitude of silvicultural possibilities is likely to be broad enough that further efforts to develop some optimum silvicultural system is normally based on analysis of the effect of all factors, natural and social.

The analytical balancing of competing demands and conflicting considerations can be defeated or paralyzed when the partisans of single uses induce legislatures to try to plan the details or manage to transfer the planning process to adversarial proceedings in courts.

Silviculture by Stand Prescription

Silvicultural treatments are best prescribed, stand by stand, by foresters on the ground. However, the general forms of silvicultural systems that may be prescribed and the basic management policies ordinarily have to be determined for the forest ownership as a whole. Some degree of standardization is necessary to ensure uniformity and continuity of action.

Too much standardization can lead to treatments that fit well in some stands but badly in others. Results can become especially poor if conformity to standard operating procedures either causes or allows field personnel to stop thinking about silvicultural problems.

The need for quasi-independent stand prescription is illustrated by the fact that it may be logical to clearcut and replace an unhealthy 35-year-old pine stand and yet thin a healthy one, even though both are located side by side on the same site within the same ownership. A standard procedure for all 35-year-old pine stands would dictate that the two be treated the same. Most such decisions are more sophisticated and complex than in this simple case, but the principle is the same.

Foresters on the ground should be able to detect matters of silvicultural significance that are not obvious to others viewing stands from roadsides, distant offices, or legislative chambers. Policies about the objectives of forest management are determined by forest owners, public or private, under the constraints of laws and regulations. The role of professional foresters is to help develop such policies and to execute them through the formulation and prescription of silvicultural treatments and systems. In some places laws require that these functions be carried out only by licensed professional foresters, usually to protect various public interests. However, this is not always the case. Forest policies are often poorly conceived and badly executed; it is the traditional responsibility of the forestry profession to correct such situations both as a matter of principle and as a way to avoid being blamed for them.

Few forms of human endeavor have more distant time horizons than those that practitioners of silviculture must strain their eyes to see and explain to those who govern forest policy. Silvicultural systems are the silvicultural programs designed for reaching the policy goals. The time periods involved, usually one or more whole rotations, are long enough that the goals and the means of achieving them usually change before they are reached. Thus the programs must be modified occasionally and must be kept flexible enough to do so. Such changes of method in time and within stands are likely to be common when a forest is, like most American forests, being brought under management for the first time. It is by means of such changes that the procedures are perfected and the pattern of stands is molded into arrangements that are rational from the standpoint of site variations and management objectives.

Naming of Silvicultural Systems

Treatments are first devised to fit the circumstances, and the naming of them is done afterward. In other words, the terminology describes treatment but does not dictate it; it is descriptive rather than prescriptive. The long-standing terminology of reproduction methods may be used to the extent of its limited capacity for providing information in terms that are meaningful to all foresters and then supplemented with adjectives that convey the additional, detailed information that is usually necessary. Methods of reproduction, kinds of stands, and silvicultural systems are classified and recognized largely because the act of doing so indirectly forces planning for the future care, development, and replacement of stands.

A cutting does not have to be planned by a forester or involve any thought whatever about the future to be recognizable as a kind of intermediate or regeneration cutting. It matters not how indiscriminately the trees are removed or how haphazardly the new growth develops. Of course, it is not necessarily easy to categorize cuttings that involve no future plans, and the act of doing so may be purely an intellectual exercise.

Sloppy use of the terminology can corrupt it to the point where the words mean little. *Selective cutting* is a potentially useful term that slovenly usage has almost destroyed as a technical term. It seldom helps to try to redefine existing technical or common terms. *Clearcutting* is really a word with many meanings as well as an ugly connotation; as a term of technical silviculture, it is an unhappy attempt at redefining a logging term and might better be replaced.

The Silvicultural System as Working Hypothesis

The practice of silviculture must be conducted in the absence of complete knowledge about the immutable natural and changing social factors that affect each stand. Furthermore, most of the treatments cannot be properly evaluated until many years after their application when the results become more evident. Decisive action cannot await absolute proof of validity, nor can it be evaded indefinitely by fence-straddling. The forester must, therefore, proceed as far as possible on the basis of proven fact and then complete plans for action in light of the most objectively analytical opinions that can be formed. This combining of fact with opinion is treacherous, especially because people easily become committed to opinions when they base actions on them; such opinions can then be mistaken for facts.

The soundest basis for action derived from a mixture of proven fact and unproven opinion is the **working hypothesis**, which is not truth but the best estimate of truth formed by analyzing all available information. It is not allowed to become a ruling doctrine but is, instead, constantly tested against new information and modified accordingly. It is not embraced so wholeheartedly that it cannot be discarded and replaced. One must always be ready to admit, at least inwardly, that an earlier decision was wrong and correct the procedures accordingly. It is important to monitor the results as objectively as possible. This process of action based on analysis of results is sometimes called **adaptive management.**

The silvicultural system is logically based on a working hypothesis and is altered through adaptive management as it becomes necessary to change the hypothesis. Lest they become ruling doctrines, existing procedures should be constantly examined to determine whether they have outlived their time or have become inconsistent with new information. It is, for example, logical to consider whether silvicultural practices developed for manual logging remain valid under mechanization. Similarly, it is well to question whether thinning policies should not be altered when it is found that the rate of diameter growth does not, as was once believed, control the properties of wood. It may be necessary to initiate prescribed burning if previous policies of fire exclusion have caused dangerous accumulations of forest fuels. Most good ideas survive to be overdone, so they should continually be modified by new ones.

Radical changes and excessive fluctuations in silvicultural procedures lead to confusion and lost motion. If the silvicultural system for dealing with a particular local kind of stand has been kept in conformity to circumstances by occasional modifications, disruptive reforms are unlikely to be necessary. It is prudent to avoid the petulant temptation to discard tested procedures in favor of radically different untested ideas because of problems of the moment. If once promising plantations become riddled with root rot, one should not necessarily swear off planting. Neither should the lapse of nine years between good seed crops cause a wholesale shift to artificial regeneration.

Forestry is almost unique in the extent to which the actions of the present govern those of future generations of practitioners. Any treatment that is applied to a stand now is likely to restrict the choices available in subsequent treatment. In a sense, the forester conducting a treatment in a stand is entering into a pact of mutual understanding with succeeding foresters about the stand. We of the present are entitled to expect that the plans we put into effect now will be given the benefit of all doubts by our successors, but not to the extent of unquestioning adherence.

The results of treatment that are most difficult to change are species composition and age-class structure of stands, so changes in these attributes should be approached with deliberation. It must be recognized that the period of regeneration is nearly the only one during which major changes can be made in stand characteristics and silvicultural systems. The period of intermediate cutting is more one of modification than of change. We do the best we can with stands inherited from the past; those that we start anew provide the best opportunity to inflict new ideas on the future (Fig. 11.4).

The silvicultural system should be built where it is to be used, not prefabricated and brought from some other kind of forest. Furthermore, silviculture has too often been conducted based on the view that each method or system constitutes a rigid set of procedures, usually defined in quantitative terms, which, if religiously followed, will produce optimum results. There has even been the view that all silvicultural practice in a given locality

Figure 11.4   One-and-a-half centuries of evolution in silvicultural practice in Bavaria illustrated in three stands at the Forest of the University of Munich. The Scotch pine stand in the background was established by direct seeding 150 years ago on soils degraded by long periods of overuse. On the left is a Norway spruce plantation established 70 years ago where the initial pines had been harvested and the soil had improved enough to allow the spruces to grow. On the right is a planted stand of mixed hardwoods, 30 years old, which represents the reestablishment of a forest resembling the original one and believed to be resistant to many of the maladies that pure spruce plantations sometimes suffer. *(Photograph by Yale University School of Forestry and Environmental Studies.)*

should follow a single method or system. These rigid views have too often collided with natural, economic, or political disaster. The present volume is dedicated to the presentation of as many possibilities and alternatives as its authors can discern.

No one method or system has any single ideal routine of application akin to the precise movements of parade-ground drill. There is, for example, no one ideal mode of shelter-wood system compared with which all others are necessarily inept, imperfect, or "less intensive" imitations. If they have these attributes, it is not because of departure from some preordained ideal treatment program. None of these methods is a schedule or routine that needs only to be copied to produce success. Furthermore, it cannot be presumed that any of the listed methods can be safely applied to any kind of forest just because it is a "recognized standard method."

This chapter has been an attempt to describe the considerations that enter into the construction; the remaining chapters include discussions of most of the general categories of systems that foresters have devised. Examples of methods of reproduction and particular silvicultural systems are presented, as is the reasoning underlying their development. These examples are chosen mainly to illustrate application to specific kinds of stands or management objectives. In application, silvicultural systems include a myriad of additional variations designed to accommodate differences in objectives of ownership, accessibility, site quality, and all the other factors that make every stand at least slightly different from all others. Many of these kinds of variations are described and discussed in collections of accounts of silviculture applied to different forest types (U.S. Forest Service, 1983) and regions (Barrett, 1995; Society of American Foresters, 1986) of the United States.

BIBLIOGRAPHY

Bacon, W. R., and A. D. Twombly. 1980. National forest landscape management. Vol. 2, Chap. 8, Timber. *USDA, Agr. Hbk.* 559. 223 pp.

Barrett, J. W. (ed.). 1995. *Regional silviculture of the United States.* Wiley, New York. 643 pp.

Burschel, P., and J. Huss. 1987. *Grundriss des Waldbaus.* Verlag Paul Parey, Hamburg. 352 pp.

Burns, R. A. (ed.). 1989. The scientific basis for silvicultural and management decisions in the National Forest system. USFS Gen. Tech. Rept. WO-55. 150 pp.

Champion, H. G., and S. K. Seth. 1968. *General silviculture for India.* Govt. of India Press, Delhi. 511 pp.

Crowe, S. 1978. The landscape of forests and woods. U.K. For. Comm. Booklet 44. 47 pp.

Davis, L. S. and K. N. Johnson. 1987. Forest management. 3rd ed. McGraw–Hill, New York. 790 pp.

Dengler, A. 1990. *Waldbau, auf ökologischer Grundlage. II. Baumartenwahl, Bestandesgrundung und Bestandspflege.* 6th ed. revised by E. Rohrig and H. A. Gussone. Verlag Paul Parey, Hamburg. 314 pp.

Evans, J. 1984. Silviculture of broadleaved woodlands. UK Forestry Commission Bul 62. 232 pp.

Eyre, F. H., and P. Zehngraff. 1948. Red pine management in Minnesota. *USDA Circ.* 778. 70 pp.

Florence, R. G. 1979. The silvicultural decision. *FE&M* 1:293–306.

Forest Ecosystem Management Assessment Team. 1993. *Forest ecosystem management: an ecological economic, and social assessment.* USDA Forest Service Govt. Printing Office, Doc. 1993-793-071. 250 pp.

Jones, G. 1993. *A guide to logging aesthetics. Practical tips for loggers, foresters, and landowners.* Northeast Regional Agricultural Engineering Service, Ithaca, NY. 28 pp.

Leuschner, W. A. 1984. *Introduction to forest resource management.* Wiley, New York. 298 pp.

Marquis, D. A., R. L. Ernst, and S. L. Stout. 1992. Prescribing silvicultural treatment in hardwood stands of the Alleghanies (revised). USFS Gen. Tech. Rept. NE-96. 101 pp.

Matthews, J. D. 1989. *Silvicultural systems.* Oxford University Press, Oxford. 284 pp.

Mayer, H. 1977. *Waldbau auf soziologisch-ökologischer Grundlage.* Gustav Fischer, New York. 500 pp.

Ohmann, L. F., et al. 1978. Some harvest options and their consequences for the aspen, birch, and associated conifer forest types of the Lake States. USFS Gen. Tech. Rept. NC-48. 34 pp.

Savill, P. S. 1991. *The silviculture of trees used in British forestry.* CAB International, Tucson. 142 pp.

Society of American Foresters. 1981. *Choices in silviculture for American forests.* Society of American Foresters, Washington, D.C. 80 pp.

Society of American Foresters. 1986. Silviculture: the next 30 years, the past 30 years. *J. For.,* 84:

I. C. D. Oliver. Overview. (4):32–42; **II.** J. C. Tappeiner II, W. H. Knapp, C. A. Wierman, W. A. Atkinson, C. D. Oliver, J. E. King, and J. C. Zasada. The Pacific Coast. (5):37–46; **III.** S. G. Boyce, E. C. Burkhardt, R. C. Kellison, and D. H. Van Lear. The South. (6):41–48; **IV.** R. S. Seymour, P. R. Hannah, J. R. Grace, and D. A. Marquis. The Northeast. (7):31–38; **V.** J. W. Benzie, A. A. Alm, T. W. Curtin, and C. Merritt. The North Central Region. (8):35–42. **VI.** J. N. Long, F. W. Smith, R. L. Bassett, and J. R. Olson. The Rocky Mountains. (9):43–49.

U. S. Forest Service. 1983. Silvicultural systems for the major forest types of the United States. *USDA, Agr. Hbk.* 445. 191 pp.

Williams, M. R. W. 1981. *Decision-making in forest management.* Wiley, New York. 143 pp.

CHAPTER *12*

THE SILVICULTURE OF PURE EVEN-AGED STANDS

The next several chapters describe, in sequence, stands of successively increasing complexity starting with those of one age class or cohort and then those with more than one. All but the last of these five chapters are arbitrarily limited mainly to considering pure stands so that the complexities of mixed stands, dealt with in Chapter 16, the last of the series, will not complicate consideration of the development of different kinds of pure stands.

Pure uniform even-aged stands regenerated by clearcutting are the kind that are best understood and simplest to manage. They start with the growing space almost free of trees and of most other potentially competing vegetation. The regeneration is established *after* the regenerative disturbance, usually within a very short period of years, and the new trees all grow up together in single-canopied stands. The new trees are regenerated artificially or naturally. The development of the stands is so uniform that the even-aged yield tables discussed in connection with thinning can be used to predict their development. Some foresters, for good reason or poor, know no other kind of stand and practice no other kind of silviculture.

Monocultures of sprouting species are often as simple and uniform. In fact, no kinds of stands can be regenerated more dependably. These will be described in Chapter 13 on vegetatively regenerated stands, which include some moderately complicated kinds.

The silviculture of planted monocultures is very simple and often dependable, but may be both intensive and expensive. Naturally regenerated monocultures that depend on post-disturbance regeneration from seeds are also easy to understand. However, their establishment by silvicultural treatments can be difficult and uncertain.

Regardless of whether the stands are artificial or natural, it should be noted that pure even-aged stands usually persist only where the site conditions are so difficult that only one tree species can endure. This is usually true of sites that are too dry, too wet, too cold,

316

or too hot. In many localities, there are scarcely any other kinds of forests. On better sites, severe disturbances may create conditions during the establishment period that are favorable to only one hardy species. Some pest or damaging agency may allow but one species to endure. It is also possible to create such stands by silvicultural treatment, although it is not necessarily easy to maintain them as such. On highly favorable sites, true clearcutting may open the way for invasions by jungles of mixed species that are virtually impossible to control.

Such simple kinds of stands are regenerated by true clearcutting, which simulates the natural catastrophic events that have led to regeneration of individual species that are shade-intolerant and exposure-tolerant. Creation of the necessary conditions for establishment often requires site preparation measures such as those described in Chapter 8. Clearcutting, instead of regeneration methods involving partial cutting, is logical when residual trees are not worthy of retention for further increase in value, source of seed, protection of the new crop, wildlife habitat, amenity, or other useful purposes. There are also some very intolerant pioneer species that not only endure such conditions but also, for practical purposes, require them.

Clearcutting should also be seriously considered if the methods of partial cutting are substantially more costly or if this method provides the best means of replacing a poor stand with a good one. If such conditions exist and the desirable species is ecologically adaptable to the conditions created, clearcutting may be distinctly superior to any other method of regeneration. On the other hand, some species are so sensitive to exposure that they must be regenerated in shaded conditions with some sort of partial regeneration cutting. On some extreme sites the planting of seemingly resistant seedlings grown in nurseries is not enough to overcome this problem.

Plantation Silviculture

The least complicated kinds of stands are pure, even-aged plantations (Fig. 12.1) promptly established with nursery-grown plants after clearcutting or some other event has eliminated virtually all of the preexisting vegetation. It is possible, though not necessarily desirable, to grow planted stands of almost any species, except for those that are too sensitive to exposure. The silviculture is simple in concept but sophisticated in detail. Many of the details have already been described in the sections on site preparation, planting, direct seeding, and operations for the release of established trees.

These techniques evade many restrictions on silvicultural treatment, although the freedom conferred also entails an increased opportunity to make serious mistakes. The size and pattern of cutting areas are not limited by the necessity of reserving sources of seed. There is no need to modify procedures to protect the seedlings from those hazards of the natural environment that are bypassed in the nursery. Harvesting operations are not hampered by the need to protect residual vegetation, except during thinnings. No residual trees or desirable advance growth obstruct the use of fire, herbicides, or machinery during site preparation. It is usually much easier, cheaper, and more dependable to dispose of preexisting undesirable plants before the new stand is established than it is afterward. However, the disturbances related to the site preparation can open the way for invasion by aggressive undesirable vegetation that may overwhelm the planted trees, especially on good sites that are favorable to many species.

Usually, the stocking of the new stands is fuller and more uniform than is possible with any other mode of regeneration. This is especially advantageous if the objective is to

Figure 12.1   Thinned plantation of slash pine in southern Alabama.

maximize the yield of wood in trees of relatively uniform size. However, the stands can become so crowded with trees of equal development that thinning is very necessary for good diameter growth and salvage of the production of trees that die from intense competition. Pure-stand yield tables can be used to guide and predict stand development. Opportunity exists for the use of species not found on the site or for establishing improved genetic strains.

The establishment of new stands starts with complete removal of the previous crop. The growing space is made vacant by the use of machinery, herbicides, fire, or combinations thereof in site preparation. One of the common goals of site preparation is to reduce physical impediments to planting. If soil-moisture conditions can be improved by reshaping the surface terrain, this may be done also. The trees that are then planted can represent the best genetic material available and be carefully fitted to the sites. The growth of the planted crop of trees can often be improved by fertilization or by reducing any competing vegetation that survived the site preparation or came in after the planting. However, fertilization is usually postponed until after the stands have closed. Thinning is an important part of any kind of truly intensive plantation silviculture, although the objectives and methods can vary widely. When the stand is deemed to be mature, ordinarily on financial grounds, it is replaced by a new planted stand and the whole cycle is repeated.

As little as possible is left to chance or nature. Planting plays a key role because it is a much more dependable kind of regeneration than that from natural or artificial seeding. Planting represents one way of shortening the period during which the stand does not occupy the site fully and has thus not resumed full production. The loss of even one year's growth can be a significant fraction of the total production of a rotation, especially if the rotation is short. Some gain in forest production, or shortening of the rotation, may be made possible if the seedlings are already one or more years old when moved from the nursery. The gain of one year with a 50-year rotation is a 2 percent increase in the production secured on each unit of land area. Prompt regeneration also ensures capture of the benefits of site preparation. Delay in regeneration increases the risk that the growing space will be usurped by undesirable vegetation.

With planting it is possible to govern the density, spacing pattern, species composition, and genetic constitution of the new stand more precisely than is possible with any other regeneration method. There is the greatest possible freedom for the use of various kinds of machinery, often powerful, in harvesting, site modification, and planting in ways that are similar to or borrowed from those used in agriculture. Deliberate efforts can be made to reduce the need for tinkering with undesirable vegetation or erratic stocking after stand establishment. Sometimes there is even the hope that everything can be set up so well that the stand can be left alone, except for protecting it, until the clearcutting and replanting at the end of the rotation. An effort is often also made to secure the closest possible approach to uniformity of species and tree size so as to reduce the cost and complications of harvesting and processing the timber crops.

Policies about Application of Plantation Silviculture

In most localities throughout the world, plantation silviculture starts as a means of afforesting degraded lands at times when society first accepts the idea of sustainable forestry. This kind of forestry is so direct and uncomplicated that it is often next applied to the rehabilitation of degraded forests and then to the replacement of good stands. It is so simple to administer that it has sometimes become the standard operating procedure of large governmental agencies and corporations. Among the accounts of plantation forestry are those of Hibberd (1991), Savill and Evans (1986), Nwoboshi (1982), Ford, Malcolm, and Atterson (1979), and many listed in Chapter 10 on artificial regeneration.

Plantation silviculture is often associated with some forthright ideas about achieving definite improvement upon nature. One manifestation of this approach, though not a necessary part of it, is the planting of exotics (Zobel, van Wyck, and Stahl, 1987). The concept is a very positive one with a powerful appeal. It seems to be the only possibility to those who believe that useful plants will not appear unless planted; there are certainly times and places where this view is correct. The policy also has certain psychological dimensions. Anyone who tries to regenerate forests by planting trees cannot be accused of having been lazy or of leaving anything to the caprice of nature.

Furthermore, there is something heart-warming about planting trees regardless of financial or ecological attitudes about the practice. The actions of people who are dedicated to planting can be ascribed to this attitude as much as to either cold calculation or sheer impatience with nature. Public opinion vacillates between extremes about the whole matter. Initially, it enthusiastically supports the view that trees grow only if planted because it perceives that this is true with denuded landscapes. Later, when existing forests are regenerated by planting, the public typically sees only the bad appearance of clearcut areas and not the newly planted trees.

For some foresters and in some circumstances, this is the only kind of silviculture worth considering; in some circles it is an anathema. Matters of opinion aside, the ideas underlying this approach must be recognized and understood. Neither this nor any brand of silviculture is a complete, unified system or ritual that must be applied as a whole or not at all. The individual pieces can be applied selectively.

Regeneration of Pure Stands from Natural or Artificial Seeding

It is possible to secure natural regeneration after complete clearcutting, although only under quite special circumstances is there much assurance of success. This way of applying the clearcutting method has a few attributes in common with that involving planting. In both kinds of application, the idea is to establish the regeneration as promptly as possible on some area that is nearly free of vegetation and exposed to sun and sky. Each also allows much latitude for disturbance of soil and vegetation during harvesting and site preparation. However, clearcutting with natural regeneration is usually associated with comparatively extensive, low-investment silviculture.

Real clearcutting exposes the site so much that it is compatible with natural (or artificial) seeding only when the species are capable of enduring exposure. Thus it is not suitable for certain shade-tolerant but exposure-intolerant species unless they are planted (and sometimes not even then). This attribute of clearcutting can, however, be of some advantage if the object is to encourage species of early-successional status and discourage seed regeneration of shade-tolerant species that happen to be undesirable. This effect is usually perceived only when one examines clearcut areas and observes what has not regenerated well; the unsuccessful species probably disappear when they are too young and small to be noticed.

The prospect of success with natural regeneration after clearcutting and the ways of securing it vary widely depending on the source and means of dispersal of the seeds. These distinctions are so important that each will be considered separately here.

Clearcutting with Seeding from Adjacent Stands

Sometimes it is possible to clearcut a stand and depend on seeds dispersed from some nearby, untreated stand to provide the regeneration. With such an approach, the agency of dispersal is all-important. In the temperate zone, the pioneer species best adapted to this kind of silviculture are disseminated by wind. In the tropics, where it is less windy, birds, rodents, and bats may play a more important role, but there are still a number of wind-dispersed pioneer tree species such as Honduran mahogany. There are even species of river flood-plains, such as the tupelo gums of the bottomlands of the Southeast, that can be regenerated after clearcutting by floating seeds.

Wind, water, and animals can carry a few seeds over astonishingly long distances. However, in regenerating forests, it is usually necessary to remember that the density of seeds deposited on a unit of ground surface varies inversely as some power of the distance from the source. The rest of this discussion about dispersal of seed into clearcut areas involves wind, but some of the same ideas may apply to dissemination by other agencies.

Wind-dispersed seeds usually have wings, and moving air is fortunately turbulent enough that some are wafted aloft rather than dropping only downward after being released from the cones or fruits. Where surrounding trees are the only source of seed, the clearing must be sufficiently small (usually long and narrow) to allow for adequate dissemination to all points. Safe widths for clearings to be stocked by wind-disseminated seed are likely to range, depending on species, from one to six times the height of the adjacent timber from which seed will be obtained. The normal direction of the wind during the season of seed dispersal should be known, and the clearing should be so located that its long axis is at right angles to this direction. Unless the clearcut area is very narrow, the distribution of regeneration is likely to be uneven.

Most dissemination occurs during dry, sunny weather when the winds are brisk and gusty. The most effective winds are frequently those that blow out of the dry interior regions of continents; they are not necessarily of the same direction as the so-called prevailing winds. In rugged terrain, it is best to have the long axes of the openings run at right angles to the contour lines because the winds responsible for dispersal of seeds are usually altered in direction so that they blow up or down valleys. Where other considerations dictate that the clearings be oriented in a manner less favorable from the standpoint of seed dispersal, the dimensions of the openings should be correspondingly reduced. The width of the clearcut areas is often restricted by distances of effective seed dispersal but sometimes also by the desirability of retaining enough side shade to keep out pioneer weeds or to favor species not fully resistant to exposure. Marquis (1966) has shown how control over the width of openings can determine whether yellow, paper, or gray birch dominates regeneration in some northern hardwood forests.

As will be described in Chapter 14, the difficulties of securing seed dispersal over long distances can be mitigated by reserving individual trees or uncut strips or groups of trees as sources of seed within the clearcut area.

Clearcutting with Regeneration from Seed on the Site

The restrictions set on the size of clearcut areas by seed-dispersal distances can clearly be avoided if one can rely on a supply of seed stored on the trees or in the soil. With almost any species, a substantial amount of seed may come from the trees removed in clearcutting, provided that the cutting is made just before or after the release of the seed of a good seed year.

With most species, the only seed from the cut trees that is useful for regeneration is that produced in the current year. Among the exceptions are certain conifers that have stored seed in serotinous cones that open gradually over a period of years, as is true of black spruce, or that rarely open to any extent except under the influence of high temperatures. The most prominent species in the latter group are the jack and lodgepole pines of the northern United States and Canada (Fig. 12.2). Many other localities that are subject to hot fires also have pines that reproduce in nature after crown fires. Seed stored in serotinous cones remains viable for many years, but special measures are often necessary to secure the release of this vast quantity of seed after cutting.

Some eucalypts have seeds stored on the trees that are released in abundance after hot fires. This is true of the second tallest tree species on earth, the mountain ash of southeastern Australia (Hillis and Brown, 1984). Investigations in Tasmania have shown that it is necessary to create substantial amounts of dry fuel for prescribed burning to prepare seedbeds after clearcutting. The fuel includes not only the logging debris from the mountain ash but also that from the deliberate felling of small noneucalypts of the understory. The resulting hot fires induce mountain ash regeneration without destroying much of the organic matter incorporated in the mineral soil. Less severe fires induce regeneration of less desirable eucalypts.

Adequate amounts of seed of a few species may remain viable in the humus layers beneath uncut stands for periods longer than one year. One example is Atlantic white-

Figure 12.2 Young stand of naturally regenerated lodgepole pine with older stands adjacent to it. *(Photograph by U.S. Forest Service.)*

cedar, the seeds of which are stored in large quantities in the poorly aerated peats on which this species grows (Little, 1950). Another is yellow-poplar (Clark and Boyce, 1964), which seems to require high temperature or exposure to light to germinate. The seeds of the ashes remain stored in the forest floor for one year because it takes them that long after they fall from the trees to mature. With such species it is often possible to expect seemingly miraculous regeneration after removing the entire standing source of seed. However, it is well not to proceed without being sure that the seeds are there.

Unfortunately, long-term storage of seeds in the forest floor is more characteristic of undesirable pioneer trees, annuals, and shrubs than of commercial tree species. Many of these weeds have seeds with hard, nearly impermeable coats that allow the embryos to survive for the many decades that may elapse between major fires or similar regenerative disturbances. They germinate or flourish long enough to produce seeds and then die; they may also lapse into the understory. Examples of such species are the genera *Rubus, Ribes,* and *Ceanothus,* as well as the short-lived pin cherry.

As described in Chapter 10, direct seeding is sometimes used after clearcutting to establish new stands of species under the somewhat unusual circumstances in which the technique can be made to work.

Applications of Pure-Stand Silviculture

West Coast Douglas-fir Region

Most regeneration associated with clearcutting simulates the effects of fires that do not actually or immediately kill all of the trees on the burned areas. The most common criterion of adaptability to clearcutting is the capacity of a species to grow faster in height than undesirable competitors. This is why clearcutting is not well suited to those kinds of shade-tolerant species that grow slowly in the juvenile stages, even if they are planted. They grow so much more slowly than the associated pioneer vegetation induced by clearcutting that they require much release work.

The evolution of silviculture in the West Coast Douglas-fir Region provides a good illustration of the development of the silviculture of pure even-aged stands. Under entirely natural conditions, coastal Douglas-fir regenerated after lightning fires; the sources of seeds were scattered seed trees or patches of trees in swales that escaped fires or were not promptly killed by them. Douglas-fir, which is normally the fastest growing conifer of these forests, tended to dominate for the first century of stand development. The Douglas-firs then dwindled from the attrition of various lethal agencies and were gradually replaced by western hemlocks, true firs, and western red-cedar that became established in the understories at the time of the initiating fires or afterwards.

Early in the twentieth century, these forests were being liquidated by railroad logging in combination with cable yarding. As a result, clearcut areas were extended progressively along the logging railroads. Even though only the best trees were harvested, the shifting of the yarding cables usually pulled over the rest creating huge volumes of slash. Society then had no concern for forest regeneration, but it did have the pathetic hope that laws requiring broadcast burning of slash might forestall the dangerous conflagrations of the time.

Foresters had little to say about what was going on and, even at that, tended to be complacent about the very large size of the clearcut areas. There was the mistaken view that the seeds of Douglas-fir from the old stands would remain stored and viable in the forest floor for many years, even through clearcutting and slash burning. This view was based on the observation of the fine Douglas-fir regeneration that had appeared after fires in uncut stands. Foresters had discounted the importance of wind dispersal of seeds from adjacent stands or from scattered survivors. They did correctly perceive that a quick repetition of tree-killing fires would eliminate the conifers because the seed supply, whatever its source, would be destroyed. Unfortunately, the otherwise good research that had been done about the regeneration process had omitted any good test of the long-term viability of Douglas-fir seeds.

By the 1930s it was shown (Isaac, 1943) that retention of living seed sources was crucial. Foresters began to have some modest influence on harvesting practices, and interest arose in regeneration provided that it cost little. For the first time there were tractors and trucks powerful enough to move the huge old-growth trees that could previously be logged only with cumbersome steam-driven machines. However, only the biggest and best trees had enough value to be harvested.

These circumstances set the stage for ill-fated efforts to prolong the lives of old-growth stands by light partial cuttings referred to as "selective logging." The cuttings were intended to operate as selection cuttings to create uneven-aged stands. The only trees cut were commonly those roughly 3 feet in diameter; these were mostly the scattered surviving Douglas-fir, which, in spite of their great size and age, were still the most durable trees in the old stands. Unfortunately, most of the remaining trees were western hemlocks and true firs already weakened by age and decay. The usual result was an acceleration of the process by which wind, fungi, and bark beetles break up ancient stands. By the 1950s the whole episode was recognized as a fiasco to be quietly forgotten.

During the next phase, clearcutting returned to fashion. By then it had been found, at least on the most common kinds of sites such as those in western Washington, that reasonably good natural regeneration could be obtained from windborne seed if the clear-cuttings were less than about 100 acres in size. It was also found that if a few seed trees were left within the cutting areas, the regeneration was better than when reliance was placed on adjacent uncut stands.

By the 1960s, however, attitudes toward forest regeneration had changed to the point that state laws required it. The exigencies of cable logging in old-growth timber also made it desirable to clearcut areas too wide for natural seed dispersal. There was, therefore, a period during which aerial direct seeding was a common way of regenerating after clear-cutting and broadcast burning of slash.

Planting (Fig. 12.3) has subsequently become very common because of the willing-ness to invest money to ensure the prompt regeneration of properly spaced trees of the best available genetic qualities. Nevertheless, the evolution of silvicultural practice does not stop. It was soon found that the routine of clearcutting, burning, and planting was not a universal solution for the silviculture in the West Coast Douglas-fir Region. It tended to fail when efforts were made to extend it to atypical sites.

The most complete failures were at high elevations in the Cascade Range, where there are severe microclimatic extremes and true firs abound. Plantation failures were also com-mon on certain sites in southwestern Oregon where the rainless summers are longer than they are farther north (Seidel, 1979). True clearcutting with site preparation often produces eruptions of red alder and undesirable shrubs when it is extended westward into the narrow coastal fringe; there fog-drip precipitation supports luxuriant stands of western hemlock and Sitka spruce, which are advance-growth species.

Another general problem associated with plantation silviculture in the Northwest, especially on public lands, has been popular displeasure over the ugly appearance of clear-cut areas. Although the use of shelterwood methods has increased somewhat, there is a remaining tendency to cling to clearcutting and planting as standard operating procedure. Popular acceptance of plantation silviculture might be greater if it were simply limited to the numerous cases where it is essential.

Southern Pines

Somewhat the same sequence of events has taken place in the development of the silvi-culture of southern pines on the southeastern Coastal Plain (Williston, 1987). Initially, there was widespread devastation from the combination of clearcutting and annual burning to improve grazing. Most first attempts at long-term silviculture took the form of selection cutting. These efforts proved to be fairly effective as a means of managing existing growing stocks but were not very compatible with the hardwood control measures necessary to establish pine regeneration.

The period from 1950 to 1970 witnessed, mainly on industrial forests, the same quick progression from one means of establishing even-aged stands to another. Initially, the seed-tree method was the solution, but this method was generally replaced by aerial seeding when it was concluded that each technique required the same amount of site preparation. Then, when industrial owners became willing to spend money for prompt regeneration with well-spaced pines of good genetic qualities, planting became the common technique (Fig. 12.1) (Wahlenburg, 1965).

One useful illustration of the way silvicultural prescriptions should be fitted to the circumstances comes from observations (Lotti, 1961) in the well-watered loblolly pine stands of coastal South Carolina. Here it is possible for systematic programs of thinning and prescribed burning to enable natural regeneration by clearcutting throughout more than half of almost any year. The development of good seed-bearers by thinning provides ample seed for regeneration almost annually. Regeneration is then effective if litter and

Figure 12.3   Clearcutting of over-mature old-growth and planting of Douglas-fir, Willamette National Forest, Oregon. The upper picture was taken 4 years after planting when most of the trees were still overtopped by the herbaceous plants that appeared after the slash burning done just before the planting. The lower picture shows the same place when the plantation was 12 years old. *(Photographs by U.S. Forest Service.)*

understory hardwoods are sufficiently reduced by prescribed burning. Harvests between April 1, the time of germination, and August 1, when the seedling stems harden, are the exceptions; during these five months it is necessary to reserve seed trees to allow for losses from logging damage to succulent seedlings. During the remainder of the year, either the newly fallen seeds or the hardened first-year seedlings provide enough regeneration to allow clearcutting.

Actually, almost any method of regeneration can be and is used in regenerating southern pines. Other methods are used for meeting various management objectives best achieved by partial cutting systems. Regeneration of the southern pines can be obtained by almost any method that provides for controlling competition from hardwoods if there is enough sunlight and pine seeds or seedlings are on the ground (Hu, Lillieholm, and Burns, 1983).

The silviculture of the southern pines is very similar to that employed in other parts of the world for many other two- and three-needled pines. Among these are *Pinus radiata*, native to a few thousand acres on the Monterey Peninsula of California and grown on millions of acres in Mediterranean climates around the world (Lewis and Ferguson, 1993), and *Pinus sylvestris*, the most important European pine.

Closed-Cone Pines

The species of pines that have serotinous cones are the one important group that is generally more satisfactorily regenerated by clearcutting than by any other method. If the cones are truly serotinous, the seed crops of many years are stored in the crowns of the trees and there is no necessity of reserving standing trees. The main problems are to get the cones to open and to expose enough mineral soil. If all of the cones on standing trees remain tightly closed, it may be futile to reserve any trees strictly for a source of seed. Because it is rarely feasible to duplicate the severe crown fires that bring these stands into being in nature, the effects of the fires must be simulated by other means.

Reasonably successful techniques for this purpose have been developed for jack pine in the Great Lakes Region (Smith and Brown, 1984; Benzie, 1977) and sand pine on old beach deposits in Florida. The lodgepole pine of the Rocky Mountains is quite variable in the extent to which it exhibits the closed-cone habit, and so it must be handled accordingly (Baumgartner et al., 1985; Lotan and Perry, 1983).

The first prerequisite is exposure of the mineral soil by mechanical scarification. If most of the cones are tightly closed, cone or cone-bearing slash must be scattered over the area in such a manner that a sufficient number of cones lie exposed to the sun within 15 centimeters of the ground. Within this zone of sluggish air movement temperatures may increase to about 50°C, which is the melting point of the resin that seals the cone scales. The necessity of such high surface temperatures poses a dilemma because they cause heat injury to the seedlings. Consequently, some of the seeds must fall into small spots shaded by debris.

It would be ideal if the scarification could be done before clearcutting, but it is usually more expeditious to do it afterward or by appropriate modifications of skidding procedures. The most common technique is to disk the areas after cutting and then lop and scatter the slash. If these steps are reversed, the cones are likely to be buried unopened. Success can also be achieved if the slash is bunched with bulldozer root-rakes, provided that enough cones are broken off and deposited on scarified soil. Sometimes the dense little clusters of seedlings that arise from the opening of cones on the ground pose a difficult kind of precommercial thinning problem. If slash burning is necessary, it must be limited to the part of the slash that has been bunched into concentrated areas; broadcast burning destroys too many seeds.

These species display variation in the degree to which the closed-cone habit is manifest. The habit is most pronounced where forest fires have been most common in nature. If the cones do not remain closed and store seeds, it is necessary to employ some sort of partial cutting or to clearcut in narrow strips and thus depend on seeds from current production. The assumption that these species are always serotinous has led to regeneration failures in cases where reliance was placed on stored seeds that did not exist (Wyoming Forest Study Team, 1971). If some cones open and some do not, it is possible to compromise between the two approaches. Variations in cutting practice, degree of scarification, and slash treatment can be employed to reduce the risk of getting excessively dense natural regeneration that is likely to stagnate. However, if all the cones are truly serotinous, definite efforts must be made to open them; the difficulties are compensated only by the fact that the size, shape, and arrangement of clearcut areas can be governed entirely by considerations other than seed dispersal.

An important incidental advantage of clearcutting in the management of lodgepole pine is that it provides one of the surest means of eradicating infestations of the debilitating parasitic seed plant, dwarf-mistletoe (Hawksworth and Sharf, 1984). Because the seeds of the pathogen must come from living trees, it is necessary only to prevent slow reinfestation from adjoining stands adjacent to the clearcut areas. With partial cutting, the mistletoe is very likely to spread from the old stand to the new unless great care is taken to cut all the infected trees of the previous crop (Baumgartner et al., 1985).

The seeds of serotinous conifers rarely exhibit dormancy, so there is little natural control over the season of germination. Therefore, the release of seed should be timed so that the seedlings will germinate at the proper season. With jack and lodgepole pine, the seeds must germinate before summer so that the seedlings will harden before the first frost. With sand pine in Florida, however, it is best to schedule scarification and slash treatment so that seedlings commence development during late fall or early winter when rainfall is adequate and when the risk of heat injury is lowest. It is indeed remarkable how much careful effort is necessary to obtain natural regeneration of these species that are adapted to reproduce themselves so easily and abundantly after catastrophic fires.

Although most hardwoods naturally grow in mixed stands, some that are pioneers in succession can form pure stands and be regenerated by clearcutting. Among these are the birches (Densmore and Page, 1992; Safford, 1983; Marquis, 1966), red alder (Hibbs, DeBell, and Tarrant, 1994), and most eucalypts (Hillis and Brown, 1984).

The Semantics of Clearcutting

It might seem that no term associated with cutting practice could be less ambiguous than *clearcutting*. Actually, it is a semantic morass made even deeper by the negative connotation sometimes attached to the word. As a technical term of silviculture, clearcutting refers to treatments in which virtually all vegetation is removed and almost all of the growing space becomes available for new plants. Timber harvesting alone is seldom sufficient to achieve such complete removal of vegetation; that is one of the main roles of the site preparation measures that were described in Chapter 8. If this kind of clearcutting is meant, it is prudent to refer to it as *silvicultural*, *true*, or *complete* clearcutting; sometimes the term *clean cutting* is also used.

The use of *clearcutting* as a forester's technical term sometimes extends to operations in which heavy removal cuttings release new stands already established beneath them as advance growth. Use of the term *one-cut shelterwood cutting* (discussed further in Chapters 14 and 16) has the virtue of focusing attention on the source of the regeneration. This is crucial in those situations in which new seedlings of the desired species are poorly adapted for establishment after the exposure resulting from true clearcutting. The simple coppice method, considered in Chapter 13, depends on regeneration from vegetative sprouting, but the degree of removal of the old stand is so nearly complete that the operation very much resembles clearcutting.

One kind of semantic confusion is perhaps inevitable in this terminology. In some mixed stands, it is entirely possible to carry out heavy removal cuttings that lead to regeneration arising from advance growth, sprouts, and new seedlings germinating after the cutting. In a certain sense, these are combinations of the shelterwood, coppice, and clear-cutting methods. Such regeneration is very common in mixed hardwood forests. The non-committal term *heavy removal cutting* provides a useful solution to this dilemma.

The sloppiest usage of the term *clearcutting* is with reference to logging operations in which only the merchantable trees are cut. If only a few trees are cut from a stand, the result should often be termed *selection thinning* or should be regarded as some crude variant of the selection method for creating uneven-aged stands.

BIBLIOGRAPHY

Baumgartner, D. M., R. G. Krebill, J. T. Arnott, and G. F. Weetman (eds.). 1985. *Lodgepole pine: the species and its management.* Wash. State Univ. Coop. Ext. Service, Pullman. 381 pp.

Benzie, J. W. 1977. Manager's handbook for jack pine in the North Central States. USFS Gen. Tech. Rept. NC-32. 18 pp.

Brissette, J. C., and J. P. Barnett (eds.). 1992. Proceedings of the Shortleaf Pine Regeneration Workshop. USFS Gen. Tech. Rept. SO-90. 236 pp.

Clark, B. F., and S. G. Boyce. 1964. Yellow-poplar seed remain viable in the forest litter. *J. For.,* 62:564–567.

Densmore, R. V., and J. C. Page. 1992. Paper birch regeneration on scarified, logged areas in southcentral Alaska. *NJAF,* 9:63–66.

Evans, J. 1992. *Plantation forestry in the tropics.* 2nd ed. Clarendon, Oxford. 416 pp.

Fedkiw, J., and J. G. Yoho. 1960. Economic models for thinning and reproducing even-aged stands. *J. For.,* 58:26–34.

Ford, E. D., D. C. Malcolm, and J. Atterson (eds.). 1979. *The ecology of even-aged plantations.* HSMO Publications, London. 382 pp.

Hamilton, L. S., and S. C. Snedacker. 1984. *Handbook for mangrove area management.* East-West Center, Honolulu. 123 pp.

Hawksworth, F. G., and R. R. Scharf (eds.). 1984. Biology of dwarf mistletoes. USFS Gen. Tech. Rept. RM-111. 131 pp.

Hermann, R. K., and D. P. Lavender (eds.). 1973. Even-aged management, symposium, 1972. *Oreg. State Univ., Sch. For. Paper* 843. 250 pp.

Hibberd, B. C. (ed.). 1991. Forestry practice. *U.K. For. Comm. Hbk.* 6. 11th ed. 239 pp.

Hibbs, D. E., D. S. DeBell, and R. F. Tarrant (eds.). 1994. *The biology and management of red alder.* Oregon State Univ. Press, Corvallis. 256 pp.

Hillis, W. E., and A. G. Brown (eds.). 1984. *Eucalypts for wood production.* Academic, Orlando, Fla. 2nd ed. 434 pp.

Hu, S. C., R. J. Lillieholm, and P. Y. Burns. 1983. Loblolly pine: a bibliography, 1959–82. *La. State Univ. Sch. For. & Wildlife Mg Mt., Research Rept.* 2. 199 pp.

Isaac, L. A. 1943. *Reproductive habits of Douglas-fir*. Charles L. Pack Forestry Foundation, Washington, D.C. 107 pp.

Laderman, A. D. (ed.). 1987. *Atlantic white cedar wetlands*. Westview Press, Boulder, Colo. 401 pp.

Lewis, N. B., and I. S. Ferguson. 1993. *Management of radiata pine*. Butterworth Heinemann, North Ryde, New South Wales. 404 pp.

Little, S., Jr. 1950. Ecology and silviculture of whitecedar and associated species in southern New Jersey. *Yale Univ. Sch. For. Bul.* 56. 103 pp.

Lotan, J. E., and D. A. Perry. 1983. Ecology and regeneration of lodgepole pine. *USDA, Agr. Hbk.* 606. 51 pp.

Lotti, T. 1961. The case for natural regeneration. In: A. B. Crow (ed.). *Advances in southern pine management*. Louisiana State University Press, Baton Rouge. Pp. 16–25.

Marquis, D. A. 1966. Germination and growth of paper and yellow birch in simulated strip cuttings. USFS Res. Paper NE-54. 19 pp.

Nwoboshi, L. C. 1982. *Tropical silviculture, principles and techniques*. Ibadan Univ. Press, Nigeria. 333 pp.

Oliver, C. D., D. Hanley, and J. Johnson (eds.). 1986. Douglas-fir: stand management for the future. *Univ. of Washington College of Forest Resources Contribution 55*. 388 pp.

Roach, B. A. (ed.). 1975. *Clearcutting in Pennsylvania*. Penn. State Univ., Sch. For. Resources, University Park. 81 pp.

Ruark, G. A. 1990. Evidence for the reserve shelterwood system for managing quaking aspen. *NJAF*, 7:58–62.

Russell, C. E. 1987. *Plantation forestry: Case Study No. 9: the Jari Project, Pará, Brazil*. In: C. F. Jordan (ed.), *Amazonian rain forests: ecosystem disturbance and recovery*. Springer–Verlag, New York. Pp. 76–89.

Safford, L. O. 1983. Silvicultural guide for paper birch in the Northeast. Rev. ed. USFS Res. Paper NE-535. 29 pp.

Savill, P. S., and J. Evans. 1986. *Plantation silviculture in temperate regions with special reference to the British Isles*. Clarendon, Oxford. 246 pp.

Schmidt, W. C., R. C. Shearer, and A. L. Roe. 1976. Ecology and silviculture of western larch forests. *USDA Tech. Bull.* 1520. 96 pp.

Seidel, K. W. 1979. Regeneration in mixed conifer clearcuts in the Cascade Range and the Blue Mountains of eastern Oregon. USFS Res. Paper PNW-248. 24 pp.

Smith, C. R., and G. Brown (eds.). 1984. Jack pine symposium. *Canadian Forestry Service COJFRC Symp. Proc.* O-P-12. Sault Ste. Marie, Ontario. 195 pp.

Wahlenberg, W. G. 1960. *Loblolly pine: its use, ecology, regeneration, protection, growth, and management*. Duke Univ., Sch. For., Durham, N.C. 603 pp.

Wahlenberg, W. G. (ed.). 1965. A guide to loblolly and slash pine plantation management in southeastern USA. *Ga. For. Res. Counc. Rept.* 14. 360 pp.

Whitehead, D. 1982. Ecological aspects of natural and plantation forests. *For. Abst.*, 43:615–624.

Williston, H. L. 1987. *Southern pine management primer*. Vantage Press, New York. 167 pp.

Wyoming Forest Study Team. 1971. *Forest management in Wyoming*. USFS, Washington, D.C. 81 pp.

Zobel, B. J., G. van Wyck, and P. Stahl. 1987. *Growing exotic forests*. Wiley, New York. 508 pp.

CHAPTER 13

VEGETATIVELY REGENERATED STANDS

Silviculture that depends on vegetative sprouting is not only remarkably simple and successful but is also the most common and ancient kind of deliberate forest regeneration. With the right species and sizes of trees, the old crop is merely cut down and a new one dependably sprouts back. The practice started with the dawn of civilization; because of its widespread use in the less developed countries, it is employed on more forest areas than any other form of silviculture (Panel on Firewood Crops, 1980, 1983). Vegetative regeneration is so easy, at least where it is feasible, that special measures are seldom necessary to secure it; in fact, it is difficult to prevent where it is not wanted.

A **low forest** is forest that originates vegetatively from natural sprouts or layered branches as contrasted with a **high forest**, which develops from seeds or planted seedlings. The ancient Germanic origin of these words refers to the fact that most sprout forests in Europe were grown on short rotations and the trees were not tall. It could not be known at the time that the coast redwood, the tallest of all plants, commonly reproduces by sprouting (Fig. 13.1). The term **coppice** denotes a stand arising primarily from sprouts. In the simple coppice method, all standing trees are cut at the end of each rotation, and a perfectly even-aged stand springs up almost immediately. A special form of coppice management is **pollarding**, in which the tops of trees are removed, thus inducing them to sprout at points above the reach of browsing animals. The planting of rooted cuttings or other asexually regenerated material is not construed as a coppice method of regeneration.

Vegetative Regeneration

The ability to reproduce vegetatively is of tremendous survival value for many species of perennial plants. It is the normal regeneration mode for most shrub species and grasses,

Figure 13.1 A ring of very large stump sprouts of coast redwood surrounding the charred remains of a stump about 8 feet in diameter near Santa Cruz, California.

many herbaceous forest plants that have perennial roots and annual shoots, and some species of trees. For such species, sexual regeneration from seed may be a safety factor that persists and preserves capacity for genetic change and adaptation to new conditions. For many other species, mostly angiosperms, sprouting is just an additional mode of regeneration. For others, mostly conifers, it occurs never or infrequently.

Vegetative regeneration produces good grass for grazing and much growth of shrubs for browsing animals and other purposes. It is a traditional mode of fuelwood production throughout much of the world. It is reputed to be a poor means of timber production, but this opinion may derive from cases in which the technique has been overused. The sprouts of desirable tree species often produce stems that are not as straight or well formed as plants of the same species grown from seeds.

As far as most woody plants are concerned, vegetative sprouting is a natural adaptation to fires that are hot enough to kill the above-ground parts. Only by this means could a species maintain itself in the face of catastrophes that occur so often that no individuals reach seed-bearing age. Almost every forest region has areas where fires have occurred so often that sprouting species predominate.

Sprouting is also a response to sudden death caused by cutting as well as damage resulting from disease, wind breakage, serious physiological disorder, or other injury. The appearance of sprouts of any form near the base or on the bole of a tree is frequently the first sign of unhealthiness. The process is triggered by interruption of the flow of auxins that are produced in the top of the tree and normally inhibit the development of basal buds.

The shoots that arise from portions of an established tree grow much more rapidly in youth than seedlings of equal age. This tendency is to some extent the result of inherent physiological differences that are not clearly understood. However, much of the unusual rapidity of growth of young sprouts can probably be accounted for by the inheritance of an extensive root system and supplies of carbohydrate from the parent tree.

Trees cut during the dormant period sprout much more vigorously than those cut during late spring and summer. This is because the reserves of carbohydrates in the roots are at a maximum during the winter and at a minimum immediately after the formation of new leaves and shoots (Kramer and Kozlowski, 1979). If cuttings are made late in the growing season, sprouts sometimes appear almost immediately, but they rarely harden before the first frost. Therefore, coppice cuttings are best done in late fall or winter.

If full advantage is to be taken of vegetative reproduction in renewing a stand, all the trees of the species to be reproduced should be cut in one operation to stimulate sprouting to the utmost. Trees of the same species in a single stand are often interconnected by root grafts. Therefore, if a few scattered trees are left behind, they may draw on the food supply of the common root system or contribute hormonal inhibitors of sprouting and thus reduce the vigor of the new sprouts. The reduction is often greater than could be caused by the shade of the residual trees.

Although trees of vegetative origin grow faster during early life than those developing from seed, they do not retain this advantage on long rotations. Vegetative reproduction generally attains larger size on a short coppice rotation than trees grown from seed would in the same time. That is the primary advantage of coppicing stands on short rotations, although the difference in rates of growth appears to reverse at greater ages. If the rotations are extended to produce sawlogs, there appear to be no consistent differences in size that can be related to the mode of origin (Leffelman and Hawley, 1925). McIntyre (1936), however, found that oak sprouts grown beyond a century in age on good sites tended to be smaller than trees originating from seed. He found no evidence that oak sprouts were inherently more short-lived than oaks grown from acorns.

The foregoing generalizations apply to most stands that are regenerated vegetatively, but there are some important differences between several forms of vegetative regeneration.

Stump-Sprouts

Stump-sprouts arise from dormant buds at the root collars of tree stumps (Fig. 13.2). These shoots almost invariably develop from lateral buds that were originally formed on the leading shoot of the seedling; these buds did not develop into lateral shoots but remained dormant, growing outward with the cambium to maintain positions just beneath the bark. The pith of dormant buds can be traced all the way back to the pith of the original stem. If this connection is broken during the growth of the tree, the bud becomes incapable of developing into a new shoot. If the bark over the dormant bud becomes too thick, it may be impossible for the bud to break through and develop into a sprout. Both the

Figure 13.2 Two desirable stump sprouts of low origin on an oak stump with one large and two small undesirable ones on top of the stump. *(Photograph by U.S. Forest Service.)*

thickness of the bark and the possibility of interruption of the bud trace increase with age. Therefore, the sprouting capacity of the tree tends to decline with age.

Dormant buds can give rise to new shoots at any level on the bole, although only those that arise at or very close to the ground line have much potential for developing into good trees. Trees that develop from sprouts that originate too far above ground are often crooked, forked, or subject to breakage and decay. Epicormic branches are shoots that develop even higher on the living boles of trees.

Stump-sprouts usually cluster in dense rings around sprouting stumps. Their numbers usually dwindle rapidly because of severe competition between them. The sprouts begin from lateral buds, so they must curve upward as they develop into main stems. Because the stems of many species also bend toward the light, it is not easy to produce straight trees.

The risk that rot will spread from the old stumps to the heartwood of the new shoots varies with species. It does not appear to happen in such diffuse-porous species as the maples in which any infected interior of the trees is "compartmentalized" or sealed off from the sapwood by gum formation (Shigo, 1985). This does not seem to help with another diffuse-porous species, yellow-poplar. With the oaks, the incidence of such rot increases with the diameter of the parent stump and the distance of the sprout above ground (Fig. 13.2) (Roth and Hepting, 1943). However, American oak forests have many large, sound trees that grew so rapidly when young that they must have been of sprout origin

(Lamson, 1976). There seems to be no problem with coast redwood, which produces sound, straight sprouts from gigantic stumps (Fig. 13.1).

The sprouts of conifers come from dormant buds that form at the base of either primary or secondary needles. Among the conifers that can be usefully regenerated by stump-sprouting are coast redwood, bald-cypress, and several pines, including shortleaf, pitch, pond, and certain Mexican species.

Stool-Sprouts

Stool-sprouts, usually crooked and slow-growing, develop from stumps that have been cut so often over so many decades that they have callused over with cambium and bark (Fig. 13.3). Such stumps or **stools** may be several feet in diameter and resemble the mysterious burls that sometimes form on tree stems (Rackham, 1980). It is not clear whether the sprouts arise from dormant or adventitious buds, or from both. They are often less than

Figure 13.3 Crooked stool-sprouts, both about 80 years old, on an ancient stool in a stand in Connecticut that had been repeatedly coppiced or burned for at least two centuries. New sprouts have started on the stool. The other trees in the picture originated either from stump-sprouts or seedling-sprouts.

an inch apart when they start, and the competition between them is so intense that none can support themselves adequately; their height growth is stunted, and they remain crooked even after natural thinning has greatly reduced their numbers.

The dismal reputation of the coppice system in countries where stands have been coppiced at short intervals for very long periods has come chiefly from the fact that most of the stumps developed into burl-like stools. The rare cases in which such stumps have developed in the United States, such as the dwarf pitch pines of the Pine Plains of New Jersey, have resulted more from frequent forest fires than from fuelwood cutting.

Seedling-Sprouts

Sprouts that come from small stumps usually have the same anatomical and physiological origin as stump sprouts, but they also have so many attributes of seedlings that they are called **seedling-sprouts**. They are defined as coming from stumps less than 2 inches (5 cm) in diameter. This generally means that the stumps may have only sapwood and may quickly be covered over with callus tissue; both factors greatly reduce the risk of heart-rot spreading from stump to sprout. They are often superior to seedlings as sources of regeneration because the new shoots grow more rapidly as the result of having established root systems to nourish them. The stems tend to be straighter simply because the faster a tree grows through a certain zone of height, the less time there is for the operation of agencies that create deformities.

Although almost any sprouting species can produce seedling sprouts, they are especially important with many large-seeded species such as the oaks, hickories, chestnut, and walnuts. The seedlings of these species are seldom numerous, but, once established, they are very persistent and may survive for many years in the shade of older trees. There is an ample supply of sprout buds at the ground line even in old trees. The seedling tops continually grow up and get killed back, but the carrotlike roots often survive and grow larger. When some permanent release takes place, these kinds of plant grow rapidly; they often are a chief source of regeneration for the species involved. The mechanism enables a kind of long-term storage of advance growth, especially of species that produce small numbers of large seeds. On the other hand, this particular source of regeneration is much more important with the so-called high-forest systems than with the coppice method. Simple coppice rotations are often too short to be conducive to the kind of seed production that would, over some years, allow this kind of regeneration to develop.

Seedling sprouts that develop after injury or fire are sources of regeneration in shortleaf pine and the Montezuma pine of Mexico, as well as some other pine species.

Lignotubers

Many of the Australian eucalypts, which are planted widely in the warmer parts of the world, have such an affinity for fire that they have some special modes of sprout regeneration (Hillis and Brown, 1984). The mode that is largely limited to this genus is the **lignotuber**. Lignotubers are paired globular masses of tissue, full of buds, that form in the axils of the opposite leaves of small seedlings. These tissues grow much larger but usually become buried at the base of the tree stems. If the stems do not get more than about 8 inches (20 cm) in diameter, sprouting from the lignotubers is prolific when the tops of the trees are cut, killed, or damaged.

Many species of eucalypts also have typical dormant buds that form in the leaf axils and produce stump-sprouts or epicormic branches. However, some of the most important members of the genus do not sprout but respond to infrequent catastrophic fires by mass seeding. Significantly, another nonsprouting eucalypt, *Eucalyptus deglupta*, occurs naturally only in New Guinea and the southern Philippines and in rain forests where there has not been enough fire to call forth adaptations to it.

Sprouts from Adventitious Buds

Sprouts can also arise from adventitious buds that sometimes develop in the callus tissue that forms over wounds or in the root cambium of a limited number of tree species. Sometimes, if sprouts from dormant buds are cut, new sprouts can arise from adventitious buds that form anew in the callus tissue of the wounds.

The most important role of adventitious buds as constructive sources of forest regeneration involves several hardwood species that produce **root-suckers** or **root-sprouts** along the roots of trees that are killed or even slightly damaged. This is so nearly the normal mode of regeneration in the aspen poplars that true seedlings are rare in many localities (Schier, 1978). It is common with sweet-gum, black-gum, American beech, black locust, and sassafras. Trees that arise from root-suckers can be much better than those from stump-sprouts. The shoots are much more likely to be free of rot, as well as evenly spaced and straight, because they are not clustered around any stumps.

In a few species, including American beech, adventitious buds produce sprouts that are called **stool-shoots**. These sprouts arise between the bark and wood and develop into peculiar, usually short-lived, rings of sprouts around the tops of stumps. They are not the same as the stool-*sprouts* that come through the bark.

Layering

This peculiar kind of vegetative reproduction arises from living, low-hanging branches that have been partially buried in moist organic matter. Layering of tree species occurs mainly in peat bogs where sphagnum moss tends to overgrow the lower branches of trees in open stands. However, many forest shrubs spread by layering when branch ends are buried beneath leaf litter. Artificial layering is one of the techniques of inducing branches to form roots so the branches can be severed and planted.

Stump-Sprout Stands

The most common kinds of vegetatively regenerated stands are those that arise from clustered sources of sprouts such as stumps or stools. These typically arise as absolutely even-aged stands after simple coppice cutting in which all the trees are cut so that nothing is left to reduce the vigor of sprouting (Fig. 13.4). If such coppice cutting has been repeated at short intervals for many decades, a single species usually comes to predominate, but if, as in the eastern deciduous forests of the United States, the cuttings have not been frequently repeated, the stands are of mixed species.

The congested nature of sprout clumps (Fig. 13.5) restricts what can be done with the simple coppice system. The rotations must usually be shorter than those contemplated for the production of good sawlogs. Ordinarily, the objectives are to grow fuelwood, pulpwood, animal browse, and small wood products.

The spacing of a coppice stand derived from stump-sprouts is controlled by the arrangement of the parent trees. If the rotation is extended sufficiently to produce sawtimber, many of the original sprout clumps will die of suppression. If dependence is then placed on reproduction by the simple coppice method, the new sprout clumps will grow up so far apart that production suffers. Best results are obtained when the rotation is so short that the final spacing of sprout clumps is acceptable as an initial spacing for a new coppice stand. The vigor of sprouting from stumps also declines rapidly with increasing age and

Figure 13.4 Simple coppice stands of mixed hardwoods in Connecticut. The stand in the background is 30 years old. That in the foreground is part of the same stand that was cut for fuelwood during the previous winter; the new sprouts have not yet finished the first year's growth in height. *(Photograph by Yale University School of Forestry and Environmental Studies.)*

diameter. The period of satisfactory sprouting is usually coincident with that of most rapid growth and may end before the trees become effective seed-bearers.

The incidence and extent of decay in stump-sprouts also increase with age, although the rate of deterioration varies greatly with species and climate. Some species are affected very slowly and to an insignificant extent; in others, decay spreads quickly and may be so injurious as to preclude use of the method.

Attention must be given to suitable preparation of the stumps. Saws are better for the cutting than axes or other devices that are likely to loosen the bark of the stumps. It is reputed to reduce incidence of rot if the cut surfaces are slanted so that water runs off them. Normally, it is best to cut the stumps as low as possible so that the new sprouts will not have to grow on the side of a decaying and unstable stump (Fig. 13.2). Sometimes the deliberate use of slash burning to kill the sides of the stumps is useful in restricting the sprouting to buds at or below the ground line. These steps, which have been found useful for oaks, are not advantageous for the eucalypts because the best sprouts of eucalypts are those that come from sides of undamaged stumps 4 inches (10 cm) tall (Jacobs and Métro, 1981).

If repeated crops are to be secured from stump-sprouts, the rotations are ordinarily less than 35 years and sometimes only a year or two long. The logical rotation length is that which will produce the maximum mean annual increment of fuelwood and other small products. The form of the stems is usually so poor that these are the only products inten-

Figure 13.5 Stand of 8-year-old red maple stump-sprouts in Connecticut showing the typical form and spacing of a short-rotation coppice stand. *(Photograph by Yale University School of Forestry and Environmental Studies.)*

tionally grown under the simple coppice method. It is desirable to obtain some reproduction from planting or natural seeding to replace decadent root-stocks and improve the spacing. This kind of restocking can sometimes be encouraged by thinning the coppice stand late in the rotation.

Another important objective of any thinning is the reduction of the number of sprouts in a clump; the faster this number can be reduced to one, the better is its growth. In places, such as certain sal forests of India, where there is overwhelming demand for fuelwood, this procedure can be used to reserve some stems to grow to pole sizes for structural material.

Root-sucker Stands

The coppice system functions at its best in the management of species that can regenerate as stands of rather uniformly spaced sprouts from root-suckers. Regenerating species that produce root-suckers is even simpler than creating a new crop from stump-sprouts; in fact, it is usually too abundant. The stands are not only even-aged but also pure and can be managed as such. Sprouting is not confined to the stumps of the original stand so that spacing is not a problem; no care is necessary in preparing the stumps. The tree stems are usually straight, and there is no evidence that the ability to produce root-suckers declines with age. The maximum length of the rotation is determined by the same criteria that might apply to a pure plantation.

Among the species that regenerate from root-suckers are the aspen poplars, sweet-gum, beech, and black locust, as well as many species of shrubs and other trees that are usually regarded as undesirable. By far the most important are the quaking and bigtooth aspen poplars, especially in the Great Lakes Region (Brinkman and Roe, 1975; Graham, Harrison, and Westell, 1963) (Fig. 13.6). Vast aspen forests replaced virgin stands of pine and hardwoods after they were devastated by heavy cutting and fires near the beginning of the present century.

A large and diversified wood industry now subsists on coppice management of aspen. This works best on the better soils where the trees can grow to middle age and good diameters before the heart-rotting fungi harm them. Soils that are excessively wet or dry are best converted back to conifers. Aspen regeneration from root-suckers depends mainly on being sure to cut all the trees down, preferably during the dormant season (Bates, Blinn, and Alm, 1993). As is true of other sprouting species, the regeneration step is simple and almost magically automatic.

Planted Coppice Stands

There is no fundamental reason why stands intended for coppice management cannot be created partly or completely by planting. This actually has two very important advantages. It enables the growing of successive crops from one investment in plantation establishment, and it preserves the genetic constitution of the planted trees for many rotations. Some stands that have been coppiced for centuries in England were originally planted, and the initial root-stocks have survived that long (Rackham, 1980).

Figure 13.6 A well-stocked stand in Chippewa National Forest, Minnesota, of 4,800 aspen root-suckers per acre, which originated after a clearcutting made during the dormant season 11 years previously. *(Photograph by U.S. Forest Service.)*

Coppice with Standards

The most ancient example of stands in which different species are grown on rotations of different lengths are those managed on the **coppice-with-standards system** in which small numbers of scattered promising trees, or **standards**, are reserved above sprout-coppice stands and left to grow for two or more short coppice rotations. The standards are usually of species different from those grown as coppice sprouts, but they can also be of the same species. Ordinarily there are several age classes of standards constituting an uneven-aged stand (Fig. 13.7); such stands are sometimes called **compound coppice stands**. This system evolved in places such as Europe before the Age of Coal, where it was important to combine the production of large amounts of fuelwood with that of some larger trees.

The trees reserved for standards are usually dominants of good form and species, preferably of seedling origin. The sprouts that arise after the cutting form a distinct story beneath and between the standards. The number of standards in a given age class is gradually decreased through successive coppice rotations. It is possible to develop standards gradually by successive thinnings in clumps of sprouts.

When the old standards are finally harvested, their stumps may be too large to sprout, and they may have to be replaced by planting or natural seeding from the remaining standards. Cleanings are sometimes necessary to bring young prospective standards of seedling origin up through the competition of the faster-growing stump-sprouts. The growth of well-established seedlings can also be accelerated by cutting them back to the ground and converting them into seedling sprouts. Plants of this type have the rapid initial growth of true sprouts, combined with the good form and rot resistance of trees of seedling origin. If the seedling sprouts spring up in pairs or clusters, all but the best one should be eliminated.

Intolerant species with light foliage are usually favored as standards, but trees capable of thriving under partial shade are desirable in the coppice. The species of the coppice must sprout readily and produce wood that is merchantable in small sizes. The standards

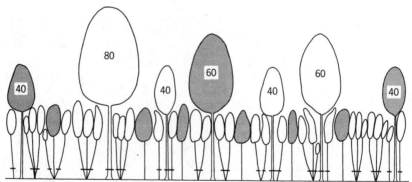

Figure 13.7   A compound coppice, with three age classes of standards, at the end of a 20-year coppice rotation. All trees of the coppice stratum will be cut except those *(shaded crowns)* chosen for new standards. The 80-year-old standard has reached the end of its assigned rotation and will be cut. The cutting of three younger standards is an operation analogous to thinning by which the number of standards in a particular age class is gradually reduced with advancing age, just as in a balanced, uneven-aged stand.

do not have to be of species that sprout; therefore, a conifer can be grown over a hardwood coppice, being introduced by planting if necessary.

Generally, standards are sufficiently isolated to develop deep crowns so that their rather short boles grow rapidly in diameter. This habit of growth resembles that of the emergents that soar above the main canopy strata of even-aged stratified mixtures, except that the standards probably grow even faster in diameter.

If the crowns expand rapidly enough, this usually inhibits the development of epicormic branches which can cause problems after any kind of partial cutting with most hardwoods and some conifers. However, if the coppice sprouts grow fast enough, they may eliminate any epicormic branches that have formed and may help keep the boles of the standards free of new ones. Dominant trees with full, vigorous crowns are less likely to develop epicormic branches after exposure than those that have unhealthy crowns or are from lower crown classes.

Individual trees with stems of high quality can be grown rapidly in open stands without risk of creating large vacancies in the growing space. The invasion of undesirable vegetation is prevented by the rapid development of the young coppice, which occupies the soil continuously.

In coppices with standards, much less of the production goes into fine logs than into small sprouts that are useful only if there is a strong demand for fuelwood, pulpwood, or other small wood products. This is why stands of this kind fell into disfavor in Europe and are grown only in rare cases in America (Fig. 13.8). Nevertheless, there may be important use for them in cases in which demands for fuelwood, pulpwood, or animal browse impinge on the use of forests to grow larger timber, especially on small holdings.

Coppice stands are almost always dealt with as being even-aged, but there is no basic reason why patches of sprouts of different age should not be intermingled in uneven-aged stands by the **coppice selection method**. If the object is to include production of some large trees, then the coppice-with-standards system is more appropriate.

The Role of Coppice Stands

The chief role of the coppice system is in growing fuelwood, pulpwood, and other products of small trees from sprouting species of angiosperms. For centuries it has been the most efficient way of growing crops of fuelwood. Because this has always been the most important human use of wood, it may be presumed that the coppice system has always been the most common kind of silviculture. In times and places in which fossil fuel is cheap and available, the method has usually been out of favor.

Provided that the species is one that sprouts, no regeneration method has a higher certainty of success. There is no delay except until the start of the next growing season; the growing space of the soil remains almost continuously and completely occupied. The new stems grow astonishingly fast in height, so that the above-ground growing space is quickly reclaimed. The rotations are generally confined to the short periods during which accumulation of total dry matter is most rapid. The optimum rotation length is usually regarded as that which maximizes the mean annual increment of dry matter. Although the trees are seldom impressive in size or appearance, the return on the small investment in growing stock is very high and it may be obtained in a short period.

The method has always been very useful in situations in which the continuing existence of a human community depends on the present and future supply of fuelwood. This was true for many centuries in much of Europe and for about two centuries in parts of the

Figure 13.8   A 30-year-old coppice of beech root-suckers beneath a fine red oak standard about 60 years old on a farm woodlot in southern Connecticut. The coppice, which has been cut periodically for fuelwood, has been used to control the form and pruning of the standard. *(Photograph by Yale University School of Forestry and Environmental Studies.)*

Atlantic seaboard of the United States. The availability of cheap coal and, later, oil caused the situation to change during the first half of the twentieth century. In both Europe and the United States, the modern concern about the coppice method has mostly taken the form of replacing coppice forests with high forests. There is, however, one noteworthy difference between the continents.

In Europe and elsewhere in the Old World, the coppice system developed a bad reputation because repeated coppicing from the same root-stocks with short rotations was associated with declines in the quality and vigor of successive crops. Coppicing was repeated so long and so often that all the stands in vast areas degenerated to grotesque clumps of crooked stool-sprouts. Some degradation also came from depletion of the inorganic nutrient capital of the soil that resulted from the nearly complete utilization of the forest production, including litter gathering, that is associated with coppicing. However, if repeated vegetative reproduction was inherently deleterious, deterioration would have been observed in certain varieties of domesticated plants that have been perpetuated entirely from rooted cuttings for centuries.

Foresters in the United States were for some years convinced that what they saw in the so-called Sprout Hardwood Region (Fig. 13.5) of the Boston–Washington megalopolis was the same sort of degraded coppice forest about which they had read in the European literature. However, coppicing in the American cases had usually not been repeated more than two or three times before the collapse of the fuelwood market caused it to cease. Early in the twentieth century, it was possible to see many potential deformities in the asymmetries of multiple-stemmed sprout clumps in the aging stands once coppiced for firewood. However, unlike Europe, true coppice stools clothed with deformed sprouts were so rare as to be curiosities (Fig. 13.3). Now that most of these trees have grown to sawlog size and the old sprout clumps have thinned themselves down to single stems, it takes close examination to detect what once caused concern. The fine standard shown in Fig. 13.8 was actually of stump-sprout origin.

The coppice method is associated with the kind of close utilization and frequent cutting that is most likely to remove chemical nutrients from the site and thus cause their depletion. This is true whether twigs are being removed for fuel by labor-intensive means or whether they are chipped for pulpwood under circumstances in which the labor needed for manual delimbing is too costly. The losses can be reduced to some extent by measures such as harvesting trees when they are leafless, or delimbing or debarking by the stump and thus leaving the twigs, leaves, and branches on the site. If the depletion becomes too great, it may be necessary to apply fertilizer, just as is the case in the essentially depletive practice of agronomy.

Another difficulty historically associated with the coppice method is that of changing from short to long rotations if the coppice method has to be replaced with high-forest methods. If, as was once the case in southern New England, large areas are coppiced every 30 years and it becomes necessary to switch to sawlog management with 80-year rotations, the situation becomes very awkward. During the 50-year transition period, it is possible to have large areas of forests but few trees large enough to harvest. Circumstances of this kind were part of the reason why European foresters have cursed the coppice method for so long. Foresters who have inherited such situations are likely to take dismal views of short rotations and proposals for departing from the objective of sustained yield.

Modern Use

One virtue of the coppice method is that it thoroughly preserves the genetic attributes of any trees grown by it. This attribute is often used in the culture of various poplars not only by coppicing but also, at least with the cottonwood poplars, by the planting of cuttings that easily take root. If a genetic form of chestnut could be developed to replace the ill-fated American chestnut, it would be logical to grow it by the coppice method to perpetuate whatever planted genetic combination was necessary. Fortunately, this species sprouted vigorously even from large stumps and grew rapidly into stems suitable for sawlogs and rot-resistant telephone poles.

It would be theoretically possible to use the coppice system to grow almost any American hardwood and the few sprouting conifers listed earlier. It might be done in cases representing special or particular kinds of circumstances. It would be the logical system to follow in order to obtain domestic fuel for a single household entirely from a woodlot of several hectares. A paper company might use it as a means of providing a supply of hardwood pulpwood that was always readily available, regardless of mud conditions, from lands close to the mill. Coppicing is a very useful way of producing browse and low cover for some wildlife species, especially those herbivores that are such important game species. One important and traditional use of the coppice system is for the production of round posts and small poles from such species as black locust.

With regard to most sprouting species of North American trees, the most important benefit from sprout regeneration is as an important supplement to seedling regeneration (Wendel, 1975). The sprouts of many hardwood species, coast redwood, and even the sprouting pines often develop as well as or even better than associated individuals of seed origin in naturally regenerated stands. Sometimes during regeneration cuttings it is desirable to cut small malformed trees back to the ground so that their large root systems can send up straight, fast-growing sprouts.

From the aesthetic standpoint, the coppice cutting can be even less beneficial than clearcutting; the cutover areas have the same appearance, and the harvests are repeated more often than is the case with true clearcutting. On the other hand, the new vegetation covers the land more quickly. Nevertheless, the stands are generally low in stature, monotonously regular, and often full of poorly formed trees; sometimes they seem little different from brushfields.

Conversion of Coppice Stands to High Forests

Although the coppice methods have an important place in silviculture, much of the forester's concern with coppice stands is to replace them with stands of seed origin. The very same sprouting habit that makes coppice stands easy to regenerate also makes them hard to eradicate. The degenerate coppice stands of Europe have been a classic problem.

The most serious North American problem of this kind has to do with stands of sprouting woody plants that are not normally regarded as coppice stands. These are the extensive brushfields of shrubs or poor hardwood species that have developed after heavy cutting and repeated fires on soils that are best adapted to the growing of nonsprouting species of conifers. On many of these sites conifers grow well and broad-leaved plants grow badly but the conifer seed source has been eradicated. Fortunately, the development of herbicides and mechanical methods of site preparation has made it much easier and cheaper than it once was to eliminate the sprouting species and restore the better ones by planting. In these situations, prescribed burning by itself is seldom helpful, although it can be useful in conjunction with the chemical and mechanical methods.

In the case of formerly coppiced hardwood stands in the eastern United States, there has been reason for replacement chiefly on sites where conifers would grow better. If the soils are good enough for the hardwoods to grow well, it has generally been appropriate to let the stands grow and to thin them while the crop trees grow to sawlog size. Shelterwood cuttings can be used for ultimate regeneration of the stands, although both seedlings and sprouts can be anticipated in the ultimate regeneration. However, some of the regeneration will also come from stump-sprouts. As is the case with vegetative regeneration in general, the question of whether this is good depends on the characteristics of the sprouts and not on some ancient dicta about the relative merits of sprouts and seedlings.

Stands Reproduced by Natural Layering

In the peat swamps of the northern forest, regeneration of black spruce and northern white-cedar by layering is an important supplement to regeneration from natural seeding (Morin and Gagnon, 1992; Paquin and Doucent, 1992; Johnston, 1977). Although these species bear seed in abundance, a high proportion of the seedlings are suppressed by the relatively fast-growing sphagnum moss, which is the main constituent of the ground cover. The seedlings have the best chance of survival on scattered, elevated hummocks and on decaying wood, a seedbed that should not be abundant in the managed forest. As time progresses, the peat grows up and engulfs the bases of the trees, covering any low-hanging branches that remain alive. The adventitious buds of buried portions of the branches develop into roots, while the living ends of the branches turn upward and grow into separate trees. Young shoots of this kind are sturdier and more vigorous than seedlings, but their form is not as satisfactory.

Reproduction of black spruce by layering is most readily achieved in peat bogs by the use of a modification of single-tree selection cutting (Johnston, 1977). If continued reproduction from layers is to occur, the uneven-aged form must be maintained so that the living crowns of the trees will remain in contact with the ground long enough for layers to become firmly rooted. It is significant that the residual stands are surprisingly resistant to wind because the peat is so resilient that the force of the wind tends to be dissipated in agitating the soil itself rather than in damage to the trees. However, black spruce growing on the firmer soils of uplands lacks windfirmness and rarely layers; therefore, it is best reproduced by clearcutting after advance regeneration has become established or by planting.

In northern white-cedar, reproduction from layers is fully as important as that from seed. Although normal layering from low-hanging branches is the most common form of vegetative reproduction, new, upright stems can also develop from the branches of wind-thrown trees. Appreciable amounts of reproduction may also rise from the rooting of small branches that become partially buried after being severed from the parent tree by browsing animals or during logging. The species is important not only for the production of posts and poles but also as a favored winter food of deer.

BIBLIOGRAPHY

Adams, R. D. (ed.). 1990. Aspen Symposium, '89, Proceedings. USFS Gen. Tech. Rept. NC-140. 348 pp.

Bates, P. C., C. R. Blinn, and A. A. Alm. 1993. Harvesting impacts on quaking aspen regeneration in northern Minnesota. *CJFR*, 23:2403–2412.

Brinkman, K. A., and E. I. Roe. 1975. Quaking aspen: silvics and management in the Lake States. *USDA Agr. Hbk.* 486. 52 pp.

Brown, D. 1994. The development of woody vegetation in the first 6 years following clear-cutting of a hardwood forest for a utility right-of-way. *FE&M*, 65:171–181.

DeByle, N. V., and R. P. Winokur (eds.). 1985. Aspen: ecology and management in the western United States. USFS Gen. Tech. Rept. RM-119. 283 pp.

Graham, S. A., R. P. Harrison, Jr., and C. E. Westell, Jr. 1963. *Aspens: Phoenix trees of the Great Lake Region.* Univ. of Mich. Press, Ann Arbor. 272 pp.

Greaves, A. 1981. *Gmelina arborea. For. Abst.,* 42:237–258.

Hillis, W. E., and A. G. Brown (eds.). 1954. *Eucalypts for wood production.* Academic, Orlando, Fla. 427 pp.

Jacobs, M. R., and A. Métro. 1981. Eucalypts for planting. *FAO Forestry Series* 11. 677 pp.

Johnston, W. F. 1977. Manager's handbook for black spruce in the North Central States. USFS Gen. Tech. Rept. NC-34. 18 pp.

Jones, R. A., and D. J. Raynal. 1988. Root sprouting in American beech *(Fagus grandifolia)*: effects of root injury, root exposure, and season. *FE&M* 25:79–90.

Kemperman, T. A., and B. V. Barnes. 1976. Clone size in American aspens. *Can. J. Bot.*, 54:2603–2607.

Kennedy, H. E., Jr. 1975. Influence of cutting cycle and spacing on coppice sycamore yield. USFS Research Note SO-193. 3 pp.

Kramer, P. J., and T. T. Kozlowski. 1979. *Physiology of woody plants*. Academic, New York. 811 pp.

Lamson, N. I. 1976. Appalachian hardwood stump sprouts are potential sawlog crop trees. USFS Res. Note NE-229. 4 pp.

Leffelman, L. J., and R. C. Hawley. 1925. Studies of Connecticut hardwoods. The treatment of advance growth arising as a result of thinnings and shelterwood cuttings. *Yale Univ. School of Forestry Bull.* 15. 68 pp.

McIntyre, A. C. 1936. Sprout groups and their relation to the oak forests of Pennsylvania. *J. For.*, 34:1054–1058.

Morin, H., and R. Gagnon. 1992. Comparative growth and yield of layer- and seed-origin black spruce *(Picea mariana)* stands in Quebec. *CJFR*, 22:465–473.

North Central Forest Experiment Station. 1972. Aspen: symposium proceedings. USFS Gen. Tech. Rept. NC-1. 154 pp.

Panel on Firewood Crops. 1980, 1983. *Firewood crops, shrub and tree species for energy production*. 2 vols. National Academy Press, Washington, D.C. 319 pp.

Paquin, R., and R. Doucet. 1992. Productivité de pessières noires boréales régénérées par marcottages à la suite de vieilles coupes totales au Québec. *CJFR* 22:601–602.

Peterken, G. F. 1993. *Woodland conservation and management*. Chapman and Hall, New York. 2nd ed. 374 pp.

Rackham, O. 1980. *Ancient woodland, its history, vegetation and use in England*. Edward Arnold, London. 402 pp.

Roth, E. R., and G. H. Hepting. 1943. Origin and development of oak stump sprouts as affecting their likelihood to decay. *J. For.*, 41:27–36.

Schier, G. A. 1978. Vegetative propagation of Rocky Mountain aspen. USFS Gen. Tech. Rept. INT-44. 13 pp.

Shigo, A. L. 1985. Compartmentalization of decay in trees. *Sci. Amer.*, 252(4):96–103.

Wendel, G. W. 1975. Stump sprout growth and quality of several Appalachian hardwood species after clearcutting. USFS Res. Paper NE-239. 9 pp.

CHAPTER 14

DOUBLE-COHORT PURE STANDS REGENERATED BY PARTIAL CUTTING

One step up in complexity from pure even-aged stands are those in which some members of an older cohort are retained for variable periods of time. There may be many purposes in such retention. The most common purpose is to preserve a source of seed until the new cohort of advance regeneration is safely established. Sometimes it helps to keep the new seedlings partly shaded in the early stages. Trees may be left because they are still growing well, they support certain wildlife species, or they serve aesthetic purposes.

Two cutting methods have long been associated with stands of these kinds. In the **seed-tree method**, a few trees are left primarily for seeds. In the **shelterwood method**, more trees are left not only for seeds but also for additional growth and shelter for the new stand. In the traditional mode of application, the reserved trees were supposed to be removed as soon as the regeneration was established and the new stands were regarded as even-aged. The two methods are so nearly similar in basic principles that they will be considered simultaneously.

In each method, fundamental reliance is placed on inducing advance regeneration, although advantage may be taken of any such regeneration that is already established. Any site preparation measures are done before, rather than after, all the old trees are removed. The transition from the old stand to the new is acccomplished without losing complete command of the growing space. The advantages of the even-aged condition and of partial cutting can be combined in varying ways and degrees when these two methods are applied.

Almost any tree species can be regenerated in partial shade. In fact, the majority of them actually require at least some shade; only a few very intolerant ones require full exposure to regenerate. Pure stands established by these methods usually develop as even-aged stands or single cohorts and conform to pure-stand yield tables as long as they remain closed.

This chapter and that on pure uneven-aged stands which follows it are restricted to pure stands mainly because certain concepts can be described more clearly without bringing in the added complications of mixed-stand structures.

Both regeneration methods involve a series of partial cuttings and other treatments applied in the following order:

1. **Preparatory treatments,** which set the stage for regeneration.
2. **Establishment** or **seeding cuttings,** which induce the actual establishment of seedlings and are often associated with treatments for site preparation.
3. **Removal cuttings,** which release the established seedlings.

This classification has usually been applied only for shelterwood systems but it fits seed-tree systems as well.

Preparatory Cuttings

Preparatory cuttings, very similar to thinnings, may be conducted before the real regeneration cuttings in order to strengthen and improve the vigor of trees destined to be left in the establishment cuttings. In cold, dry, or very wet climates where too much humus accumulates and seedbeds become too thick, preparatory cuttings may also be done to accelerate litter decomposition. However, one should avoid creating the kind of environment that will favor the establishment of undesirable species that are more shade-tolerant than those ultimately desired. If such species are likely to invade after light cutting, it is normally best to omit such cutting. If undesirable shade-tolerant trees and shrubs that will interfere with the desired regeneration are already established, it may be important to kill them.

In most cases preparatory cuttings are not necessary, particularly where their purpose has been largely accomplished by thinnings prior to the reproduction period. In fact, reproduction of some species may become established as a result of thinnings, or even in the absence of cutting. Herbicides, prescribed burning, and other active measures of site preparation are usually more expeditious than preparatory cutting as a means of reducing the accumulation of humus and controlling undesirable vegetation beneath the old stands.

Establishment or Seeding Cuttings

The establishment cutting is generally the first step in shelterwood and seed-tree cutting. The purpose of this cutting stage is to open up enough vacant growing space in a single operation to allow the establishment of regeneration. The ideal objective is to open the canopy enough to allow only the desirable species to become established. Any site preparation that is actually intended to lead to the establishment of seedlings is usually done just before or simultaneously with the establishment cutting, because any favorable surface conditions thus created may not last long. If the canopy is opened too much or if any site preparation is too heavy, unwanted intolerants may appear; too little of either may invite undesirable tolerant species.

It is at this stage that the seed-tree and shelterwood methods differ the most. The seed-tree method usually fits species that are more intolerant than potential competitors. The shelterwood method usually provides just enough shade that it is better for species that are at least moderately tolerant compared with competitors that are intolerant pioneers. Much depends on the kind and degree of any previous site preparation and the relative desirability of the species favored by it. Because this general approach to controlling species composition is seldom perfect, post-regeneration release measures are usually necessary if the objective is to create pure stands.

Use of the terms *establishment* or *seeding* denotes that these cuttings are the true regeneration cuttings of shelterwood and seed-tree systems. As is the case with any regeneration cutting, the cutting should ideally be done just after good seed crops have fallen. However, because a source of seed is retained, it is possible to do the cutting and then await a good seed year; in such cases, any site preparation can be postponed until the good year arrives.

Characteristics of Reserved Trees

Provision for sources of seed obviates the need to restrict the sizes of cutting areas. Vacancies hospitable to regeneration are created in the growing space beneath the reserved trees. It is not really possible to distinguish seed-tree cutting from shelterwood cutting on the basis of some arbitrary number of reserved trees per acre. However, because the production of seeds depends mainly on the number of trees, it helps to think in these terms with respect to the seed-tree method. With shelterwood cutting, plenty of potential seed trees are left, and the main considerations are usually the degree of shelter and the volume of wood left to grow for later harvests. These variables are correlated with the basal area reserved on each acre.

With the seed-tree method, the criteria followed in selecting trees to retain are crucial; with the shelterwood method, enough trees are reserved that they are less important but the principles are the same.

The best trees to reserve are those members of the dominant crown class that have wide, deep crowns, relatively large live crown ratios, and strong, tapering boles. Such trees are the ones most likely to combine windfirmness with the health and vigor that lead to good seed production. Trees with short, narrow crowns are, in spite of the smaller area on which the wind may act, vulnerable to breakage and uprooting because they have weak roots and slender boles without much taper. The occasional production of large "distress crops" of seeds by damaged trees is not a sufficiently dependable phenomenon to make it wise to use such trees for seed sources. Neither method is suitable for species that have extremely weak resistance to wind (or ice damage) or to soils that are conducive to shallow-rootedness.

Seed-source trees must be old enough to produce abundant fertile seed; the age at which seed bearing begins in closed stands is the safest criterion. Early and abundant production of seed can sometimes be encouraged by thinning out the crown canopy or fertilizing the soil around the chosen trees. Anything that can be done to protect the seeds on such trees from insects, arboreal rodents, or similar pests will enhance production. The seed production of good trees is often quickly increased by their release in the cuttings. A few good seed trees sometimes produce as much seed per acre as the full stand did before the cutting.

Because of both hereditary and environmental factors, the trees that have been good seed producers in the recent past are very likely to be the most dependable. The remains of cones or fruits on the trees or on the ground beneath them are often useful evidence of fecundity. The kind of trees used for seed trees depends to some extent on the intensity of silviculture practiced. If a second cutting can be made within a few years to utilize the seed trees, there is little reason not to pick fine, healthy dominants, presumably containing valuable timber. When the seed trees must be abandoned or are left for purposes of wildlife management after they have provided seed, individuals of lower commercial value may be appropriate. In stands that are distinctly even-aged, the reservation of the smaller trees is likely to be a futile gesture.

The seed-tree and shelterwood methods theoretically provide opportunity for more rigorous selection of the parents of the new crop than is attainable with any other method of natural regeneration. It is rational to assume that undesirable characteristics are heritable and to attempt to confine choices to the best phenotypes. If there is some compelling reason why the most valuable trees cannot be reserved for seed, it is wise to try to evade the problem. The most important way that this can be done is to select trees that have developed imperfectly because of strong and obvious environmental influences that are likely to have operated at random. Trees that have developed poor form because of lack of competition would be logical choices in such circumstances. Trees that are poor because of apparent susceptibility to damaging agencies that operate in immature stands should be rejected. There is probably no reason to avoid trees damaged by heart-rot, which characteristically infects trees only when they are older than the normal rotation age.

Arrangement, Number, and Distribution of Seed Trees

With the seed-tree method, the number of trees to be reserved depends on factors such as the amount of seed produced per tree, the prospective survival of the trees, and the number of seeds required to yield an established seedling with the seedbed conditions that will prevail.

If the seed trees are uniformly scattered, the prospective quantity of seed needed seems to be a more critical factor than the distance of dissemination. Satisfactory dissemination by wind can usually be counted upon for distances of at least three times the height of the trees. With species that are disseminated by winged or terrestrial animals, the distances depend on the habits of these animals. Even these distances can be great enough that regeneration success depends more on the quantity of seed and the nature of the regeneration microenvironments than on limitations of dispersal distances.

The number and distribution of seed trees influence the pollination of female flowers and stroboli and, therefore, the number of viable seeds produced. A few species, such as white ash and cottonwood, are dioecious; both male and female trees of such species must be left regardless of the method of regeneration employed.

In monoecious species, self-fertilization usually results in the production of a high proportion of inviable seeds as well as progeny with the inferior genetic characteristics associated with inbreeding, but the male and female flowers of the same tree rarely become mature simultaneously. The possibility of self-fertilization is also reduced by the tendency for female flowers to develop at the top of the tree, above the male flowers; the wind is not often turbulent enough to cause much pollen to rise straight upward. However, the concentration of pollen grains decreases rapidly with distance from the source because of atmospheric turbulence; therefore, the proportion of fertile seeds probably decreases as seed trees become more widely separated.

It may, therefore, be assumed that isolated seed trees produce lower percentages of *viable* seeds than they would if they were adjacent to other trees of the same species, but the *total* seed production is usually increased by releasing trees from competition. Thus the fruitfulness of trees in closed stands is clearly not a good criterion of the number of isolated seed trees required.

The proper number of seed trees to leave per acre depends on knowledge of seed production and on how many seeds are required to yield one established seedling on the kinds of seedbeds that will exist. A goal must be set for the number and spatial distribution of established seedlings that constitute a satisfactorily stocked stand. For example, this minimum might, at the third year, be 1000 trees per acre, with 40 percent stocking of a sample of randomly chosen milacre plots. Detailed studies of loblolly pine on the Atlantic Coastal Plain indicated that this goal could ordinarily be reached during a mediocre seed year with four dominant or codominant pines, 16 inches D.B.H., per acre on sites prepared by prescribed burning after logging (Wenger and Trousdell, 1957). However, as part of this example, it was estimated that 12 pines per acre would be required if they were 12 inches D.B.H., and, without the burning, the number of seed trees would have to be increased by 50 percent.

The risk of loss to lightning may influence the number of trees reserved. Trousdell (1955) found that in northeastern North Carolina the annual losses to lightning occurred at the rate of 5 trees per 100 hectares, regardless of how many were reserved. An increase in the number of trees left would reduce the percentage, but not the number, that were struck. Because the number of thunderstorms there is close to the national average, one might estimate prospective lightning losses in other localities from the relative frequency of such storms.

Sporadicity of Seed Crops

Regardless of the regeneration method, success ordinarily comes only when there are good seed crops. Except in the case of species with very small seeds, and sometimes even with those, almost all of the seeds produced by trees are consumed by rodents, birds, insects, fungi, or other pests. Only a small fraction remains to renew the forest, and that small amount is enough to restock the forest only when the total seed crop is large.

Unfortunately, the planning of natural regeneration is seriously hamstrung by poor understanding of the factors that control seed production and its timing. If there is any important tendency toward regular periodicity of seed crops, it is so obscured by disrupting influences that it is no basis for reliable predictions.

Sometimes the number or proportion of female reproductive buds, often a limiting factor, may be enhanced by improved tree vigor or nitrogen nutrition. On the other hand, there is also evidence to suggest that the most critical question is whether various damaging abiotic agencies or biotic pests interrupt one of the many steps that lie between bud formation and seed production. Often the best crops of seed come after years of total failure. With such episodes the lack of seeds in one year may cause collapse of populations of seed predators during the next year.

With most species good seed crops can be predicted only a few months in advance, when the fruits or cones have formed and the time during which damage can occur begins to dwindle. The earliest times for assessing the prospects come when, usually a year in advance, the buds first differentiate into vegetative or reproductive. If development of cones or fruits takes two years, as is true with most pines and red oaks, useful predictions can be made rather far in advance. However, with a process that has many steps from bud formation to seedling establishment, it is important to remember that failure at any one step can defeat the whole process.

In practice, it is normally best to reserve enough trees to restock the area in one moderately good seed year. Excellent crops can be depended upon only if they are already on the trees; meager crops are usually consumed by rodents or birds.

Both the seed-tree and shelterwood regeneration methods have an inherent safeguard. The seed source is retained on the site until the new stand is established. If regeneration fails after one good seed crop, it is possible to wait for the next with no cost other than loss of time and the need to repeat any necessary site preparation.

Severity of Cutting

The trees removed in the establishment cuttings of the shelterwood method come mainly from the lower crown classes and should include undesirable species and trees suspected of being inferior genetically. However, consideration is sometimes given to reserving trees that are poor for timber and good for wildlife. The temptation to eliminate all coarse, branchy dominants should be restrained if these are the only good seed-bearers.

The establishment cutting is best confined to a single operation if the new crop is to be uniform in age and size. In this way, the severity of the cutting can also be adjusted so as to create a range of environmental conditions that is restricted as closely as possible to that which is optimum for germination and establishment of the desired species. In the case of the shelterwood method, a series of cuttings of ever-increasing severity might involve the reservation of an overstory that was at first too dense and later too thin to provide the proper environment.

Where moisture deficiencies are likely to limit regeneration, an excessively dense overwood may cause too much root competition and interception of precipitation. If heat injury to succulent young seedlings is the primary difficulty, the establishment cutting should be regulated so as to increase the amount of diffuse light admitted from the sky as much as possible without allowing much direct sunlight to reach the forest floor.

The appropriate density of the overwood differs within wide limits, depending on the requirements of the species sought and the site factors; it may also be as important to make conditions unfavorable to undesirable species as to create conditions favorable to the desirable (Ashton, 1992; Marquis, 1966). If the species that appear are more shade-tolerant than those desired, it is usually necessary to increase the severity of the establishment cutting. If, for example, the existing species to be reproduced is pine, but a more tolerant species burgeons after shelterwood cutting, it is probably a sign that the cutting was not heavy enough or that site preparation was inadequate. The opposite is true if unwanted grasses or other pioneer vegetation appears. The unintended establishment of mixtures of species is usually not of concern if the desirable species outgrows the undesirable. Various indexes, such as stand basal area, related to crown cover can be used to regulate and guide the severity of establishment cuttings and thus enable transfer of experience from one case to another. Because the extraction of trees felled in the subsequent removal cuttings can cause damage to the new seedlings, it may help to take off as much of the old stand as possible in establishment cutting.

It is useful to take advantage of all groups of advance growth that may have started naturally or as an unintentional result of preparatory cuttings or thinnings. Over such groups the cutting is heavy and may remove all the old trees.

If satisfactory regeneration does not become established as a result of seed present at the time of cutting or from that produced in the next few years, two courses of action are available. The inadequate natural reproduction may be supplemented by planting, or measures of site preparation may be applied during a subsequent seed year.

Arrangement of Seed Trees

In most seed-tree cutting, the seed trees are left isolated and rather uniformly distributed whereas the rest of the stand is removed, as shown in Fig. 14.1. Another option is to leave the seed trees in groups, strips, or rows, rather than having them scattered uniformly. Concentrating seed trees on restricted areas makes them easier to protect during the initial cutting and also to harvest when they have served their purpose. Cross-pollination might be increased, although seed dispersal would be less efficient and uniform. Sometimes it is difficult to get new regeneration established under such concentrations of seed trees, but this difficulty can sometimes be mitigated by applying localized shelterwood cutting to

Figure 14.1 Two winter views of the same spot in a 95–year-old stand of loblolly pine in North Carolina, immediately before and after a cutting in which eight seed trees were reserved on each acre. The larger hardwoods were killed with herbicides and prescribed burning was done in the late summer to kill small hardwoods, dispose of slash, and prepare the seedbed. *(Photograph by U.S. Forest Service.)*

them at the time of their reservation. There is no evidence that groups of seed trees are significantly more wind-resistant than comparable trees left isolated.

Removal Cuttings

Removal cuttings have, in varying degrees, the objectives of gradually uncovering the new crop and of making the best use of the potential of the remaining old trees to increase in value. There may be one or, in intensive management, several removal cuttings, the last of which is called the **final cutting**. With the seed-tree method there is usually only the final cutting, although some trees may be kept for long periods as insurance against the loss of the new stand, for additional increment of valuable wood, or to favor wildlife dependent on old trees.

In the shelterwood system, the ideal objective of the removal cutting process is to proceed so that the new stand fills the growing space as fast as the old stand is caused to relinquish it. This approach enables regeneration to be accomplished without reducing total wood production as far below the capacity of the site as is the case with clearcutting in which a new crop of seedlings, natural or planted, is left to regain full occupancy alone. If the growing space is kept filled, there is not much opportunity for undesirable species to invade.

The largest and most vigorous trees usually have the greatest capacity to increase in value, and so are the trees most logically reserved until the final cutting (Fig. 14.2) unless they happen to be ones that interfere unduly with the new crop. After the reproduction is established, it is desirable to watch for signs of poor growth. If the young trees develop unhealthy foliage, bend aside toward the light, or fail to maintain a satisfactory rate of growth in height, the competition of the overwood should be eliminated or reduced. If the development of the new stand is not uniform and if there is more than one removal cutting, it may be necessary to vary the severity of a given cutting operation in different parts of the area. Some patches may need full release, others may require partial release, and still others that remain unstocked may need a repetition of the establishment cutting.

It is important to distinguish between the conditions suitable for the establishment of regeneration and those necessary to make the new trees grow satisfactorily in height. Establishment cuttings often provide ideal shelter for new seedlings, but much heavier removal cuttings are necessary to get them to grow well. It is important that removal cuttings open stands rapidly to allow desirable species to outgrow more tolerant competitors. After they are well established, intolerant species tend to increase in height growth with each added increment of light right up to full sunlight, whereas the tolerant species attain a modest but maximum height growth at some intermediate light intensity.

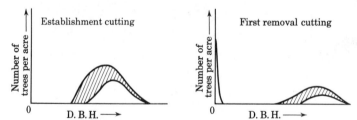

Figure 14.2　Diameter distribution curves showing (by cross-hatching) the relative numbers of trees of different diameters removed in a typical establishment cutting and the first of two removal cuttings. In the establishment cutting (left) the smaller and less desirable trees, representing about a third of the volume, are cut. In this example, the first removal cutting (right) takes place after the trees left in the establishment cutting have increased substantially in diameter and after some of the reproduction resulting from that cutting has grown to measurable D.B.H. The trees reserved in the first removal cutting are the biggest and best in the stand; they will all be harvested in the second and final removal cutting.

The time sequences of cutting range from those in which old trees are left only until the new crop is established to those in which some of the trees of the previous crop are deliberately retained as a part of the new one. Except for the need to regulate the composition and rate of development of the new crop, there is little reason why the timing cannot be fitted to any management consideration. The patterns of cutting in space include those in which the residual trees are scattered uniformly throughout what will become one large, even-aged stand. There may also be patterns of cutting in which trees are cut or left in strips or groups; sometimes the resulting openings are gradually enlarged in sequences of cuttings. The more complicated seed-tree and shelterwood systems differ from the selection system mainly because the resulting stands can be counted as essentially even-aged (or of single cohorts) for purposes of forest planning and in regulating harvests for sustained yield.

Because of the principle of leaving the best and most vigorous trees until the end, the shelterwood and seed-tree systems usually provide the best way of applying the concept of financial maturity to management of the growing stock of even-aged stands. Efficient use can thus be made of the capacity of the better trees to increase in value without sacrificing the advantages of concentration of operations associated with even-aged management. The sequence of distinctly separate operations is more visible, more systematic, and simpler to administer than is the case with uneven-aged stands.

The shelterwood and seed-tree methods can, however, be applied intensively only if there are markets for trees of small size or poor quality, unless such trees are killed rather than harvested. The principle of cutting the poor trees before the best can make the first cuttings of the shelterwood sequence financially unattractive, especially if the cost of making the stands accessible is high.

The removal cuttings, if uniformly applied over the regeneration area, are almost certain to cause some injury to the new stand. This injury can be reduced, but not eliminated, by care in logging. The least damage is likely to result if the overwood is harvested while the seedlings are still flexible. The greatest difficulties result when it is necessary to fell trees with broad crowns into stands of saplings. The damage usually appears more serious than it really is, and it is sometimes even a benefit in disguise. Both shelterwood and seed-tree cutting often lead to the development of grossly overstocked patches of reproduction that look handsome when young but are apt to stagnate in the sapling stage. These clumps can be crudely thinned by the skidding or felling of trees across them. It is usually best to direct the inevitable logging damage toward the densest parts of the new stand or those that are entirely unstocked and away from the sparsely stocked portions.

Variations in Spatial Patterns of Stand Structure

Most of the preceding discussion of systems of stand regeneration and management has been limited to patterns in which both the new trees and any older ones are uniformly distributed over the whole area of the stand in what is called the **uniform** arrangement. If fact, the shelterwood method is sometimes called the **uniform method** because commonly the residual trees are uniformly distributed.

Sometimes there is reason to depart from patterns of stand structure in which trees are distributed uniformly in the horizontal dimension. The openings left by regenerative disturbances and the new aggregations of trees established in them can be in the form of groups, patches, strips, wedges, circles, ovals, squares, rectangles, or simply of irregular shapes. The same is true of the spatial arrangement of seed trees or other trees reserved in partial cutting. Sometimes these shapes are simply inherited from past stand structure; sometimes they are deliberately created for significant reasons.

If the units are small enough, they are arbitrarily regarded as parts of the same stand. If the time interval between the regeneration of each unit is short enough, the stands are regarded as either even-aged or of the same cohort. If these intervals are so long that the trees of different units are clearly not of the same total height, the stands are regarded as uneven-aged, provided there are at least three such units.

Such terms as strip-shelterwood, patch clearcutting, group-selection, or the strip-seed-tree method may be used to describe these patterns of regeneration cutting.

Strip Arrangements

There can be reasons to apply almost any regeneration method in strips that progress, quickly or slowly, across the stand area. The usual purpose of leaving seed trees or clear-cutting in strips is to provide sources of seed for regenerating the cleared areas. The reserved strips are generally no wider than they absolutely must be and the cleared strips must be, narrow enough to permit adequate seed dispersal. When seedlings have become established on the clearcut strips, the originally reserved strips may be regenerated from seed on the trees that are cut or from artificial regeneration, although the seed-tree or shelterwood methods can also be employed there. The clearcut strips may be oriented so that they are at right angles to the direction of seed-dispersing winds or to that of the midday sun. However, if the new seedlings require protection from the sun, it should be noted that this does not extend beyond the shadows of the reserved trees.

Such cutting is sometimes done in **progressive strips** that advance in some direction chosen so that timber is extracted through uncut strips and not across areas under regeneration. Sometimes the strips are advanced from leeward to windward against the direction of the most dangerous storm winds. This creates a new streamlined, uneven-aged stand that may shunt the force of the wind harmlessly up and over the stand (Fig. 19.2).

Shelterwood cuttings may also be applied in progressive strips. Such systems can be used to manage pure stands, but they are more common with mixed stands and will, therefore, be described in Chapter 16 .

Application of Seed-tree Systems

The seed-tree method can provide good regeneration of wind-disseminated species that are naturally adapted to become established after severe fires, provided that appropriate site preparation exists and the desired species is not overwhelmed by even more intolerant weed species. The method has, for example, been used with moderate success for various southern pines (Wenger and Trousell, 1957) (Fig. 14.1), coastal Douglas-fir, some euca-lypts, and forms of jack and lodgepole pine that have nonserotinous cones. In many such cases, however, the technique has been replaced by that of clearcutting and artificial re-generation because it has often been necessary to employ planting or direct seeding to take appropriately prompt advantage of site preparation. The desire to achieve genetic improve-ment has also given impetus to planting, although the high initial investment of that ap-proach deters some owners. Seed-tree cutting can be used instead of the shelterwood method in cases where the supply of seed might be super-abundant and lead to overstocking.

Unfortunately, the seed-tree system carries a burden of guilt by association with disappointment. It has too often been a half-hearted gesture toward regenerating forests during the early stages of deliberate forest management. There have been misleading cases in which crops of new seedlings ascribed to reserved seed trees were, on closer examination, found to be from advance growth already present before cutting. Frequently, the number or vigor of seed trees was inadequate. Another problem has sometimes been that the number of trees left were too few to justify returning to harvest them; leaving more trees in shelterwood cutting has often been the solution to that problem.

Application of Shelterwood Systems

The shelterwood systems have many more variants and wider applicability than the seed-tree systems. Within the framework of the shelterwood method, there can be variation in the relative degrees of shelter and exposure in both space and time. Adjustments can be made to meet the environmental requirements of almost all species except those that are exceedingly intolerant of shade or root competition (Hannah, 1988; U. S. Forest Service, 1979). It is particularly useful for species that do not initiate rapid height growth until they have slowly developed the deep root systems necessary to sustain such growth. It also fits the numerous species that are tolerant in youth but intolerant after reaching the sapling stage.

Shelterwood methods simulate the effects of those windstorms and similar enemies of large trees that kill forests from the top down and release small trees of species adapted to start as advance regeneration. Conversely, when used with light surface fires or similar measures of site preparation, both shelterwood and seed-tree cutting can be used to secure the establishment of seedling regeneration of many fire-following species.

The term *shelterwood* denotes one purpose succinctly, although sheltering seedlings from damaging agencies is not the *only* one. The shelterwood method is also employed partly, even primarily, to get efficient use of the productive capacity of the growing stock. Trees that are no longer capable of much further increase in value are the ones cut to make space for regeneration.

Advance regeneration secured by shelterwood cutting can seem stunted and be mistakenly dismissed as hopeless. It is best to base opinions about response to release on direct observations of the development of the small trees. If many older trees that have grown well have annual rings that reveal very slow growth in the early stages, it can be assumed that advance regeneration of the species involved can be depended upon to respond well to release. Sometimes such opinions are best formed by examination of ring patterns right at the ground line. Among the observations of good response to release are those of Douglas-fir by Tesch et al. (1993) and longleaf pine by Boyer (1993), as well as of conifers in Quebec (Doucet, 1988), California (Helms and Standiford, 1985), and Idaho (Ferguson and Adams, 1980).

The only common problem associated with natural regeneration from advance growth is that it tends to be too dense and thus to develop into overcrowded thickets. It may help to leave shelterwood trees to induce irregularity of height growth in the new stand. Such trees can also become an important part of the new stand.

Residual trees are sometimes retained to favor wildlife species that depend on trees older than normal rotation age for food or shelter. Well-chosen residual trees also make a stand under regeneration look less devastated than is the case with clearcutting.

In the case of most shelterwood cutting, some other potential advantages can be used in varying degrees. When one seeks natural regeneration from seed with clearcutting, seed-tree cutting, or even some forms of group-selection cutting, it is easy to say that the methods lead to regeneration if one cuts in a good seed year. This is not much help for the normal situation in which good seed years come sporadically and seldom. Most kinds of shelterwood (and selection) cutting provide a basic way around this dilemma.

The significant point is that the main harvest cuttings follow, rather than precede, the establishment of seedlings. The ecological adaptations of most species are flexible enough that shelterwood cutting can usually be applied with some degree of latitude about timing. Even an establishment cutting can often be made in any year with the idea of simply waiting for the next seed crop. The procedure that is most crucial is any site preparation for regeneration that may be necessary; this may call for a special campaign to treat many stands in some year when a good crop of seed is due to fall.

Most species also have seedlings tolerant enough of shade that they do not have to be released in removal cuttings on some precise time schedule. As a result, it is sometimes possible to store the seedlings that become established in one spectacular seed year and then release them, first in one stand and then in another, over a period of years or even of decades. This latitude is not available with all species, but it can be very advantageous if it does exist.

Most shelterwood cutting also has the very important advantage that, if there are enough residual trees, the old stand can continue to make economically effective use of the growing space while regeneration is awaited. This eases the cost of waiting for good seed crops. Even more importantly, it can provide for the "telescoping" of one rotation into another. If the new stand is, in effect, six years old at the time of the final removal of a rather full shelterwood that is 60 years old at the time, the average rotation may, for practical purposes, be shortened by 10 percent. The gain can be greater yet if the alternative is a clearcutting procedure in which seeding or planting, by either accident or design, is delayed for a year or more after cutting.

If the new stand gets well started before the old one is removed, the inevitable loss of production during the transition is kept under some control; this is because the desired vegetation comes close to retaining full command of the growing space. With true clear-cutting, for example, there is an inevitable hiatus between the time the old vegetation is cleared away and the time when the new trees have developed sufficient foliage and roots to make full use of the growing space. This hiatus is ordinarily several years long. It is also a time when unwanted vegetation can usurp some of the temporarily vacant growing space.

The presence of seed trees or a shelterwood can make it difficult to burn slash. Conversely, the shade that a shelterwood casts could, in some climates, restrict drying or speed decomposition enough to make slash disposal unnecessary. The presence of shade often eliminates the necessity to expose mineral soil as a means of getting light-seeded species to germinate. Shaded litter is more likely to remain cool and moist.

The shelterwood system is superior to all others except the selection method with respect to protection of the site and aesthetic considerations.

Shelterwood management that involves sequential harvest cutting ordinarily does not develop until conditions favor silviculture of at least moderate intensity. Often it develops where it becomes necessary to regenerate stands that have been thinned and were otherwise well cared for. However, this development usually depends on the existence of other conditions, including the desirability of establishing even-aged stands by natural seeding and the existence of terrain and machinery suitable for partial cutting. Another common and important impetus is the opportunity to reap financial advantage from the continued growth of the trees retained as the shelterwood (or as seed trees).

Matters of aesthetics and public relations also play an important role. The residual trees of a shelterwood are not only attractive but also provide, at a glance, evidence of concern for the future.

The use of shelterwood and seed-tree cutting to secure natural regeneration is common among owners who have a long-term outlook but wish, for any reason, to avoid the large out-of-pocket investments that artificial regeneration requires. The validity of this policy depends on the extent to which the gains may be counteracted by the potentially higher costs of logging or post-regeneration treatments. As is the case with other matters, many forest owners will follow courses of action that require little investment even if the results are less than ideal.

The situation in which this policy of seeking low-cost regeneration fits best is with those species, usually rather tolerant of shade, that grow slowly when young but faster later in life. For these species the rapid compounding of high regeneration costs can be a more serious financial impediment than that of costs of precommercial thinning of over-dense shelterwood regeneration undertaken later in the rotation.

The most common American use of shelterwood cutting for the regeneration of pure stands is for many pines and other conifers. It has been used successfully for loblolly, shortleaf, and pitch pines when coupled with adequate control of hardwoods. It is now regarded as a good way of regenerating longleaf pine, which is difficult to plant and has very slow-growing seedlings (Balmer, 1979; Boyer, 1979). When ponderosa pine is grown under even-aged management, the regeneration is often accomplished by shelterwood cutting. In the case of the Black Hills of South Dakota, which have summer rainfall and abundant regeneration, this is fairly easy; in other parts of the rather arid range of the species, it may be necessary to wait patiently or plant (Barrett, 1979). Red pine has also been successfully managed under the uniform shelterwood system, although it has sometimes been difficult to secure prompt regeneration (Benzie, 1977; Eyre and Zehngraff, 1948).

With lodgepole and jack pines, shelterwood cutting is often used in stands where the cones are not closed or serotinous; even if they are, the reservation of a shelterwood may provide some shade for seeds from cones on the slash and also impose the kind of variations in seedling height that mitigate overstocking.

In the cases of the moderately shade-tolerant eastern (Fig. 14.3) and western white pines the shelterwood system is advantageous for pest management and is ideal for natural regeneration. The main problem associated with growing eastern white pine is usually the white pine weevil (Funk, 1986). This insect often kills the terminal shoots of trees with tops in the sunshine. The replacement branches are usually the laterals of the previous year, and, when they turn upward, they form crooks and forks that badly degrade the stems. One solution is to to use the shelterwood overstory to keep the young trees shaded until at least one log-length has formed. Older white pines provide the best kind of shade because the weevils are lured away by their sunlit tops. The overstory must remain for such a long time that it should be composed of pines that can produce value rapidly and thus compensate for the slow growth of the deliberately stunted regeneration.

With western white pine, for which the introduced blister rust fungus is the major problem, shelterwood cutting plays a role in keeping the understory *Ribes* shrubs, the alternate hosts of the fungus, in check. The partial shade that is conducive to pine seedlings may be too much to allow the *Ribes* to survive and produce seeds. There may be possibilities of building rust-resistant natural populations if rust-free parent trees are retained as the shelterwood seed source. With both eastern and western white pines, it is necessary

Figure 14.3   A 74–year-old stand of eastern white pine in the Pack Forest in eastern New York being regenerated by the shelterwood method. The establishment cutting was carried out 12 years previously and the partial shade of the overstory is being used to reduce damage to terminal shoots by the white pine weevil. *(Photograph by State University of New York, College of Environmental Sciences and Forestry.)*

that shelterwood cuttings be severe enough to allow the pine seedlings to grow faster than those of hemlocks and other associated species that are more shade-tolerant than the pines (Boyd, 1969).

Douglas-fir and many other conifers of western North America are ecologically well adapted to shelterwood cutting (Fig. 14.4) (Ruth and Harris, 1979; Atkinson and Zasoki, 1976; Harris and Farr, 1974; Williamson, 1973). In the past, the use of the method did not

Figure 14.4 Regeneration, 7 years old, of Douglas-fir from uniform shelterwood cutting on a forest of the U.S. Bureau of Land Management in the Oregon Coast Range. *(Photograph by Yale University School of Forestry and Environmental Studies.)*

develop because of the difficulties of doing partial cutting, especially in ancient unstable stands and on the steep terrain that is so common. In such places as western Washington, the climate and soils are favorable enough that clearcutting and planting is a dependable solution; shelterwood cutting may be used to meet various management objectives, but partial shade is not necessary for regeneration.

Most of the region is so mountainous that elevational differences and rainshadow effects create an intricate pattern of variations in site factors to which silviculture must be adapted. One large locality where shelterwood cutting often seems desirable is southwestern Oregon, which has rainless summers that are longer than those farther north. The problem sites there include south-facing slopes, very coarse-textured soils derived from physically weathered granite, and low elevations where the rainfall is small. The difficulties can also be overcome by thorough site preparation and planting of well-developed seedlings (Tesch and Mann, 1991). On such sites partial shade often appears to reduce direct evaporation and extremes of surface temperature enough to allow seedlings to become established. In addition, the retention of some shade may reduce the development of ceanothus and manzanita shrubs, which are very prevalent and can easily exclude trees. Sometimes very large, valuable, old-growth Douglas-fir have been left to provide shade. Adroit use of madrone or other trees of low value, as well as nonliving shade, can alleviate the problem.

At high elevations, both incoming solar radation by day and outgoing radiation by night are so great that extremes of surface temperature are very wide (Tranquillini, 1979); shelterwood cutting is often necessary to secure establishment of regeneration.

Except for true pioneers, most broadleaved species are adapted to start as advance regeneration, and their seedlings require partial shade. Because most of these hardwood species grow in mixed stands, they will be considered in Chapter 16. However, if they are to be grown in pure stands, it is usually necessary to employ shelterwood cutting or the selection system that is described in the next chapter.

BIBLIOGRAPHY

Ashton, P.M.S. 1992. Establishment and early growth of advance regeneration of canopy trees in moist mixed-species forest. In: M. J. Kelty, B. C. Larson, and C. D. Oliver. *The ecology and silviculture of mixed-species forests.* Kluwer Academic, Boston. Pp. 101–122.

Atkinson, W. A., and R. J. Zasoki (eds.). 1976. Western hemlock management conference. *Univ. of Wash. Inst. For. Products Contrib.* 34. 317 pp.

Balmer, W. E. (ed.). 1979. Proceedings, Longleaf Pine Workshop. *USFS Tech. Publ. SA-TP-3.* 119 pp.

Barrett, J. W. 1979. Silviculture of ponderosa pine in the Pacific Northwest: the state of our knowledge. USFS Gen. Tech. Rept. PNW- 97. 106 pp.

Benzie, J. W. 1977. Manager's handbook for red pine in the North Central States. *USFS Gen. Tech. Rept.* NC-33. 22 pp.

Benzie, J. W., and A. A. Alm. 1977. Red pine seedling establishment after shelterwood harvesting. USFS Res. Note NC-224.

Boyd, R. J. 1969. Some case histories of natural regeneration in the western white pine type. USFS Res. Pap. INT-63. 24 pp.

Boyer, W. D. 1993. Long-term development of regeneration under longleaf pine seedtree and shelterwood stands. *SJAF*, 17:10–15.

Brender, E. V. 1973. Silviculture of loblolly pine in the Georgia Piedmont. *Ga. For. Res. Counc. Rept.* 33. 74 pp.

Doucet, R. 1988. La régénération préétablie dans les peuplements forestiers naturels au Québec. *For. Chron.*, 64:116–120.

Eyre, F. H., and P. Zehngraff. 1948. Red pine management in Minnesota. *USDA Circ.* 778. 70 pp.

Farrar, R. M., Jr. (ed.). 1990. Proceedings of the symposium on the management of longleaf pine. USFS Gen. Tech. Rept. SO-75. 293 pp.

Ferguson, D. E., and D. L. Adams. 1980. Response of advance grand fir regeneration to overstory removal in northern Idaho. *For. Sci.*, 26:537–545.

Funk, D. T. (ed.). 1986. Eastern white pine: today and tomorrow. USFS Gen. Tech. Rept. WO-51. 124 pp.

Hannah, P. R. l988. The shelterwood method in northeastern forest types. *NJAF*, 8:99–104.

Harris, A. S., and W. A. Farr. 1974. The forest ecosystem of southeast Alaska. 7. Forest ecology and timber management. USFS Gen. Tech. Rept. PNW-25. 109 pp.

Helms, J. A., and R. B. Standiford. 1985. Predicting release of advance reproduction of mixed conifer species in California following overstory removal. *For. Sci.*, 31:3–15.

Korpela, E. J., S. D. Tesch, and R. A. Lewis. 1992. Plantations vs. advance regeneration: height growth comparisons for southwest Oregon. *WJAF*, 7:44–47.

Lipscomb, D. L. 1989. Impact of feral hogs on longleaf pine. *SJAF* 13:177–181.

Marquis, D. A. 1966. Germination and growth of paper and yellow birch in simulated strip cuttings. USFS Res. Paper NE-54. 19 pp.

Tesch, S. D., K. Barker-Katz, E. J. Korpela, and J. W. Mann. 1993. Recovery of Douglas-fir seedlings and saplings during overstory removal. *CJFR*, 23:1684–1694.

Tesch, S. D., and J. W. Mann. 1991. Clearcut and shelterwood reproduction methods for regenerating southwest Oregon forest. *Oregon State Univ., For. Res. Lab., Res. Bul.* 72. 43 pp.

Tranquillini, W. 1979. *Physiological ecology of the alpine timberline.* Springer-Verlag, New York. 137 pp.

Trousdell, K. B. 1955. Loblolly pine seed tree mortality. Southeastern For. Exp. Sta., Sta. Paper 61. 11 pp.

U.S. Forest Service. 1979. *The shelterwood regeneration method.* USFS Div. Timber Management, Washington, D.C. 221 pp.

Wenger, K. F., and K. B. Trousdell. 1957. Natural regeneration of loblolly pine in the South Atlantic Coastal Plain. USDA Production Research Rept. 13. 78 pp.

Williamson, R. L. 1973. Results of shelterwood harvesting of Douglas-fir in the Cascades of western Oregon. USFS Res. Paper PNW-161. 13 pp.

Youngblood, A. 1990. Effect of shelterwood removal methods on established regeneration in an interior Alaska white spruce stand. *CJFR*, 20:1378–1381.

CHAPTER *15*

SILVICULTURE OF PURE UNEVEN-AGED STANDS

Uneven-aged stands, like pure plantations, have staunch adherents and scornful critics. Patchy regenerative disturbances have commonly created them both in natural forests and in forests that have been subject to partial cutting. It is often better and easier to preserve the preexisting age-class structure than to attempt to change it. Where some management objective requires that a stand always have some large trees, it is necessary to develop or maintain some sort of uneven-aged structure. It is claimed by some that uneven-aged stand structures are essential to the well-being of forest ecosystems.

There are several kinds of uneven-aged stands, and conflicting ideas about how to manage them. Some foresters have devoted much attention to quantitative methods of molding these stands into self-contained units that will supply sustained yield of timber. Others have been content simply to develop stands with several age classes and to let yields of timber fluctuate in varying degree. Sometimes the quantitative methods are used to mold stands into arrangements of diameter classes that some believe to be characteristic of old-growth stands.

In order to simplify the initial presentation of these ideas, this chapter deals with comparatively simple, pure uneven-aged stands; discussion of more complex stands that are mixtures of species or are both uneven-aged and mixed is postponed until a later chapter.

The term **selection system** applies to silvicultural programs that are used *to maintain* uneven-aged stands; **the selection method** is employed to *regenerate* such stands. An uneven-aged stand contains at least three well-defined age classes; ''well-defined'' means differing in total height and age, not just in stem diameter. It is not necessary that it be a self-contained sustained-yield unit or even an approximation of it.

As far as the discussions in this book are concerned, the even-aged aggregations of which uneven-aged or multicohort stands are composed are small patches. This qualification is applied because there is not, in the strictest sense of the term, any such thing as an uneven-aged stand. Even when a single large tree dies, it is replaced not by one new tree but by many that appear nearly simultaneously. This is true even if the new trees are from advance regeneration.

The uneven-aged stand is an artificial entity invoked to help understand what might otherwise be a chaos of little "stands." The question of how large an even-aged aggregation must be to represent an individual stand depends entirely on the particular context in which the forest is viewed at the moment. Someone concerned about silviculture or logging, for example, may see many little even-aged stands. In contrast, someone concerned about forest administration, wildlife, or watershed management may see these same stands as a homogeneous entity characterized by much internal variation. Whereas the first party would map the little units separately, the second would regard that as a useless complication.

An essentially ecological definition would make the minimum size of a stand equal to that of the largest opening that was fully under the microclimatic influence of adjacent mature trees. An opening of this critical size would, at the very center, have the same temperature regime as that which prevailed over a large clearcut area. Such an opening would be about twice as wide as the height of the mature trees. The effects of shade and root competition with adjacent older trees would be significant.

Some observers believe that application of the selection systems requires that each stand be made into a self-contained sustained-yield unit. This condition is one that can be approached but is almost never attained in practice; even approximations of the condition are difficult to maintain. In fact, single-minded efforts to mold stands into sustained-yield units often produce results that are illogical in the light of other considerations. Nevertheless, the essentially mathematical manipulations involved in these efforts are introduced in this chapter because they have a powerful appeal to many and also because, as also described in Chapter 17, they provide one means of monitoring programs for achieving sustained yield in whole forests.

In the selection method of reproduction, the mature timber is removed either as single scattered trees or, more often, in small groups at relatively short intervals to open growing space for regeneration. Such cuttings are repeated indefinitely; the cuttings may be at regular intervals or carried out whenever the condition of the stand or other considerations dictate. The whole flowing process depends not only on periodically establishing reproduction but also on making it free to grow so that the continuing recruitment of new age classes is achieved. The method is usually associated with natural reproduction but can also be used with planting or artificial seeding. True reproduction cutting is typically accomplished by cutting the oldest or largest trees, *and, if necessary, enough smaller trees to ensure that enough trees of needed age classes are free to grow.* The seed and any protection necessary for natural reproduction come from the trees that remain around the openings.

The regeneration can arise from any source—that is, from new seedlings, advance regeneration, sprouts, planted stock, or combinations of all of these. In a certain sense, all of the methods of regeneration previously described can be applied in small units of area. In some patches, all the trees can be eliminated and replaced by planted trees or new seedlings as with clearcutting. The equivalent of shelterwood cutting can be used to establish advance regeneration. Reliance on sprouts emulates the coppice method. There can also be mixtures of species as described in Chapter 16.

Intermediate cuttings may be made among the younger trees at the same time that the older or larger trees are removed. Each immature even-aged aggregation is treated essentially as if it were a stand by itself. Within the whole uneven-aged stand, the periods of regeneration and intermediate cutting may thus extend through the entire rotation and be indistinguishable from one another.

The Place of Uneven-aged Stands

There are two common reasons for developing such stands. The first is simply that the stands were inherited in that condition and cannot be replaced with even-aged or double-cohort stands without prematurely cutting too many young trees.

The second is the existence of management objectives requiring that a stand always have some large trees. Not the least important are aesthetic considerations, especially in roadside strips and parts of forests that are in public view. Any approach, even a crude one, to achieving sustained yield within small holdings requires that trees arrive at mature size at short intervals even if they are sporadic. The juxtaposition of old trees with younger ones is frequently important in wildlife management or for other purposes that require diversity of habitats and of the species of plants and animals therein. If natural regeneration is difficult, as on adverse sites, there may be reason to maintain a permanent source of seed. Uneven-aged stands may be essential on steep slopes that are either geologically unstable or subject to avalanches.

The selection method is often said to favor regeneration only of shade-tolerant species or to be unusable for intolerant ones. This really is not true. As has been pointed out before, tree seedlings are small, and their survival is governed by characteristics of microenvironments with dimensions measured in centimeters. The shelterwood methods can be modified to regenerate even the most shade-requiring species in even-aged stands; the selection methods accommodate intolerant ones in large openings. However, the single-tree selection method, which is described next, is associated mainly with shade-tolerant species or stands on sites so dry that crown closure is incomplete.

Single-Tree Selection System

In the classic form of the selection system, each little even-aged component of the uneven-aged stand occupies a space equal to (or somewhat less than) that created by the removal of a single mature individual (Fig. 15.1). Theoretically, single mature trees are harvested at short equal intervals of time, and each group is thinned artificially or naturally so that only one tree is left at the end of the rotation. The development of even-aged groups of trees in the very small, scattered openings thus created is the main characteristic. The only reason for making the cuttings at equal intervals of time would be to make the stand into a perfect sustained-yield unit.

The species most likely to be perpetuated are those that are very tolerant, although the opening left by the removal of a single large tree will, if site conditions are otherwise favorable, allow the establishment of a few seedlings representing early successional stages. However, if the opening is not soon enlarged, the crowns of adjacent older trees will almost certainly widen and overtop all the new regeneration. The difficulty of carrying out light harvests often enough to accomplish this has usually made the single-tree selection system very difficult to conduct.

The removal of single trees in thinnings or the choice of trees to harvest on the basis

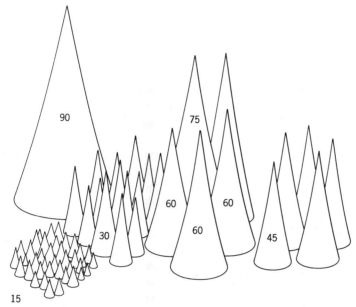

Figure 15.1 Schematic oblique view of a 1/10 hectare segment of a balanced uneven-aged stand being managed by the single-tree selection system on a 90-year rotation with a 15-year cutting cycle. Each tree is represented by a cone extending to the ground; the numbers indicate the ages. Each age group occupies about 1/60 hectare. The 90-year-old tree is now ready to be replaced by numerous seedlings while the numbers of trees in the middle-aged groups are appropriately reduced by thinning.

of individual analysis constitutes use of the single-tree selection method only if it results in the establishment of regeneration that is free to grow.

Group-Selection System

A much more feasible way of managing uneven-aged stands is the group-selection system under which the final age-class units consist of two or more single mature trees (Fig. 15.2). If the regeneration openings are made larger, it becomes possible to accommodate the ecological requirements of almost any tree species. As previously stated, if the matter is viewed from the standpoint of ecological site factors, the maximum width of what are called "groups" could be set at approximately twice the height of the mature trees. With such a classification limit, the same environmental conditions that exist in a huge clearcut area could be made to exist but only near the center of the opening. Regeneration of some very intolerant species might thus be virtually impossible.

As indicated earlier, this semantic matter may also be viewed from the standpoint of someone making administrative maps of forests for such purposes as determining areas of different age classes. Under this interpretation, the smallest areal unit of a given age class that one was ready to delineate on a map would be the smallest recognizable "stand." Anything smaller would be deemed a "group" and part of a larger "stand" with two or more age classes. With aerial photographs and satellite-referenced position locators, it is

Figure 15.2   A stand of ponderosa pine in the Fort Valley Experimental Forest, Arizona, managed under the group-selection system. The saplings form a group that resulted from one unusually favorable regeneration year. The group that occupies the foreground has been thinned twice to favor trees of best bole form such as the ones now remaining. The small cards mark the stumps of medium-sized trees cut 32 years earlier; the large card marks the position of a large member of the group harvested 2 years ago. *(Photograph by U.S. Forest Service.)*

possible to identify "stands" very much smaller than was the case when foresters could map stands only by measuring distances on the ground. In other words, it is now much easier to recognize small even-aged stands, and there is less reason to lump them into larger uneven-aged stands.

The group-selection method has other advantages (Roach, 1974) over single-tree selection cutting. The older trees can be harvested more cheaply and with less damage to the residual stand. The trees develop in clearly defined even-aged aggregations; this characteristic is of substantial advantage in developing good form in many species, especially in hardwoods.

This modification is also more readily applied to stands that have become uneven-aged through natural processes because such stands are more likely to contain even-aged groups of mature trees than a mixture of age classes by single trees. Furthermore, the openings created are large enough that the progress of regeneration and the growth of younger age classes are more readily apparent.

It is important to consider the significance of the effects involved in having trees growing on the edges of even-aged aggregations. The total perimeter of such edges is greater the smaller the size of the groups. The effects are both beneficial and harmful but vary according to the circumstances. Influences on reproduction are beneficial to the extent that side shade protects young seedlings. However, the competition from the older trees for soil moisture can be serious. The roots of the older trees are capable of spreading out into the newly created openings. They may even do so automatically if they are already attached to the roots of the cut trees through intraspecific root grafts.

On the other hand, this lateral expansion of the roots and crowns of older trees along the edges of groups may make it possible to allocate more area to the growth of large trees than is the case with forests of even-aged stands (Fig. 15.3). This enables slightly more of the total production of the stand to be shifted to the larger and more valuable trees, thus at least theoretically increasing the yield of the stand in terms of merchantable volume added to such trees. The smaller the groups, the greater would be the advantage. However, because of the uncertainties involved in studies of forest yield, this advantage is not known to have been verified experimentally.

The juxtaposition of different age classes provides an opportunity to release the young trees along each edge once during each rotation when the older adjoining groups are harvested. This takes the place of one heavy thinning and is of some advantage when ordinary thinning must be long delayed or is not feasible. The effects of this phenomenon on the form of the edge trees vary considerably depending on the species and circumstances. The development of large branches is forestalled as long as there are larger trees at the side, but the effect is reversed when the larger trees are removed. Distinctly phototropic species, such as most of the hardwoods, do not fare well because the stems tend to bend outward toward the light or to become crooked if the terminal shoots are battered

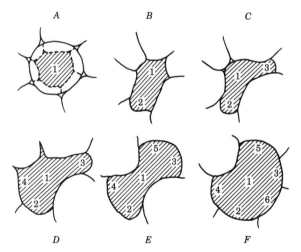

Figure 15.3 The initial contraction and subsequent expansion of area occupied by the trees of one small age-group in an uneven-aged stand. *A* shows the circular outline of the area previously covered by a single mature tree that has just been removed; the smaller cross-hatched area 1 is that to which the new age-group of seedlings is confined by expansion of the older adjacent trees. Sketches *B* to *F* show how this group of trees would re-expand as periodic, sequential cutting of areas 2 to 6 successively removed adjacent trees during the remainder of the rotation.

by the crowns of taller trees. This is not so true of distinctly geotropic species like conifers, which tend to grow straight under all circumstances. However, in climates subject to wet, clinging snows, any trees growing beside taller conifers are likely to be damaged badly when accumulations of snow suddenly slide off the higher crowns. Some wildlife species profit from the combination of environmental conditions existing along the boundaries between very young groups and older trees, but it is harmful to others. If the advantages outweigh the disadvantages, an effort may be made to increase the length of group perimeter per acre by creating smaller groups.

The simplest kinds of group-selection cutting create definite gaps in the forest canopy. Some ecologists regard ''gap-phase'' regeneration as the common way that forests are renewed without departing from the ''climax'' or late-successional kinds of species composition. The gaps are construed as being caused by wind or maladies associated with old age. If the gaps are large enough, they will, for the reasons described in Chapter 7, provide microenvironmental conditions suitable for the regeneration of species from almost the whole spectrum of local vegetation. This may help explain why climax vegetation is regarded as having much higher species diversity than that of earlier successional stages. If so, this is because the environment of the gap is not uniform but covers a wide range of microclimates. Neither the forester nor the ecologist should regard the entire area of small openings as conducive to only one category of vegetation.

Another ecological attribute of small openings that must be anticipated is their tendency to become pockets of hot air by day and of frost and dew by night. This is because they are not well ventilated by the wind but are open to the sky and, if large enough, to the sun. At night plant surfaces cooled by radiational heat loss may not be adequately reheated by contact with warm air wafted down from above; dew or frost may form on the plants. This may cause risk of damage by frost or increase the risk of infection of leaves by fungus spores, such as those of the stem rusts of conifers. The frost effect has been suggested, along with deer browsing, as a means of controlling unwanted true firs in the regeneration of pine forests in the Sierra Nevada (Gordon, 1970).

The effects of hot-air accumulation in sunlit openings on plants are more obscure; humans find them very pleasant in winter and oppressive in summer. Studies of German spruce forests have shown that the risk of frost is most extreme in openings that are one and a half times the height of the surrounding trees (Geiger, Aron, and Todhunter, 1995). Wider openings are better ventilated and smaller ones more shaded, especially at high latitude. If these effects are undesirable, one can wholly or partly substitute the effects of more diffuse cover of the kinds associated with, but not limited to, shelterwood cutting. Partial cover combines some shade with more opportunity for ventilation by the wind, which may be thought of as entering the forest from above in the degree that canopy openings allow.

Not all the mature trees of a group need be cut at a single time. If desirable because of ecological requirements for reproduction, the need to leave some trees for additional growth, or some other consideration, the groups can be reproduced by small-scale applications of the shelterwood and seed-tree methods. There is no necessity that the individual groups in one stand be uniform in size, shape, or arrangement, nor that one age class be represented by a single group. In fact, if proper advantage is taken of the existence of groups of different ages when an irregularly uneven-aged stand is placed under management, such regularity is unlikely. A systematic arrangement of individual groups would be convenient but is usually almost impossible to secure.

Strip-Selection System

The components of an uneven-aged stand can be created in slowly advancing strips. This arrangement enables the transportation of logs through the next strip to be harvested. There is opportunity to obtain advance reproduction in the side-light adjacent to the most recently cut strip. Similarly, if the progression of cutting is directed toward the equator, regeneration of species that require partial shade may be encouraged. If successive cuttings advance against the direction of the most dangerous winds, the stand gradually becomes streamlined so that the main force of strong winds is diverted up and over the stand (Fig. 19.2). Strip and patch selection cutting have sometimes been effective in creating snowdrifts and thus improving the yield of water from snowmelt (see Chapter 18).

Quantitative Methods of Managing Uneven-aged Stands

This section starts with a discussion of the concept of the perfect sustained-yield stand as a series of little even-aged stands representing all age classes. It then considers another theory, driven by a simple mathematical assumption about numbers of trees in diameter classes, that proposes somewhat different structures that are claimed either to be more efficient in timber production or to simulate some perpetual equilibrium alleged to characterize old-growth stands. It concludes with accounts of more flexible quantitative methods of regulation that are not based on simplistic rigid assumptions about desirable stand structures.

The Concept of the Balanced All-Aged Stand

One of the most attractive and idyllic concepts of forestry is that of the theoretical, balanced, all-aged stand (Davis and Johnson, 1987). It continually yields benefits and regenerates itself steadily; it is a dynamic system that is always the same and always in equilibrium. The adjective "balanced" means that every age class up to that of rotation age is represented by an equal area so that the stand is a perfect sustained-yield unit. If these conditions are met, *and only if they are met,* one could annually cut whatever amount of wood the stand produced in a year and count on doing so indefinitely if the cutting was done every year and the schedule was not disrupted by unplanned destructive disturbances.

Even though this condition has a powerful naturalistic appeal, it does not come into existence in nature but would have to be an essentially artificial creation. Some foresters and ecologists take it as axiomatic that virgin or old-growth forests are in a state of self-maintaining equilibrium. Even if such stands have several cohorts and biomass production is perfectly balanced by decay and mortality, this does not mean that stand even approximates a balanced all-aged stand. One would have to regenerate the same amount of forest area each year for about a whole rotation to create the condition exactly. It is a simple encapsulation of good ideas toward which foresters strive by various means but should not expect to achieve perfectly, even in whole forests.

The theoretical all-aged stand would have trees of every age class from one-year seedlings to veterans of rotation age, but there is no clearly good way of specifying how many there should be in each age class. The idea that each age class should have foliage covering an equal area has always provided the soundest theoretical basis for sustained-yield management of forests. Unfortunately, it is not feasible to regulate a balanced uneven-aged stand directly on this basis because the necessary measurements of areas of small aggregations of trees and determinations of their ages would be too intricate. Therefore, it is necessary to employ surrogates for foliage area and resort to modifications and approximations of the idea.

Simplifying Approximations

The simplest modification has to do with the impracticality of operating in each stand every year. Instead, it is presumed that the stand would be treated periodically, with each period being called a **cutting cycle**. The uneven-aged stand would have as many age classes as there were cutting cycles during the rotation. If the result were a stand with an arrangement of age classes that approximated the all-aged stand, it would be regarded as a balanced uneven-aged stand. Figure 15.4 is a schematic map of such a stand, and Fig. 15.5 shows how such stands might be combined into a whole forest managed on the basis of some perfect selection system.

In order to avoid direct determinations of tree ages and the areas occupied by each age class, it is commonly assumed that tree diameter can be taken as the index of age. The distribution of diameters in the stand is used to assess and control the allocation of area of growing space to each age class. Total tree height, which has a much higher correlation with effective age, would be better but is more difficult to measure or even to estimate.

Much confusion and error can be avoided if it is recognized that the best indicator of the rate of future growth of a tree is the amount of foliage that it bears. Just because the

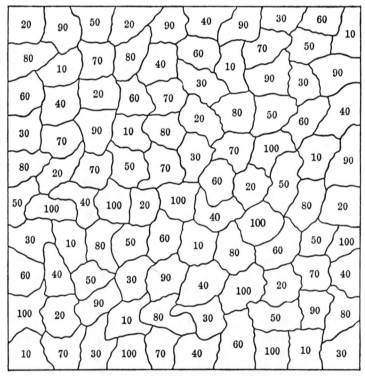

Figure 15.4   A 1-acre portion of a fully balanced selection stand managed on a rotation of 100 years under a 10-year cutting cycle. Ten age classes are represented, each occupying approximately one-tenth of the area. The numbers indicate the ages of the individual groups of trees.

Stand 1 contains age classes: 1, 11, 21, 31, 41, 51, 61, 71, 81, and 91	Stand 2 contains age classes: 2, 12, 22, 32, 42, 52, 62, 72, 82, and 92	Stand 3 contains age classes: 3, 13, 23, 33, 43, 53, 63, 73, 83, and 93	Stand 4 contains age classes: 4, 14, 24, 34, 44, 54, 64, 74, 84, and 94	Stand 5 contains age classes: 5, 15, 25, 35, 45, 55, 65, 75, 85, and 95
Stand 6 contains age classes: 6, 16, 26, 36, 46, 56, 66, 76, 86, and 96	Stand 7 contains age classes: 7, 17, 27, 37, 47, 57, 67, 77, 87, and 97	Stand 8 contains age classes: 8, 18, 28, 38, 48, 58, 68, 78, 88, and 98	Stand 9 contains age classes: 9, 19, 29, 39, 49, 59, 69, 79, 89, and 99	Stand 10 contains age classes: 10, 20, 30, 40, 50, 60, 70, 80, 90, and 100

Figure 15.5 Diagram of a selection forest managed on a rotation of 100 years with a cutting cycle of 10 years. The forest contains ten stands, one of which is cut through each year, thus giving equal annual cuts. Each stand contains ten age classes, and together the age classes in the ten stands form a continuous series of ages from 1 to 100 years.

diameter of a tree is used as a surrogate for its total leaf area does not mean that the diameter of the tree controls the rate of future growth; it is a result of growth and not a cause of it. Neither can it be assumed that every 9-inch shortleaf pine will increase in diameter at the same rate as others of the same diameter in the same stand on the same soil, especially if the positions in the crown canopy are different.

Regulation of Cut by Diameter Distribution

A series of little stands, each of the same area, conforming to a yield table for pure stands of the same species, is the fundamental basis for defining the mathematics of development of a balanced uneven-aged stand. It is because the yield tables represent observational evidence that predict how pure even-aged aggregations of trees, if they are free to grow, actually develop over time. They show the decline in numbers from competition or thinning as well as the increase in the average diameter of each age class with the advancing years. The second column of Table 15.1 shows an appropriate diameter distribution for a pure, unthinned, balanced, uneven-aged stand of loblolly pine grown on a 40–year rotation and based on a yield table of Meyer (1942) that presents diameter distributions for even-aged stands. It is very important to note that trees of a given diameter class will not grow at the predicted rates if their crowns are shaded from above.

The characteristics of the proper diameter distributions are depicted in Fig. 15.6. Because it requires many saplings to cover the space eventually occupied by a single mature tree, the distribution should approximate the smooth, "reverse-J-shaped" curve of Fig. 15.6*A*. This curve represents the collective total of the diameter distributions of a series of little even-aged, pure groups of trees covering equal areas and separated by equal intervals of age, as shown in Fig. 15.6*B*. The objective is to harvest and replace trees in such patterns that this diameter distribution is the same either just before or just after each harvesting operation.

Assuming that the appropriate distribution of age classes has been converted to a

Table 15.1 Diameter distributions exemplifying the results of different criteria[a] used to determine these distributions for various kinds of pure uneven-aged stands

D.B.H.	Yield table	BA 128 $q = 1.7$	BA 80 $q = 1.3$	BA 80 $q = 1.7$	BA 60 $q = 1.7$
(in.)	(1)	(2)	(2)	(2)	(2)
6	138	90	35	76	57
8	84	53	27	44	33
10	44	31	21	26	20
12	22	18	16	15	12
14	10	11	12	9	7
16	7	6	10	5	4
18	2	4	7	3	2
20	1	2			
Total	308	215	128	178	135

(1) Based on a series of even-aged stands in a yield table for old-field stands of loblolly pine in northern Louisiana managed on a 40-year rotation; site index of 90 feet, and total basal area of 128 ft^2; based on Meyer, 1942.

(2) Derived entirely from chosen values of q, basal areas per acre, and maximum D.B.H.
[a]See discussion in text for explanation of these criteria.

diameter distribution and created in a stand, it is necessary to consider the small but crucial changes that will be made in the distribution at the time of each cutting (Fig. 15.7). Selection cuttings in such stands involve successive harvests of the largest trees in a stand. A diameter limit (point x in Fig. 15.7) is established as an indication of age, with the understanding that trees below this size are in general to be reserved and those above cut. This diameter limit should be regarded as a flexible guide rather than as a rigid dividing line. Depending on their silvicultural condition, especially capacity for further increase in value, a few trees above the limit should be left and some below the limit should be removed. Trees that are surplus in a given diameter class are usually harvested, except where they are needed to compensate for deficiencies in the next higher or lower classes. Commonly, there are surpluses in many diameter classes larger than the smallest that are measured; this is clear evidence of deficiencies in seedlings or other small trees.

True thinning may also be conducted simultaneously with the harvest cutting. Such thinning must be distinguished from the harvest of the oldest trees and should take place in clumps of trees that are too young for final harvest cutting. The thinnings should be guided by the estimates of required numbers of trees shown in a graph such as that of Fig. 15.7 with surplus trees removed if they are not needed to remedy deficiencies in other diameter classes. The methods of thinning followed are the same as those employed in even-aged stands. Unlike the final harvest cuttings, the thinnings should not remove the largest trees in the various clumps, unless they are of undesirable form or species. The primary objective of the thinnings should be to anticipate the inevitable reduction of numbers denoted by the steep slope of the J-shaped curve (Fig. 15.7), so as to salvage prospective mortality and allow the remaining trees space for more rapid growth.

The whole program of thinning and stand-density regulation is the key to the crucial question of how rapidly the growing and aging trees flow through the diameter distribution. The process also depends on the continual regeneration of stands.

It has been said that a forester cannot cut more than is grown or grow more than is cut. If, and only if, the stand is of one species, has the perfect balanced diameter distri-

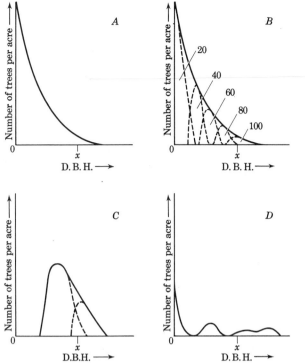

Figure 15.6 Several types of diameter distributions found in uneven-aged stands. *A* shows the distribution curve of an all-aged stand of 1 acre containing sufficient trees in each diameter class to produce an unvarying number of trees of optimum size (of diameter *x* or larger) at rotation age. *B* indicates how a balanced uneven-aged stand of the same diameter distribution may be composed of five age classes, each occupying an equal area, with a 100-year rotation. *C* represents a stand with two closely spaced age classes and no advance reproduction. Uncritically supervised selection cuttings in stands of this kind will produce abnormally high yields of timber until all existing trees reach optimum size; thereafter they will yield nothing until the new growth reaches merchantable size. It is difficult to create the balanced distribution in such stands in one rotation period, especially if the smaller trees do not respond to release after selection cuttings. *D* shows one of the many kinds of irregular uneven-aged distributions that may be found in virgin stands. This one contains four well-distributed age classes, one of which is well beyond optimum age for an economic rotation. Such a stand could be gradually converted to the balanced form provided that the intermediate age classes did not deteriorate after partial cutting.

bution, and is cut on the precise schedule just described, the allowable annual cut under sustained yield is equal to the *periodic* annual increment in the same unit of measure. It seems logical that one should safely be able ''to harvest the annual growth every year,'' but this is true for a stand or forest only if the stated conditions of perfection have been created and are rigidly maintained. It is prudent to regard the *mean* annual increment at rotation age for an otherwise similar even-aged stand as the best estimate of the allowable

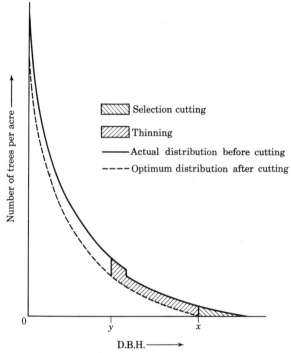

Figure 15.7 Diameter distribution of a balanced uneven-aged stand under intensive management, indicating the number of trees of different diameter classes theoretically removed in a single cutting. All trees larger than diameter *x*, which has been set as the index of rotation age, are removed. The trees of smaller diameter may also be reduced in number by thinning, provided that they are larger than the tree of lowest diameter *(y)* which can be profitably utilized. Under extensive practice there would have been no cutting of trees less than diameter *x* and most of those represented above as being cut in thinnings would have been lost through natural suppression.

annual cut in a balanced uneven-aged stand. It must also be kept in mind that the appropriate diameter distribution is not the cause of the desired regime of growth and harvest but an indicator and result of the process.

Negative Exponential Distributions of Diameters as Guides

A different hypothesis about the proper diameter distribution uses a mathematical distribution that has a general resemblance to reverse-J-shaped curves but was not based on ideas about having all age classes equally represented. Late in the nineteenth century, deLiocourt (1898) observed that some very old forests in Europe often had diameter distributions in which the number of trees in each diameter class was some multiple of that of the next larger diameter class. Meyer (1952), in some very old natural stands (usually of mixed species) in North America, concluded that this situation prevailed and that the multiple, termed the *q*-**factor**, might vary from 1.2 to 2.0. In a stand with a factor of 2.0, each diameter class would have twice as many trees as the next larger class. They assumed *(but never verified)* that any stand with any value of *q* had arrived at a stable equilibrium and would remain the same perpetually if the effects of periodic harvests or mortality kept returning the diameter distribution to that defined by the same q-factor. This assumption of a stable dynamic equilibrium was taken as justification for assuming that the periodic annual increment was the amount that could be harvested under sustained yield forever or at least into some indefinite future. They also assumed that appropriate numbers of new seedlings would materialize at the right times.

Diameter distributions based on q-factors fit negative exponential curves; these convert to straight lines if numbers of trees are converted to logarithms and they are plotted over arithmetic tree diameters; the appropriate value of q defines the slope of the straight line. The allure of this neat mathematical relationship has led many foresters to conclude that q-factors define mathematical ''laws'' governing the growth of trees and stands, although no biological basis has been advanced for the idea. The relationship does at least define some kinds of reverse-J-shaped curves.

The q-factors that are chosen depend on the species and the assumptions that are made about rates of growth and mortality as well as stand basal area and the sizes to which trees are to be grown (Fiedler, 1995). Some choices of the slopes and intercepts of the lines are shown in Fig. 15.8. The slope of each line is the q-factor. In customary usage, if the q-factor is 1.4, the number of trees in a given 2–inch D.B.H. class is 1.4 times that of the next larger 2–inch class.

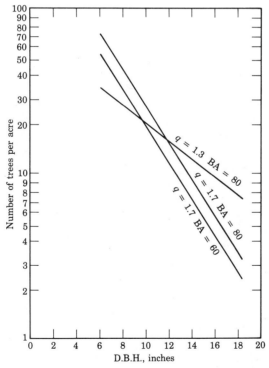

Figure 15.8 Diameter-distribution curves converted to straight lines by plotting logarithms of numbers of trees over arithmetic values of D.B.H., showing the effects of some changes in values of the q-factor and stand basal area (sq. ft. per acre) that are also shown in Table 15.2. The values of q are for 2-inch classes and the stands defined would have no trees larger than the 18-inch class.

Table 15.2   Table from Tubbs and Oberg (1978) showing values of a coefficient, *K*, for determining the number of trees to be grown in uneven-aged stands with diameter distributions with different amounts of basal area per acre and various *q*-factors

Maximum D.B.H.	*q*-Factor						
(in.)	1.1	1.2	1.3	1.4	1.5	1.6	1.7
24	18.5	24.8	34.3	48.1	68.0	97.3	139.3
22	14.0	18.2	24.0	32.1	43.4	58.9	80.1
20	10.3	13.0	16.4	21.1	27.2	35.2	45.6
18	7.4	9.0	11.0	13.5	16.7	20.6	25.5

Divide the chosen value of stand basal area by the *K*-value for the desired combination of *q*-factor and maximum D.B.H. to determine the number of trees in the largest diameter class. To determine the number of trees in each 2-inch class, multiply the number of trees in the next class with larger D.B.H. by the *q*-factor, as has been done for Table 15.1.

Table 15.2 by Tubbs and Oberg (1978) can be used to guide the calculations, which are based on the fact that basal area, tree numbers, and diameters are all perfectly related. Table 15.1 shows examples of the effect of some of the choices on diameter distributions with different values of *q*, maximum D.B.H., and basal area. For comparison, as previously mentioned, it also shows a diameter distribution for a balanced uneven-aged stand constructed from diameter distributions for known even-aged stands of loblolly pine.

Possibilities with More Large and Fewer Small Trees

Diameter distributions defined by the *q*-factor do not fit those distributions deduced from yield tables for pure even-aged stands, as shown in Fig. 15.6*B* and Table 15.1. The discrepancy lies in the fact that, if the relationship specifies the logical numbers of trees in the larger sizes, then there seem to be far too few in the smaller diameter classes. For some, this casts doubt on the validity of using this mathematical relationship in managing stands or forests for sustained yield. Sometimes provision is made to increase the number of small trees by having the *q*-factor high in the small diameter classes and lower in the middle and upper ranges.

For advocates of *q*-factors, the small numbers of small trees are regarded as evidence of some advantages deemed to be inherent in the uneven-aged arrangements defined by *q*-factors. As a consequence, efforts to deduce appropriate diameter distributions for sustained yield have been diverted into schemes for allocating large amounts of growing space to trees in the middle and large sizes that increase in *merchantable* volume rapidly (Cochran, 1992; Adams and Ek, 1974). Allocation of space to seedlings, saplings, and other submerchantable trees is reduced either deliberately or unwittingly, with or without use of *q*-factors. Some of the advocates of such action contend that sustained yield is not an objective or that somehow the use of any sort of *q*-factor magically takes care of it.

The surest way to maximize *short-term* periodic annual increment of board-foot volume would be by having whole forests with nothing but *even-aged* stands in the diameter classes around 13 inches where such volume approximately doubles with each 2-inch increase in diameter. Unfortunately, if one harvested trees continually from such a forest, the recruitment of new 13-inch trees would soon collapse while the large trees were whittled away. Trees have to be small before they can be large.

There are ways, consistent with sustained yield, for shifting production onto large stems by diminishing the proportion of small ones. Low and crown thinning are the most important ways.

If the regeneration units of a selection stand are small, the area that was previously occupied by one or two mature trees is not recolonized fully by seedlings. Some is taken over by the horizontal expansion of the crowns and roots of adjacent trees. When they are cut in their turn, the regeneration group can, in its turn, take over part of the newly vacated growing space. In this way, as shown in Fig. 15.3, the amount of area used by a given age-class group increases each time an adjacent older group is removed. This effect must have some reality in stands with small age-class groups and, therefore, much area of boundary zones between groups. The effect is negligible if the groups are large.

Another potential source of gain of efficiency lies in the "advance regeneration effect" described in connection with shelterwood cutting in Chapter 14. If small trees can be grown under large ones, it is not necessary to allocate the space to them that would be required if they were grown in the open. As will be described in Chapter 16, this same general effect can be obtained by releasing species of the lower strata of stratified mixtures, including those that are of single cohorts.

If this effect were perfectly achieved, one could simply allocate all of the "space" in the regulated diameter distribution to the released trees and none to those too small to release. However, trees of sufficient size to be released would have to be available and they would not grow to this size anywhere near as fast as they would in an open-grown even-aged stand. This source of efficiency, to the extent that it exists, is equally attainable with shelterwood cutting.

In stands on very dry sites, there can be another effect that diminishes the required numbers of small trees. In such cases, as in some ponderosa pine stands in the interior of the western United States (Fig. 15.2) (Baumgartner and Lotan, 1988), groups of trees can have root systems that fully occupy the soil but crowns that do not come close to touching because the possible amount of foliage is so restricted. This effect is somewhat the same as that involving tree crowns and expanding groups of trees depicted in Fig. 15.3. In such situations, it may not take many small trees to restock an area, and often the effect of root competition seems severe enough to restrict branch size and stem taper without any obvious crown competition (Pearson, 1950). It is also possible that small, slow-growing, but viable trees can be stored for long periods in small openings if they are kept alive by being root-grafted to their larger neighbors.

Other Approaches to Regulation of Stands for Sustained Yield

The various diameter distributions just discussed are best viewed as ways of relating existing stand structures to those presumed to be desirable. They are useful chiefly in determining which classes are deficient and conversely which are over-represented. They do not dictate whether to work toward the balanced structure or, if so, how fast and which trees to harvest in the process.

Techniques of predicting the development of various alternative kinds of uneven-aged stand structure by computer simulation are being developed. The soundest of these calculations are based on relationships between amounts of foliage or crown sizes on the accretion of wood (O'Hara, 1995). Although the results of such work may be translated into diameter distributions to guide partial cuttings, they are based on some fundamental ecophysiological factors and not on the idea that the diameter distributions actually control anything.

A less rigorous and reasonably successful solution to the problem is to be content with maintaining uneven-aged stands that fluctuate rather widely around the diameter distributions that are tentatively deemed to be appropriate. This process is apparently followed in long-continued management of selection stands in Europe. During the process, changes in distribution and stand volume are monitored, and all reasonable efforts are made to readjust them at harvest times. If the larger size classes seem to be over-represented and the periodic annual increment is harvested, it is recognized that the rate of cutting is faster than can be permanently sustained (Burschel and Huss, 1987; Davis and Johnson, 1987). If the periodic annual increment decreases, the diameter distribution can be examined to determine why and also to deduce what should be done to adjust the situation. If the recruitment of medium-sized trees begins to fail and the growth of the stand declines seriously, steps are taken to start over again with an irregular cohort of trees.

With any approach to uneven-aged management, it is crucial for sustained yield to keep cutting openings in the stands for the recruitment of new regeneration and to set goals in terms of the areas opened rather than the numbers and sizes of trees cut for the purpose. If the rotation is 100 years and the cutting cycle is 10 years, then the regenerated area should at least approximate a tenth of the total.

Pitfalls

Regimes of harvests in uneven-aged stands that allocate inadequate space to small trees can sometimes continue for many decades, with rates of cutting greater than can be sustained in the very long run. Forest managers who fall into this trap have sometimes done so by assuming that any diameter distribution that is reverse-J-shaped or approximates some *q*-factor in the merchantable size classes is *ipso facto* a balanced uneven-aged stand. The next false premise is that the conditions have been met under which the allowable annual cut under sustained yield is equal to the periodic annual increment; this harvest rate would be unsustainably high if too much space were allocated to the larger size classes.

The problem is aggravated in situations where whole forests consist of just one age class because all stands arose after a single devastating event. If one counts only trees large enough for sawlogs, a middle-aged stand can sometimes seem to have a reverse-J-shaped diameter distribution curve. One could then leap to the conclusion that the whole forest is and should be composed of all-aged stands in which the allowable annual cut is equal to the current annual increment. This expression of stand growth rate is unsustainably high in stands that are not yet at the rotation of maximum mean annual increment. If this situation is allowed to continue too long and new age classes are not recruited, some future generation will run out of trees to harvest, and it may take at least one human generation of diminished timber harvests to correct the situation.

Creation of Balanced Uneven-aged Stands

Theoretically, an absolutely even-aged stand can be replaced by a balanced uneven-aged stand during the period of one rotation. If the new stand is to have five age classes with a rotation of 100 years, all that is necessary is to conduct cuttings leading to the establishment of reproduction on one-fifth of the area at 20-year intervals. Thus some parts of the stand must be cut before or after the age of economic maturity. The disadvantages of such awkward timing of the harvests can sometimes be mitigated if any parts of the stand that mature early are replaced first and those that continue to increase in value longest are cut last. The conversion will be successful only to the extent that the cuttings are appropriately timed and heavy enough to lead to the establishment of the age classes and species of reproduction desired. It is usually a fallacy to pretend that the large trees of essentially even-aged stands are older than the small ones and to attempt to make the conversion by successive cuttings of dominants.

The creation of balanced uneven-aged stands can be carried out more rapidly and easily in stands that already contain several age classes. Ordinarily these will be in irregular distribution, having resulted from a history of high-grading or patchy natural disturbances. It will usually be necessary to make the sacrifice represented by cutting some trees earlier or later than might otherwise be desirable. Sometimes small, relatively old trees of the lower crown classes can be substituted for nonexistent dominants of the same size and younger age, thus inventing ''age'' classes that might otherwise take decades to grow from seedlings. This shortcut is useful to the extent that the trees involved can withstand exposure and return to full vigor. However, it would be a mistake to indulge in this procedure to the exclusion of the creation of the new age classes of seedlings that must be continually recruited and made free to grow if the balanced distribution of age classes is to be completely developed.

The age-class distribution of a stand is likely to be readjusted with least sacrifice of other objectives if the changes are made very gradually. Several rotations might well be required to transform essentially even-aged stands to the balanced uneven-aged form.

The degree of precision of regulation of the cut necessary for maintaining or creating any real approximation of a balanced, uneven-aged stand cannot be achieved without making an inventory of the stand before marking it for cutting. Usually, this involves determining the distribution of diameter classes by basal area or numbers of trees per acre. The prospective cut is then allocated among the various diameter classes on the basis of a comparison between the actual distribution and that which is assumed to represent the balanced condition. Leak and Gottsacker (1985) described an expeditous way of guiding the marking by using prism point-sample plots to assess stand conditions and having only three broad classes of tree diameter.

Irregular Uneven-aged Stands

There are many valid reasons other than sustained yield for deliberate maintenance of uneven-aged stands. These other objectives can be achieved without the balanced arrangement.

Long-continued application of the concept of financial maturity in efforts to extract optimum value from the growing stock is very likely to cause the development of uneven-aged stands. If this concept is the best key to intelligent management of an uneven-aged stand, it is as logical to use the selection system as it would be to employ the shelterwood system under similar circumstances in an even-aged stand. However, the sequences of cutting necessary for true sustained yield and those guided by the concept of financial maturity are usually compatible only in uneven-aged stands that are already balanced and composed of good trees. If the diameter distribution is unbalanced, it can be brought into balance only by cutting some trees long before or long after they are financially mature.

Most reasons that exist for having uneven-aged stands and using the selection system are not concerned with manipulating the growing stock of stands in keeping with the concepts of either sustained yield or financial maturity. Rather, they may have to do with nothing more than the preexistence of the uneven-aged condition. The purposes may be wholly or partly matters of beauty, wildlife habitat, seedling ecology, or protection of stand, soil, or site against damage. Such objectives may require uneven-aged stands but not balanced ones. In fact, the warping of patterns of harvest removals to fit some preordained diameter distribution commonly interferes with fulfilling many other logical purposes.

If an uneven-aged stand is to be allowed to remain irregular and not be refitted to some diameter distribution, it is not necessary to take the costly step of determining the diameter distribution before the stand is marked for partial cutting. It is feasible to maintain an unbalanced uneven-aged stand indefinitely and reap any benefits, other than sustained yield, that may be induced by the uneven-aged arrangement.

The distribution of age and diameter classes can be allowed to fluctuate almost at random, except to the extent that the cutting or reservation of particular classes must be adjusted to balance the books of sustained yield for the whole forest. In this kind of management, the allowable cut and its approximate distribution among diameter classes are determined for the entire forest, and then harvests are made in various stands on the basis of other considerations, silvicultural and otherwise, until the scheduled volume and kinds of trees are removed.

Theoretically, the true reproduction cuttings of the selection system involve removing the oldest and largest trees, with some predetermined diameter limit being taken as a definition of the smallest trees old enough for such cutting. However, this should be only a first approximation of the characteristics of trees selected for harvest. Trees above the limit may be left because they are increasing in value with unusual rapidity or for reasons such as their capacity to provide seed and protection for reproduction. The trees designated for cutting, even though they are below the limit, are those that increase in value too slowly because of low quality or poor growth; those that interfere with logging or the growth of better trees; or those that are likely to be lost after the cutting. In other words, many of the trees to be removed are chosen based on the same principles that are followed in thinning.

The guidance of such complicated timber marking is sometimes facilitated by tree classifications. This is especially true if the various age groups are small, because the standard Kraft crown classification (Fig. 2.5) is most useful in rather large, pure, single-canopied, even-aged aggregations. In other kinds of stands, the Kraft classification must be supplemented by other information, although the matter of position in the crown canopy is always important. The other characteristics important in deciding whether trees should be cut or left are (1) age or size, (2) quality, and (3) vigor. Tree vigor is usually assessed by observations of the foliage, which is the most crucial productive machinery of the tree. The condition of the foliage can be categorized in terms of its amount, color, density, and leaf size. If attention is given to the quality of the foliage, the live crown ratio is a very useful index of tree vigor.

Evolution of Selection Methods

The first timber harvests in most forests typically involve crude, unconscious applications of the selection method in which only the biggest and best trees are removed. At the other end of the spectrum are some of the most intensive and intricate kinds of silviculture. Although one may gradually develop from the other, the early crude selection cuttings are most commonly superseded by some other method of cutting.

The extensive kinds of selection cutting are often diameter-limit cuttings. The diameter limits can be set with varying degrees of sophistication. In mixed stands, for example, it may be beneficial to set a low limit for species of little future promise but higher limits for the more valuable species. Such extensive applications may be practiced indefinitely or may be used as temporary expedients until conditions become more favorable. The cutting cycles are usually long, and there is seldom anything akin to intermediate cutting; there is little control over stand density or species composition.

In pure stands, an extensive application of the group-selection method may prove reasonably successful when the desirable species represent a physiographic or climatic climax.

Economic Selection Methods and "Selective Cutting"

In situations where landowners have no interest beyond liquidation of existing merchantable timber, foresters have used one important principle of extensive application of the selection method as a way of encouraging retention of growing stock. There is always a limiting diameter below which trees cannot be profitably utilized for a given purpose, because the handling costs exceed the sale value of the small material produced. The tree that can be harvested with neither profit nor loss is referred to as the **marginal tree** of the stand or forest in which it occurs; a cutting that removes all trees that can be utilized without financial loss is called a **zero-margin selection cutting**. If a landowner can be induced to recognize the existence of the marginal tree, it becomes clearly desirable to leave smaller trees in logging. If the presently unmerchantable trees are sufficiently large and numerous to provide the basis for another harvest in the near future, there is an additional incentive to protect the smaller trees and to shift away from a policy of short-term liquidation.

This approach is effective in getting the practice of forestry established, but is not an appropriate basis for long-term practice. In fact, as economic factors become more favorable, the size of the marginal tree often drops to the point where virtually every tree in a stand can be utilized at a profit. Under these conditions, zero-margin selection cutting may verge on clearcutting.

The primary drawback of basing the selection method on the concept of the marginal tree is that the potentialities of different trees for future growth are not taken into account. Although a tree may be logged profitably at a given time, it often can be harvested at an even greater profit at some later date. Once a landowner is committed to retaining small trees for growing stock, it becomes desirable to know which trees are best reserved to the next cutting cycle. This can be done by using the concept of financial maturity to detect those trees that will earn an attractive rate of compound interest on their own present value if left to grow. This procedure usually results in setting a guiding diameter limit that is substantially higher than the one that would apply to zero-margin cutting. If the total return to be obtained during another cutting cycle is high enough, the practicability of continuing operations at least to the end of that cycle is demonstrated. This line of financial logic is often sufficient to guide forest practice up from mere mining to a level of extensive practice likely to ensure continuity of management.

If practice can be intensified still further to secure optimum production and sustained yield, the suitability of the selection method should be reexamined. During the period 1930 to 1950, many authorities regarded extensive applications of the selection method, described as "selective cutting," as the panacea of American forest management. The times were peculiarly appropriate for such proposals. The development of trucks and tractors suitable for use in logging big timber had begun to displace railroad logging, thus making partial cutting possible in old growth. The lumber markets of that period were so badly depressed that only the biggest and best trees could be harvested profitably. Most owners were not willing to make significant investment in regeneration. In some cases, foresters were not sure that they knew how to regenerate the stands if they had funds to do so. Partial cuttings provided a means of postponing regeneration until the necessary funds and knowledge were available. Selective cutting played a highly important role in demonstrating that partial cutting was feasible and that residual stands of usable timber could be profitably left for future harvests.

Unfortunately, selective cutting also proved to be a tool that could be used for evil as well as good (Isaac, 1956). Much of it was simply high-grading that resulted in neither maintenance of the uneven-aged form, establishment of desirable reproduction, nor preservation of residual stands. Selective cutting, and indeed much of the idea of uneven-aged stands in general, fell out of fashion during the 1950s when many owners overcame their reluctance to invest in regeneration and various forms of even-aged silviculture were found to be more effective. As usual, the pendulum swung too far. It has not stopped swinging.

Examples of Management of Pure Unven-aged Stands

Restrictive Sites

Pure uneven-aged stands are most commonly found and easiest to maintain on sites that have such pronounced seasonal deficiencies of available water that natural monocultures are favored. These deficiencies may result simply from drought or physiological dryness caused by low soil temperature or the accumulation of poorly aerated water. On such sites, conditions are only sporadically conducive to regeneration, so it helps to retain trees that can produce seeds whenever circumstances allow. In such cases, tree growth is usually too slow to justify the high cost of planting, so one is restricted to natural regeneration. Another reason for uneven-aged stands in these circumstances is the desirability of trying to keep nearly every seedling that becomes established. If they come in sparsely and infrequently, the inevitable result is likely to be an uneven-aged stand.

The evolution of this approach in the dry ponderosa pine forests of the interior of the western United States (Cochran, 1992) illustrates some of the considerations involved in maintaining uneven-aged stands of this kind. Many stands are composed of small even-aged groups of trees that have arisen when trees were killed by bark beetles, lightning, dwarf-mistletoe, or the irregular lethal effects of fires; group-selection management has long been common.

The species has a broad geographical range, and the patterns of seasonal dryness to which it is exposed differ considerably, as do the management problems. The only part of the range in which regeneration is easily secured is a region of summer rainfall that lies east of the Rocky Mountains and includes the Black Hills of South Dakota. Most of the stands in those localities are even-aged, and, because of the relative ease of regeneration, they are typically managed that way.

The widest part of the ponderosa pine range lies between the Rocky Mountains and the mountain chain formed by the Cascades and Sierra Nevada to the west. At the north there is a long, dry summer but with just enough extension of the winter rains into the spring that it is only moderately difficult to obtain regeneration. The selection system is commonly used but has evolved as conditions have allowed the intensity of practice to increase.

Before the 1940s, the old-growth forest was still ruled by *Dendroctonus* bark beetles and the limitations imposed by railroad logging. It was anticipated that this cumbersome mode of log transportation would dictate a long cutting cycle of about 30 years. Only fine old trees more than about 20 inches D.B.H. were worth cutting. The Keen Tree Classification (Fig. 19.4) was devised as a basis for predicting the survival prospects of various categories of trees for the long cutting cycles that were envisioned. It was used as the basis of financial maturity analysis (see Chapter 17), which also guided the choice of trees for harvest or retention. In fact, this approach, termed the **maturity-selection system** (Munger, 1941), was perhaps the most successful selective logging scheme of the era because it closely fit existing developmental processes in truly uneven-aged stands.

With the advent of tractor logging and much more favorable markets, it became possible to shorten the cutting cycles and to produce thinning effects by cutting in a broader range of diameter classes. Problems with the bark beetles subsided as large, old, susceptible trees were eliminated. It has gradually become more customary to prescribe treatments for individual stands on the basis of their prevailing condition rather than following some standard system. At one time, the term **unit area control** (Hallin, 1959) was used to denote this shift in emphasis. The "unit" was any homogeneous part of a large, heterogeneous stand, but it gradually came to be recognized as a small, separate stand in its own right. As problems such as dwarf-mistletoe and the need for site preparation to deal with brush competition got increased attention, even-aged systems, including those involving planting, became more prominent among the alternatives used. Concern about the invasion of tolerant conifers and increases in the amount of forest-fire fuel have also made it desirable to institute prescribed burning beneath the stands.

Farther south, in Arizona and New Mexico, severe drought in spring makes regeneration of the forest a rare event. These areas have summer showers and thunderstorms, but they do not occur in time to induce seedlings to germinate early enough to harden before frost. In 1919 the combination of an abundant seed crop, the existence of much vacant growing space from previous fires or heavy grazing, and unusual spring rains brought abundant pine regeneration to northern Arizona. The 1919 class filled so much growing space that for some decades it was logical to turn to other silvicultural management problems. Since bark beetles were not a major problem, the selection cutting could be aimed at developing trees with good bole form and natural pruning. A program for doing this, called **improvement-selection cutting** (Fig. 15.2), and many other aspects of ponderosa pine silviculture were described by Pearson (1950). The kinds of pines with small crowns and branches that he found capable of responding to release and turning into fine trees were among those that were highly vulnerable to beetle damage farther north.

In most of the ponderosa pine types, the selection system has been used to take advantage of rare regeneration episodes. In a different general case, the selection system is often used to take advantage of easy regeneration coupled with weak competition from undesirable species. Such circumstances exist on specific sites in many regions where there is good rainfall at the season of seedling establishment but such poor moisture supply at other times that only drought-tolerant species can endure. Various species of pines have this adaptation. Under just the right soil-moisture regime, the regeneration phase of silviculture may require little more than drifting with the tide of natural processes. This situation can develop on soils of deep sand, such as glacial outwash or old coastal beach deposits, in humid climates. If the stands are already uneven-aged, there is little reason to convert them to the even-aged condition. The dry soils usually make the logging easy and reduce problems with undesirable vegetation. Any large vacancies created in the growing space by cutting usually fill up promptly and almost automatically with dense regeneration. The main problem is ordinarily that the natural regeneration is far too dense and precommercial thinning may be required.

One famous and influential case involved Scotch pine on some deep sands south of the Baltic Sea. Because of the almost effortless regeneration, a mode of management called the *Dauerwald* or "continuous forest" developed early in the present century. This was a revolt against policies of strict regimentation of the forest, fixed rotation lengths, and plantation silviculture. Most efforts focused on tending individual trees and harvesting them only when they interfered with better trees. This policy tended to accentuate any variations in age-class structure that already existed in the stands.

This general approach is so easy that it is useful to recognize the narrow range of site conditions that make it feasible. They occur in the eastern and southern parts of the United States only on deep, sandy soils, which are often among the poorer soils for tree growth. They are, for example, the easiest places to grow longleaf pine, eastern white pine, and some other pines because hardwood competition is feeble. Such sites are more common in the western interior; examples are the kinds of ponderosa pine sites where regeneration is easy but brush competition is not serious. Another important case is the mountains of northern Mexico, which have some uneven-aged, easily regenerated forests of *Pinus durangensis* and *P. arizonica* in a summer-wet, winter-dry climate (Fig. 15.9).

Whether this dry-site regeneration is difficult or easy, it is often associated with some

Figure 15.9   An uneven-aged stand of Durango pine on a soil in northern Mexico that is moist during part of the summer but very dry the rest of the year.

peculiar phenomena of root competition. The supply of available water is often small enough that the invisible root systems can be fully closed but unable to support a closed canopy of foliage. The stands easily develop persistent gaps that appear to be unstocked but are actually being fully utilized by the roots of adjoining trees. If root-grafting is well-developed, as it often is with pines, partial cuttings may merely donate the use of living roots of the cut trees to the remaining ones. This is good for the remaining trees but can defeat efforts to create real soil vacancies for regeneration.

Deep sands in humid climates can present an additional peculiar phenomenon. The problems with moisture deficiency may be mostly near the surface. Although trees may grow slowly at first, they can grow at accelerating rates for long periods as their roots become increasingly extensive. What may seem to be poor sites initially become much better. If the trees of crowded groups differ enough in height, the leading trees will forge ahead even without the benefit of thinning as their root systems become deeper.

In the poorly aerated peat swamps of the northern forest, regeneration of black spruce and northern white-cedar by natural layering is an important supplement to that from natural seeding and is encouraged by developing uneven-aged stands. Because the layering takes place only when the ends of live branches are overgrown by sphagnum moss, it is essential that tree crowns touch the ground in a sufficient number of places. This condition can be maintained only in uneven-aged stands (Johnston, 1977). The sites are poor enough that the stands often do not close, and the rise of water-table levels that would result from reduced transpiration after clearcutting might harm the trees. The partially cut stands are surprisingly resistant to wind because the peat is so resilient that the force of the wind tends to be dissipated in agitating the soil itself rather than in damage to the trees.

Good Sites

If enough effort is dedicated to controlling unwanted species, it is possible to grow pure uneven-aged stands of many species on any kind of site. One well-documented case history (Reynolds, Baker, and Ku, 1984) involving such stands started in southern Arkansas with loblolly and shortleaf pine. These species characteristically grow together in even-aged stands so their successful culture in uneven-aged stands is proof of the versatility of the method. The basic objective of the procedure is to develop good sawtimber growing stocks as swiftly as possible from the remnants of high-graded even-aged stands. Good new stands could be created by liquidating the remnants and replacing them with even-aged regeneration, but this might cause a temporary suspension of harvests when the old trees were gone. Furthermore, there would be needless sacrifice of many small- or medium-sized trees of good potential.

The guiding principle followed in this kind of cutting is the concept of financial maturity which is applied with careful attention to the quality of increment as well as to volume. Full advantage is taken of the excellent natural pruning of some of those remnants of earlier stands that are capable of regaining full vigor (Fig. 15.10). This procedure rehabilitates the stands and preserves enough of the irregularity of the initial stands that some intermingling of age classes remains.

This approach has helped meet the need for low-investment silviculture on small, private, nonindustrial ownerships that comprise almost two-thirds of the Southern forest (Murphy, Baker, and Lawson, 1991; Williston, 1978). The chief emphasis tends to be on manipulating existing growing stocks in order to extract optimum advantage from them over the longest possible time. Although regeneration is seldom perfect, it is often possible to get sufficient natural regeneration if fire, herbicides, browsing, and harvesting of hardwoods are used to keep the hardwoods under some degree of control.

Figure 15.10   A stand, predominantly of loblolly and shortleaf pine, being managed under the selection system in southern Arkansas. The large-branched pine in front of the man is to be cut to foster the growth of the small-branched one behind him. *(Photograph by U.S. Forest Service.)*

BIBLIOGRAPHY

Adams, D. M., and A. R. Ek. 1974. Optimizing the management of uneven-aged stands. *CJFR*, 4:274–287.

Alexander, R. R., and C. B. Edminster. 1977. Uneven-aged management of old-growth spruce—fir forests: cutting methods and stand structure goals for the initial entry. USFS Res. Paper RM-186. 12 pp.

Baker, J. B. 1986. The Crossett farm forestry forties after 41 years of selection management. *SJAF*, 10:233–237.

Baumgartner, D. M., and J. E. Lotan (eds.). 1988. *Ponderosa pine: the species and its management.* Washington State University Dept. of Nat. Resource Sci., Pullman. 281 pp.

Burschel, P., and J. Huss. 1987. *Grundriss des Waldbaus.* Paul Pary, Hamburg. 352 pp.

Coates, K. D., S. Haeussler, S. Lindeburgh, R. Pojar, and A. J. Stock. 1994. *Ecology and silviculture of interior spruce in British Columbia.* Forestry Canada and Brit. Col. Ministry of Forests, Victoria. 182 pp.

Cochran, P. H. 1992. Stocking levels and underlying assumptions for uneven-aged ponderosa pine stands. USFS Res. Note PNW-RN-509. 10 pp.

Davis, L. S., and N. Johnson. 1987. *Forest management.* 3rd ed. McGraw-Hill, New York. 790 pp.

deLiocourt, F. 1898. *De l'amenagement des sapinieres.* Bul. Soc. For. Franche-Compte Belfort. Besancon, France. Pp. 396–409.

Dolph, K. L., Mori, S. R., and W. W. Oliver. 1995. Long-term response of old-growth stands to varying levels of partial cutting in the Eastside pine type. *WJAF*, 10:101–108.

Fiedler, C. E. 1995. The basal area—maximum diameter—q (Bdq) approach to regulating uneven-aged stands. In: K. L. O'Hara (ed.), Uneven-aged management: opportunities, constraints, and methodologies. *Montana For. and Cons. Exp. Sta. Misc. Pub.* 56. Pp. 94–109.

Geiger, R., R. H. Aron, and P. Todhunter. 1995. *The climate near the ground*. 5th ed. Vieweg, Braunschweig, Germany. 528 pp.

Gingrich, S. F. 1978. Growth and yield. In: Uneven-aged silviculture and management in the United States. USFS Gen. Tech. Rept. WO-24 . Pp. 115–124.

Gordon, D. T. 1970. Natural regeneration of white and red fir: influence of several factors. USFS Res. Paper PSW-58. 32 pp.

Guldin, J. M., and M. W. Fitzpatrick. 1991. Comparison of log quality from even-aged and uneven-aged loblolly pine stands in Arkansas. *SJAF*, 15:61–67.

Hallin W. E. 1959. The application of unit area control in the management of ponderosa—Jeffrey pine at Black Mountain Experimental Forest. *USDA Tech. Bull.* 1191. 96 pp.

Hansen, G. D., and R. D. Nyland. 1987. Effects of diameter distribution on the growth of simulated uneven-aged sugar maple stands. *CJFR*, 17:1–8.

Hickman, C. A. (ed.). 1990. Evaluating even and all-aged timber management options for southern forest lands. USFS Gen. Tech. Rept. SO-79. 149 pp.

Hornbeck, J. W., and W. B. Leak. 1992. Ecology and management of northern hardwood forests in New England. USFS Gen. Tech. Rept. NE-159. 44 pp.

Hotvedt, J. E., Y. F. Abernethy, and R. M. Farrar, Jr. 1989. Optimum management regimes for uneven-aged loblolly-shortleaf pine stands managed under the selection system. in the West Gulf Region. *SJAF*, 13:117–122.

Isaac, L. A. 1956. Place of partial cutting in old-growth stands of the Douglas-fir region. USFS, PNW Res. Paper 16. 48 pp.

Johnston, W. F. 1977. Manager's handbook for black spruce in the North Central States. USFS Gen. Tech. Rept. NC-34. 18 pp.

Knuchel, H. 1953. *Planning and control in the managed forest*. Transl. by M. L. Anderson. Oliver & Boyd, Edinburgh. 360 pp.

Leak, W. B., and J. H. Gottsacker. 1985. New approaches to uneven-aged management in New England. *NJAF*, 2:28–31.

Long, J. N., and T. W. Daniel. 1990. Assessment of growing stock in uneven-aged stands. *WJAF*, 5:93–96.

Lilieholm, R. J., L. S. Davis, R. C. Heald, and S. P. Holmen. 1990. Effects of single tree selection harvests on stand structure, species composition, and understory tree growth in a Sierra mixed conifer forests. *WJAR*, 5:43–46.

Meyer, H. A. 1952. Structure, growth, and drain in balanced uneven-aged forests. *Jour. For.*, 50:85–92.

Meyer, W. H. 1942. Yield of even-aged stands of lobolly pine in northern Louisiana. *Yale Univ. Sch. For. Bul.*, 51. 39 pp.

Munger, T. T. 1941. They discuss the maturity selection system. *J. For.*, 39:297–303.

Murphy, P. S., J. B. Baker, and E. R. Lawson. 1991. Selection management of shortleaf pine in the Ouachita Mountains. *SJAF*, 15:61–67.

Nyland, R. D. (ed.). 1987. Managing northern hardwoods: proceedings of a silvicultural symposium. *State Univ. of N. Y. Coll. Envir. Sci. & For. Misc. Publ.* 13. 430 pp.

O'Hara, K. L. (ed.). 1995. Uneven-aged management: opportunities, constraints, and methodologies. *Montana For. and Cons. Exp. Sta. Misc. Publ.* 56. 166 pp.

Pearson, G. A. 1950. Management of ponderosa pine in the Southwest. *USDA, Agr. Monogr.* 6. 218 pp.

Reynolds, R. R., Baker, J. B., and T. T. Ku. 1984. Four decades of selection management on the Crossett farm forestry forties. *Ark. AES Bul.* 872. 43 pp.

Roach, B. A. 1974. What is selection cutting and how do you make it work—What is group selection and where can it be used? *State Univ. of N. Y., Coll. of Env. Sci. and Forestry, Appl. For. Inst. Misc. Rept.* 5. 9 pp.

Scott, V. E. 1978. Characteristics of ponderosa pine snags used by cavity-nesting birds in Arizona. *J. For.*, 76:26–28.

Tubbs, C. H., and R. R. Oberg. 1978. How to calculate size-class distribution for all-age forests. USFS, North Central Forest Experiment Station, St. Paul, Minn. 5 pp.

U.S. Forest Service. 1978. Uneven-aged silviculture and management in the United States. USFS Gen. Tech. Rept. WO-24. 234 pp.

Williston, H. L. 1978. Uneven-aged management in the loblolly—shortleaf pine type. *SJAF*, 2:78–82.

CHAPTER 16

STANDS OF MIXED SPECIES

The concept of stratified mixtures was introduced in Chapter 2, and the means of maintaining and regenerating them are considered in this chapter. First, however, it is important to state why this kind of complexity must be introduced and how it relates to the simpler kinds of stands described in the four previous chapters.

Mixed stands are common in nature, especially where soil and climate are not restrictive. Under favorable site conditions it is often difficult, expensive, or even impossible to maintain purity of stand composition, so it may be prudent to work at least partly with natural tendencies rather than row upstream against them. There can be other reasons for maintaining mixed stands. Sometimes they are more resistant to damage or more productive than pure stands. In some cases they provide better protection of the physical and mechanical properties of soils. Usually, they are more attractive and support a wider diversity of wildlife.

Stratified mixtures are not only most common on sites where soil moisture or other site factors are not limiting but are often most appropriate there. Although such sites have great biological potential to produce wood, the stands can easily get out of silvicultural control. In the early stages particularly, they can be the confused tangles for which the term *jungle* was invented. Much of the gross production may be diverted into species of low economic value, so that the very sites that are fundamentally most productive may be regarded as poor from the economic standpoint. This view is often applied to moist tropical forests or the bottomland hardwood forest of the southeastern United States. In the latter region, the adjacent dry uplands may be deemed more productive only because it is easier to control hardwoods and grow pines.

It is ironic that the most successful silviculture is conducted not on the very best sites, but in areas where the soil factors are just restrictive enough that species composition can be kept under control. On the best sites, it may be easier and cheaper to learn to understand and live with mixed stands than to destroy them and substitute pure plantations.

The concept of the stratified mixture is one way of systematizing, understanding, and harnessing the development and productivity of complex forests; it provides a way of inflicting order on chaos. The concept may play somewhat the same role for mixed stands that conventional yield tables do for pure stands in predicting stand development. The results will be less precise and the predictions will depend mainly on computer simulation of the growth of the crowns of trees and stands. It may help to think of each stratum as a separate stand with one stand on top of another.

Not the least of the problems is that efficient silviculture usually depends on being able to make economic use of the wood of all species. Mixed stands are often clogged with unmerchantable species. Even where most species can be used, there can be complications about sorting them.

Most mixed stands develop stratification by species because one species usually gets ahead of another sooner or later. If all the losing species die, the result is ultimately a pure single-canopied stand. If, on the other hand, the laggard species are adapted to the shade of some subordinate position and thus survive, the result is a stratified mixture. In other words, after the different competing species have sorted themselves out according to a pattern analogous to a pecking order in chickens, mixture in the vertical dimension develops.

Single-cohort Stratified Mixtures

For purposes of simplicity, this account starts with single-cohort stratified mixtures because their structure and development are less complicated than those of mixed stands that have several cohorts or age classes. The results of partial disturbances and partial cutting in the past have actually made multicohort mixed stands more common than the simpler kind. However, the developmental processes of multicohort stands are most easily understood if it is recognized that the patterns within each cohort are usually the same as they would be in a whole stand composed of that cohort. The difference is that they take place in patches.

Single-cohort mixtures usually exist only where some lethal disturbance has eliminated all trees larger than saplings and some new combination of advance regeneration, sprouts, or newly germinating seedlings forms a new cohort. The elimination of the larger trees makes it comparatively easy to see how the new cohort develops. Such stands have often appeared after very heavy cuttings of mixed stands for fuelwood or other small products. They may result from severe blowdowns or, much less commonly, from the attacks of relatively unselective pests. Fires can create them if the sprouts or seeds of more than one tree species remain viable.

Relation Between Development and Treatment

The developmental pattern may be obscure in the early stages; chaos may appear to reign. It has been common to attempt to bring order out of this chaos by employing the concept of the balanced uneven-aged stand. This approach seldom works because the large trees of mixed stands are not necessarily slower-growing and older than their neighbors; the smaller trees of the lower-stratum species do not exhibit the rapid growth ascribed to small trees under the uneven-aged concept.

Often the best course of action in the early stages is to let the various species fight it out among themselves. Early release work can be difficult, costly, and futile. Usually, it is best to wait until people can walk beneath the new stands to determine the outcome of competition and to kill any plants that are causing trouble. What may seem to be overcrowding in the sapling stage may actually be creating straight, branch-free boles in trees that ultimately become the leading ones.

Sooner or later, those long-lived species, regardless of their degree of shade tolerance, that have become the biggest and best trees in natural mixed stands on the same sites can be depended on to ascend to similar status. This may not happen, however, if some aggressive exotic species has joined the cast or some previous member has found a new role as an aggressive stump-sprouter. The species that persist in successively lower strata are generally of greater shade tolerance the lower the stratum. The lowermost are often shrubs that never grow very tall. The stems of negatively geotropic species remain straight and vertical in the lower strata; those that are more phototropic bend toward the light and often become crooked as the sources of light from canopy openings shift position.

The early development of stratified mixtures can be even more chaotic if much mineral soil is laid bare by the regenerative disturbance. Fast-growing weedy pioneer vegetation may seem to overwhelm everything and hide the tree seedlings that have started beneath it.

Vines are apt to get started at this stage and are dangerous competitors if they actively climb or ride up atop the desirable trees. In many cases, these "support parasites" will form the top canopy and suppress the trees unless active measures are taken to kill them. Usually, the best way to control them is to exclude them at the start by avoiding exposure of mineral soil and keeping the site fully occupied by preexisting woody vegetation. Vines are most likely to be problems on good soils in warm climates.

Herbaceous annuals seldom cause problems because they usually vanish after the first year. Their seeds may remain stored in the forest floor and be ready to germinate in some major disturbance decades hence. Perennial herbaceous plants, including grasses, can impede development for long periods, especially if they become tall. Neither these nor shrub species will maintain dominance for long provided that some tree species is able to grow up through them.

If pioneer tree species have germinated, they usually spurt above everything else and form an overstory that may remain for some decades (Fig. 16.1). In fact, their shade may shield and nurse the development of some slower-growing shade-tolerant species that are longer lived. Such pioneers may die young or slow down in height growth and become suppressed by other species that overtake them. Usually their main effect is to postpone the development of the more shade-tolerant species beneath them. However, if either they or some other overstory shade persists, some relatively shade-tolerant species may overtake a less tolerant one that would otherwise ascend to a leading position. In other words, shading effects or other factors controlling interspecies composition can be regulated to favor some and discriminate against others. Some of these intolerant species are long-lived; yellow-poplar, yellow birch, and certain hard pines, for example, can continue to grow rapidly in height and even endure for centuries. The adaptations of species are endless.

The key to guiding the development of such stands lies in knowing how the constituent species interact with one another in their height growth on specific sites. Fortunately, the rates of height and diameter growth tend to be well correlated so that much can be learned from examining annual growth rings on stumps. Unfortunately, moist tropical

Figure 16.1 A simple kind of even-aged stratified mixture, consisting mainly of aspen over balsam fir, but with one white spruce and one white pine, remnants of an earlier stand, projecting above the aspen stratum. Superior National Forest, Minnesota. *(Photograph by U.S. Forest Service.)*

forest trees do not have annual rings; thus reconstruction is not possible in the very kind of stands for which such information would be the most helpful.

In some cases, a whole stratum can logically be harvested in one operation. Those species that race ahead and soon die must be harvested early or allowed to go to waste. It is indeed possible that one might harvest successive overstory layers as each matured, but provision would have to be made for the fact that the seed source of each species might also be successively eradicated. The advisability of working down to the lower strata and leaving them to continue development usually depends on the potential quality and prospective growth of the trees that are released. Figure 16.2 shows a kind of shelterwood cutting sequence in which the strata are removed from top to bottom but each species is retained long enough to ensure its regeneration.

If the uppermost species are the most desirable, as is usually the case, stratified mixtures have some useful developmental attributes that are not found in pure stands. If these species can be depended upon to get on top, it is not necessary to have anywhere near as many of them in the initial regeneration as is the case with pure stands. The rest of the growing space is filled with trees of the species that lag or fall behind. These trees of the subordinate strata are usually tall enough to provide the crowding necessary for the natural pruning and other kinds of training of the taller trees but without restricting the horizontal expansion of the upper parts of the crowns (Fig. 16.3). The congestion that is common in the crown canopies of pure stands is thus avoided.

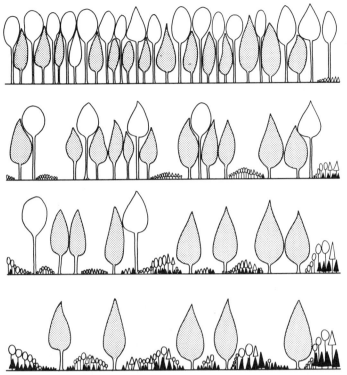

Figure 16.2 Regeneration of an even-aged stratified mixture of eastern hemlock (dark crowns), hardwoods (rounded white crowns), and white pine (pointed white crowns) by the irregular shelterwood method. The upper sketch shows a 75-year-old stand prior to treatment. Successive sketches, from top to bottom, show the stand at 15-year intervals, just before each cutting. They indicate how the upper stratum is opened up rapidly, releasing the hemlock but leaving some of the upper stratum for seed and extra increment. The hemlock stratum is opened more slowly and only to the extent necessary to allow regeneration of the intolerant species of the upper stratum. The bottom sketch shows the stand at 120 years, just before the last of the old hemlocks are removed; this leaves a new stratified mixture of somewhat broader range of age than the previous stand. One could accelerate the whole process and make the new stand more uniform by eliminating the hemlock stratum first.

If the uppermost trees ascend increasingly farther above the lower strata, the freedom of their crowns to expand becomes greater, so that one may get the effect of a thinning without having to conduct one. Upper-stratum trees that develop in this manner may have such large crowns that they can continue good diameter growth to advanced ages, and they often have fat boles of high quality, even in untreated stands. It may nevertheless be useful to set some maximum length for the amount of clear bole that is desirable and to plan to release the tree enough that branch dying can be halted and the expansion of the live crown can be promoted when its base has retreated to that level. Any thinning nec-

Figure 16.3   The pure stand on the left is of the same age as the mixed stand on the right and the trees with open crowns are of the same species. The two emergent trees on the right have larger crowns and greater stem diameter than those of the comparatively congested pure stand. The trees with cone-shaped crowns in the mixed stand have acted as trainers of the taller emergents.

essary to achieve this effect would, as shown in Figure 16.3, involve mostly the removal of trees of the same species, with crowns competing in the same level as the released trees.

When one sees fine, wide-crowned trees such as cherry-bark oaks in stratified mixtures of bottomland hardwoods in the southern United States, it is prudent to resist the temptation to assume that a pure plantation of the species would be better yet. The result of a pure plantation would likely be a congested stand of slender oaks. Whenever a forester finds a large, fine tree in the woods and ponders the ways of growing similar ones, it is sobering to measure the great width of the crown and consider the implications of that for stand management. Sometimes it is difficult to thin single-canopied stands hard enough to produce such results without major waste of growing space.

When stratified mixtures are thinned, the relationships among tree crowns should be examined carefully and in three dimensions. The diameters and spacings of the tree stems are not as reliable as indicators of crown relationships as they are in single-canopied stands. With stratified mixtures, it is necessary to note that the crowns of two trees may touch above those of several trees that have stems that stand between them. Much craning of the neck is required.

It often helps to look upon each stratum as a separate stand and thin it as if it were (Fig. 16.4). Sometimes one can discern, within each stratum and if it is viewed separately, the same crown classes that are recognized in pure stands. If a lower stratum is ultimately to be released and made into the main top-canopy stand, it is desirable to anticipate this possibility and thin it down to an appropriate spacing in advance.

The question of whether one or more of the lower strata can be released and made into successive main-canopy stands depends on the feasibility of appropriate logging and on the qualities of the subordinate trees. Sometimes they have grown in height such that the tops of their crowns have battered against the bottoms of the crowns above. This can leave the crowns and stems deformed, but the damage is usually not of long-term consequence. Stem malformations from phototropic leaning are more common. The length of time that the subordinate trees linger in the lower strata increases the period during which they can suffer accidents or pest attacks. Negatively geotropic species such as conifers or sugar maple, sweet-gum, and hickories tend to remain straight and are thus the most promising kind to release.

It is not always possible to make a "new" stand by removing the top layer and releasing the next. If it is, however, one can get two or more stands for the costs of money, difficulty, and lost production that are normally paid for regenerating one. The countervailing matters to consider are mainly those of more complicated logging and the question of whether the production by the released stand is worth as much as that of an entirely new one.

It should not be presumed that the uppermost stratum is always the one to remove first. The trees of any stratum can be removed whenever they are economically mature or when this will facilitate regeneration or accomplish some other purpose. Reasons could be found in different cases to remove the strata in almost any sequence, including starting in the middle. When eastern hemlocks grow under hardwoods, they can actually become larger in D.B.H. than the trees above and, furthermore, attain optimum size for sawlogs without ever reaching the upper stratum. Because such hemlocks also inhibit regeneration, it helps to take many of them out before the hardwood stratum above them.

As is always the case, new trees will not appear in the ground story unless adequate growing space is made vacant for them. In stratified mixtures, this is most likely to be accomplished by removing or thinning one or more of the lower strata. Because advance growth is usually important for the regeneration of the desirable species of stratified mixtures, regeneration cuttings commonly take the form of some sort of shelterwood cutting. However, precautions are necessary to maintain sources of seed of the desired species

Figure 16.4   A stand in southern Maine, similar to that of Figure 2.8, which has just been thinned with removals from each of the strata and sufficient opening of the lowermost to allow establishment of new advance growth. *(Photograph by U.S. Forest Service.)*

if reliance is ultimately to be placed on natural regeneration. When a given stratum is removed, it would be easy to eliminate a species rather permanently.

Stratified mixtures can be manipulated in many ways. Figure 16.5 shows some of the pathways of stand development that might be followed, starting with a single-cohort strat-ified mixture. These include the creation of the initial kind of stand by shelterwood cutting as well as two pathways that start with the removal of some large emergent species. One pathway proceeds down the blind alley called "high-grading" in which shade-tolerant lower-stratum species are left and tend to block new regeneration. On another pathway this degradation is partly halted by thinning all closed strata to allow regeneration of all remaining species, although the original emergent species may be lost. Several other op-tions, including the creation of a stand with two age classes, are also possible develop-mental pathways. A sketch (Fig. 16.9) in a later section shows how a multi-cohort stand of essentially unchanged composition might be created starting with the same kind of single-cohort stand.

The silvicultural management of mixed stands is easiest and most profitable econom-ically where stands are readily accessible to markets that can make use of trees of all species and qualities. In fact, finding or developing such versatile markets is usually the key to managing mixed stands successfully. The sorting of logs and products is expensive enough that it should be carefully planned and coordinated by each party to a harvesting operation. It may be desirable to focus a given cutting step on the removal of a limited number of species, sizes and qualities of trees, or stratum in order to simplify later steps in harvesting and processing. It helps if there are diversified markets for pulpwood and fuelwood as well as for sawlogs. The conservatism of markets and manufacturers about unfamiliar species is often a stumbling block. It can be useful to advance the point that a good log is usually a good log regardless of its species.

The concept of the single-cohort stratified mixture represents an interpretation of the natural development of most, but not all, mixed stands. The fact that it appears to "follow nature" does not mean that nature needs to be followed. The forester can imitate the parts that meet management objectives and, provided that the health of the forest does not suffer, omit parts that do not. It is well to remember that the pure even-aged plantation also imitates a kind of natural development and provides a highly predictable pattern of stand development. The development of mixed stands is less predictable, but knowledge of the stratification process helps.

If stands are so remote or in such steep terrain that they are expensive to access, only the biggest and best trees of a few species are valuable enough to return the high costs of getting them to market. Such circumstances are conducive to high-grading. Often the best remedy for that is killing enough residual trees to provide some vacant growing space for either regeneration or the enhanced growth of desirable trees that have been reserved.

Irregular Shelterwood Method

The **irregular shelterwood system** is usually associated with the management of mixed stands, although it can also be applied in pure stands. It differs from the uniform application of the shelterwood method in that the regeneration period is extended so long that the new stand may be of a single cohort but is not absolutely even-aged. This does not mean that it has the three or more age classes that denote the uneven-aged condition. It does, however, mean that the stand will include two age classes for long periods and sometimes even for a whole rotation. The adjective "irregular" refers mainly to the variations in tree heights

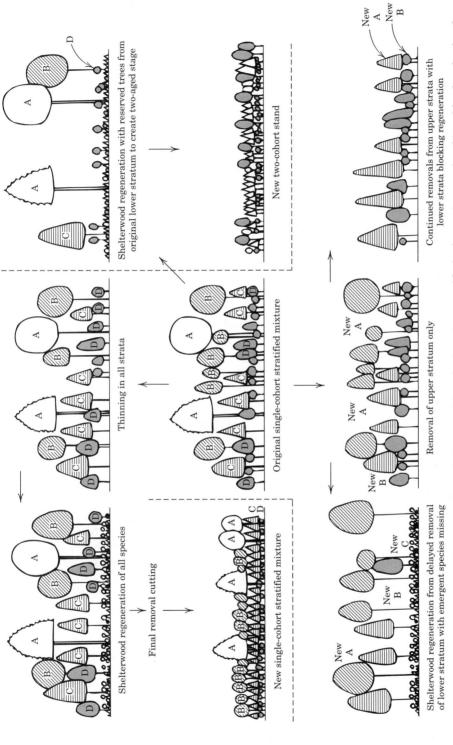

Shelterwood regeneration of all species

Final removal cutting

Thinning in all strata

Shelterwood regeneration with reserved trees from original lower stratum to create two-aged stage

New single-cohort stratified mixture

Original single-cohort stratified mixture

New two-cohort stand

Shelterwood regeneration from delayed removal of lower stratum with emergent species missing

Removal of upper stratum only

Continued removals from upper strata with lower strata blocking regeneration

Figure 16.5 Various pathways of treatment that might start with a single-cohort stratified mixture (center sketch) and lead to the development of various other kinds of single- or double-cohort stands. (*Adapted from Smith, 1989.*)

within the new stand. The partial cutting procedures depicted in Fig. 16.5 are often examples of the irregular shelterwood system.

"One-Cut" Shelterwood Method

Single-cohort mixed stands are sometimes created by very heavy removal cuttings of stands that have ample advance regeneration that has started by accident or without deliberate effort to establish it. Fires or other damaging agencies as well as the canopy gaps that develop from twig breakage caused by tree-sway and other effects in aging stands often allow advance growth to become established. If soil-moisture conditions are favorable, advance growth of species that are only moderately shade-tolerant can appear as a result of canopy opening that may seem insignificant. It is entirely possible for stands of some early successional species to have desirable advance growth of a later successional stage beneath them. In the Great Lakes Region, for example, red pine may follow jack pine in this manner.

If satisfactory advance growth is already established, there is no fundamental reason to refrain from releasing it with a single removal cutting. This sort of "one-cut" shelterwood cutting, often done unintentionally, has been remarkably effective in the early stages of forest management in many places, especially in eastern North America. It is a useful procedure for rehabilitating many kinds of degenerate old-growth stands or stands that have been high-graded. The best results in such circumstances have often been associated with the development of good markets for defective or small trees. One drawback of the one-cut method is that it gives little control over the nature of the regeneration; it seldom produces pure stands. Furthermore, substantial losses of volume are sometimes sustained from the processes that led to establishment. The one-cut shelterwood system has a place in the early development of silvicultural practice, but it does not lend itself well to regenerating stands that are kept well stocked and on shortened rotations.

Blunders have sometimes resulted from failing to recognize one-cut shelterwood cutting for what it is. These mistakes have occurred when heavy removal cuttings mistakenly identified as "clearcutting" were thought to have led to new regeneration when they were only releasing advance growth. It is easy to overlook small seedlings or to underestimate the age of small trees a few years after a harvesting operation. The blunders come about when the "clearcutting" is shifted to stands that are too young or too tightly closed to have advance growth or when the logging machinery is changed to kinds that destroy it.

One instance of this chain of events was in northern hardwood stands on the Allegheny Plateau in northwestern Pennsylvania (Fig. 16.6) (Marquis, Ernst, and Stout, 1994). These stands had been subjected to successive high-gradings during the late nineteenth century. Early in the twentieth century a wood-distillation industry developed in the locality and all remaining trees larger than about 3 inches D.B.H. were removed. Fine new stands burgeoned and were dominated by good, fast-growing black cherry along with sugar maple, beech, and other species in the lower strata. In the keen vision of hindsight, it can now be seen that this all came from the advance growth that had been unintentionally established during the era of high-grading. What foresters mistakenly thought they knew was that the fine black cherries had germinated after "clearcutting" and possibly even depended on the exposure.

The heavy chemical-wood cuttings had taken place during the short span of a couple of decades so a large forest mostly of one age class was created. Not surprisingly, there was eagerness to start regenerating some stands very early in order to get started with readjusting the age-class distribution for sustained yield. Because what was deemed to be clearcutting seemed to have worked before, some of it was tried in stands only 40 years

Figure 16.6   Two stages, 43 years apart, in the development of a stratified mixture of black cherry (rough-barked in lower picture) and sugar maple (smoother bark) in northwestern Pennsylvania. The small pole-sized trees left after the heavy cutting shown in the upper picture are of the kind shown in the upland hardwood portion of Figure 16.13; they are obscured by younger trees in the later picture. *(Photographs by U.S. Forest Service.)*

old. The new vegetation commonly consisted of expanses of fern and grass with a few beech root-suckers. Initially, the excessive deer populations common in the locality were blamed, but it soon became apparent that the lack of advance growth was the real reason. Shelterwood cutting has now been successfully substituted (Marquis, 1979).

Planting of Mixed Stands

Stratified mixtures can be created by planting (Montagnini et al., 1995; Cole, 1993). The most successful mixtures are composed of faster-growing intolerant species above slower-starting tolerants (Fig. 16.7). If the trees of the uppermost species are not too dense and numerous, they grow more rapidly in diameter than they would if crowded into the single canopy of a pure plantation; the lower-stratum species can provide much the same training effect of a pure stand. The upper-stratum species should not comprise more than a quarter of the total number of trees. It is rare that much advantage can be gained by having more than three species.

Properly selected species can be planted simultaneously at random or according to a variety of patterns, and the stratified mixture will develop automatically. Two-species mixtures in alternate rows are common, although they usually have too many of the leading species. Often the fastest growing can be planted several years after those destined for the lower strata. Pioneer species are the most satisfactory overstory species because they grow

Figure 16.7 A Connecticut forester in a stratified mixture of conifers that he planted 33 years earlier, with equal numbers of four species distributed at random. The species, in order of decreasing present height, are European larch, red pine, white pine (mostly removed in thinnings), and Norway spruce. The larches, such as the one beside the forester, form a topmost story and each has had plenty of space for growth of large crowns and stem diameters. *(Photograph by Yale University School of Forestry and Environmental Studies.)*

rapidly in height and cast light shade over the more tolerant species beneath. Species of intermediate tolerance, such as Douglas-fir and soft pines, can be used for any story depending on the associated species. A species may be a leader on poor sites but a follower on good sites.

If two species are nearly the same in rates of height growth and shade tolerance, one will usually suppress the other because the small differences in growth gradually add up to differences in height large enough to cause one species to shade the other. The resulting self-thinning effect enhances the diameter growth of the ascendant species (Fig. 16.3), but the other becomes a rather costly filler and trainer if it does not grow to useful size. If mixtures of closely similar species are desired, they should be separated into groups and strips wide enough to allow space for mature individuals of the particular species. If horizontal mixtures of this sort are planted, it may be well to arrange then in ways that will expedite future operations or fit variations in site and existing vegetation.

Alternatives to the Concept of Stratified Mixtures

The general idea expounded in most of this chapter is that the intermingling of species occurs much more commonly in the vertical than the horizontal dimension and that this is the key to understanding their silviculture. However, there are always exceptions as well as alternative hypotheses. Figure 16.8 is a schematic illustration of the differences between three possible structures of mixed stands.

Much thought about mixed stands has been predicated on the idea that they are comprised of small, pure groups of trees in which the species are intermingled in the horizontal dimension. This condition seems to exist more often in plantations than in nature but is certainly possible, especially where there are soil differences within stands. Where mixtures occur in pure groups, each can be managed as if it were a pure stand but with adjustments to deal with edge effects.

It is possible to have different species intimately intermingled in the same canopy stratum although this condition is uncommon, temporary, or maintained only with repeated thinnings. Most mixed stands in Europe are thinned in ways that hold back species that tend to surge ahead but stimulate the growth of those that lag behind. However, the stands involved are often mixtures of spruce, true fir, and beech which are all shade-tolerant species with nearly the same rates of height growth.

If it is found that two or more species are of the same age (or cohort) and indeed remain in the same stratum through a whole rotation, the stand might as well be managed as if it were single-canopied. While it is contended here that such a condition is not very common, stands should be managed for what they are and not be warped to fit to some preconceived theory about what they ought to be.

Mistaking Size for Age in Mixed Stands

There is a different interpretation of mixed stands that clings to the tacit assumption that large trees are older than smaller ones, so that mixtures of species can be treated as if they were always uneven-aged. This misperception is then extended to the fallacious view that the small trees can be depended on to grow just as rapidly as they would if they were young dominants in some little group of trees with their tops open to the sky. This optimistic view can, however, hold true for small overtopped trees if they are of some shade-tolerant species that are capable of rapid acclerations in height growth when released from trees above them. However, that development is the result of an attribute of many kinds of stratified mixtures rather than an advantage somehow conferred only by an uneven-

Figure 16.8 Three possible structures of even-aged mixtures of the same four species. The upper sketch shows a young stand consisting of pure groups. The middle sketch shows an older stand in which deliberate treatments have caused all four species to form a single canopy stratum; this structure would take place naturally if all four species grew in height at the same rate. The lower sketch is of an old stand in which the four species have become sorted into four different strata because of their different rates of growth in height.

aged condition. In fact, more advantage can often be taken of it where the single-cohort condition is maintained.

Sometimes it is arbitrarily assumed that all mixed stands are not merely uneven-aged but that all age classes needed to make the stand a sustained-yield unit have somehow magically appeared. These assumptions are usually based on the existence of reverse

J-shaped curves of diameter distribution among trees of merchantable size and the failure to notice shortages of smaller trees. This fallacy is discussed in detail in Chapter 15 in connection with uneven-aged management for sustained yield. The q-factors discussed in that chapter were actually first based on observations in old mixed stands and there has been much research with efforts to use them for regulating diameter distributions of mixed stands. There is no evidence that such efforts have made stands into sustained-yield units. The correlation between diameter and age is even poorer in mixed stands than it is in pure stands. It is difficult to see any theoretical basis for the idea that conformity to diameter distributions defined by J-shaped curves or q-factors confers magical properties on mixed stands. The false assumptions involved can lead to high-grading and dangerous overestimates of allowable harvest rates, both of which can defeat the long-term sustainability of forestry programs.

Mixed, Multi-cohort Stands

Mixed stands with basically stratified structure commonly include truly different age classes or cohorts. In fact, some of the constituent sub-stand groups may even be pure, even-aged, and single-canopied. The uneven distribution and sporadicity of forest disturbances have made this kind of complexity very common. Even where the tendency for the species to segregate into different strata is strong, the pattern will be obscured if the initiating disturbance is not complete and scattered older trees have remained.

If a species is found in some stratum higher than the one that it normally occupies, it is very likely to be of an older cohort than trees of equal height. Such trees are often shade-tolerant species left by some earlier cutting or lethal disturbance, but they can be of almost any origin. Within a species and single age class, trees of sprout origin may occupy a higher stratum than those originating from new seed. The presence of these kinds of aberrant trees is not necessarily undesirable, but it can mask the processes leading to development of stratified mixtures. It has also been an important reason why the concept of the uneven-aged stand has been inflicted on mixed stands without regard for the actual age-class structure.

Although it is technically possible to convert complex forests to pure stands, it can be costly if natural tendencies resist the conversion. There can also be a variety of unfortunate consequences of an ecological nature. Simple silviculture has important virtues, but the forest should not be torn down and replaced with simple stands just because they are easy to understand.

If the irregularity of uneven-aged mixed stands and the unused growth potentialities of immature trees are of no important future benefit, they can sometimes be converted to single cohorts. This is simple and easy if the whole stand area is well stocked with advanced regeneration, but sometimes the cost of killing unmerchantable immature trees makes such action an expensive way of securing simplicity.

Often the best course of action is to live with the complexity and attempt to take advantage of it. Such advantages may be extracted from making the most of the timber-producing potentialities of each cohort or of superiority of the complex structure for wildlife and other nontimber benefits.

The kind of partial cutting that plays on the advantages of such complexity is that in which each component of the stand is recognized for what it is and treated accordingly. Provided that the seed sources of all desirable species are maintained, it may be possible to select trees to be retained or removed mainly on the basis of their degree of financial maturity or their capacity to continue to increase in value at an acceptable rate. It is also important to ensure that established regeneration is made free to grow by making large enough openings when the time comes to release it. It is easy to succumb to the temptation to leave too many trees for continued growth. If openings are too small, the regeneration of shade-intolerant species is likely to fail.

It is also possible, however, to convert single-cohort mixtures into multi-cohort stands by the very gradual process depicted in Fig. 16.9. Such action would usually mean cutting some parts of a stand prematurely and holding others to advanced ages. Most management purposes of this procedure, such as those of wildlife management and aesthetics, can

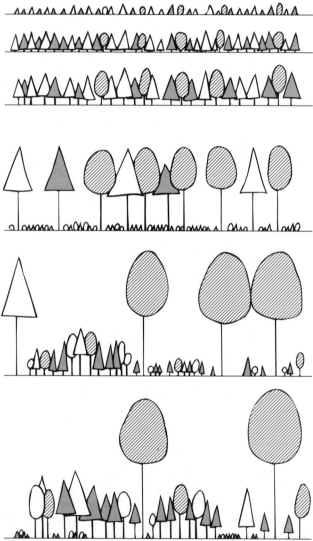

Figure 16.9   Treatment pathway leading from a young single-cohort mixed stand to the initiation of the first new cohort of an uneven-aged stand in the left side of the diagram. The trees with the shaded, rounded crowns are of a shade-intolerant species for which a seed-source must be retained in the stand until the new trees of that species reach seed-bearing age. The species with the black, conical crowns is the most shade-tolerant of the four species.

usually be achieved by proceeding only part of the way toward creating the balanced, all-aged condition.

Strip Arrangements

The attributes of the uneven-aged stand structure and of regeneration by the shelterwood method can be combined in spatial patterns in which shelterwood cuttings are made in successive strips that are advanced gradually across the area. At any given time, cutting is concentrated in certain strips, and the rest of the stand remains temporarily untouched.

After the pattern and direction of progression of the advancing strips is determined, a seeding or establishment cutting is conducted on the first strip. After a few years, a removal cutting is made on the first strip, and the next strip is given a seeding cutting. A few years later the first strip is subjected to a final cutting, the second strip to a removal cutting, and a third strip to a seeding cutting. In this way (Fig. 16.10) the series of cuttings progresses strip by strip across the stand. The cuttings usually progress toward the direction from which either dangerous winds or midday sun come so that the new stand can either start in the shade or be streamlined so that the wind is shunted up and over it (see Fig. 19.2). If the strips advance slowly enough, the thinning may cause some of the final crop trees to develop some degree of windfirmness before they are completely isolated.

In the most intensive applications, such as some that are common in Central Europe, the cuttings are applied at short intervals so that there is a gradual transition from final removal cutting to light preparatory cutting across strips about 30 meters wide (Fig. 16.11). Within these strips there are preparatory, establishment, removal, and final removal cuttings, but each in zones that are scarcely distinguishable from each other. The whole strip is advanced slowly across the stand, at average annual rates of 2 to 6 meters. Several years normally elapse between harvesting operations in a given stand; any advantages that might be gained from annual fine-tuning of the regeneration process would be outweighed by the nuisance and disruption of very frequent cuttings. One of the objectives of strip-shel-

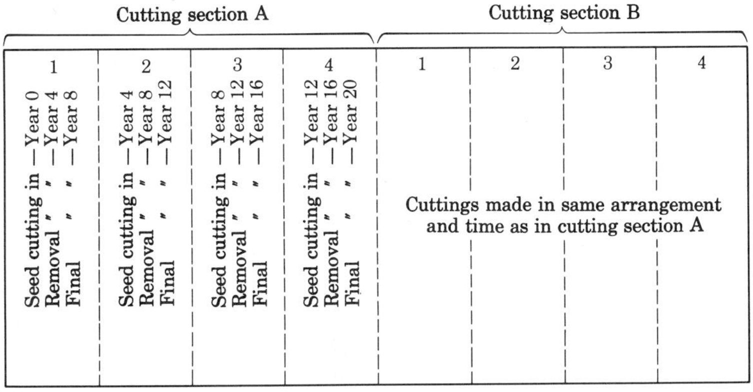

Figure 16.10 Diagram of a stand replaced in two cutting sections by the strip-shelterwood method. The period of regeneration is about 20 years long. By operating simultaneously in two or more contiguous cutting sections it is possible to build a new stand that could be regarded as of one cohort. In this case, the rotation is 100 years and preparatory cutting is deemed unnecessary. Figure 19.1 shows the same kind of stand in vertical cross-section and how damaging winds might be diverted over it.

Figure 16.11   Regeneration of spruce, fir, and beech obtained by the use of the strip-shelterwood system, with cutting advancing from north to south (left to right) in a forest in southern Bavaria. The timber is extracted to the right. If the cutting is done before the advance regeneration is established, dense grass invades and hampers tree seedlings. *(Photograph by Yale University School of Forestry and Environmental Studies.)*

terwood cutting in Central Europe is prevention of invasion by coarse grasses that thrive on overexposure and hamper seedlings of spruce, fir, and beech.

The strips need not be straight unless this would be of real advantage in systematizing or concentrating operations. An irregular or undulating line may even expose a longer frontage for regeneration and thus hasten the advancement of the strip cuttings.

Perhaps the greatest advantage that the advancing strips have over the uniform method is the opportunity to extract the felled timber through the old stand rather than through the regenerated belts. The possibility of doing this is greatest if the stands have been appropriately thinned before the start of regeneration cuttings so that there is enough space between the old trees. Both strip cutting and thinning require that the stand be served by a good transportation system. If logs are best skidded downhill, it is logical that on hillsides the strips advance either downhill or along contour lines but not uphill. One approach developed in southwestern Germany involves patterns in which strip shelterwood cuttings start at the center of a stand and advance, sometimes during the course of a whole rotation, to the roads around the edges of the stand. The purpose is mainly to reduce logging traffic through regeneration areas.

Group Arrangements

If different cohorts are arranged in groups, it may be possible to develop the kind of regularized uneven-aged stand of stratified mixtures depicted in Fig. 16.12. In this variant of the group-selection method, each small sub-unit could be regenerated from seed pro-

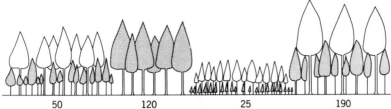

50	120	25	190

Figure 16.12 Part of an uneven-aged stand being managed by the group-selection system and composed of a stratified mixture of two conifers arranged in even-aged groups. The trees with the light crowns are of an intolerant species that grows comparatively rapidly when young and attains greater size than the more tolerant (dark-crowned) species of the mixture. The numbers indicate the ages of the groups. The upper stratum is ready for the first thinning at 25 years of age and, at 50 years, most of the crowns of this stratum have been freed around most of their perimeters. This stratum is harvested soon after 90 years; this releases the lower stratum, which has been left untreated in the meantime, to fill any gaps. The remaining stratum is managed for the rest of the rotation as a simple, single-canopied, even-aged group. After 120 years, it is ready for uniform shelterwood cuttings designed to replace the mixture. The trees reserved in these cuttings provide the seed of the tolerant species; that of the intolerant species comes from adjacent groups. There is no silvicultural reason why either stage of the rotation could not be prolonged or terminated earlier; the timing indicated here is purely an example.

duced by the seed sources of the adjacent groups if seed-bed conditions and other factors were favorable.

Sometimes it may be useful to apply the scheme of strip-shelterwood cutting in patterns of expanding groups or patches. This can be done to induce patches of advance regeneration or to take opportunistic advantage of those that have arisen from thinning or natural disturbances. Sometimes this cutting procedure is regarded as a variant of the irregular shelterwood system.

Although the groups may be expanded uniformly in all directions various problems often make it more logical to advance the cuttings mostly in one direction and turn the patches gradually into advancing strips. The chief advantage of such a shift would be the systematization of timber harvesting and administration. The direction of progression can also be determined by the need to protect seedlings from direct sunlight and logging damage or residual stands from wind damage.

The little openings may become frost pockets, and the question of whether this is undesirable depends on one's attitude about whatever frost-sensitive species might appear. Regeneration groups or patches are sometimes deliberately started in conjunction with shelterwood cutting as a means of enriching mixed stands with intolerant species or giving slow-growing species a head start over more aggressive species.

Two-aged Stands

There can be a variety of reasons for reserving scattered trees when the rest of the trees of a stand are removed to make way for a new even-aged stand or cohort. Such trees can be reserved for additional growth of wood, for wildlife management purposes, or for aesthetics. This can be done in any kind of stand, mixed or pure, provided that the reserved trees can be depended upon to remain standing long enough to accomplish the purpose of leaving them.

This may be a common state of affairs in nature. Veirs (1982) showed that coast redwoods may endure for more than a dozen centuries, while Douglas-firs come and go around them every century or two. Mahogany trees in Yucatan can endure repeated hurricanes and the hot fires that may follow them through several natural rotations (Snook, 1993). In that case, the mahoganies also germinate after the severe fires and grow up as straight stems clothed with compound leaves instead of woody branches and tightly surrounded by less enduring species. When the mahoganies are about 7 meters tall, they begin to forge ahead of the other species and change their crown form to one in which long strong branches spread above the other species.

Such scattered large trees are sometimes left to provide nesting sites for cavity-inhabiting organisms or sources of food for others. In many cases, malformed trees or those with heart-rot are often better for such purposes than those of good timber quality.

The trees left to form two-aged stands can also be of shade-tolerant species that can lapse into long stages of postponed development such that it may take them twice or even three times as long to grow to mature size as some more quickly growing associates. This is not necessarily bad and may even represent a common natural growth habit of some species. In the Allegheny hardwood stands previously mentioned, small pole-sized sugar maples and beeches, left as being too small to cut during the very heavy removal cuttings, sometimes developed remarkably well as components of otherwise even-aged stands that came up around them (Marquis, 1981b).

As shown in Figs. 16.6 and 16.13, such trees, if of good form, can become valuable components of what might otherwise seem to be even-aged stands. The irregularity that these holdover trees imposed on the strata of black cherry that grew up around them caused broader differentiation of crown classes in the cherries and, therefore, much better diameter growth of the potential crop trees among them. The cherries also grew up rapidly enough to prevent the whippy holdover trees from becoming large-crowned wolf-trees. It has often been presumed that such seemingly stunted, small residuals are of no value and ought to be eliminated as a routine matter. As is usually the case, it is well to look for the exceptions that seem to lie behind each general rule.

The same phenomenon often occurs with shade-tolerant conifers such as certain spruces, true firs, and hemlocks. For example, in mixtures of red spruce and balsam fir, the spruces initially grow more slowly than contemporaneous firs (Fig. 16.13). Some of the spruces, when less than about a meter in height, stall in height growth. Branch elongation continues so that these saplings develop ''umbrella'' tops or the form of a candelabra. They survive in this quasi-dormant condition and are able to resume height growth when released by the cutting or death of the short-lived balsam firs. Large trees with this history of release can be identified by dense clusters of dead branches that once formed the umbrella tops (Fig. 16.14). More normal advance-growth saplings with deep, conical crowns actually respond to release more quickly, but the peculiar umbrella spruces are striking evidence of the reality of the phenomenon.

The red spruces may even go into two stages of arrest such that they complete their development during three balsam fir rotations (Davis, 1991). If they grow well after their ultimate release, their total age at that time is of no consequence and does not add to the length of the ''rotation.'' Because their size gives the effect of having planted a very tall seedling, any rotation reckoned from the time of release is actually shortened.

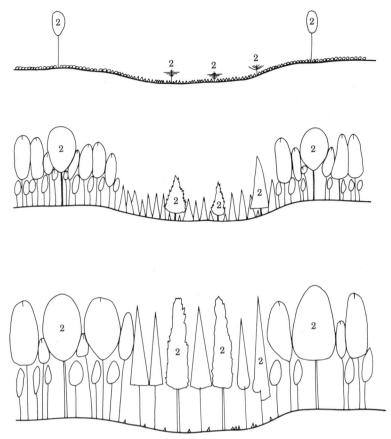

Figure 16.13 Three stages in the development of an unthinned mixed stand, on two different soils, in which trees held over from the previous rotation become major components of the next stand. The trees marked 2 are taking two rotations to make full development. Those with rounded crowns and vertical marks at the tips are the fast-growing black cherries of the case described by Marquis (1981b), while those without such marks are shade-tolerant maples or beeches. The conifers on the poorly drained site are red spruces (rough, pointed crowns) and balsam firs (smooth, conical crowns). The trees marked 2, which were subordinates in the previous stand, become the largest trees of the new stand and also prevent it from developing into excessively uniform thickets. The top sketch shows the initial condition.

Examples of Management of Mixed Stands

North American Stratified Mixtures

The most outstanding example is the eastern deciduous forest, which, with its admixture of conifers, is the most complex forest formation outside the tropics. The most complicated mixtures of conifers on earth are the coastal forests of the Pacific Northwest with their outliers in the Northern Rocky Mountains; those of the west slope of the Sierra Nevada are only slightly simpler.

Figure 16.14   The closely spaced dead branches of several red spruces in this stand of red spruce in Maine show that the trees were very slow-growing "umbrella" subordinates before they were released by the cutting of an old-growth white pine.

Mixed forests in these regions sometimes have definite emergents; distinctions can often be made between species of the upper, middle, and lower stories. A species that behaves as an emergent or upper-stratum species in one part of its range may elsewhere occupy the B or C stratum in mixture with different associates. Douglas-fir is an upper-stratum species in relation to hemlock or true firs on the West Coast, but in some places in the dry interior it is a lower-stratum species with respect to ponderosa pine and may not be able to start in the open. At high elevations, most tree species cannot become established without some shade of the sort most easily maintained in mixed stands. Similar problems exist in the Alps and in other high mountain areas throughout the world (Tranquillini, 1979).

In the Cascade Range of Oregon and Washington, such problems have been most acute in the zone where Douglas-fir tends to give way to true firs and mountain hemlock, species that, among other adaptations, have saplings that restraighten themselves better after annual bending by massive snowfalls. At these high elevations, so little atmosphere above the terrain that seedlings in the open are defeated by the effects of heat injury by day, frost by night, and direct evaporation from the uppermost soil strata. The true firs are even less successful at surviving in the open than Douglas-fir. Shelterwood cutting has become a common solution to the problem. The maintenance of mixed stands is likely to play an increasingly prominent role on large areas of public forests in the western United States because of aesthetic considerations, the unpopularity of plantation silviculture, and the high cost of artificial regeneration, especially for sites with low or mediocre productivity. The approach can have some drawbacks. A common one is the prevalence of dwarf-mistletoe, a parasitic seed-plant that can spread from overstory trees to their progeny if they are left there too long.

Most stands of the eastern deciduous forest are stratified mixtures. The chief exceptions are some stands on very poor sites and the extensive root-sucker stands of aspen poplar of the Great Lakes Region. Except for some genera, mostly of pioneers such as *Betula, Populus, Liquidambar,* and *Liriodendron,* almost all of the important tree species must start as advance regeneration beneath partial shade so that shelterwood or selection methods of cutting are generally required for their silviculture (Kellison, Frederick, and Gardner, 1981; Kelty and Nyland, 1981; Roach and Gingrich, 1968).

The stands often regenerate well after heavy removal cuttings, such as those of the ''one-cut'' shelterwood method, provided that adequate advance growth exists. Advance growth of oaks usually develops best in partial shade and can then be stored for some years until it is well enough established to release (Holt and Fischer, 1979; Carvell and Tryon, 1961). Some valuable intolerants, such as yellow-poplar, black cherry, and black walnut in eastern hardwood forests, can scarcely exist in mixture except as emergents. Very tolerant species, such as sugar maple, beech, and hemlock, almost always form the lower strata of those stratified mixtures in which they occur.

Surprisingly, complicated stands can be reestablished with these simple procedures. In many instances, especially with hardwoods, it is often hard to know when or whether to intervene with cleanings in the early stages of stand development. Usually, little experience is available to serve as a guide. Often it is best to leave the stands alone and let the trees fight it out among themselves; some remarkably fine stands have developed from what may seem to be neglect. One advantageous kind of intervention is the removal of certain trees, vines, or shrubs that are obviously detrimental and likely to endure.

One of the well-differentiated stratified mixtures is that of eastern white pine, various hardwoods, and eastern hemlock that are occasionally found in the northeastern United States and adjacent regions. In old stands, the scattered white pines are the emergents simply because they continue growing in height after the hardwoods have stopped. The mixed hardwoods form stratum B; the very tolerant hemlock forms stratum C, so that the two layers of conifers are separated by one of hardwoods. If the stands are viewed casually from a long distance to the side, the few pines may appear to form a nearly pure stand; from above, as nearly pure hardwoods; and from underneath, as nearly pure hemlock with some hardwoods and scarce pines. Figure 16.6 shows how such a stand might be regenerated by the shelterwood method.

Of course, many stratified mixtures are less complicated. Two-storied mixtures of fast-growing pioneers, such as aspen poplars, red alder, or the cecropias of the American tropics, over more tolerant species are easily recognized for what they are and treated accordingly.

Comparatively uncomplicated, healthy stratified mixtures of conifers are common in those forests of western North America where soil moisture is adequate for them. For example, Alexander (1987) described the management of several different kinds of mixtures of Engelmann spruce over subalpine fir in the Rocky Mountains.

Moist Tropical Forests

Although the concept of the stratified mixture was originated in complex tropical forests, neither it nor any other interpretation of stand development has led to any universally satisfactory silvicultural procedure for them (Kio, 1979). With such complexity, it is logical to suppose that a large number of procedures, many of them yet unimagined, will be invented.

In both the tropics and the temperate zone, a common silvicultural response to problems of species complexity has been to wipe out mixed stands and replace them with pure plantations of fast-growing species, often exotic, of the early stages of succession. In the tropics, this approach has been most prominent on soils degraded by previous use for agriculture or grazing. On such sites it may, in fact, be the only feasible first step in the reforestation process, regardless of whether it is continued in subsequent rotations. The chief silvicultural problem associated with pure plantations is to keep them pure, especially where highly favorable soil moisture encourages classic jungles of pioneer weed trees and vines. Weeds of these kinds are usually not initially present on abandoned pastures that have become grassy, although tall grass itself can hamper plantation silviculture.

One of the first principles of silvicultural management under the concept of the stratified mixture is avoiding the kinds of severe disturbance of vegetation and soil that favor pioneer weeds. The best way to keep command of the growing space on good sites is to have it well stocked with advance regeneration or sprout sources of desirable species before doing any heavy cutting.

Much of the more theoretical kind of silvicultural and ecological thought has always gravitated toward the perpetuation of the very complicated "primary" or old-growth forests of evergreen broad-leaved species on those sites that are biologically the most favorable. It must be recognized that there is no silvicultural magic that will enable such forests to be harvested and regenerated without altering them in some degree. This has not happened with any other kind of forest. There are all sorts of reasons, including the guidance of silvicultural practice, to safeguard significant areas of old, natural forests as areas of vegetation that are simply preserved and not treated.

Few serious and sustained efforts have been made to achieve acceptable natural regeneration in these forests. The longest experience and closest approaches to success can be found in Southeast Asia (Whitmore, 1984) with the one-cut and "tropical shelterwood" systems. In the tropical shelterwood systems, one or more lower strata are removed to make way for regeneration. It is advantageous that the dipterocarps, which include some of the most valuable species of the region, are heavy-seeded and adapted to start as advance growth.

The trees of tropical forests are often overutilized for fuel or fodder where the human population is dense and underutilized where forests are difficult to access. Where there is clear purpose in maintaining forests and fuelwood is not the overriding concern, silvicultural problems are much easier to solve if uses are found for more of the species. In the past, as in many other parts of the world, the preoccupation has been with a few species highly valued for export or some other specific purpose. Although some species are always more useful than others, a weed may cease to be a weed if a use can be found for it. Furthermore, if a plant has to be eliminated in order to guide some silvicultural development, it is usually more economical to use it than to kill it.

It is commonly alleged or implied that "forestry" is destroying tropical forests. This is true only to the extent that some forestry *agencies* are involved in mining forests for one-time revenue. Sometimes the mining is guided by blind faith on the part of legislatures and some foresters that if only big trees above some diameter limit are harvested they will be magically replaced in kind by the smaller ones. Most often, however, the timber harvests are merely the first step in conversion to agricultural or grazing use. No conscious silviculture practice is possible without stable control of forests and commitment to long-term production.

The principles of silvicultural management of mixed stands in the tropics are not greatly different from those applicable to complex mixed deciduous forests in such places as the eastern hardwood forests of the United States. Regeneration of most, but far from all, important species is dependent on advance regeneration and vegetative sprouting present before any true final-harvest cutting.

The necesssary advance regeneration can usually be established by thinning in the various strata, especially the middle and lower ones, in what amounts to shelterwood cutting. After enough seedlings, saplings, or sources of desirable sprouts are found to exist, they are best released by cutting openings or strips wide enough to allow the regeneration to grow rapidly. The appropriate widths of openings are best determined by observation of results and vary depending on the light requirements of the species chosen for encouragement.

If cutting is too severe or if advance regeneration and sprout sources are lacking, pioneer trees and vines are likely to develop. In such cases, however, the longer-lived tree species may reestablish themselves beneath the pioneers and reclaim the sites after the pioneers have died. Sometimes attempts are made to improve species composition by adding desirable species in enrichment or reinforcement plantings. These are usually made along lines that have been opened through the established vegetation. Very frequent releasing operations are necessary to keep such planted trees free to grow.

In spite of all precautions, some vines or lianas usually climb above the crowns of desirable trees and suppress them. They often form interlaced networks that are so strong that felling one tree may pull over many that should be left. Enough become established with any kind of silvicultural treatment that it is usually desirable to conduct occasional climber-cutting operations as a kind of release cutting. It is especially useful to do this long enough in advance of any felling operations to give the dead vines time to decay.

Some maintain that the production of primary forests in the humid tropics is very low, and others hold the opposite view. The most plausible explanation of these contradictory views is that if only the production by a few scattered trees of cabinet wood is counted, it would be small even if the total biological production were large. However, the situation is complicated by the lack of annual rings and the difficulty of knowing the ages of trees. Stand development may occur so rapidly that the net current annual increment of a 50-year-old stand of primary or climax vegetation in Panama may be as small as that of a 350-year-old stand in Oregon.

Attention sometimes turns to managing for the "secondary" or early-successional species that are often rather easily regenerated. These may or may not be associated with or followed by regeneration of species of later successional status. The combinations and permutations of what may happen or be done are both numerous and mostly still unknown.

Most tropical regions are rather dry, and deciduous forests, savannahs, and grasslands are much more common than moist forests and certainly more so than true rain forests. Most people tend to live in the drier regions. Production of fuelwood is really more important to the population than old-style exploitation of distant humid forests for export species. Under drier conditions, the silvicultural practices are not different in principle from those applicable in most of the temperate zone. Coppice management for fuelwood may be more important, but pure stands are probably just as common.

Relative Merits of Pure and Mixed Stands

The question of whether mixed stands are superior to pure stands depends on the circumstances and not on sweeping generalizations. It takes more care and skill to treat and manage them. Their yields and patterns of development are less predictable than with pure stands. They may be safer, more productive, less costly to maintain, and more attractive, but not always.

The question of whether mixed stands are more productive than pure depends on (1) the units of volume or value on which comparison is based, (2) relative vulnerability to losses, and (3) relative degree of utilization of the growth factors of the site.

In terms of total dry-matter production, single-canopied mixtures, with species mixed in the horizontal dimension, are almost sure to give less than pure stands of the most productive member. The productive potential of the fastest growing species is normally diluted by its intermingling with the less productive. Exceptions would exist if the mixed stand suffered less loss or had some beneficial characteristic, such as improved nitrogen fixation or soil porosity, that increased growth.

Stratified mixtures, however, are likely to show greater total productivity than pure stands of the intolerant overstory species. The tolerant species of the lower strata, by their very nature, are more efficient in photosynthesis than the overstory species. Furthermore, they must to some extent use solar energy that is not absorbed by the overstory. They also use some of that part of total production that might exist in any lesser vegetation that would otherwise develop beneath a "pure" stand. It is possible, however, that a stratified mixture might be less productive than a pure stand of an evergreen, tolerant species adapted to grow as an understory. Evidence on these matters is very scanty. The situation may differ depending on whether soil factors or light are most limiting, or on whether production is measured in terms of weight or volume of wood.

Foresters are usually more concerned about merchantable yield than about total production, even though one cannot be managed well without understanding the other. One highly important consideration is the rate of diameter growth, which in turn affects rotation length. This often means that the typical overstory species may have some valuable attributes, even though it is fundamentally less productive than some associate normally found in the understory. The wood of one species is often inherently more valuable than that of another without respect to its ecological status. If the stratified mixture has virtue with respect to yield, it is often because of the opportunity to combine the quick growth of the individual trees of the overstory species with the slow-starting but high and long-sustained production that the understory species achieves after its release.

Mixed stands probably utilize the soil more effectively than pure stands, although the extent to which they do so is not clear. There is no evidence that the root systems of different species become arranged in different strata like the crowns. All tree species have most of their feeding roots close to the surface, but some also have roots that penetrate more deeply and thus extract water and nutrients from deeper levels. The nutrients ultimately return to the forest floor in fallen leaves and thus become available to the whole community. Not all tree species have the same uptake of each chemical element so that mixed stands may sometimes make more efficient use of the total nutrient capital of a site (Montagnini and Sancho, 1994). In some cases, the inclusion of nitrogen-fixing species such as legumes or alders in mixed stands enhances production.

Mixed stands can be more effective than pure stands in preventing nutrients from leaching out of the forest system. Most evidence suggests that if one component of the vegetation does not make full use of some available nutrients another will appear or expand to do so. Some nutrients leach into streams, but thus far nothing shows that the rate is faster with pure stands.

If stand composition is such that litter decomposes very slowly, much of the nutrient capital of the site may be bound in unavailable form in thick layers of humus. This is most likely to occur under very dry, wet, or cold conditions. This problem is best handled by cuttings or site treatments to speed decomposition. Sometimes, however, the decomposition of coniferous litter is enhanced by an admixture of hardwood litter, which supports a more abundant and versatile population of decomposing and disintegrating organisms.

No manipulation of vegetation can create any increase in that part of the total nutrient capital of a site that is of geological origin. Fixation of atmospheric nitrogen can be improved, but that is about all. The fact that poor soils often have pure stands, whereas mixed stands are associated with good ones, is an effect rather than a cause of the condition of the soil.

The choice between pure and mixed stands is often one between pure conifers and mixtures of conifers and hardwoods. Evergreen conifers generally produce more wood than deciduous hardwoods, and a higher proportion of it is deposited in usable boles. Furthermore, the growth of conifers is diminished less by unfavorable site conditions than that of hardwoods. Consequently, a pure coniferous stand plagued with some ills might still be superior to a conifer-hardwood mixture. Therefore, the conversion of hardwood stands on poor and mediocre sites to pure conifers is usually regarded as desirable. It is on some of the best sites, where hardwoods (and most other species) grow well and where the vegetation is hard to control, that growing quasi-natural stratified mixtures may be the best solution.

BIBLIOGRAPHY

Adam, P. 1992. *Australian rainforests*. Oxford Clarendon, New York. 308 pp.

Alexander, R. R. 1987. Ecology, silviculture, and management of the Engelmann spruce—subalpine fir type in the Central and Southern Rocky Mountains. *USDA, Agr. Hbk.* 659. 144 p.

Aust, W. M., J. D. Hodges, and R. L. Johnson. 1985. The origin, growth, and development of pure, even-aged stands of bottomland oak. In: Proceedings of third biennial southern silviculture conference. USFS Gen. Tech. Rept. SO-54. Pp. 163–170.

Budowski, G. 1963. Forest succession in tropical lowlands. *Turrialba*, 13:42–44.

Budowski, G. 1976. Distribution of tropical American rain forest species in the light of successional processes. *Turrialba*, 15:40–42.

Carvell, K. L., and E. H. Tryon. 1961. The effect of environmental factors on the abundance of oak regeneration beneath mature oak stands. *For. Sci.*, 7:98–105.

Cole, D. M. 1993. Trials of mixed-conifer plantings for increased diversity in the lodgepole pine type. USFS Res. Note INT-412. 9 pp.

Davis, W. C. 1991. The role of advance growth in regeneration of red spruce and balsam fir in east central Maine. In: C. M. Simpson (ed.), *Proc., Conf. on Natural Regeneration Management*. Maritimes Region, Forestry Canada, Fredericton. Pp. 157–168.

DeByle, N. V. 1981. Clearcutting and fire in the larch/Douglas-fir forest of western Montana—a multifaceted research summary. USFS Gen. Tech. Rept. INT-99. 73 pp.

de Graaf, N. R. 1986. *A silvicultural system for natural regeneration of tropical rain forest in Suriname*. Agricultural University, Wageningen, Netherlands. 250 pp.

Denslow, J. S. 1987. Tropical rain forest gaps and tree species diversity. *Ann. Rev. Ecol. and Syst.*, 18:431–541.

Fajvan, M. A., and R. S. Seymour. 1993. Canopy stratification, age structure, and development of multicohort stands of eastern white pine, eastern hemlock, and red spruce. *CJFR*, 23: 1799–1809.

Fiedler, C. E., W. W. McCaughey, and W. C. Schmidt. 1985. Natural regeneration in intermountain spruce-fir forests–a gradual process. USFS Res. Paper INT-343. 12 pp.

Figueroa Colón, J. C., F. H. Wadsworth, and S. Branham (eds.). 1987. *Management of the forests of tropical America: prospects and technologies*. USFS Institute of Tropical Forestry, San Juan, PR. 469 pp.

Food and Agriculture Organization of the United Nations. 1985. *Intensive multiple-use management in the tropics: analysis of case studies from India, Africa, Latin America, and the Caribbean.* FAO, Rome. 180 pp.

Fredericsen, T. W., S. M. Zedaker, D. W. Smith, J. R. Seiler, and R. E. Koch. 1993. Interference interactions in experimental pine-hardwood stands. *CJFR*, 23:2032–2043.

Gomez-Pampa, A., T. C. Whitmore, and M. Hadley (eds.). 1990. *Rain forest regeneration and management.* Parthenon, Park Ridge, NJ. 457 pp.

Hill, D. B. 1987. Stand dynamics and early growth of yellow birch and associates in the White Mountains. In: Current topics in forest research: Emphasis on contributions by women scientists. USFS Gen. Tech. Rept. SE-46. Pp. 20–26.

Hodges, J. D. 1987. Cutting mixed bottomland hardwoods for good growth and regeneration. In: *Proceedings of the 15th annual hardwood symposium.* Hardwood Research Council, Asheville, N.C. Pp. 53–60.

Holt, H. A., and B. C. Fischer (eds.). 1979. *Regenerating oaks in upland hardwood forests.* Purdue Univ., Dept. of For. and Nat. Resources, W. Lafayette, Ind. 132 pp.

Hopkins, M. S. 1990. Disturbance—the forest transformer. In: L. J. Webb and J. Kikkawa (eds.), *Australian tropical rainforests.* Pp. 40–52. CSIRO, East Melbourne. 185 pp.

Hornbeck, J. W., and W. B. Leak. 1992. Ecology and management of northern hardwood forests in New England. USFS Gen. Tech. Rept. NE-159. 44 pp.

Horsely, S. B. 1994. Regeneration success and plant species diversity of Alleghany hardwood stands after Roundup application and shelterwood cutting. *NJAF*, 11:109–116.

Johnson, P. S. 1992. Perspectives on the ecology and silviculture of oak-dominated forests in the Central and Eastern States. USFS Gen. Tech. Rept. NC-153. 28 pp.

Kellison, R. C., D. J. Frederick, and W. E. Gardner. 1981. A guide for regenerating and managing natural stands of southern hardwoods. NC AES Bull. 463. 23 pp.

Kelty, M. J. 1989. Productivity of New England hemlock/hardwood stands as affected by species composition and canopy structure. *FE&M*, 28:237–257.

Kelty, M. J., B. C. Larson, and C. D. Oliver. 1992. *The ecology and silviculture of mixed-species forests.* Kluwer Publishers, Norwell, Mass. 287 pp.

Kelty, M. J., and R. D. Nyland. 1981. Regenerating Adirondack hardwoods by shelterwood cutting and control of deer density. *J. For.* 79:22–26.

Kio, P.R.P. 1979. Management strategies in natural tropical high forest. *FE&M* 2:207–220.

Kio, P.R.O. 1981. Regeneration methods for tropical high forest. *Environmental Conservation,* 8 (2):139–147.

Kittredge, D. B. 1986. The influence of species composition on the growth of individual red oaks in mixed stands in southern New England. *CJFR,* 18:1550–1555.

Ladrach, W. E., and J. A. Wright. 1995. Natural regeneration in a secondary Colombian rain forest: its implications for natural forest management in the tropics. *Jour. Sustainable For.,* 3:15–38.

Lamson, N. I., and H. C. Smith. 1991. Stand development and yield of Appalachian hardwood stands managed with single tree selection for at least 30 years. USFS Res. Paper NE-655. 6 pp.

Loftis, D. L. 1990. Predicting post-harvest performance of advance red oak reproduction in the southern Appalachians. *For. Sci.,* 36:908–916.

Loftis, D. L. 1990. A shelterwood method for regenerating red oak in the southern Appalachians. *For. Sci.,* 36:917–929.

Loftis, D., and C. E. McGee (eds.). 1993. Oak regeneration: serious problems, practical recommendations. USFS Gen. Tech. Rept. SE-84. 319 pp.

Manuel, T. M., K. L. Belli, and J. D. Hodges. 1993. A decision-making model to manage or regenerate southern bottomland hardwoods. *SJAF,* 17(2):75–79.

Marquis, D. A. 1979. Shelterwood cutting in Allegheny hardwoods. *J. For.,* 77:140–144.

Marquis, D. A. 1981a. Removal or retention of unmerchantable saplings in Allegheny hardwoods: effect on regeneration after clearcutting. *J. For.,* 79:280–283.

Marquis, D. A. 1981b. Survival, growth, and quality of residual trees following clearcutting in Allegheny hardwood forests. USFS Res. Paper NE-277. 9 pp.

Marquis, D. A., R. L. Ernst, and S. L. Stout. 1984. Prescribing silvicultural treatment in hardwood stands of the Alleghanies. USFS Gen. Tech. Rept. NE-96. 90 pp.

Means, J. E. (ed.). 1982. *Forest succession and stand development research in the Northwest.* Oreg. State Univ. For. Res. Lab., Corvallis. 171 pp.

Miller, G. W., T. M. Schuler, and H. C. Smith. 1995. Method for applying group selection in Central Appalachian hardwoods. USFS Res. Paper NE-696. 11 pp.

Montagnini, F., E. González, C. Porras, and R. Rheingans. 1995. Mixed and pure forest plantations in the humid neotropics: a comparison of early growth, pest damage and establishment costs. *Commonwealth For. Rev.*, 74:306–314.

Montagnini, F., and F. Sancho. 1994. Above-ground biomass and nutrients in young plantations of indigenous tree species: implications for site nutrient conservation. *Jour. Sustainable Forestry*, 1:115–139.

Panayotou, T., and P. S. Ashton. 1992. *Not by timber alone: economics and ecology for sustainable tropical forests.* Island Press, Washington. 280 pp.

Oliver, C. D., and R. M. Kenady (eds.). 1981. *Biology and management of true fir in the Pacific Northwest.* Institute of Forest Resources, University of Washington, Seattle. 344 pp.

Oliver, C. D., and B. C. Larson. 1996. *Forest stand dynamics.* Wiley, New York. 518 pp.

Roach, B. A., and S. F. Gingrich. 1968. Even-aged silviculture for upland central hardwoods. *USDA, Agr. Hbk.* 355. 39 pp.

Ross, M. S., T. L. Sharik, and D. W. Smith. 1986. Oak regeneration after clear felling in southwest Virginia. *For. Sci.*, 32:157–169.

Siedel, K. W., and P. H. Cochran. 1981. Silviculture of mixed conifer forests in eastern Oregon and Washington. USFS Gen. Tech. Rept . PNW-121. 70 pp.

Seydak, A. H. W. 1995. An unconventional approach to timber yield regulation for multi-aged, multispecies forests. I. Fundamental considerations. *FE&M*, 77:139–153.

Seymour, R. S. 1992. The red spruce–balsam fir forest of Maine: evolution of silvicultural practice in response to stand development patterns and disturbances. In: M. J. Kelty, B. C. Larson, and C. D. Oliver (eds.), *The ecology and silviculture of mixed-species forests.* Kluwer Publishers, Norwell, Mass. Pp. 217–244.

Smith, D. M. 1989. Even-aged management: when is it appropriate and what does it reveal about stand development. In: C.W. Martin et al. (eds.), New perspectives on silvicultural management of northern hardwoods. USFS Gen. Tech. Rept. NR-124. Pp. 17–25.

Snook, L. K. 1993. Stand dynamics of mahogany (*Sweitenia macrophylla* King) after fire and hurricane in Yucatan. University Microfilms International No. 9317535, Ann Arbor.

Sutton, S. L., T. C. Whitmore, and A. C. Chadwick (eds.). 1983. *Tropical rain forest: ecology and management.* Blackwell Scientific, Oxford. 498 pp.

Tesch, S. D., and E. J. Korpela. 1993. Douglas-fir and white fir advance regeneration for renewal of mixed-conifer forests. *CJFR*, 23:1427–1437.

Tranquillini, W. 1979. *Physiological ecology of the alpine timberline.* Springer-Verlag, New York. 137 pp.

Veirs, S. D., Jr. 1982. Coast redwood forest: stand dynamics, successional status, and the role of fire.

Whitmore, T. C. 1983. Secondary succession from seed in tropical rain forests. *For. Abst.*, 44:767–779.

Whitmore, T. C. 1984. *Tropical rain forests of the Far East.* Oxford Univ. Press, London. 2nd ed. 352 pp.

Wilson, R. W. 1953. How second-growth northern hardwoods develop after thinning. USFS, Northeastern For. Exp. Sta., Sta. Paper 62. 12 pp.

Wormald, T. J. 1992. *Mixed and pure forest plantations in the tropics and subtropics.* FAO, Rome. 152 pp.

PART 5

SILVICULTURAL MANAGEMENT OBJECTIVES

CHAPTER 17

TIMBER MANAGEMENT

The most important product of forest management is timber. Many times, however, the words ''timber management'' and ''silviculture'' are used incorrectly as if they were synonymous. Actually, the roots of Western European forestry are as firmly entrenched in the aristocracy's desire for wild game management for hunting as in the production of fuelwood and construction timber. It can also be argued that a major factor in the introduction of scientific forestry in the United States was the concern for clean water. Nevertheless, the production and harvest of trees for wood products has been the primary objective of American silviculture. Silviculture has often been likened to agriculture, and this association has reinforced this perceived single-minded objective inasmuch as the goal of agriculture is producing a crop for consumption. Foresters have felt comfortable with this objective in the belief that ''good forestry'' (the production of timber with concern for preserving the productivity of the soil for future crops) would result in strong, viable wildlife populations, clean water supplies, and ample recreational opportunities.

The introduction of linear programming techniques into contemporary forest management institutionalized this single-purpose perception. The mathematical optimization equations used a single ''objective function'' combined with a host of constraints. Managers strove to maximize the economic value of the forest, which was soon described as the Net Present Value of the forest (described later in this chapter). Because wood products were either the only or the dominant sources of income, maximizing the timber value of the forest became an exercise in scheduling harvests (income) and stand-regenerating and stand-tending operations (costs) in a way that optimized the cash flow discounted to the present. Concerns about wildlife and aesthetics were reduced to constraints on timber management, such as the size and location of cutting areas and the minimum age of trees at time of harvesting.

423

The fundamental objective of silviculture is better thought of as the tending of trees in forest stands to produce stand structures over the course of time that can be used to provide a variety of goods and services. The production of wood products will generally remain the dominant, if not sole, income source. Therefore, timber management should remain in the forefront of the silviculturist's concerns in most settings, but production of other goods and services and protection of the forested environment also need to be considered as reasons for creating certain types of stand structures.

If timber management is the landowner's primary objective, the regeneration system that is used must be matched not only with the species and site, but also with the rotation length, product markets, and rate of stand development. Too often it is assumed that managing for timber means clearcuts and monocultural plantations. Monocultures are often used, not for biological efficiency, but because of real or perceived market constraints. Industrial forest companies usually own land for a specific purpose, such as supplying wood to a Kraft pulp mill. In this case it might be advantageous (and more profitable for the company as a whole) to produce a crop with very uniform wood fiber characteristics. This is why most industrial landowners in the southeastern United States grow loblolly or slash pine. Except on harsh sites, monocultures of any but the most shade-tolerant species are unnatural and difficult to maintain.

Because of the investment in growing stock and its creation, silvicultural prescriptions need to be sensitive to disturbances that may affect the stand. Silviculture procedures directly related to damaging agents are specifically addressed in Chapter 19, but all plans must take disturbances into account. The two prime factors affecting prescriptions are those disturbances that affect the stand during the initiation or regeneration phase and those that destroy trees before the final expected merchantable product is achieved. A fire may destroy the advance regeneration that was going to be used to regenerate the stand, and an ice storm may destroy the entire stand just before the trees reach sawlog size. Knowing the likelihood of different types of disturbances is crucial in making the silvicultural prescriptions for timber management.

The Role of Wood Utilization

Although forest stands can produce a multitude of goods and services, the objective of timber production should not be underestimated. Wood for timber products, including paper and fuel, is, in terms of tonnage, the most important raw material of civilization except for nonrenewable concrete. With proper soil protection and wise management, forests are an infinitely renewable resource. Silviculture directed at wood production must therefore be aimed at growing useful trees, setting the stage for efficient harvesting, and simultaneously maintaining the capacity of the site and vegetation to produce more.

The utilization and harvesting of wood are generally more costly than growing trees; they also have a certain immediacy in satisfying human wants. Consequently, it is not easy to keep them from becoming such dominating considerations that they defeat the long-term objectives of stand management. Many of the solutions devised to strike balances between long-term goals and immediate wants are put into effect by choices among different silvicultural procedures. Because silvicultural practices determine what kinds of stands and trees will exist in the future, much, perhaps too much, of the ethical concern of forestry for future productivity has traditionally been viewed as a silvicultural consideration.

Mainly because it can truly control some remote future, silviculture for wood production is both blessed and cursed with the most distant planning horizons that govern present actions. Most foresters harvest trees that they did not grow and grow trees that they are not likely to harvest. They must attempt to predict what kinds of trees will be useful in the future and then start to grow them now.

Predictions of future wants are always fraught with uncertainty. Fortunately, the technology of wood utilization seems to develop faster than trees can be grown. The trend is toward enabling the use of poorer and smaller trees of an increasing variety of species. This makes the silviculture easier but can also cause harm and generate complacency. If more kinds of trees can be harvested economically, the greater will be the opportunity for silvicultural manipulation of stands. However, if everything can be harvested, the greater is the temptation to damage forests by harvesting everything.

If everything can be harvested, it is usually because markets have been developed for small or poor trees for some single product such as pulp or fuel. This can lead to the temptation to grow trees only for such purposes. This policy may be justified in places with an overriding local demand for pulp or fuel or on sites that are too poor to produce anything else. It is sometimes argued that one could grow fine trees of traditionally valuable species and then find that no one wanted them. The instances in which this has actually happened are rare and, even then, they involved such specialized products as curved oak timbers for the ribs of wooden ships.

Trees that are large enough to saw or slice into solid or laminated wood products command higher prices than those that are simply chipped or burned. Furthermore, because of the high energy requirements of substitute materials (mostly from nonrenewable resources), solid-wood products also represent the best way to use wood for energy conservation. Ideally, wood fuel and products from chipped wood would come only from mill or logging residues and the large numbers of trees that never become large or good enough for any other use. There are plenty of circumstances in which this ideal is unattainable, so the silviculture is modified accordingly.

It is generally prudent to grow trees of species, sizes, and qualities that will make them usable for a variety of products. Because of market uncertainties and pathological risks, it is better to grow the species that are best adapted to a given site than to try to force the site to produce the species that sells best at the beginning of the rotation. Even though logging machinery changes, it is usually well to assume that comparatively simple and uniform stands will be easier to log than those with intricate spatial variations. It is generally best to start by formulating silvicultural procedures that will create good, productive, healthy stands and then modify the actual practices to meet nonsilvicultural problems. Difficulties often arise when planning starts from some goal of utilization that does not fit the natural circumstances.

Sustained Yield

Consideration of timber management is entwined in the concept of sustained yield. Wise silvicultural planning should provide for a sustainable yield, but it is not always clear what product definitions should be used for the consideration of yield. Most often sustained yield has meant the sustainable production of wood volume. This idea has been enhanced to include optimization such as the mandates to produce the maximum sustained yield from areas designated for timber management. This rather simple order becomes quite complicated because wood yield can be differentiated into either total volume or merchantable volume, and often only the volume of certain species is considered. The time period also needs to be considered; what appears to be sustained yield over the next 500 years may not appear sustainable if only the next 50 years are considered. Similarly, what is sustainable over the next 50 years may not be sustainable over the next 500 years. The area that is to be included is also important. Should the flow of wood products from each acre be continuous, or is the stand, forest, or region the appropriate scale?

Most of these policy questions are best left in the broader realm of forest management or forest policy rather than silviculture, but the silviculturist needs to be fully aware of how each stand fits into the overall management picture. Forests are made up of individual stands. Because the labor and capital inputs available for management are not completely flexible, it should be obvious that the optimal management solution for a forest is not the simple addition of the optimal solutions for each stand. It is also important to realize that each acre cannot be expected to provide maximum output of every good and service. On the other hand, a system of "zoning" in which timber management is completely separated from other outputs such as wildlife or biodiversity usually does not provide a satisfactory outcome either.

Viewpoints about sustained yield vary from unquestioning worship to derision. Close adherence tends to prevail in places such as Western Europe, where society has not forgotten times of timber famine. In localities with much old-growth timber, the importance of the idea has traditionally seemed remote. It is often attacked on the ground that only short-term economic considerations count and that technology can always find substitutes for depleted resources.

Some critics of sustained yield equate the term with *maximum* attainable sustained yield. Maximum sustained timber yield has sometimes been used as an excuse not to manage for other benefits. The overriding consideration of timber in the name of maintaining sustained yield is often the result of flawed economic analysis in which timber is deemed to be the only value in the stand.

It is important for the followers of the concept to note, however, that rigid adherence to annual sustained yield would force timber onto the market when demand was low and withhold it when demand was high. The purpose here is only to show how silvicultural systems are modified to provide for sustained yield and also how these modifications can be applied in ways that either restrict or liberate silvicultural practice.

Over the centuries, nothing has come closer to putting silvicultural practice in a straightjacket than attempts to develop sustained yield by simple, arbitrary procedures. Modern practices for producing a steady wood supply are much less restrictive and increasingly effective. It is easier to include other objectives than during more traditional times when silviculture could not be as flexible. The idea of the silvicultural system itself emerged mainly as a device to ensure that the cutting practices followed in stands would be orderly enough to fit into the old, simple plans for sustained-yield management for whole forests. In fact, the reputation of the silvicultural system as a highly rigid procedure comes more from the requirements of this ancient kind of management than from any that have directly to do with growing stands of trees.

The basic site factors and the disturbance regime of the area need to be considered in any plan for sustained yield. More recently, the idea of **sustainable management** has been used. This term goes beyond the basic accounting system for wood that has been used in the more traditional sustained yield calculations and is extended to the regime of operations in a stand that are adapted to the site, the social needs, and the range of benefits that can be utilized from the stand.

The first planned forestry in the Western world developed in Europe late in the Middle Ages (Knuchel, 1953). The main objective was to guarantee the perpetual supply of indispensable fuel from sprouting hardwoods. Transportation was so primitive that the forest area tributary to a community was rigidly circumscribed. Sustained yield was achieved by

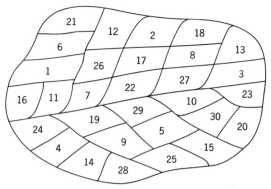

Figure 17.1 Schematic diagram of a forest divided into 30 equally productive stands so that the cut can be regulated by area to provide a sustained yield under purely even-aged management on a 30-year rotation. The numbers indicate the sequence of cutting according to an arrangement designed to avoid having contiguous areas under regeneration. Note that stand boundaries conform to the terrain of a typical stream drainage rather than to any arbitrary pattern.

what has come to be called the **area method of regulation** of the cut. This consists basically of dividing the total forest area into as many equally productive units as there are years in the planned rotation and harvesting one unit each year (Fig. 17.1). Such management worked very well under the coppice system, because there is no kind of reproduction more certain or prompt than that which comes from stump sprouts. It was almost equally successful in growing conifers by clearcutting and planting, a system that imitated agriculture and represented the next step toward more efficient timber management. With such a scheme, reproduction was also reasonably sure and could be obtained without delay.

If sustained yield were the only objective or if the coppice system and clearcutting and planting were the only silvicultural systems, there would be no need of any other methods of regulating the cut to secure sustained yield. However, this regimented kind of forestry with its fixed rotations and annual cutting areas does not provide adequate latitude for dealing with the kinds of nonuniform stands that are either encountered or else created by partial cutting for the attainment of other management objectives. The method of regulating the cut by area can be applied without great difficulty if the stands are essentially even-aged and partial cutting is limited to measures such as thinning and most forms of uniform shelterwood cutting. However, this method becomes very difficult to apply when the individual age-class units become too small for practical area measurement and some kind of uneven-aged condition prevails or is being created.

It may be necessary to work with the uneven-aged condition simply because the particular stands involved happen to be uneven-aged when placed under management. In other instances, the ecological requirements for natural regeneration, wildlife, and watershed management, or protection of the forest necessitate diversity of age classes within stands. Efficient use of growing stock requires recognition of the fact that not all the trees of a stand are of the same age or even of the same size when the best time comes to cut them. This point tends to argue for some departure from the even-aged condition.

The area method is not applicable to those unhappy situations in which severe economic limitations dictate "high-grading," the kind of partial cutting in which only the biggest and best trees can be cut. This kind of cutting may have to be applied to even-aged as well as uneven-aged stands. Although it is hardly an efficient kind of silviculture, it is one in which sustained-yield management is difficult but possible. Careful scrutiny is necessary to ensure that the rate of removal of large trees is kept in balance with the ability of the smaller trees of the residual stands to supply a flow of large trees of adequate quality in the future.

European foresters became aware of the desirability of occasional departures in the direction of uneven-aged management more than a century ago. By that time, the concept of achieving sustained yield through regulation of the cut by area was firmly entrenched not only in the tradition of forestry but also in the laws regulating forestry. Consequently, the idea of introducing more latitude into silviculture could not then be adopted without substantial concessions to the objective of sustained yield and the traditional technique of ensuring it. The method of regulating the cut by volume and probably the concept of the silvicultural system were fruits of these concessions.

In the **volume method of regulation** of the cut the basic procedure is to determine the allowable annual or periodic cut in terms of volume of wood with due regard for the rate of growth, current and potential, and for the volume of growing stock, existing and desired (Davis and Johnson, 1987). Ideally, this means that if one could ever determine and create the appropriate volume of growing stock made up of the proper distribution of sizes and kinds of trees, one could depend on having it yield annually a certain well-defined volume of growth available for cutting. If this could be done, the length of the rotation and the amount of area exposed by cutting each year would not have to be known.

The volume method of regulation is a sophisticated and indirect way of applying the area method. In fact, it depends so often on regulating diameter distributions that it might better be called **regulating the cut by diameter distribution**. Instead of working directly with the idea of cutting equal areas each year under a definite rotation, one determines the distribution of number of trees with respect to D.B.H. that would prevail in the forest at hand if it were composed of a series of even-aged stands with each age class, from age 1 to that of rotation age minus 1, represented by an equal area. The resulting diameter distribution defines the growing stock that should be left after each year's cutting if sustained yield is to be assured; the allowable annual cut is whatever volume this growing stock will yield in annual growth. The process of determining the appropriate growing stock and the allowable annual cut were discussed in more detail and more critically in Chapter 15 on uneven-aged management.

It is easy to get the impression that the volume method replaces the tyranny of the area method with the hard labor of creating and maintaining balanced uneven-aged stands by the selection system. It is true that this technique can be used to make a single stand into a self-contained unit of sustained yield. However, there is seldom any point in going to this extreme because the stand itself rarely needs to be a sustained-yield unit. Usually, whole forests of many stands are the units for which foresters attempt to develop sustained yield.

The main virtue of the volume method is that it frees the forester of the rigid restraints of either extreme. It enables one to make a sustained-yield unit out of a *forest* composed of almost any combination of patterns of even-aged, two-aged, balanced uneven-aged, and irregularly uneven-aged stands that might be desirable to satisfy other objectives. If it is most feasible to keep certain kinds of stands in the even-aged condition, it is still possible to have others elsewhere in the forest in an irregularly uneven-aged status without disrupting a sustained-yield program. The chief requirement is that all the stands of the whole sustained-yield unit in combination conform to the appropriate distribution of diameter classes. Although it calls for skillful forest management, this approach is better than trying to fit the forest into a highly arbitrary pattern merely to simplify the bookkeeping. Usually, it also enables conversion of an existing growing stock into a properly distributed one much faster than would be the case if each stand had to be made perfectly even aged or uneven aged and balanced.

Many of the rigid ideas about regulation of the cut have been handed down from bygone times when transportation was poor and sustained yield was a more crucial necessity than now. The techniques of both area and volume regulation were so crudely developed that the stands had to be uniform if they were to work at all. The age distribution and basic program of management had to be sufficiently simple and sharply defined to be clearly comprehensible to anyone concerned with regulating the cut. The silvicultural system was the vehicle of terminology by which this information was conveyed. In a sense, the silvicultural system, once established, was regarded as a binding contract between silviculture and forest management.

This viewpoint still finds expression in the idea that there is some sort of requirement that a whole forest should be handled according to a single silvicultural system. Another idea that has often became entrenched is that whole forest types, regardless of ownership and objectives, should be managed using the same silvicultural system. The same tradition also makes it easy to conclude that stands must be reproduced at the end of the planned rotation regardless of market conditions or any other consideration. It also leads to the idea that stands must be either distinctly even-aged or definitely balanced and uneven-aged, and cannot be anything in between. There are plenty of good reasons for imposing some degree of uniformity and homogeneity on forest stands, but these are more likely to involve the objectives of increasing the efficiency of logging and other operations in the stands than just those of assuring sustained yield.

Both methods of forest regulation are simple in principle but maddeningly difficult to apply, even when ownership has embarked on sustained yield as a policy. Each method works well only when the proper distribution of age classes exists. This kind of distribution can only be an artificial creation. It could come about in nature only if regenerative disturbances occurred annually and affected equal areas. Even as an artificial creation, the balanced distribution is a goal approached only over more than one rotation; it would probably not be maintainable even if it was attainable. The forest manager is normally concerned mainly with some planned alteration of the existing distribution of age classes. This requires knowing what the distribution is and forecasting the consequences of various silvicultural alterations of the distribution.

The method of area regulation remains the most dependable technique of developing and guiding sustained yield. Mapping techniques such as aerial photography, satellite imagery, and global positioning systems have made it easier to map small areas of different age classes. Geographic information systems have made it possible to determine areas of stands accurately and to investigate the interrelationship of land attributes such as soil type, topography, and stand type. Computer simulation growth models make it possible to project future yields and to investigate variations associated with different schedules of stand replacement. Optimum harvest schedules can be determined by use of various optimization routines of operations analysis such as linear programming and, with advances in multi-objective modeling, other objectives can be included simultaneously.

Area regulation is very useful if the proportion of land in different stages of stand development is also important for many nontimber reasons. Many wildlife species require several development stages, and forest planning must be sure that adequate land areas of each stage in the right spatial proximity will be available into the future.

Modern inventory systems improve the method of volume regulation by volume or diameter distribution. However, the inventory data need to be carefully analyzed and interpreted before long-term programs of sustained yield are designed. Changes induced by growth, mortality, and harvest in the composition and structure of the growing stock of forests must be carefully monitored. Management schemes that combine use of the area regulation method for sustained-yield control and the techniques of volume regulation primarily for analysis of growing stock are normally better than reliance on either method alone. What remains for consideration here are those modifications of silvicultural treatment that might seem to make no sense except for turning forests into sustained-yield units.

Such cases are common in America because real forest management has existed for only a rotation or less and has been applied to forests that rarely have any semblance of the distribution of age classes necessary for sustained yield. Until recent decades, most forest management in the United States was focused on forests that had not been subject to much cutting or severe natural catastrophe and had a plethora of overmature stands and a paucity of any age classes younger than the ultimate rotation age. If heavy cutting has been concentrated in a brief period in the past, almost all the stands will be of one age class dating from the time of the cutting. Even where the inheritance consists of uneven-aged stands of natural or accidental origin, it is usually found that some age classes are deficient and others present in surplus.

Much of the management of old-growth forests in the western United States involved circumstances of this kind. Management might, for example, have started with a large virgin forest consisting almost exclusively of stands that originated after a single massive fire that occurred 375 years ago. If one wished to convert this to a sustained-yield forest on a 100-year rotation the simplest way would be to clearcut 1 percent of the area annually for the next century. One could accelerate this process, among other ways, by cutting over the old growth more rapidly and by planning to cut some of the first new stands before they reached 100 years. However, substantial acreages of old timber would be left standing for decades without any real increase in value. If one considered only the timber management of the individual stand, it would seem witless not to replace all the old timber with young vigorous stands immediately. In this situation it is often useful to conduct salvage cuttings to mitigate losses in overmature stands that are being stored until the time of their scheduled replacement.

In other localities, such as much of the eastern and southern United States, it is common to find whole forest regions full of middle-aged stands that arose after heavy cuttings some decades ago. In many areas the growing stock had neither timber of mature size nor stands of seedlings and saplings for decades. For a long period of time, the forester could do little but eke out meager income from intermediate cuttings that did nothing to correct the distribution of age classes. Many of the first stands that reached the condition where they could be harvested and replaced were removed long before they were mature, just to get some new age classes started and income realized. Some of the stands with which one starts were held well beyond the logical rotation age in order to have something to cut when the missing younger age classes had not reached maturity. Almost any sort of adjustment of age-class distribution is likely to involve holding some stands beyond rotation age and, at other times, cutting some before they become mature.

The same kind of manipulations may be made in more detail with particular diameter classes. Consider, for example, the case of a forest being managed with the intention of growing the crop trees to 20 to 24 inches D.B.H. An overall surplus of trees in the range from 16 to 20 inches is compensated by deficiencies in all diameter classes below 15 inches. If trees smaller than 10 inches D.B.H. are unmerchantable, adjustments must be made mainly by the cutting of trees that are larger than 10 inches. The surplus diameter classes should be picked over very carefully with several definite objectives in mind. In general, the best trees should be reserved in sufficient quantity not only to replace the trees larger than 20 inches that are harvested but also to provide an adequate volume to compensate for part of the present deficiency of trees now less than 15 inches. The removals from the surplus diameter classes should, to the greatest extent possible, come from the poorest members and provide for the early establishment of new reproduction in order to accelerate recruitment of new age classes.

While the surplus diameter classes are winnowed quite rigorously, the deficient diameter classes receive an indulgent treatment that might seem inconsistent. In these classes it even becomes desirable to try to make good trees out of poor ones. Although one would be unlikely to release a 19-inch codominant in these circumstances, it would be logical to cut an 18-inch dominant to promote two adjacent 12-inch codominants to the dominant position. It might be justifiable to release and even prune a 10-inch tree with a 25 percent live crown ratio if one could be sure that it would ultimately recover and develop into a good tree. This would not be the most logical way to grow good trees if one could start from the beginning with seedlings, but there is the necessity of creating a good 12-inch tree in 15 years rather than in several decades.

The adjustments are by no means always made with partial cuttings. In the example at hand, one might carry out the seemingly reckless and premature harvest of a fairly good stand with trees mostly in the 16- to 20-inch range even while carefully nursing a rather poor adjacent stand in which most of the trees were 8 to 15 inches.

One swift remedy for deficiencies of age or diameter classes within the area being managed for sustained yield is the acquisition of stands and forests that have them. It may, for example, be wiser to purchase a heavily cutover area than to clearcut an immature stand to balance the age distribution of the forest. Management for sustained yield is not an objective to be pursued unquestioningly, but neither can it be ignored. The imbalances that exist must be identified, and their consequences must be faced. Decisions have to be made about whether to rejuggle age-class distributions and, if so, how this will alter silvicultural treatment. Perfect solutions do not exist. Ugly surprises or hard times lie in wait for the successors of forest managers who ignore, aggravate, misunderstand, or neglect the problem.

Preoccupation with sustained yield for timber should not obscure the fact that the appropriate distribution of age classes also provides for a continuing flow of other benefits. It also brings about high diversity in those ecological factors governed by stand age. The benefits to the environment that go with using forests to sequester carbon and chemical nutrients are at their greatest if there is an equal representation of all age classes up to maturity. When annual losses of biomass to death and decay exceed annual production, the stand begins to release carbon and nutrients back into the rest of the system. Some of the benefits associated with other nontimber uses of forests require some stands older than optimum rotation age for timber. Others require an allocation of area to young stands that is too great for optimum sustained yield of timber. However, in general it can be said that age-class distribution that is best for sustained yield of timber is at least good for the sustained flow of most other important benefits of the forest.

Financial Considerations

Timber management decisions depend heavily on financial considerations. The monetary value of standing timber equals the value of the wood products that are produced from the timber minus all costs incurred during harvest, transport, and manufacture. Therefore harvesting costs are important; the value will increase if the harvesting and transportation costs are reduced. These costs can be minimized if the equipment type and size are appropriately matched with the trees and the site and the transportation system provides quick access to the manufacturing site. Decisions are also affected by the various modes of investment analysis and the way that interest rates are incorporated into it. These factors influence rotation lengths, decisions about thinning and pruning, and other silvicultural matters. Much of this book describes silvicultural methods aimed at producing high-value trees, but other financial considerations need to be dealt with.

Planning the silvicultural procedures for timber management requires knowledge of the basic site characteristics of the stand. Timber management is based on rather simple financial relationships. Managing for timber value needs to be thought of as seeking a good return on the investments (costs) made during the growth of the timber. The site characteristics will determine what timber products can be produced most "naturally." Certain species will have a competitive advantage over others on any given site. Silvicultural techniques using large inputs of machinery or chemicals can often overcome this advantage, but usually at high cost (reducing the rate of financial return) and sometimes with damage to the soil, the fauna, or water quality.

Treatment of Monetary Investments

There are two basic capital investments in forestry. The most obvious is the money actually invested out-of-pocket in the costs of growing trees and holding the land beneath them. The usual way of analyzing these kinds of investments is by using compound interest calculations to compare the **net present values** of different courses of action. In this type of analysis all costs and revenues are discounted to the present using a chosen interest rate. Such matters are treated in much more detail in various works on forest economics (Gregory 1987; Leuschner, 1984; Gunter and Haney, 1984).

One of the most important uses of this particular analytical technique is the financial comparison of different programs or systems of silvicultural treatments. It is a technique used by foresters and other planners but not in conventional accounting, which does not recognize unrealized future returns. Comparisons of present net values help with choices about regeneration techniques, rotation lengths, thinning programs, logging methods, and many other decisions made in building a silvicultural system. Such analysis always militates in favor of securing early returns and postponing costs, with these effects being more pronounced the higher the chosen interest rate.

The basic concept is sometimes thought of as an enemy of long-term forestry, but the real culprit is often the sloppy use of it. Society and forest owners seldom use such analysis to determine whether to engage in long-term silviculture. That decision is usually based on some more intuitive kind of concern for and faith in the future. When comparing two quite different land uses a different type of financial analysis, **internal rate of return** is often used. This analysis avoids the problem of using a set interest rate and compares the interest generated by the project itself. The internal rate of return is the breakeven interest rate where discounted revenues exactly equal discounted costs.

Once the decision has been made to invest in long-term forestry, analysis of present net value becomes a forester's tool for comparing different ways of executing the policy and of detecting financially optimum solutions. Before it has been decided to invest in silviculture, much is made of the difficulty of securing high rates of interest on such charges as regeneration costs. In many places regeneration after cutting has become a legal requirement. In this case the cost of regeneration can be viewed as a harvest cost for the previous stand and is usually treated as such. However, the analyses done by calculating present net values become ways of determining which mode of stand regeneration is financially best given that the stand is to be replaced. The net present values are calculated for all management alternatives, and the highest value is chosen. This is the purpose for which these methods were first developed; their use as a deterrent of forestry is an unfortunate byproduct.

Another mode of financial analysis of forestry is called **financial maturity analysis** (Davis and Johnson, 1987, Leuschner, 1984). This considers the interest that growing trees earn on their own value. The realizable value of a stand at any point in its development is treated as an investment, and the rotation is ended when the increase in value of the stand ceases to exceed some desired rate of compound interest. The return on this particular kind of investment is zero until the first day one tree in the stand is suddenly worth one penny of net value. On that day, the compound interest earned with that one penny, on an investment of zero of the day before, is infinitely high. From that day onward, except for important effects of trees improving in value because of quality improvement or becoming suitable for products of higher value, the rate of compound interest earned on this peculiar kind of investment represented by the stand sinks inexorably downward. However, this rate of decline can be slowed or temporarily reversed by commercial thinning (Fig. 17.2).

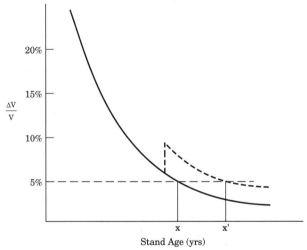

Figure 17.2 Using financial maturity analysis to set rotation length. The actual annual increase in growing stock value per acre (ΔV) divided by the current value of the growing stock (V) is the percentage of increase in growing stock value. This rate of increase would have no meaning when the stand was young and had no value. At the point when the stand just begins to have value, the rate of increase would be infinite; from this point on, the rate of increase declines each year. In this system the time to harvest (rotation length) is set as the time when the value increase rate declines to a predetermined rate (in the graph shown as 5 percent). A stand growing along the solid line would be harvested at time X. The effect of a low thinning is shown by the dotted line. The thinning causes an immediate rise in the increase rate because there is less growing stock and the more valuable trees are left. As the rate of increase again declines, the rotation age would occur at X' which is later than in the unthinned stand. The rate of increase might not decline as rapidly as shown if the thinning made the remaining trees grow valuable wood faster.

The same concept can be applied to individual trees. For example, a tree increases from 18 to 20 inches D.B.H. and from 53 to 63 cubic feet in volume. Quality and other factors associated with change in size increase stumpage value per cubic foot from 66 to 83 cents, so tree value increases from $35 to $52.20. If the tree has been growing 10 rings per radial inch and so continues, it takes 10 years for this increase to take place and the compound interest earned on the $35 would be 4 percent. However, if the tree can be made to increase in radial growth to a rate denoted by six rings per inch, the interest earned during the period shortened to six years would be 7 percent. If the goal rate was 6 percent, it would be logical to harvest the tree at the beginning of the period in the first case but to leave it if the growth acceleration of the second case seemed likely. In the first case the tree is economically mature, given the choice of interest rate; in the second case, it is not.

It should be clearly recognized that the values of the two trees were *not* the costs of growing the trees. There is not necessarily any direct relationship between this kind of value and the growing costs. The value of the growing stock is referred to as the opportunity cost because the owner has the opportunity of harvesting the tree and investing the value elsewhere. Thus the analysis of return on growing stock is different from the analysis of actual monetary investments by calculations of present net value, described in the previous section.

Choice of Interest Rate

The rate of compound interest demanded on investments will not be the same for all owners; it may also be higher for actual monetary investments than for return on growing stock. A logical rate is one realistically available to the owner from some alternative investment that might actually be made, often in some other part of the forestry enterprise. It is possible to arrive at a choice by analytically subdividing an interest rate into its logical components.

The first component is a risk-free cost of borrowing money (or investing), usually regarded as approximately 3 percent. Allowance must also be made for certain risks. One is the risk of unsalvageable loss of values from damaging agencies. Sometimes certain data about percentage rates of annual losses can be used to help with estimating the size of this component. Another kind of risk can be called the "market risk," the risk that the demand for the trees being grown will dwindle or vanish. People no longer want elm wood for wagon-wheel hubs and their current demand for Douglas-fir framing timbers might someday wane. This kind of allowance must be made, but the percentage assigned to it depends on imaginative forecasting. The combined effects of these risks is considered to amount to 1 percent to 3 percent, so 4 to 6 percent is often used for the interest rate in computing net present value rather than the risk-free 3 percent.

One offsetting component that can be kept separate from or included in the market risk is allowance for any appreciation in value. This is, in general and on the average, more real and less elusive than the risk that some kinds of trees will become technologically underemployed. The stumpage values of quality sawtimber have increased at an average rate of nearly 6 percent compound interest in terms of constant dollars for many decades, whereas values of low-quality wood such as pulpwood have remained almost constant. The overall increase in wood prices is often considered to be 1.5 to 5 percent, depending on the relative amounts of high- and low-quality wood.

If capital is limited, then the interest rate must reflect the other possible investment returns for the capital tied up in the growing stock or the cost of borrowing more capital. Most large companies have a large amount of capital available to them, but at a cost significantly higher than the risk-free cost of capital (3 percent). Small landowners may be limited in their borrowing capacity, and the interest rate may be very high.

One component of interest rates normally does not belong in the rates used for analysis of investments in forest-growing stock and silvicultural treatment. That is the allowance for inflation or other change in the value of money. If a tree is worth a pair of shoes today and it neither grows nor loses its utility during the next 10 years, it will still be worth a pair of shoes. Allowances for inflation have to be made for investments in bonds, savings accounts, or other paper promises of future payment of money. Trees are inflation-proof equity. In times and places of rapid inflation, they have long been a refuge for capital, provided that their owners are not taxed severely. In analysis of returns on investments made in growing trees or in value of growing stock, most of the effects of inflation are canceled by making the calculations with present costs and values for all investments and returns regardless of when they occur. The possibility that costs might increase faster than values is a factor included in the allowance for market risk.

Other Financial Factors
Sometimes special circumstances dictate special financial needs. Forests often can provide relief because of the unusual flexibility of the investment. One case is cash-flow constraints. Some owners, particularly small landowners, need cash even though their forest may not be financially mature. Lending sources have been reluctant to accept growing timber as secure collateral for loans because of perceived risk in timber management. Some owners have a strong aversion to assuming debt at any cost. Another example of cash-flow constraints for small landowners is the need to meet annual costs such as property taxes.

Larger owners also have cash-flow constraints. A typical problem is the large debt service after a major acquisition such as land or mills. The company may choose to liquidate growing stock timber on forest stands in order to reduce debt and increase dividends and stock prices even if this jeopardizes future yields. The opposite condition sometimes develops when a company builds up certain age classes to solidify timber self-sufficiency for their mills which can improve their bond rating and reduce their future borrowing costs.

Rotation Length
The planning of thinning schedules and the setting of rotation lengths are strongly interactive kinds of analysis.

There is one kind of biological limit on rotation length. When the dominant trees cease to grow much more in height, shoot elongation and crown expansion also become very slow. At this time the trees cease to respond much to thinning, so there is no point in continuing such treatments. This development is one of the causes of the declines in production rates that set economic limits on rotation lengths. Growth continues, but at a declining rate. Losses to rots or diseases can also increasingly offset growth. The only reason to have longer rotations is the desire to allow trees to grow very slowly to large diameters. The culmination of height growth comes very early in some species, especially those that grow fastest in height in youth; slash pine and aspen are good examples. At the other extreme are species such as eastern white pine, many western conifers, and yellow-poplar, which continue height growth to substantial heights and ages. Other factors, such as interest rate, influence the optimal rotation length, and frequently rotation lengths are chosen which are shorter than this biological limit.

Two different principles are used to set rotation length, although in practice they are interrelated. One, which may be termed the **physical rotation**, is aimed at maximizing physical yield under long-term sustained yield in terms of the mensurational units of yield.

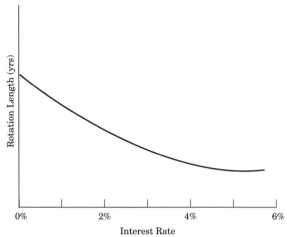

Figure 17.3   Effect of demanded interest rate on optimal rotation length. As the interest rate is increased, the rotation length is shortened (see Fig. 17.2). The longest optimal rotation is when the interest rate is 0 percent. This assumes that capital has no value and that the rotation would end when M.A.I. was maximized. Using a rate higher than 0 percent leads to rotation lengths shorter than that at which M.A.I. is maximized.

The other, the **financial rotation**, has the goal of optimizing money return on capital under the soil rent principle; calculations of it are characterized by use of compound interest. The idea of the physical rotation fits best when it is necessary to guarantee a perpetual supply of timber to some voracious local or regional forestry industry from a fixed and limiting amount of forest land. The financial rotation concept applies best when the amount of capital available for long-term silvicultural investment is the limiting factor. If the interest rate for financial rotation calculation is set to 0 percent, then the rotation length will be the same as that calculated for the physical rotation. An interest rate of 0 percent means that capital is in no way limiting (Fig. 17.3).

The physical rotation length is that of the maximum mean annual increment in terms of the mensurational unit chosen for measurement of yield. The financial rotation is that of highest present net value of all costs and returns discounted back to the beginning of the rotation at some chosen rate of compound interest. Both kinds of choices are guesses about distant future conditions. Rotation lengths can vary widely depending on which mensurational units, future prices, and interest rates are selected. The full range of possibilities can be great enough to make the difference between the two basic principles seem trivial. Nevertheless, the usual situation is that financial rotations are shorter than physical rotation lengths.

The higher the chosen interest rate, the shorter are the rotations and the greater the advantage from making commercial thinnings early in the rotation. In the case of physical rotations, the analogous variable effect comes mainly from the way in which tree diameter affects the proportion of total production that is counted as yield. The higher the minimum diameter that is set on what is reckoned as yield, the longer is the rotation and the greater it will be prolonged by kinds of thinning that increase diameter growth.

Under any given set of assumptions, the effect of not thinning is usually to shorten rotations because there is no salvage of prospective mortality and no acceleration of diameter growth is possible.

Analysis of Stand Increment

No matter what system is being used to determine rotation length and the scheduling of other timber harvest operations, it is imperative to understand stand growth rates. The bookkeeping needed to track the change in the amount of wood in stands and forests can be very tricky. In the first place there is the problem that the change or "growth" is determined usually from measurements made at two different times. Each measurement itself is subject to errors, so this "growth" (the difference between two measurements with high variance) is an estimate with even higher variance.

Second, it is important to subdivide the "growth" into its components. **Survivor growth** is that on trees present at the beginning and end of the period. Other components include the **volume harvested** during the period, and the **mortality** that took place but went unsalvaged. The growth that took place during the period on trees that died before the second measurement cannot be determined by repeated plot measurements. It may be noted that the growth loss from unsalvaged mortality counts as a heavy tax against the survivor growth of the period. This is because the loss is not merely of what grew during the period but also the loss of all the production that had previously accumulated in those trees.

This is counteracted in some degree by another important artifact of the bookkeeping, **ingrowth**. Ingrowth takes place when trees pass some arbitrary minimum D.B.H. and are counted for the first time. This contribution to production is magnified because these trees suddenly acquire a substantial volume, much of which was really laid down before the measurement period started.

In a perfectly balanced uneven-aged stand (or forest), these components would theoretically be the same in each successive period. Because most stands and forests are unbalanced, the magnitude of these components fluctuates markedly from one measurement period to another. However, by keeping track of the separate components, one can more dependably analyze the growth of forests and reduce the risk of large errors in prediction. There is even more opportunity for error if volume estimates are made in board feet rather than cubic volume. The ingrowth, a component always difficult to assess properly, is usually very large when measured in board feet.

The board-foot volume of a tree increases so rapidly with diameter that small fluctuations in the diameter distribution cause remarkable differences in the apparent growth. This phenomenon is accentuated by the use of misleading log-rules; with the Doyle Rule, for example, there are overestimates of growth in diameter classes less than 22 inches.

The choice of mensurational units is important in any calculation of stand growth. If wood is to be chipped up for bulk products such as pulp or particle board or if it is used for fuel, the appropriate units would be merchantable volume or dry tons, ordinarily with small diameter limits. With this kind of utilization, rotations would be short and any advantages of thinning small but not necessarily absent. If the trees are to be made into sawn or sliced wood products, the units are more likely to be thought of as board feet in the United States or cubic meters of sawn wood elsewhere, each with comparatively large diameter limits set on how much wood is counted. With this approach, the rotations are comparatively long, and the effects of thinning on diameter growth become crucial and controlling.

None of these approaches, however, really solves the problem of taking the economic effect of tree diameter fully into account because all lead into the trap of the simple assumption that all units of a given kind of product are equally valuable. The only way out is the difficult one of using studies of logging costs and product values to determine the true net value of units of product from trees of differing size (and, for that matter, quality). If this line of logic is followed to its end, it is seen that the most logical physical rotation would be that which maximized mean annual increment of money, which is a better measure of social utility than cubic volume. Financial rotations would be determined by the same calculations except that their length would be shortened by taking the time value of money into account.

Regardless of which economic or financial methodology or mixture thereof is followed, there is the general effect that the simpler the assumptions about mensurational units and net value the shorter will be the rotations compared with those that would prevail in a state of full knowledge. The only simple assumption that logically shortens rotations is the selection of a higher rate of compound interest under soil rent and the logic of this depends on whether the chosen rate is appropriate to the circumstances. One pragmatic way of dealing with this general situation might be to calculate a rotation from simplistic assumptions and then lengthen it on the basis of intuitive judgments about those less knowable factors that tend to lengthen rotations.

Concentration and Efficient Arrangement of Operations

A number of considerations have been presented that tend to argue more against than in favor of developing uniform stands. It has, for example, been pointed out that the objective of sustained yield is hardly cause by itself for inflicting the uniformity of the pure, even-aged stand on silviculture. In general, highly flexible procedures of cutting and arrangement of stands enable the closest approach to optimum use of both growing stock and growing space, and in lesser degree to the successful regeneration of the forest and its protection against damaging agencies. These objectives are, however, not always best attained by creating or maintaining irregular stands and there are other important reasons for resisting the temptation to do so. Most of these involve the point that all sorts of operations can become complicated and expensive in a stand composed of a mosaic of age classes or species.

Harvesting Costs

The harvesting of timber crops is usually the most expensive operation conducted in the forest. Therefore, it is important to arrange stands so that costs, per unit of volume harvested, will be kept at the lowest level consistent with other objectives. Transportation is the component of logging costs most affected by the arrangement of stands. If the merchantable age classes or species are scattered rather than concentrated in a contiguous unit, the gross area that must be covered to harvest a given volume of timber in a single operation is correspondingly increased. This is especially true if terrain is difficult, if roads must be built or improved for each operation, or if the cost of shifting heavy equipment from one operation to another is high. If the heterogeneity of the stands dictates handling a broader range of sizes, qualities, and species of trees than is possible with a single set of machinery or a single procedure, there is the additional cost of having a wider variety of equipment or of trying to handle material with equipment not suited to the purpose. The cost of supervision also tends to increase the more scattered and complicated the operation.

The ideal stand from this viewpoint would be the largest even-aged, pure stand that could be harvested during the period before the harvesting equipment had to be moved for some reason not related to characteristics of the stand. One point of departure for thinking about this matter is the size and shape of the cutting area tributary to a single log-loading point. If the costs of trails or roads to the loading points are high, there will be a tendency to make the cutting units, which often become new stands, even larger.

In considering the relationship between silviculture and logging costs, it is crucial to distinguish between area-related *transportation* costs and those of *handling and processing*, which are most strongly affected by tree size. For example, the costs of felling, bucking, loading, and manufacturing material from forty 20-inch trees are not significantly different if they come from 1 acre or from 10 acres of gross stand area. The larger the trees, the lower are these handling costs per unit of volume, at least up to the point where the trees are too large to be handled by the chosen kind of equipment.

If the only objective were to minimize transportation costs that depend on the volume of cut per gross acre, the solution would be to cut everything worth the cost of harvesting and processing. If, on the other hand, it were to minimize handling costs that depend on tree size, only the larger trees would be cut.

The question at hand, however, also involves planning for the production of material for future harvests. Not only does this mean that a host of other objectives must be considered, but also that the plan covers a series of harvests rather than just one. In principle, it is no more difficult to accelerate the rate at which trees grow to good size in uniform stands than it is in heterogeneous ones. In other words, problems involving tree size can be solved in either kind of stand, but those involving transportation can best be solved with uniform stands. It is, therefore, ordinarily better from the standpoint of a long-term program of harvests to work toward increasing the uniformity of stands.

Another cogent argument for uniform stands is the point that the forest industries are hamstrung at all stages of harvesting and manufacture by the tremendous variability of raw wood. This necessitates repeated sorting and complicates all handling operations; some of this extra work is avoided in the processing of competitive materials that are more homogeneous. To cite merely one example, the variability in size of trees dictates that logging equipment be made versatile at the expense of being completely efficient. Any silvicultural steps that can be taken to reduce variability are advantageous, although there are inherent biological limits to the uniformity that can be achieved. Variation in diameter as well as in other characteristics is bound to appear in any stand that fully occupies the site for any length of time and in stands where competitive effects are used to develop good stem form. However, it is certainly possible to counteract the heterogeneity often found in nature.

Many methods of partial cutting, especially thinning, can be applied to enable the systematic removal of various relatively uniform categories of trees at successive intervals, with a high proportion of the volume being harvested from favorable ranges of diameter. The main objective should be to obtain the widest possible margin between the costs of logging and the value of the material harvested during a series of harvests. This goal is not likely to be achieved if the sole consideration is minimizing the cost of each separate harvest without regard for those that will follow. Silvicultural measures to create trees of good size and quality, preferably in stands of high volume per acre, contribute fully as much.

Other Silvicultural Costs

Most other operations are facilitated by systematic arrangement of uniform stands. Silvicultural operations such as pruning, release cutting, and site preparation can become very complicated and expensive in heterogeneous stands. Such blanket operations as prescribed burning and application of chemicals from the air are expensive or impossible unless large areas can be treated as solid units. Timber marking is very time-consuming and expensive in heterogeneous stands, a situation aggravated by the fact that the more complicated the stand the more expensive the talent required for the marking.

Much of the cost incurred in harvesting wood is for developing access to the logs. Road-building costs are often attributed to specific timber sales without much consideration of road life or other road uses. Planning timber sales and access road building jointly can reduce harvest costs and therefore increase timber values.

Road Network and Harvest Layout Design

It is important to consider stands not only as individual units, but also as part of a forest and a landscape. Many important considerations for wildlife and aesthetics are considered in the following chapters, but the arrangement has implications for timber management as well. Stands are connected by a road network that brings the logs to the mills. Just as the value of the trees to the owner is greatly affected by harvesting costs, transportation has to be considered as well.

The forest manager can do little about the relative location of the forest and the mills, but can exert control over the road network within the forest. Many plans labeled forest management plans have really been harvest plans. Stands were selected for harvest and the road network extended from stand to stand. This system has often led to an inefficient road network and has locked the silviculturist into a less than ideal pattern of stand arrangement.

The roads are such an important determinant of timber value that their placement and construction should be considered at the outset. A poorly designed road system can leave some areas inaccessible, and poorly designed roads will slow trucks and increase costs. Roads that are badly placed can result in mass wasting, which can take valuable land out of production as well as cause damage to stream water quality and fish habitat. Similarly, poor stream crossings can lead to recurring stream bank and bridge repairs, adding significantly to road costs.

Preexisting stand boundaries should not be changed without good reason. Where it is desirable to create new stand boundaries or variations within stands, they should coincide with features such as changes in site, variations in accessibility, and barriers to damaging agents. It is especially important to try to get stand boundaries to fit major differences in site conditions. Where stand boundaries are merely artifacts of variation in disturbance history, not site conditions, it may be desirable to make some changes in the boundaries during one rotation in order to better design the road network. This might mean that part of the stand will be cut earlier or later than maturity, but conditions will be set for better forest management in the future.

The road system and the stand harvest schedule together determine the management of the landscape. To provide the most flexibility in the harvest schedule, it must be possible for logs to move quickly and easily from all parts of the forest. This will also permit timely application of intermediate operations such as thinnings and fertilization.

The access provided by the road network can be both a blessing and a curse. When forest fires occur, it makes it easy to bring people and equipment quickly to the site. On the other hand, increasing public access increases the number of fires caused by people. The network provides access for hunters and other sports enthusiasts, but may cause excess hunting pressure on small populations of vulnerable animals. Usually, the cost of controlling road use is minor compared to the costs incurred by not having roads where and when they are needed.

Although the road system should be carefully designed and constructed rather than added piecemeal, a segment at a time, it does not all have to be built at one time. Road-building costs are formidable, and the construction of a large network before harvesting is usually unwise. The road system should be designed at the outset, then a forest management plan should be developed, followed by construction of roads as needed.

Through a landscape management plan, the forest can be cut so that timber management can be carried out in an efficient and profitable way. At the same time, other goods and services can be obtained through a well-planned spatial arrangement of different species and ages. The road network is the glue that holds the plan together while increasing the income from timber.

Aesthetic Considerations

The layout of harvests should take aesthetic qualities of the landscape into consideration. People often dislike change, and the layout of new or altered stands will be accepted only if most people feel that the visual change is for the better. There are many aspects to the aesthetic quality of the landscape and, whenever possible, they need to be taken into account.

The visibility of the area will determine how much effort needs to be devoted to visual qualities. The visibility is determined by the position in the landscape (high areas are more easily seen than low areas), the steepness of the area, and the blocking effect of the stands in the foreground.

The principles used in visual design include shape, scale, diversity, unity, and the "spirit of the place" (U.K. Forestry Commission, 1992, 1994). Shape is the most important of these principles. Whenever possible straight lines, particularly those that follow ownership or administrative rather than natural boundaries, should be avoided. The more that shapes can interlock, the better. Harvested areas should blend into the landscape. The scale of the harvest is also important. Stands resulting from natural disturbances have a scale that is influenced by site and natural disturbance; harvest areas should have a similar scale. Scale is also influenced by diversity. Increased diversity will tend to reduce the scale of the harvest areas from a visual standpoint. A large cut in one continuous forest type will be more noticeable than a similar cut in a more diverse forest type. In addition, the harvest area can be designed to unify or connect other stand types in the viewshed. Effort should be made to increase the beauty of the view by putting the uncut areas into better perspective.

Landscape designers also give weight to the "spirit of the place." The altered stands need to fit people's perceptions of what the landscape "should" look like. This perception will change over time, and the harvest design should also change over time.

The view must be considered in two parts—the foreground and the distance. More attention should be paid to individual tree and edge characteristics in the foreground (close to the viewer). Designed edges need to be more sweeping in the distance because the viewer will not be able to detect subtle changes. In addition, the edges near the viewer can "lead the eye" to more appealing views in the distance.

Anything that can be done to screen cutting areas and logging debris from public view will reduce problems. In the immediate stand, where people might walk, the aesthetics of harvest operations need to be controlled (Jones, 1993). Height of the remaining slash, size and form of the uncut trees, and rutting of soil from harvest equipment are all important. Foresters and loggers need to work together to plan skid trails in ways that are not visually offensive and that will provide access rather than do environmental damage. The height of cut tops and slash should blend into the height of the understory and shrubby vegetation.

The visual beauty of a landscape is dynamic; as trees grow, the view changes. Changes in the landscape can be modeled through computer simulation and landscape visualization systems. Silvicultural activities can be planned so that the long-term effects as well as the immediate effects can be taken into consideration. Aesthetic matters in harvest planning will likely have a greater role in the future, and stands need to be organized to lead to an orderly progression of harvests.

Control of Operations

The success of silviculture depends heavily on the proper execution of the harvesting operations that are ordinarily used to carry them into effect. Cutting is the chief tool by which the forest is controlled; this is true even when the main objectives are the management of resources other than timber, such as water, forage, or wildlife.

The most important problems are encountered in the application of the numerous kinds of partial cutting that are so commonly essential to the efficient use of the forest to produce wood and other benefits. When whole stands are cut in single operations, these problems are changed in kind but not in magnitude.

The harvesting of trees for wood products is, from the silvicultural viewpoint, an indispensable means of controlling forest vegetation as well as an economic end in itself. The felling and removal of trees is the most costly forestry operation. In most instances, the objective is to proceed so that the value of the products substantially exceeds the costs of harvesting and subsequent operations. However, the opportunity to recoup even some of the removal costs through conversion to wood products can make it possible to grow forests for aesthetic or similar reasons at endurable expenses.

If a tree is to be cut and used, nothing that can be done about the fact that the distance from where it grew to the point of use is fixed. The best that can be done about cost of moving the cut product is to try to avoid incurring it while the tree is any smaller than it must be. The most favorable balance between costs and returns in harvesting a stand are more likely to be achieved not by cutting all the trees but by removing the largest and best. Divergence between the long- and short-term viewpoints arises mainly over the kind of trees cut and left and not so much over the total volume cut and left. Large trees of good quality, which are most attractive for immediate harvest, can be of sufficient vigor that they will bring a better return if left to grow. Conversely, there is a temptation to leave small or poor trees of kinds that are attractive for neither present harvest nor future growth but just impede the growth of anything that might be better.

From the silvicultural standpoint, it is best to have all operations connected with harvesting, from the tree at least to the point where forest roads leave the forest, under the direction of the forester. The prevalent practice of putting harvesting responsibilities in the purchasers or contractors, whose primary interest is necessarily in cheap logging, is not ideal.

Designation of Trees and Stands for Cutting

The trees or stands that are to be cut, left, or otherwise treated in harvesting or any other silvicultural operation can be designated by (1) specification in words, (2) physical marking, or (3) some combination of the two.

Cutting instructions specified in words are normally used when it is either unnecessary or not feasible economically to make sophisticated distinctions between trees. The simplest specifications consist of nothing more than designation of the diameter limits or species to which cutting is to be confined. This approach usually gives poor to mediocre results from the silvicultural standpoint. The more that the structure of a stand varies from point to point, the poorer are the results of diameter-limit cutting. It often leads to skipping dense patches of slender trees that ought to be thinned and to overcutting where the trees are sparse and large.

Better results can be achieved by laying out more definite rules for cutting or by marking small plots to demonstrate the type of cutting desired. This method works best with well-trained personnel who understand the objectives and have no reason, economic or otherwise, to depart from the instructions. Advantage should be taken of opportunities to train woods personnel so that the simpler silvicultural operations can be conducted in this way.

The use of such general specifications must be associated with frequent and careful instructions, particularly at the beginning of each operation.

Tree Marking

A much closer degree of control, with less dependence on inspections, can be obtained by marking the trees that are to be removed or those to be left. The workers are responsible only for removing the designated trees with due care for the site and the vegetation that is to remain. The timber marker bears a high proportion of the responsibility for planning both the logging and the future stand.

Ordinarily the trees to be cut are the ones marked, because they are normally less numerous than those to be left. However, if the majority of the trees are to be cut, it is better to mark those that are to be left; this marking should be done without wounding the living portion of their bark. Trees to be left should be marked in the case of very heavy thinnings or those kinds of reproduction cutting in which only a few trees are reserved. This has the incidental advantage of focusing the timber marker's attention on the residual growing stock. However, the cost of marking is great enough that it is always best to put the marks on whichever category of trees is in the minority. Where an area is to be cut clear, it is sufficient to mark the boundaries of the area to be so treated.

Sometimes it is possible to combine verbal specifications with partial marking. For example, in some circumstances it may be useful to reserve all trees below a certain diameter limit and those above that limit that are specifically marked. It may also be useful to specify the removal of all merchantable trees other than marked individuals of certain stated species. It may be desirable to encircle certain patches of a stand for clearcutting but limit cutting to marked trees outside those patches. The parsimony of such means of economizing on the marking effort must be balanced against the risk of causing confusion.

Mechanics of Marking

Many of the same marking procedures used for designating timber to be cut for utilization are also employed to control pruning, release cutting, and other silvicultural treatments. The marks themselves should be readily visible and durable enough to last through the period of the projected operation.

Trees are normally marked with paint because of the ease and visibility it affords, although tools for scribing the bark such as timber scribes and axes have been used. Usually, two marks are placed on each tree—one at a height where it is easily visible and another low enough on the trunk as to appear on the side of the cut stump. The mark on the stump helps determine whether the felling has been conducted according to the marking. The lower mark may be omitted when the felling crews can be depended upon to adhere to the marking.

Various schemes involving numbers and shapes of marks can be devised to stipulate such things as the direction in which a tree is to be felled, the product for which it is to be prepared, whether it is to be felled or killed, the height to which it should be pruned, or whether its utilization is at the discretion of the logger. Occasionally, distinctive marks, as permanent as possible, may be made on crop trees to expedite consistent treatment in marking for subsequent cuttings. Any special marking scheme should be so devised as to eliminate the possibility of misinterpretation. The colors of paint chosen for marking trees and for such other purposes as designating property boundaries should follow some standard system.

The best way to mark a stand is to proceed back and forth across it systematically, completing one long, narrow strip on each trip. If only one side of the tree is marked, the marks should be placed so as to face the next unmarked strip. However, it expedites logging if the trees are marked on more than one side. Marks should be sufficiently large to allow the logger to see which surrounding trees will be cut; this will result in felling directions conducive to more efficient skidding. Sometimes it is necessary to tally the trees that are to be cut and, less often, those that are to be left. Tallying can be costly enough that careful consideration should be given to the value of the data gathered and to the question of whether efficient schemes of sampling can be devised to provide necessary information at lower cost.

Costs and Benefits of Marking

The cost of timber marking is usually, but not necessarily logically, figured as a charge per unit of product designated for harvest. It varies widely, depending on the size of the timber, the data gathered during the marking, the amount cut per acre, topography, density of undergrowth, travel time, cost of materials, and the rate of pay of the personnel. It may be negligible where patches of large trees are being designated for clearcutting, or, conversely, may be a major portion of the stumpage value when small pulpwood trees are marked.

The immediate benefits are the only ones properly charged against the cost of marking the timber that is removed. These values include factors such as the information gathered about the quantity and quality of material being cut and the important opportunity to lay out efficient patterns for felling and extraction of the trees.

The essentially silvicultural benefits come mainly from increases in future production gained by exercise of discrimination among individual trees and conscious creation of vacancies suitable for establishment of reproduction. The cost of these benefits is more logically charged against the acre or hectare, and the future yield thereof, than against the timber removed. Justification rarely exists for deciding that all partial cutting will or will not be controlled by tree marking. The seemingly costly marking of trees of low value in a young stand may actually yield more future benefit than marking in an older stand of valuable trees.

Guidance of Timber Markers

Professional foresters who have attained responsible positions are usually too expensive for routine timber marking. Ironically, they do less and less of it but become more and more concerned about how it is done. They have become instructors and supervisors of timber marking; instead of doing it themselves, they usually do little more than prescribe the marking of a stand and inspect the results. This situation is one argument for developing homogeneous stands, thus avoiding needlessly complicated schemes of silviculture. Even-aged stands can be administered more expeditiously than uneven-aged stands, and the actual expense of marking them is lower because the work is more concentrated in space.

A forester who supervises timber marking must have personal proficiency and, above all, a clear idea of the ultimate objectives in the treatment of each kind of stand. These objectives must be developed in the light of a thorough understanding of local economic conditions, logging, and the ecology of all species in the stand. These ideas must be conveyed to the staff in such a manner that they will be able to apply their judgment to attaining the objectives. If the training is handled wisely, the people who do the actual marking will become more proficient than the supervising forester, who should be prudent enough to recognize the fact.

Good judgment about timber marking can be developed through observant experience in a given locality. The best test of the wisdom of a particular marking project comes a number of years afterward when the vegetation has adjusted itself to the conditions created by the cutting. The pressures of forest administration are so great that it is too easy to concentrate on setting up this year's cutting and forget to go back to inspect the results of the cutting of several years ago. It is also easy to become so obsessed with the imaginary virtues of an established marking policy as to be blind to any shortcomings that may be apparent looking at the results of past marking. The forester should take advantage of every opportunity to improve skill in marking timber for cutting and then observing the results as dispassionately as possible.

Within a given forest management organization or for a given forest, it is usually necessary to standardize practice among individuals and for various areas of the same forest type. Continuity and uniformity are needed in the treatment of individual stands, management of growing stock, development of stand structure, and pattern of logging. These prescriptions or marking rules often describe the kinds of trees to be cut or to be fostered as candidates for the final crop. The use of tree classifications, sometimes with pictures depicting crown shapes, bark characteristics (Fig. 17.4), or other indicators of tree vigor, may be helpful. It is often important to specify the kind of stand that is to be left after partial cutting. Such rules should not only describe the details of the marking itself but also indicate the objectives of stand manipulation.

Useful as such rules may be, arbitrary insistence on conformity to them can inhibit the development of skill and good judgment among individuals. The extent to which this is true depends on the intelligence and knowledge of the personnel, the nature of the rules, and the manner in which they are enforced. Furthermore, rules by themselves are no substitute for the use of demonstration plots, training sessions, and similar means of education and exchange of information among personnel.

Control of Waste and Damage in Logging

Some waste and destruction is inevitable in any logging operation, and the work is so arduous that it is too easy to become inured to the damage. Constant effort is necessary to minimize damage and to distinguish between the inevitable and the preventable. Waste of presently merchantable material comes from (1) leaving unnecessarily high stumps, (2) failure to utilize merchantable portions of felled trees, especially in the tops, (3) breakage or other damage of stems in felling or skidding, and (4) failure to utilize suitable material in dead, dying, or partly defective trees.

It may also be important to direct each part of a tree to its best use. Discriminating judgment must be exercised in determining standards of merchantability because these

Figure 17.4   Bark characteristics indicative of different rates of growth of southern red oak (from Burkle and Guttenberg, 1952), showing patterns typical of many species. As growth slows, the individual bark plates become larger and the fissures between them become deeper. A very fast-growing tree would have smooth bark composed of minute plates. Because the bark grows outward as the wood grows inward from the cambium, the outermost bark indicates the conditions of some years ago. The stems shown above are not necessarily of the same diameter. *(Photographs by U.S. Forest Service.)*

standards may vary with market fluctuations or even from the most to the least accessible parts of stands. These kinds of losses are simply evidence of inefficiency in logging and are best controlled by appropriate supervision and properly planned conduct of harvesting operations. Logging supervisors can normally be expected to cooperate in reducing waste of merchantable wood because it is in their interest to do so. Greater administrative effort is necessary to curtail damage to desirable residual vegetation, to the soil, and to other values in which loggers have no immediate economic interest.

Sometimes merchantable material is deliberately left in the woods. Many dead or dying trees have values for certain wildlife species and should purposefully be left. A certain amount of coarse woody debris left on site is desirable to maintain high levels of organic material for preservation of soil structure and to sustain many of the forest floor microflora and microfauna for amphibians and other animals.

Protecting Vegetation

Destruction of residual vegetation is a most vital silvicultural concern in those kinds of cutting in which either trees or advance regeneration are left as sources of future production. If residual trees will interfere with production, it may even be desirable to destroy them. Most damage to residual vegetation comes from the cutting, breakage, or crushing of trees that impede the felling and removal of the trees designated for cutting. The wounding of standing trees during felling and skidding is another important source of injury, especially because of the wood-rotting fungi that often invade exposed wood. Young trees are also cut for skids, corduroy roads, and other uses incidental to the harvesting operation. Supplies of such material can often be secured, with little if any added cost, from inferior species, tops, or cull material.

The greatest destruction of growing stock is likely to occur in stands of irregular form, where trees of various ages from small seedlings on up are intermingled. Greatest damage is likely to occur to large seedlings and small saplings that are tall enough to impede operations but not durable enough to resist breakage. Seedlings less than 2 feet in height are comparatively supple and are, therefore, more likely to escape destruction. Heavy cutting causes more damage than light.

It is difficult to be specific about the extent of damage that can be accepted without endangering future production in partially cut stands. Although losses are usually expressed in terms of the percentage of stems destroyed or injured, the most significant point is their distribution. For example, if each milacre in need of reproduction has one seedling, the area may be fully stocked even though 70 percent of the total number were destroyed. On the other hand, a loss of 10 percent of the total might be so concentrated as to leave large openings entirely without trees. If a thinning is being made in a stand at a stage of development when the optimum spacing of the residual trees is 12 feet ($\approx$4 m), skidding equipment that leaves a swath of destruction 10 feet ($\approx$3 m) wide can be used to advantage, with no damage of long-term consequence to the stand. However, it must be noted that root damage and soil compaction may make this path of destruction effectively wider than it superficially appears.

Damage to reserved trees or other desirable residual vegetation includes not only breakage but also the wounding of the bark, cambium, or even the wood of stems and roots. Pushing a tree into a leaning position so that it forms undesirable reaction wood is sometimes more damaging than killing it because it may become useless and yet continue to usurp growing space. The same may be true of a tree that is so badly wounded or broken that it develops heart-rot or becomes malformed. The most common damage to trees of pole size and larger is trunk wounding. A small bump against a tree that is not being cut, especially during the spring when the cambium is actively dividing and the bark is loose and easily knocked off, can result in a very costly defect. Decay will often set into the open bark wood leading to loss of the most valuable wood, the butt (Shigo and Larson, 1969).

Damage from felling can be reduced by refraining from dropping trees on advance regeneration or against trees that are to be left standing. If damage to young growth cannot be entirely avoided, the trees should be felled in such a pattern that the damage will be well distributed and have the effect of thinning the residual stand. Much difficulty can be forestalled by giving attention to the matter at the time of marking as well as during the operation itself.

In marking timber, each tree should be considered not only from the standpoint of its productive potentialities, but also with regard to the problems that might be encountered in felling it or one of its neighbors. *No tree should be marked for cutting unless a way can be seen, or created, to fell it without inconvenience or excessive damage to the residual stand.* Tree markers should take note of asymmetrical crowns and significant lean. Trees that hit leave trees on the way down usually cause serious wounds or get hung up, which necessitates additional work for the skidder, further increasing the potential of damage. Often, hung-up trees are as much the fault of the marker as they are the logger. It may then be necessary to supervise the operation closely or to train the felling crews to ensure that they follow the plan embodied in the pattern of marking. Even then some unanticipated losses may occur. In such cases, provision should be made to salvage any unmarked trees that are seriously damaged in the course of logging.

Little additional damage is likely to occur during the limbing and bucking of felled trees, if the crews refrain from swamping out any young trees that do not actually impede the operation or constitute a source of danger to personnel. In logging heavy timber or if two very different size products are being harvested, it is sometimes advantageous to finish the operation in two stages; the logs produced in the first stage are removed before cutting the remaining marked trees. This practice has the favorable effect of avoiding congestion and confusion, thereby reducing damage to the residual stand, breakage of felled trees, and the amount of labor required for both limbing and bucking.

BIBLIOGRAPHY

Davis, L. S., and K. N. Johnson. 1987. *Forest management.* 3rd ed. McGraw-Hill, New York. 790 pp.

Gregory, G. R. 1987. *Resource economics for foresters.* Wiley, New York. 477 pp.

Gunter, J. E., and H. L. Haney, Jr. 1984. *Essentials of forest investment analysis.* Oreg. State Univ. Bookstores, Corvallis. 333 pp.

Hamilton, G. J. (ed.). 1976. Aspects of thinning. *U.K. For. Comm. Bul.* 55. 138 pp.

Jones, G. 1993. *A guide to logging aesthetics: practical tips for loggers, foresters, and landowners.* Northeast Regional Agricultural Engineering Service, Ithaca, N.Y. 28 pp.

Knuchel, H. 1953. *Planning and control in the managed forest.* Transl. by M. L. Anderson. Oliver and Boyd, Edinburgh. 360 pp.

Leuschner, W. A. 1984. *Introduction to forest resource management.* Wiley, New York. 298 pp.

Shigo, A. L., and E. vH. Larson. 1969. A photo guide to the patterns of discoloration and decay in living northern hardwoods. USFS Res. Paper NE-127. 100 pp.

U.K. Forestry Commission. 1992. *Lowland landscape design.* HMSO, London. 56 pp.

U.K. Forestry Commission. 1994. *Forest landscape design.* HMSO, London. 28 pp.

CHAPTER 18

SILVICULTURAL MANAGEMENT OF WATERSHED ECOSYSTEMS

Tall forests grow where there is plenty of water; most of the fresh water of streams and lakes comes from lands that either have forests or are capable of supporting them. The quantity and quality of water that flows from forests are always matters of silvicultural concern regardless of whether the water is used by people or only by fish and other aquatic organisms. Public policy often requires that forest owners pay attention to watershed management even if they do not use the water themselves. The best kind of watershed cover is untouched, unused, ungrazed forest. However, well-conducted forestry is more compatible with water management than any other human use of land (Fig. 18.1). Many forests are created or maintained for the primary purpose of watershed protection and management. Forests are the chief sources of water for vast metropolitan areas and agricultural irrigation.

The forestry aspects of watershed management have been thoroughly reviewed by Satterlund and Adams (1992) as well as Riedl and Zachar (1984). The standard objectives of silviculture in regard to water management are erosion control and reduction of the range between extremes of streamflow. Sometimes the purpose is to increase the supply of usable water or improve its quality. The control of actual damage to forested watersheds is much more a matter of proper management of roads and trails than of modifying silvicultural treatments.

Protection of Soil and Water Quality

Forest vegetation protects soil and water by shedding the leaves and other organic matter that feed soil organisms, such as earthworms and ants, which churn the soil and keep it porous. Litter also protects the mineral soil from the impact of falling raindrops and of the even larger drops of water that fall from the leafy canopy. Water that filters into the soil

449

Figure 18.1   A 100-year-old second-growth stand of Douglas-fir prevents erosion of the slopes above this stream in the steep Oregon Coast Range. Occasional floods occur because of very high winter rainfall. The young streamside stand of red alder germinated on and now stabilizes the bare soil left by the most recent severe flood, and the stream flows clear virtually all of the time.

does not run over the surface picking up solid materials in the process of surface erosion. Erosion harms twice—first by removing soil and second by depositing it on better soils or as mud in streams. Much of the water that sinks into the soil emerges later in springs of clear, filtered water.

Erosion is a natural phenomenon by which land is continually worn down; it cannot be prevented completely. Most erosion of mineral materials in natural forests originates from flowing water cutting into the beds and banks of streams during times of heavy runoff. It is very important to maintain or augment networks of roots of woody plants along the banks of streams in order to reduce or prevent undercutting of them.

What actually moves in water through forest soils on its way to streams is limited to dissolved substances and some particles of clay or very fine organic matter. However, various disturbances can induce *accelerated* erosion of coarser materials, and this is what needs to be controlled in silvicultural practice.

Porous soils are not only resistant to erosion but they also have substantial capacity to detain water temporarily in the larger pore spaces. Water in such **detention storage** is ultimately released to streams, but any delay has the virtue of making streamflow and water supply more even through the year. Every unit of soil has capacity for **retention storage** in pore spaces so small that water cannot flow out of them; it can only evaporate or be taken up by plants. Water will not start flowing to streams until the capacities of both detention and retention storage are filled.

The cutting or killing of trees by itself does not impair the porosity and infiltration capacity of the soil, provided that the supply of food to the soil organisms continues. Porosity may diminish slowly if the supply of new organic matter is halted, but this will not happen if vegetation regrows. The soil organisms that feed on forest litter are usually much more effective than plows or similar mechanical equipment in keeping the soil loose and porous. Furthermore, they do not have wheels or hooves that compact the soil.

Silvicultural cutting is aimed at establishing new vegetation and not at preventing it. Vegetation of almost any kind will produce organic matter and maintain soil porosity regardless of whether it is the species envisioned in the regeneration cutting. However, actions that scrape, gouge, or pack the mineral soil and litter layers do impair the infiltration capacity. As was described in Chapter 8, these actions, particularly scraping, can also reduce the supply of chemical nutrients in the soil. The main problems with soil damage, erosion, and stream sedimentation come from roads, log transportation, and excessive site preparation.

Prevention of surface erosion from overland flow is the first line of defense, but it is seldom perfectly effective. Measures to filter turbid water before it flows to streams constitute the second line of defense. Roads and barren or compact soil surfaces are the main sources of turbid water. The filtration can often be accomplished by diverting such water onto filter strips which are areas of porous undisturbed forest soil; porous barriers such as bales of hay or artificial settling basins are sometimes necessary. Many such measures are not permanently effective and must be repaired frequently until the eroding surfaces are stabilized.

The areas of soil within a stand that are most crucial for watershed management are the **source areas** that are so nearly saturated that water can flow from them beneath or on top of the soil surfaces. Most source areas are not permanent. After heavy rains or the melting of deep snows, almost all soils are source areas. However, gravity soon removes all the water so loosely held by the soil that it can flow; transpiration from vegetation makes the soil even drier. Therefore, the only areas that are permanent source areas are likely to be found along streams or at the bottoms of slopes and in other places where water collects. Such places are usually indicated by the species composition of shrubs and other vegetation.

It is while a soil unit is functioning as a source area that it yields water and is also most likely to suffer compaction or surface erosion. When soil is wet, it lacks cohesion and resistance to compaction such as that caused by heavy equipment. A major portion of the soil and water damage associated with harvesting and other uses of forests comes from damage to source areas either while they are saturated or become so before the soils have recovered from the damage.

Landslides on Steep Slopes

While cutting *per se* is not likely to start surface erosion, it may induce landslides or other deep-seated earth movements on extraordinarily steep, geologically unstable slopes. Most of these kinds of sites in North America are found in small portions of the very steep and geologically very young Coast Ranges along the Pacific Ocean (Swanston, 1974). However, similar conditions can exist in other steep terrain. The causative conditions are not steepness alone but the existence of discontinuities of materials within a few meters of the surface that can become planes of slippage, especially when lubricated by water. If the roots of large trees extend downward across these slippage planes, the risk of landslides is reduced. The action of webs of tree roots can bind the surfaces together well enough to reduce the tendency for lenses and slabs of material to slide or slip down the slopes. If too many trees are killed by cutting, the web of tree roots is likely to decay after several years and cause delayed-action landslides. Whereas any surface erosion is most serious immediately after soil disturbance, this kind has the insidious characteristic of occurring a few years later.

As in the case with surface erosion, these destructive events are much more likely to be caused by roads and road building than by the cutting pattern. If landslides are likely to occur on a particular steep slope, there will usually be leaning trees or other evidence of their having taken place on that slope without any human disturbance. However, good knowledge of the subsurface geological structures is often necessary to detect dangerous situations. Susceptible areas are best logged with helicopters or balloons, without roads, or not logged at all. For most such slopes it is not a question, from the geological standpoint, of whether there will be landslides but only of when and at what rate. If such areas are going to be subjected to cutting at all, it is logical to employ kinds of selection or shelterwood cutting that will encourage the permanent maintenance of a network of living roots.

Effect of Cutting Trees

There is a deeply rooted and erroneous popular myth that trees make it rain and that cutting them causes drastic alternations between drought and flood. The water that falls from the sky comes mostly from air that has been moistened over warm oceans and is lifted by convection over warm surfaces or over mountain and cold air masses. Transpiration from forests by itself rarely, if ever, moistens air masses enough to make it rain. Forests are an effect of abundant precipitation rather than the cause of it. In the vast majority of cases, the cutting of living trees, by reducing losses of water to transpiration and interception, actually *increases* the amount of water that reaches streams. The myth probably originated from the observation that true deforestation associated with agriculture or grazing often causes rain to run off so quickly from compact soils that streams flood and then dry up.

The same exposure to sun and moving air that makes tree leaves so efficient at changing carbon dioxide and water to sugar also causes them to transpire large quantities of water back into the atmosphere. Trees also intercept appreciable amounts of precipitation when rain or snow collects on leaves and branches of trees. Some of this precipitation is detained temporarily before falling to the forest floor, but much of it is directly evaporated into the atmosphere. Water losses from transpiration are generally more significant than those from interception. Both kinds of losses reduce the amount of water that reaches soil, groundwater, and streams.

The effects of silvicultural treatments on streamflow are well exemplified by the results of comprehensive, long-continued studies conducted in Appalachian hardwood forests on steep terrain in a high-rainfall area of North Carolina (Swank and Crossley, 1988). The gains in stream runoff from tree removal were roughly proportional to the reduction of tree cover and sensitive to small variations. After clearcutting, they amounted to as much as 10 acre-inches annually for a few years after the cutting. With lighter cutting, the runoff increases were not only proportionally smaller but also far more temporary because the tree crowns expanded again quickly.

Both transpiration and interception reductions are greatest when warm weather accelerates evaporation rates. There are virtually no reductions in transpiration when trees are leafless or are so short of water that they do not transpire very much.

With these kinds of reductions in transpiration and interception, the higher the precipitation the greater is the absolute amount of gain in streamflow. Such increase of soil water seldom adds to the peak runoff associated with floods. In fact, it is ordinarily beneficial because it is most important at the very times of year when transpiration is greatest and streamflow the most feeble. However, if a flood-inducing rain came soon after one of these times, the floodwaters would be increased by the comparatively small amount of water that cutting had caused to be left in detention storage. It should be noted that increases in streamflow may sometimes be of no advantage unless some way can be found to store water in reservoirs until it can be used.

Evergreen stands intercept and transpire substantially more water than stands composed of deciduous trees. The North Carolina studies found that replacing the deciduous hardwoods with white pines caused annual reductions of streamflow of about 10 acre-inches for at least 10 years after the pine stands had fully closed. These reductions extended throughout the year and were greatest during winter and spring. It may be anticipated that any cutting in evergreen forests will have a greater effect than similar cutting in deciduous forests.

The effects of tree removal are somewhat different in the case of snow, which can accumulate in wind-created drifts behind barriers or in narrow openings. Drifts provide useful ways of storing water on the land and delaying runoff in climates, such as those of the Rocky Mountains (Fig. 18.2), where most snow melting is caused by sun rather than rain. Cuttings made to create narrow east-west strips can be used to cause drifts to form in shaded areas (Leaf, 1979). They represent one of the most effective ways in which silvicultural treatment can be used to increase water supplies. On the other hand, in climates where snow can be melted rapidly by warm rain at the wrong times, this kind of water storage can add to flood peaks.

Ordinarily the chief effect of cutting on the water regime is to provide more water for the vegetation of the site. Cutting is done for the deliberate purpose of increasing downstream water supplies only where such supplies are critical. Experiments with **maximizing** streamflow by repeated mechanical or chemical soil treatments to prevent regrowth of vegetation have been found to cause soil erosion or loss of nutrients to runoff. These effects and treatments are not those of silvicultural clearcutting, which is not aimed at denuding the landscape. Replacing forests with grasses will produce more streamflow and cause less erosion and nutrient loss, but watershed damage may result if any subsequent grazing is not stringently regulated.

The increases in streamflow resulting from cutting can sometimes be great enough to accelerate the erosion of streambanks. For this reason, it is sometimes necessary to restrict the amount of clearcutting done at one time on a given watershed.

Silvicultural treatments designed to maintain the quality and temperature of water for healthy populations of fish and other organisms in streams are considered in Chapter 20.

Fog-drip

Fog-drip is a kind of precipitation that is reduced or prevented by clearcutting. It forms when the minute, suspended water droplets of clouds are blown through leaf canopies, swept out of the air, and coalesced into droplets large enough to fall to the ground. This kind of precipitation by interception can be very significant in some ''cloud forests'' at high elevations and in some of the remarkable kinds of forests that occur along the foggy western shores of continents at middle latitudes. The coast redwood and western hemlock–

Figure 18.2   Early July storage of snow in drifts in openings in a forest of Engelmann spruce and sub-alpine fir in White River National Forest, Colorado.

Sitka spruce forests of the western edge of North America are heavily watered from this source. The degree to which heavy cutting harms such forests is not known, partly because of the difficulty of measuring fog-drip precipitation accurately. Conventional rain gauges exposed beneath the open sky do not collect fog-drip. It is likely that the most serious harm is to germination and establishment of seedlings starting in the open.

This kind of precipitation can also cause damage. When supercooled cloud droplets impinge on solid objects and freeze, they form a white ice called **rime**. Because of their large surface area, these droplets also absorb large amounts of sulfur dioxide and other atmospheric pollutants. These are the cause of much of the damage from acid ''rain'' in forests that are frequently enshrouded in fog or cloud.

Wetland Problems

The most common case where clearcutting alone can cause true site degradation is on soils that are continually wet and poorly aerated, particularly those with *Sphagnum* moss. These are usually on the flattest kinds of terrain; however, they can also be on hillsides where precipitation grossly exceeds evapotranspiration. Mossy hillside moors are common in many areas of Britain and Ireland where forests were cleared and repeatedly burned for grazing hundreds or thousands of years ago. Problems of excess water there were aggravated by the development of podzol hardpans induced by strong leaching. Reforestation often depends on deep plowing done to break the hardpans and allow water to drain from the surface soil layers.

In many forested swamps and bogs, transpiration by trees plays a major role in keeping water tables low enough for the roots of the trees to survive. If clearcutting turns off the ''transpiration pump,'' the level of poorly aerated water may rise enough to retard or even prevent regrowth of trees. This is one of the few cases in which complete clearcutting by itself can actually defeat growth of trees. Partial cutting is the best solution to such problems unless the purpose is to eliminate trees and increase yield of water.

Storage and Leaching of Nutrients

Soil nutrients as well as water can flow out of forests, but one objective of silvicultural practice is to restrain such losses. As is the case with natural erosion, they cannot be entirely prevented; in fact, aquatic life would suffer from the loss of chemical nutrients if it were eliminated. Chemical nutrients also flow off forest sites with the wood and other materials removed from them.

If nearly all of the vegetation is temporarily eliminated, as with true clearcutting, the risk that soluble substances, especially chemical nutrients, may leach out of the forest system at inordinate rates must be considered. The best defense against such loss is the continued maintenance of some sort of vegetative cover even if the trees or other woody plants are eliminated. The next best defense is prompt reestablishment of vegetation to continue the cycling of nutrients and other soluble substances on the site.

Living vegetation is not the only buffer against nutrient loss because the organic and inorganic colloidal surfaces that make up the cation exchange capacity of the soil also play a highly important role. Soils that are high in organic matter or clays (mineral colloids) may strongly resist loss of cations such as calcium, potassium, or ammonium nitrogen, but this does not help retain anions such as nitrate or sulfate ions. The retention of nitrates (and the less important sulfates) depends mostly on living vegetation and conservation of soil organic matter. The question of whether phosphate anions leach away depends more on the vagaries of their binding with iron and aluminum compounds than on anything altered by cutting practices.

Most investigations of the matter have not shown major losses of nutrients in association with regeneration by clearcutting. There was, however, a significant case involving sandy soils and the northern hardwood forests of the granitic White Mountains of New Hampshire (Bormann and Likens, 1979). Here the warming of the forest floor induced by clearcutting caused the quick decomposition of the thin leaves of the litter of maple, beech, and birch and, more importantly, nitrification. In this process, nitrifying bacteria convert soil ammonium compounds, which can be retained on the cation exchange surfaces of colloids, to nitrate anions, which are free to move out in the soil solution. If, as is the case late in the growing season, the pioneer vegetation did not develop fast enough to take up the nitrates and some other highly mobile nutrients, then there was a period of some months during which accelerated losses to streamflow took place. The fact that very large losses were caused by repeated annual applications of a soil sterilant aimed at maximizing the yield of water in streamflow has to do with the prevention rather than the establishment of regeneration by clearcutting. As far as regeneration cuttings in these particular forests are concerned, the important point is to keep their widths narrow and place them in areas already well vegetated with advance growth. Similar losses from nitrification have not been observed in experiments in conifer forests or in deciduous forests farther south or west.

The retention of leaves by evergreens enhances the capacity of such species to store nutrients above ground (Cole, 1986). The thin leaves of deciduous species usually decompose rapidly enough that they support comparatively rapid nutrient cycling, but the high mobility of the nutrients also exposes them to somewhat greater losses by leaching. Losses of nutrients in hot, moist tropical forests are so notoriously rapid that most storage of them is limited to the wood.

Association with Close Utilization of Resources

The most important risk of nutrient loss comes from removing them in wood, leaves, or litter. Some of the practices associated with marginal agriculture in less developed countries and formerly in Europe are very deleterious. Repeated removal of litter from forests for mulch or fuel is perhaps the surest way to ruin the soil (Pritchett, 1979); it often leads to the development of very slow-growing forests. Excessively close utilization of foliage or twigs for fodder, fuel, or various simple structures also places a nutrient drain on forests. Regardless of what defects it may have, it is better for people to use the coppice method to grow fuelwood than to ruin the soil by scraping up litter and animal dung to use for cooking their food.

The *concentration* of chemical nutrients in leaves and twigs of trees is much higher than in the main-stem wood and bark, but roughly half the total *weight* of nutrients is in the main stem. Sometimes the heartwood of trees has lower concentrations of nutrients than the sapwood, and the proportion of heartwood increases as trees grow older (Raison and Crane, 1986). If removals are limited to pieces of wood 10 centimeters or larger in diameter, the nutrient losses are usually offset by supplies of nutrients from (1) rock weathering or (2) nitrogen fixation and (3) those that come from the sky in such forms as desert dust, evaporated ocean spray, nitrogen fixed by lightning, and the substances of atmospheric pollution.

Somewhat the same problems associated with removing twigs for fuel by labor-intensive means could also arise where highly mechanized harvesting chips up whole trees for pulp or fuel under circumstances in which the labor needed for manual delimbing is too costly. The losses can be reduced considerably by growing trees to larger size and lengthening intervals between the removals (Raison and Crane, 1986). They can also be reduced by measures such as harvesting trees when they are leafless or by delimbing at the stump and thus leaving the twigs, leaves, and branches on the site. If the depletion becomes too great, it may be necessary to apply fertilizer, just as is the case in the essentially depletive practice of agronomy. It is more effective and cheaper to maintain optimum amounts of organic matter, especially that incorporated in the soil, than to resort to fertilization. Fertilization is better viewed as a means of improving the productivity of undamaged soils than as a remedy for sick ones.

Defense by Revegetation

There is no better means of conserving the renewability and productivity of terrestrial ecosystems than keeping the growing space full of vigorous forest vegetation. As long as the vegetation of a given forest age class continues to accumulate biomass, it is in what has been called the aggrading phase of its development. This means that it is not only adding to the standing crop of organic substance but also accumulating nutrients, including carbon, on the site. When it is destroyed or reaches the stage at which decay equals or exceeds the accretion of biomass, it enters upon the degradation phase in which nutrients and carbon cease to accumulate and begin to be lost to the atmosphere or waters moving through the system. In other words, nutrients may leak out of the forest system not only when stands are destroyed but also when they become biologically overmature.

Except for a few special kinds of sites, any cutting pattern that results in immediate revegetation is as protective of the watershed environment as any other. Most of the real problems have to do with roads and other disturbances that denude the soils.

Construction and Maintenance of Roads and Trails

Although the cutting of trees rarely induces any erosion, the building and use of roads and trails almost inevitably causes some degradation. Trails constructed for the foot-traffic of people and animals as well as those left by the skidding of logs over the ground can cause problems. Roads can be worse because they are wider and cut more deeply into the terrain. The crucial consideration is planning the disposition of water. A forest road is really a water-control device designed to provide a cohesive but well-drained surface strong enough to support rolling vehicles (Fig. 18.3).

The construction and maintenance of roads and trails are matters of engineering dealt with much more completely in books by Satterlund and Adams (1992), Stenzel, Walbridge and Pearse (1985), and Simmons (1979). Only the basic principles and some details about simpler kinds of roads and trails are considered here.

The basic solution to these problems involves measures that conflict with each other. The first is to get the water off trails and roads quickly. The second, which is much more difficult to achieve, is to prevent sediment-laden water from reaching streams. If at all possible, such water should be made to infiltrate into the soil before it reaches streams. Where that is not possible, the objective should be to impose barriers in the streams that will slow the water down so that the sediments will be deposited close to the source and the cutting of streambeds will be reduced.

Any action that removes or compacts any mineral soil reduces its capacity to absorb and store water. The removal of unincorporated organic matter not only exposes soil

Figure 18.3   A well-graded and properly drained all-weather logging road through a partially cut spruce—fir forest on poorly drained soil in eastern Maine.

particles to erosion but also impairs the ability of these materials to arrest any particles that start to flow sideways. Therefore, whenever mineral soil is exposed, it becomes important to prevent water from flowing for long distances over such surfaces and to divert it into filtering materials or settling basins before it reaches streams.

Erosion along roads and skid trails can be reduced by locating them so as to avoid steep grades and places where extensive cutting and filling will be necessary. It is desirable to avoid wet soils or very deep ones that are easily rutted when wet. On steep slopes, skid trails that converge downward should be avoided as much as possible. The location of roads close to and parallel with streams increases siltation. Proper attention to drainage of roads not only reduces erosion but also prolongs their usefulness. This applies to roads in current use as well as roads that are not used between logging operations. If the drainage systems of temporarily abandoned roads cannot be cleared of debris periodically, they should be opened enough at the end of each period of use that they will function without attention.

It is frequently advantageous to suspend log transportation during periods when roads are muddy, because the risk of creating conditions favorable to erosion is then at a maximum. Precautions of this kind often help reduce logging costs by preventing the damage and delay that occur when equipment gets stuck in the mud. Some readily accessible stands on dry, well-drained soils should be reserved for cutting in muddy weather.

After the location of a road has been determined, all trees and saplings in the way of the road should be cut before any earth moving is done. Stumps that cannot be deeply buried in place must be removed and deposited in appropriate places. Merchantable material should be converted to logs and put aside for removal when the road reaches the storage points. Burning or other disposition of unmerchantable material should be one of the first steps. It is often very advantageous to form the bed and side-ditches of a road some months before the final surfacing material is applied and the road is put to use. This gives the bed time to settle, become firmer, and thus less subject to the rutting and other difficulties that aggravate erosion and damage to equipment. Even though it is porous, the topsoil must be removed or buried beneath the roadbed because it usually turns to mud if used as surface material for transportation ways.

The next step is applying any needed load-bearing surfacing materials which are usually crushed rock or gravel. Ideally, the surface should be porous enough to absorb some water but also strong enough to support vehicles. Water-bars, dips, or culverts should be placed at strategic points along roads and trails; the spacing should be closer the steeper the slope and the more erosive the materials. It helps if roads have at least enough slope to prevent water from accumulating on them. Saturated surfaces usually lose their bearing capacity.

Paving prevents erosion of road surfaces but has the insidious effect of transferring the problem downstream. Fast-moving clear water, such as that which flows off paved surfaces (or even out of hard-bottomed streams) has tremendous erosive power because it will cut into streambeds until its total capacity to carry sediment is filled. Another deceptive kind of hard surface is that created by using bulldozers to create "hard-pan" roads. These are sunken ways more like ditches than roads. They are built by scraping down to some hard layer of subsoil. Because there is no place for diverting water to the sides, such roads usually become muddy, rutted sources of sediment during use. However, such roads can be acceptable if the hard surface is either natural rock or solidly frozen materials, provided that care is taken for disposition of the water that flows off them. Use of any sort of frozen roads should cease before thawing starts; their surfaces usually become supersaturated while they are frozen, so they become exceedingly muddy when they thaw.

Roads should be kept as narrow as possible in order to restrict the amount of exposed erosive surface. The temptation to use heavy earth-moving machinery to make roads wider than planned often needs to be curbed. Measures that make roads more suitable for swift vehicular movement have the unfortunate effect of increasing the amount of soil disturbance.

Concerns about drainage of transportation ways change if the soil materials are highly porous sands. Erosion becomes a problem only during periods when the subsoil is frozen solid but the surface has thawed. Such soils lose their cohesion and bearing capacity if they are either too dry or supersaturated. At other times, they can often be used as roads without any ditches or excavation.

Culverts or bridges should be installed over most kinds of definite streams. Fords can be used only where streambeds with permanently hard, rocky bottoms exist or can be created. Small intermittent streams can be crossed by "corduroying" them—that is, by laying logs in them side-by-side and parallel to the streams; the logs should be removed when the use has ceased.

The drainage of roads must be maintained. Ruts and potholes should be filled. This is usually done with scraping equipment that pulls material out of the side ditches and returns it to the middle of rounded road surfaces. However, the reopening of some drainage-ways can best be done with hand tools, preferably when water is actually running in them and blockages can be easily detected.

Many forest roads are used only for brief periods at intervals of some years. When such roads are taken out of use, measures must be taken to ensure that they will remain drained and uneroded without much attention. This is often done by digging "fail-safe" drainage ways that will carry water across and away from roads in deep ditches during heavy rains or similar events. High berms left beside ditches that cross roads discourage damaging vehicular use of the roads. It is also important to see that exposed soil surfaces are promptly revegetated with grasses or other herbaceous vegetation (which may have some incidental value for wildlife). These measures not only reduce stream sedimentation but also help preserve the roadbed for future use.

Management of Road Networks

From the standpoint of watershed management, the objective of forest road planning should be to make a whole forest accessible with the least possible amount of exposed road surface. This will also minimize the investment in roads in the long run. To this end, it is well to plan road networks that will serve the whole area that needs to have roads and to extend them as they are needed. Too often the networks grow in an unplanned fashion as roads are built to whatever areas need harvests in the immediate future.

Because roads that are not in active use produce comparatively little sediment, there are advantages in some degree of concentration and consolidation of the harvesting activity that takes place in any given year. To illustrate this point with an extreme case, if a fixed area of land is to be harvested in a given year, it would, from this standpoint, be best to concentrate it in one big clearcut area. This area might be served by one road that was open for use during that one year and then closed for a whole rotation. Such a plan would disturb far less road surface than the same area of cutting distributed around the forest in openings created by the group-selection system. The fact that large clearcut areas require less road use than partial cuttings means that they actually contribute less of the main form of environmental damage associated with timber harvesting. However, cutting reduces loss of water to transpiration, so that the concentration of too much heavily cut area in one watershed may sometimes induce excessive erosion of steambeds by allowing too much water to reach the streams.

The planning of any system of forest roads should be done with thorough attention to topography, soils, and geology. In steep terrain, it is very important to avoid putting roads on geologically unstable slopes subject to landslides.

The desirable widths of filter strips between roads and streams or lakes depend on the slope of the terrain. Recommendations by Simmons (1979) for the eastern United States range from 25 feet for flat ground to 45 feet for 10 percent slopes, 65 feet for 20 percent slopes, 105 feet for 40 percent, and 165 for 70 percent, with double these widths for municipal watersheds.

On very steep terrain, heavy cuttings may require networks of skid trails and haul roads that expose as much as 15 to 25 percent of the soil surface. The area that is rather permanently removed from production by haul roads varies significantly, depending on the kind of skidding equipment used. It is very high if most of the movement of logs within the cutting area is done after logs are loaded onto trucks. The disturbance of the soil on skid trails is mostly temporary and so much less serious than that on haul roads for trucks that it is desirable to accomplish as much movement of logs as possible in the skidding phase. It may therefore be desirable to use fast or powerful skidders rather than those with a small economic operating radius. Although the powerful skidders may cause greater damage to skid trails, their use does not necessitate putting so much area into even more destructive truck roads.

Effect of Different Kinds of Logging Equipment

The kind of equipment used for moving logs to roads has an important effect on soil, water, and any vegetation being deliberately left to grow. There are two basically different means of log transportation: (1) those that lift the logs partly or completely off the ground and (2) tractive equipment that drags or carries logs on the ground (Stenzel, Walbridge, and Pearce, 1985). Those that lift do the least damage to the soil; their effect on residual trees depends on how they are used.

Cable yarding with stationary engines involves lifting the logs into the air and moving them to a landing on a road. Dragging logs may cause some gouging and scraping action, but the heavy machines do not move over the soil and compact it. However, some modes of cable yarding involve so much sideways shifting of the cables that almost all residual trees are knocked over; this kind of equipment is usually compatible only with clearcutting. On the other hand, with skyline systems in which cables are suspended above the stand canopy, it is sometimes even possible to lift logs out of the stands vertically in thinning operations. With such equipment it is necessary to move the logs over well-defined corridors so that any partial cutting must be modified accordingly. Cable yarding is generally adaptable only to steep terrain because substantial differences in elevation are required to place the cable high enough to lift the logs off the ground.

One significant advantage of cable yarding is that the machines do not burn fuel merely to pull themselves over the ground. In tractive skidding, the power source expends many times more energy in moving itself than in pulling its load. This is one reason why logs are generally pulled downhill in tractive skidding, while it is both feasible and desirable with cable skidding to pull them uphill. However, this also means that the road systems are often very different for the two general methods. With cable yarding, the goal is to put the roads as high as possible on the terrain, but with tractive skidding they can be on lower ground. For these and other reasons tractive skidding is usually cheaper than cable yarding.

Helicopter logging is very costly, but it reduces road building as well as damage to soil and the residual stand. With respect to these problems, balloon logging is much like skyline yarding except that the cableway is supported with a balloon rather than a tower. The use of these aerial systems in old-growth timber leaves no permanent transportation system. This can be a blessing for watershed management but a curse for any other form of management that requires good access to stands.

Large mobile cranes with long booms and grapples are sometimes used to lift logs and then advance them by depositing them in piles that are two boom-lengths closer to the loading point. The crane is then moved (or "leap-frogged") to the other side of the pile to repeat the process which is sometimes called "shovel-yarding." This technique reduces the area of soil disturbance, but the swinging action of the boom requires that all sizable trees be removed from broad avenues through the stands.

Where heavy tractive machinery is used, the choice becomes one between skidding out material as (1) logs bucked into short lengths, (2) whole delimbed stems, or (3) whole trees minus the roots. The longer the lengths pulled out of the woods, the greater is the risk of damage to the residual stand and the greater the need to arrange for movement along straight lines. When whole trees are dragged out of the woods little gouging of the soil occurs, but there can be a sweeping of the litter that may or may not be desirable. If the crowns are large, they may damage residual trees; such damage can be reduced by partial delimbing before the skidding starts. Carrying off the small branches and foliage can have serious consequences for the nutrient capital of the site. The practice of doing the delimbing at the log landings and leaving their vicinity choked with debris simply impairs the productivity of even more land. Sometimes this debris is carried back into the woods on return trips by grapple skidders.

Crawler tractors are less likely to cause rutting than rubber-tired devices because their weight is spread over much more area. With any such machines, it is silviculturally desirable that they not be any heavier or larger than is necessary. It can help to use one device for assembling batches of material to be hauled out of the stand with another kind of device, especially when it is advantageous to collect small stems into bundles. Sometimes damage can be reduced if trees are cut into log lengths at the point of cutting and left in piles to be brought out by large specialized trucks called "forwarders."

Certain kinds of feller-bunchers, especially those with long arms that can reach out between the trees and bring in trees from several meters away, can reduce the amount of ground that needs to be run over in partial cutting operations. Some of these machines are designed to hold the tree stem while it is severed at the base, delimbed, and cut into logs left in piles for forwarders to pick them up. Sometimes the severed limbs can be left scattered on the ground in such ways that the machinery moves over them rather than on the soil. Some of these machines were actually designed for partial cuttings and are not just construction machinery with special attachments for logging.

Skidding done with draft animals generally causes the least damage to residual trees and to the soil within the stand. However, because animals tire, skidding distances must be short, and this usually increases the amount of area in roads, which are the chief source of soil and water damage. The main problems with animals are that they are slow and require much care.

Many difficulties can be avoided if both timber markers and loggers plan for the felling and removal of trees. It usually helps to arrange for logs to be moved out in straight lines. If they have to be pulled around curves, it must be anticipated that any adjacent trees will be damaged; they can either be protected with buffers or simply harvested as a final step of the operation. Trees should not be felled parallel with skid roads or in such other orientations that the logs from them will have to be pivoted around good trees to move them to the roads. If trees are felled in a herringbone pattern in relation to the road, they can be pulled out with a minimum of turning. If they can be felled so that their tops point toward where they are to be moved, that reduces the total ground-skidding distance. However, it is often necessary to move the logs out butt first.

It takes cooperation and understanding between foresters and loggers to do good logging. The loggers must have adequate incentives. The cheapest possible logging is inevitably poor logging and ultimately invites cumbersome public regulation. Almost any kind of equipment can produce good logging jobs if supervisors make it clear that good work is expected. The presumption that logging damage is inevitable becomes a self-fulfilling prophecy. It helps tremendously if all parties concerned plan and think through what they do at each step, whether it be deciding which trees to cut or how to move a skidder around in the woods to collect a load of logs.

BIBLIOGRAPHY

Anderson, H. W., M. D. Hoover, and K. G. Reinhart. 1976. Forests and water: effects of forest management on floods, sedimentation, and water supply. USFS Gen. Tech. Rept. PSW-18. 115 pp.

Bengtsson, J., and H. Lundkvist (eds.). 1994. Ameliorative practices for restoring and maintaining long-term productivity in forests. *FE&M*, 66:1–199.

Bormann, F. H., and G. E. Likens. 1979. *Pattern and process in a forested ecosystem.* Springer-Verlag, New York. 253 pp.

Brown, T. C., and D. Binkley. 1994. Effect of management on water quality in North American forests. USFS Gen. Tech. Rept. RM-248. 28 pp.

Cole, D. W. 1986. Nutrient cycling in world forests. In: S. P. Gessel (ed.), *Forest site and productivity.* Martinus Nijhoff, Boston. Pp. 103–115.

Farrell, P. W., D. W. Flinn, R. O. Squire, and F. G. Craig. 1986. Maintenance of productivity of radiata pine monocultures on sandy soils in Southeast Australia. In: S. P. Gessel (ed.), *Forest site and productivity.* Martinus Nijhoff, Boston. Pp. 127–136.

Gessel, S. P., D. S. Lacate, G. F. Weetman, and R. F. Powers. 1990. *Sustained productivity of forest soils: 75th North American Forest Soils Conference.* Univ. Brit. Col., Faculty of Forestry, Vancouver. 525 pp.

Harr, R. D., R. L. Frederickson, and J. Rothacher. 1979. Changes in streamflow following timber harvest in southwestern Oregon. USFS Res. Paper PNW-249. 22 pp.

Hewlett, J. D. 1982. *Principles of forest hydrology.* Univ. of Georgia Press, Athens. 183 pp.

Leaf, A. L. (ed.). 1979. *Impact of intensive harvesting on forest nutrient cycling systems.* State Univ. of N.Y. Coll. of For., Syracuse. 421 pp.

Likens, G. E., and F. H. Bormann. 1995. *Biogeochemistry of a forested ecosystem.* 2nd ed. Springer-Verlag, New York. 176 pp.

Naiman, R. J. (ed.). 1994. *Watershed management: balancing sustainability with environmental change.* Springer-Verlag, New York. 554 pp.

Patric, J. H., and J. D. Helvey. 1986. Some effects of grazing on soil and water in the eastern forest. USFS Gen. Tech. Rept. NE-115. 25 pp.

Perry, D. A., M. Meurisse, B. Thomas, R. Miller, J. Boyle, J. Means, C. R. Perry, and R. F. Powers (eds.). 1989. *Maintaining the long-term productivity of Pacific Northwest forest ecosystems.* Timber Press, Portland, Oreg. 256 pp.

Potts, D. F., and B.K.M. Anderson. 1990. Organic debris and the management of small stream channels. *WJAF*, 5:25–28.

Pritchett, W. L. 1979. *Properties and management of forest soils.* Wiley, New York. 500 pp.

Raison, R. J., and W.J.B. Crane. 1986. Nutritional costs of shortened rotations in plantation forestry. In: S. P. Gessel (ed.), *Forest site and productivity.* Martinus Nijhoff, Boston. Pp. 117–125.

Reidl, O., and D. Zachar. 1984. *Forest amelioration.* Elsevier, Amsterdam. 623 pp.

Satterlund, D. R., and P. W. Adams. 1992. *Wildland watershed management.* 2nd ed. Wiley, New York. 436 pp .

Simmons, F. C. 1979. *Handbook for eastern timber harvesting.* USFS, Northeastern Area State and Private Forestry, Broomall, Pa. 180 pp.

Stenzel, G., T. A. Walbridge, Jr., and J. K. Pearce. 1985. *Logging and pulpwood production.* 2nd ed. Wiley, New York. 358 pp.

Stickney, P. L., L. W. Swift, Jr., and W. T. Swank. 1994. Annotated bibliography of publications on watershed management and ecological studies at Coweeta Hydrologic Laboratory, 1934–1994. USFS Gen. Tech. Rept. SE-86. 115 pp.

Swank, W. T., and D. A. Crossley, Jr. (eds.). 1988. *Forest hydrology and ecology at Coweeta.* Springer-Verlag, New York. 469 pp.

Swank, W. T., and D. H. Van Lear (eds.). 1992. Multiple-use management: ecosystem perspectives of multiple-use management. *Ecological Applications,* 2:219–274.

Swanston, D. N. 1974. The forest ecosystem of southeast Alaska. 5. Soil mass movement. USFS Gen. Tech. Rept. PNW-17. 22 pp.

Swift, L. W., Jr. 1988. Forest access roads: design, maintenance, and soil loss. In: W. T. Swank and D. A. Crossley, Jr. (eds.). *Forest hydrology and ecology at Coweeta.* Springer-Verlag, New York, 1988. Pp. 312–324.

Tarrant, R. F., J. M. Trappe, and J. F. Franklin. 1988. From the forest to the sea: a story of fallen trees. USFS Gen. Tech. Rept. PNW-229. 153 pp.

U.S. Forest Service. 1989. Fire and watershed management. USFS Gen. Tech. Rept. PSW-109. 153 pp.

Vitousek, P. M., and P. A. Matson. 1985. Disturbance, nitrogen availability, and nitrogen losses in an intensively managed loblolly pine plantation. *Ecology* 66:1360–1376.

Weetman, G. F., and B. Webber. 1972. The influence of wood harvesting on the nutrient status of spruce stands. CJFR, 2:351–369.

CHAPTER 19

SILVICULTURAL CONTROL OF DAMAGING AGENCIES

Measures for controlling or managing sources of damage to forests are vital parts of any silvicultural system. The damaging agencies are as much a part of forest ecosystems as the trees; in fact, the killing of trees is the chief tool of silviculture. It is better to learn how to live with and manage the sources of damage than to attempt the impossible task of eliminating them.

The best approach to such management involves combinations of tactics that come in steps. The first steps involve building the health of stands and their resistance to attack as well as creating environments unfavorable to the sources of damage. Any feasible measures to prevent attack before it occurs come next. If attack takes place, combative treatments may be applied to suppress the damaging agency. Finally, dead or damaged trees can be salvaged when all else has failed.

The control of damage usually requires that stands be kept readily accessible, normally by roads. Good access permits for such measures as speedy fire suppression, good surveillance, application of pesticides, timely salvage, prescribed burning, and those kinds of partial cuttings that improve the vigor or health of stands.

It is far easier to generalize about damage by nonliving or abiotic agencies, which are governed mainly by comparatively simple physical processes, than about the highly variable biological processes involved in damage from biotic agencies.

Forest protection is so important that it is usually dealt with in academic courses in fire control (Brown and Davis, 1973), forest pathology (Tainter and Baker, 1996; Manion, 1991), and forest entomology (Coulson and Witter, 1984; Knight and Heikkenen, 1980). This chapter deals only with examples of modifications of silvicultural systems that are designed to cope with insects, fungi, fire, and other injurious biotic and physical agencies. The modifications are mostly specific steps taken against specific damaging agencies and cannot be safely based on generalities applicable to all forests. Appropriate procedures

464

must be based on the details of knowledge about the sources of damage; they cannot be reduced to sweeping generalizations.

PROTECTION AGAINST ABIOTIC AGENCIES

Fire Control

Any dead plant material can be fuel for wildfire if it becomes dry enough. Many of the silvicultural measures taken to reduce problems with forest fires involve the reduction of the fuels on which the fires feed. Some of the techniques of site preparation discussed in Chapter 8 are more important for fuel reduction than for facilitating regeneration. Silvicultural measures can be taken to modify the kind, amount, or distribution of fuels.

The risk of damage from fire varies widely between different sites locally and between broad climatic regions. Almost any kind of forest vegetation will burn if it is ignited under the right conditions and the dry fuel is continuous enough to enable fires to spread. Before there were people, there were lightning fires. Almost every habitat has plant species adapted to fill growing space left vacant by fire. Some of the most humid regions usually have areas of dry soils, such as deep sands, that are subject to fire. Even swamps that always have standing water will burn if there is a continuous fuel blanket composed of dead, dry grass or similar herbaceous vegetation.

In many humid regions, wildfires are not important enough to warrant modifying the silviculture, although it is still necessary to suppress them. In some localities with long dry seasons and highly inflammable vegetation, it may not be possible to do anything silviculturally except for salvage and reforestation after fires. There is a middle category of places in which fuel management measures will reduce the problems; sometimes it is not easy to decide where these places are or how far to go with the measures. Fire itself is the most common and effective means of fuel reduction.

All of the plants that grow in forests are potential fuels, so complete elimination of all fuels is simply out of the question. Decisions must be made about the amounts and kinds of fuel that will be left or allowed to develop in stands. Among the factors to consider are the degree of risk of serious fires, their anticipated damage, and the prospective effectiveness of fire suppression. In localities where very damaging fires are common, these decisions become questions about what degrees of calculated risk to take.

The most far-reaching modifications of silviculture for dealing with wildfire risks are those in which fire-resistant species are grown and comparatively light periodic surface fires are used to keep dangerous amounts of fuel from accumulating. For example, ponderosa, red, loblolly, slash, and longleaf pines and some other two- or three-needled pines are commonly grown as monocultures because of their fire resistance. Their tolerance of dry soils is also a factor in their adaptability to fire-prone sites. Because seedlings and saplings usually have yet to develop roughened fire-resistant bark, it may be necessary to avoid having stands with intricate intermingling of age classes; making fire-lines around many small units of area is very costly and difficult to administer. Sometimes the fires result in the development of a ground stratum of grass. Where this silvicultural policy is being followed, the fires must be frequent enough to prevent too much fuel from accumulating. It may be especially important to prevent vertical curtains of potential fuel from developing.

Policies of fire exclusion have sometimes set the stage for crown fires which are dangerous and almost impossible to suppress. In stands of slash pine on the Atlantic and

Gulf Coastal Plains, "ladder" fuels of draped pine needles frequently develop on understory shrubs with waxy inflammable leaves if there is no prescribed burning. Serious problems have arisen in some localities where there is enough precipitation in one season to support the development of luxuriant stratified mixtures of species, usually conifers, coupled with a long extremely dry fire season. Large amounts of vertically distributed ladder fuels develop, especially if fires have been excluded for many decades. The situation can be aggravated if there are too many trees for the soil to support during drought years and many succumb to insect attack as a result. The resulting dead standing trees add to the supply of fuel and invite ignition by lightning strikes.

Stands of this sort are common in the Sierra Nevada, the Rocky Mountains, and in other places where fire exclusion has allowed fire-sensitive Douglas-fir and true firs to invade the lower strata of stands of ponderosa pine. The condition is not always the result of excessively successful fire control but was also common in earlier times. In nature such stands regenerate slowly after the severe fires that occur at long intervals. However, neither the dangerous conflagrations nor the delay in restocking can be tolerated except in some remote wilderness areas. The best solutions are to reduce stocking by heavy partial cuttings to thin the stands and eliminate ladder fuels. After this has been done, it may or may not be necessary to carry out prescribed burning to achieve further fuel reduction or prepare for new regeneration. This procedure has been used, with remarkably hot fires, to induce regeneration of the big-tree sequoia, the world's largest organism, in the mixed conifer forests of the Sierra Nevada.

Not all natural forest types adapted to regenerate after wildfires are places where prescribed burning under existing stands is feasible. For example, in many stands of lodgepole and jack pine, the fuels are almost always so abundant and inflammable that most fires develop into uncontrollable crown fires. The usual solution in such cases is to create firebreaks to keep fires from getting large and then regenerate stands by planting or direct seeding after clearcutting and broadcast burning of slash. Similar policies are followed with particularly inflammable stands of eucalpyts.

Some forests include dry areas with low shrubby or grassy vegetation that is highly inflammable yet adapted to resprout after fires. These are common in Mediterranean climates such as that of California with very long dry summers (Moreno and Oechel, 1994). About the only silvicultural treatment that can help on such areas is the reestablishment of vegetation after fires have occurred.

Dealing with Dead Wood and Slash

Forestry started in North America during a period when sporadic conflagrations had destroyed many sawmill towns with terrible loss of life in the Great Lakes Region and the West. These disasters were blamed almost entirely on accumulations of slash. Laws were commonly passed, requiring that slash be burned in the vain hope that cutover land could be made fireproof. This view has persisted in the policy of some forest-fire control agencies and has been a major reason for the practice of clearcutting, broadcast burning, and planting. Sometimes slash disposal has become an end in itself, although extreme views favoring slash disposal collide with equally extreme opposition to any activity that generates smoke.

Forest fires spread chiefly through the dead, dry leafy materials of the forest floor. Dead wood may be ignited by the burning leafy materials but is not the continuous fuel blanket through which fires spread. However, woody materials may burn hotly enough that burning embers from them may be wafted aloft and spread on the wind for substantial distances. They may spread across fire-control lines, and the woody materials themselves can impede the construction of such lines even if they are not burning.

Standing dead trees (**snags**) pose yet another dilemma for fire management. They are easily ignited by lightning, and, because they are dry, they can burn high above ground and spread sparks for long distances across fire-lines. Felling them is dangerous work, but they can be even more hazardous if the wind determines the time and direction of their fall. The dilemma is that they also provide food and shelter for some of the fauna associated with old forests. It is wise to fell them if they are in places where they are apt to fall on people. Sometimes dead wood lying on the ground plays an important role in the life cycles of many forest organisms, both large and small.

In planning any silvicultural system, it is necessary to decide what to do about slash left in harvesting operations. The beneficial effects of slash must be weighed against the risks and problems it creates. In many humid climates, little or no slash disposal is necessary for fire control.

Many problems associated with smoke can be reduced by avoiding burning at night or under other conditions when temperature inversions prevent upward dispersal of smoke. Allowing large pieces of fuel to smolder for days often causes smoke problems. The ideal slash disposal fire is one that burns only during the daytime and quickly consumes the fine flashy fuels that carry fires but under conditions in which the larger fuels are too moist to burn.

Sometimes concern about smoke has led to the disposal of slash by burying it, a practice that seems hardly consistent with soil and water protection. On the other hand, if logging machines run over and crush slash that has been deposited on the ground, this usually protects the soil from rutting and also renders much of the slash uninflammable. In fact, anything that causes slash to lie close to the ground will usually speed its decomposition.

Another way to reduce forest fuels is by utilizing them, although it often requires undesirably close utilization to achieve much benefit. The grazing and browsing of animals can achieve some reduction but must usually be too heavy for the health of the vegetation to break the continuity of fine fuels. The same kind of problem exists with wood utilization. Ordinarily, the wood that can be utilized comes from diameter classes larger than those of the branches and leafy materials through which forest fires actually spread. Reducing the heavy fuels may, however, cause the smaller material to be consolidated close to the soil surface where it decomposes faster than when they are supported above ground. If partial cutting keeps shaded forest floor materials moist enough, the slash decomposes more rapidly than it would in hot, dry openings. However, the opposite effect can prevail in boreal climates as a result of low temperature.

Sometimes the continuity of slash and other forest fuels can be interrupted by establishing firebreaks along which preplanned fire-lines can be built quickly to stop fires after they start. Snags and other kinds of fuel that might emit sparks that could blow over the lines should be eliminated close to such firebreaks. Ridgetops, wetlands, and broad roads can often be developed for firebreaks which usually become stand boundaries. Fires will often stop or slow down at interfaces between radically different kinds of forests. For example, spring fires racing through light, fluffy fuels beneath leafless deciduous forests will usually slow down when they burn into dense needle litter in the shade of evergreens. Conversely, later in the growing season after the needle litter has dried and the deciduous trees have come into leaf, a fire burning in the opposite direction will become less intense when it starts to spread in the heavy shade of the deciduous trees.

Wind Damage

Trees grow tall and the destructive power of wind increases exponentially with increasing distance above the frictional effect of the ground surface. The force exerted by wind also increases exponentially with wind speed. Wind damage can be reduced by (1) thinning to increase the taper and strength of individual tree stems as described in Chapter 3, (2) avoiding the creation of local accelerations in wind speed, and (c) using strip-cutting schemes in which successive cuttings advance from leeward to windward (Fig. 19.1). Many aspects of wind damage to forests have been reviewed by Counts and Grace (1995).

Wind causes damage to trees by either breakage or uprooting of stems. In either case, the cause is compression failure of stems or roots on the *leeward* sides of trees. Such failures form because wood is weakest in its resistance to splitting. The failures appear along horizontal or diagonal planes on the leeward sides when tree stems are bent beyond the limit of their elasticity and a crushing action takes place. If compression failures have formed, the stem or root may later break at the uppermost compression failure (Mergen, 1954). If the failure is in the stem, it often breaks about halfway between the base of the crown and the ground; if the uppermost failure is in the supportive roots, the tree will uproot if subsequent winds push it over. The compression failures impair the elasticity of the wood and cause more strain on the half of the tree stem that is under tension. If the wind force becomes great enough, the whole stem breaks, leaving a splintery tension failure on one side of the break and the brash compression failure on the other.

In planning for silvicultural measures to reduce wind damage, it is very important to know the directions from which the damaging winds are likely to blow. These directions are not necessarily the same as those of the prevailing winds, although it helps to recognize that tree stems tend to strengthen themselves against the winds from the directions that predominate during the seasons when wood formation takes place. In temperate North

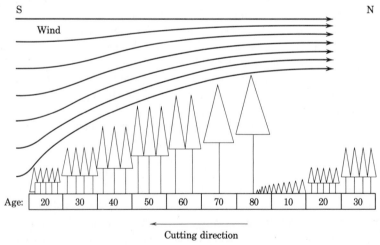

Figure 19.1   Use of the strip-selection system to divert damaging winds up and over a stand. The cross section is along a north–south axis through part of a stand managed on an 80-year rotation. The stand is streamlined against southerly winds and regeneration is also protected from the direct rays of the sun. Cutting progresses from north to south and the 80-year-old strip is about to be replaced.

America, the most common direction of winds during the growing season in major storms is from the southwest. The main exception is found along the Atlantic Coast where southeast or southerly winds blowing by the shortest distance from sea surfaces during hurricanes are most dangerous, even though the prevailing winds are westerly. The very destructive winds associated with tornadoes may blow from almost any direction. They are so powerful that there do not appear to be any measures, other than post-storm salvage, that can be taken to mitigate their effects.

The stands that are most endangered by wind are those growing on slopes exposed to winds that blow unobstructed across broad expanses of water or level terrain. Stands on very shallow soils are also vulnerable to uprooting. Although heavy, early, and frequent thinning may help, it is sometimes necessary to grow trees on short rotations so that they do not get very tall. It also helps to use species with relatively strong wood because these resist not only stem breakage but also the failures of supportive roots that lead to uprooting.

When the pathways of moving air are constricted in either the horizontal or vertical dimension, the wind speed is increased because unchanged volumes of air are forced into narrowed passages. Small accelerations in wind speed can be significant because the force of moving air varies with the third or fourth power of the speed. Wind speeds of 90 miles per hour (16 m/sec) represent the approximate threshold of severe damage but some damage can occur at much lower speeds; few trees can withstand speeds in excess of 120 miles per hour (21 m/sec), such as the winds of tornadoes.

When a stream of air passes up over a ridge or mountain range, its passageway is constricted from below and its speed is accelerated. The same effect takes place when the air moves through any sort of horizontal gap, whether it be a gap in the terrain or a gap in the margin of a stand of trees. The terrain cannot be changed, but it is silviculturally possible to shape the edges of stands in ways that reduce the possibilities of dangerous accelerations of the wind.

In most kinds of terrain, the wind damage associated with topography is most common and worst on the windward slopes where one would intuitively expect it. However, if the slopes are exceedingly steep, as in some parts of western North America, it is often worst just leeward of the ridge crests. This is because of gusty downbursts of air that take place in a turbulent zone on the lee sides of the crests. The same effect can result when storm winds collide with tightly closed vertical stand edges; the edge trees may withstand the wind, but the weaker ones behind them are hit with downbursts. One way of dealing with borders of stands that are undergoing partial cuting is to leave strong, scattered trees in broad belts as in uniform shelterwood cutting. This dissipates some of the force of the wind gradually rather than abruptly. However, the main consideration is avoiding the creation of patterns of stand arrangement that create or accentuate constrictions that cause wind speeds to accelerate.

Sequences of cutting in successive strips that advance from leeward to windward can create the streamlined pattern of stand arrangement shown in Fig. 19.1 and reduce the risk of downbursts. In this sort of **strip-selection cutting**, appropriately long periods of years must be allowed to elapse between the cuttings, and it is necessary to adhere to the schedules. This approach can be combined with efforts to protect tender seedlings against exposure and the invasion of undesirable pioneer vegetation in **strip-shelterwood cutting** as shown in Fig. 16.11. Where combinations of protection from sun and wind are sought, the strips must advance in a direction that is a compromise between the direction from which dangerous winds blow and the direction of the midday sun. In many cases in the Northern Hemisphere, this means that the strips advance from northeast to southwest. Quick replacement of stands by clearcutting in narrow, quickly advanced strips does not produce streamlining effects, but it can help to advance them from leeward to windward. In any sort of strip cutting, it is highly desirable that the windward edge of the last strip that is to be cut be composed of trees that are strong enough or deeply rooted enough to be unusually windfirm.

Protection from Ice, Snow, and Frost

The best way to keep trees from breaking under the weight of adhering snow and ice is to develop strong trees with symmetrical crowns. Where either people or the forest itself are threatened with snow avalanches, permanent selection forests can be maintained so that the snow is kept anchored in place or diverted after it starts to slide. Small trees growing in gaps in stands may suffer frost damage, because the small openings become frost pockets. If there is wet snow, it may accumulate on evergreen branches around the openings and crush small trees when it suddenly cascades down onto them. In extreme cases shelterwood cutting or the use of nurse crops may be necessary to protect regeneration from frost damage.

BIOTIC AGENCIES

Consideration of biotic enemies best starts from recognition of the fact that the trees of any kind of forest represent a source of food to a wide variety of organisms. Owing to the availability of food, organisms ranging from minute viruses to large herbivorous mammals have evolved that are adapted to feed on plants. These organisms are so dependent on the vegetation that changes in the forest cause changes in the populations of dependent organisms. Changing the stands does not get rid of parasites; it merely exchanges one group of parasites for a new set that may be more or less harmful and difficult to manage.

Fortunately, only a very few of the dependent organisms are harmful. In a sense, the parasite that kills its host and thus its supply of food is a poorly adapted one. This is why introduced parasites often cause much more economic loss than the native parasites. The well-adapted parasites sometimes cause so little damage that they go almost unnoticed. However, some of these cause substantial economic damage without threatening the life of the tree. For example, the heart-rots, which attack the nonliving wood inside a tree, can ruin the utilizable wood without significantly harming the vital processes of the tree. Need for modifying silvicultural systems to reduce losses to biotic enemies can normally be confined to those few organisms that can cause serious loss. The vast majority that merely feed on the trees without important damage do not require control measures.

Direct control of pests involves attacking the pests themselves either with pesticides or with various methods known as **biological control**. Biological control involves the introduction or encouragement of biotic agencies that combat the damaging organisms. Suitable agents include fungi antagonistic to damaging ones, parasites of insects, or predators of herbivores. **Indirect control** refers to measures that make the circumstances less favorable to the pests or more favorable to their hosts. This consists mostly of **silvicultural control** which involves the creation of forests and forest environments that resist either damaging agencies or the effects of damage by them. These distinctions are not necessarily perfect; silvicultural measures can, for example, be used to encourage the biotic enemies of pests or to eliminate trees that are sources of infection.

It must be recognized that control programs are seldom completely successful, and that a good outcome at one time and place does not guarantee similar results elsewhere. Such programs, like so many things in silviculture, are best regarded as the continuous application of adaptive management. Successful programs of forest pest management usually involve several kinds of control measures and are so tailored to each complex biological situation that it is difficult to fit them into any simple categories.

Most of the classic generalizations about the damaging biotic agencies of the forest are more nearly true than false, but they cannot be accepted as a basis for silvicultural procedure without being scrutinized for applicability in each instance. Among these generalizations or propositions are:

1. Vigorous, fast-growing trees are more resistant than less thrifty, slow-growing trees.
2. Mixed stands are safer than pure ones.
3. Multi-cohort stands are more resistant than even-aged stands.
4. Duplication of natural conditions will safeguard against difficulties.

Exceptions are numerous. For example, the stem rusts of conifers and the white pine weevil are more serious pests of vigorous trees than of the less thrifty. Pure stands of spruce are much less susceptible to the spruce budworm than are mixtures of spruce with highly susceptible species. In uneven-aged stands, there is ample opportunity for infection of young age classes from older trees by pathogens such as dwarf-mistletoe that attack trees of all ages, thus enabling an infestation to remain established in a stand indefinitely (Fig. 19.2). Highly artificial stands of well-tested exotics from faraway continents are sometimes less subject to damage than those in the native habitat. Although the cases that

Figure 19.2 A forester examining a 25-year-old stand of Montana lodgepole pine that has been infected with dwarf-mistletoe from infected residual trees, such as the two taller ones in the center of the picture, that had been left when the new stand was regenerated. *(Photograph by U.S. Forest Service.)*

fit the broad generalizations are more numerous than the exceptions, action is best guided by open-minded analysis of each case.

By far the most important silvicultural approach to reducing losses to damaging agencies is simply the evasive action of avoiding conditions conducive to damage. It is remarkable how much damage from pests or nonbiotic agencies can be traced to encouraging species or strains thereof that are not adapted to the sites or are simply exotics. Many root diseases are the result of attempts to grow a given species on soils that are too wet or too dry.

It is important to watch for syndromes of successive interdependent causes of damage or mortality. For example, a defoliator, such as the gypsy moth, weakens trees; then a drought year or a cold winter makes them weaker yet; finally some root-rot or bark-boring insect kills them. If one step is omitted, the trees may not die. If someone sees only one step, there is risk that inappropriate action will be taken or that counterproductive argument will ensue with those who see only another step.

If there is adequate knowledge of the situation, it may be possible, by silvicultural measures, to create forests that are resistant to damage. In dealing with this or any other matter relating to damaging agencies, it clarifies analysis to distinguish between (1) **susceptibility to attack** and (2) **vulnerability to damage**.

For example (Doane and McManus, 1981), oak stands on very dry soils where the litter gets hot during the day are very susceptible to gypsy moth attack because the larvae do not descend to the ground where mice can feed on them during the day. However, the defoliated oaks on those soils are not as vulnerable to mortality because the root-rot fungus, *Armillaria mellea,* is not as prevalent as it is on more mesic sites. On the better soils, the mice reduce the defoliation, but, if the trees are defoliated, they are more vulnerable to being killed by the root-rot. This can make it appear that the gypsy moth problem is confined to the better soils. The possibility that the insect might spread from the dry soils to the good is obscured by the fact that the female moths of the most common U.S. strain do not fly. However, some small larvae are dispersed by wind on silken threads. Furthermore, the risk of defoliation on the mesic sites is heightened by the tendency of birds to leave these sites and flock to feed on the dry sites where caterpillars are more abundant. This explains why outbreaks can start on dry ridgetops and later develop on the better soils, even though the biotic movement between the sites is more of birds than of insects. This is one of many cases in which the development of appropriate control measures depends on detailed knowledge of the actions of the damaging agencies.

Examples of damage control by modifying silvicultural systems are many and varied. Among these examples are thinnings and other removals of old, damaged, or infested trees in southern and western pine stands to reduce damage by *Dendroctonus* bark beetles (Thatcher et al., 1981; Sartwell and Stevens, 1975; Belanger et al., 1993). With species that are seriously threatened by heart-rot, it is logical to terminate rotations before the trees become old and highly susceptible.

The blister rust of the five-needled pines is susceptible to some degree of silvicultural control through modification of stand structure. The spores that move from the alternate hosts, *Ribes* shrubs, cannot move over long distances and can germinate only on moist needle surfaces. Anything that forestalls the formation of dew on the foliage reduces the possibility of infection. This can be done by avoiding the creation of small gaps that are open to the sky and employing the sheltering effect of shelterwood cutting. Valley bottoms and areas close to large bodies of water are places where infection rates are high. In the case of western white pine and sugar pine, most losses result from infections of small trees so that preventing infection during the early stages is crucial. Eastern white pine trees can be infected at any height but are not as susceptible, so that avoiding conditions suitable for dew formation usually suffices. In earlier years, much of the control effort centered on eradicating *Ribes,* but this approach proved expensive and not as effective as hoped. Sometimes the germination of stored *Ribes* seeds can be reduced by using shelterwood cutting and avoiding soil disturbance. The partial shade that is conducive to pine seedlings may be too much to allow the *Ribes* to survive and produce seeds. Recent evidence also suggests that rust-resistant natural populations may be built if rust-resistant parent trees are retained in the shelterwood seed source.

The main problem with growing eastern white pine is usually the white pine weevil. This insect often kills the terminal shoots of trees with tops in the sunshine. The replacement branches are usually the laterals of the previous year, and, when they turn upward, they form crooks and forks that badly degrade the stems. One solution is to use the shelterwood method to keep the young trees partially shaded until at least one log-length has formed. Older white pines provide the best kind of shade because the weevils are lured away by their sunlit tops. The overstory must remain for such a long time that it should be composed of pines that can produce value rapidly and thus compensate for the slow growth of the deliberately stunted regeneration. The *Hypsipyla* borer that damages the terminal shoots of mahogany and other members of the Meliaceae in the tropics behaves in a similar manner.

Role of Mixtures of Species and Age Classes

In the majority of cases, mixed stands probably suffer about the same amount of damage from pests as pure stands. In many other cases, however, mixed stands suffer less damage than pure. Cases in which pure stands suffer less biotic damage are not rare.

If the food supply of some pest does not include all of the species of a mixed stand, then it is diluted in ways that may inhibit buildup of the pest population. This is probably the chief advantage of mixed stands in pest management. The physical separation of susceptible plants also inhibits the spread of many pests, especially those that disperse slowly or with difficulty. Fungi that spread through the soil, such as those causing most root rots, are often less common and damaging in mixed stands. The long-term dispersal of insect pests of low mobility can be impeded by inedible plants. No real physical separation exists between species, however, in a stratified mixture if there is an essentially pure stratum above or below ground.

If a given mixture is more vigorous than pure stands, then it is more likely to be resistant to those pests that attack weakened trees. A mixture is probably more secure against pests than an uneven-aged pure stand because a given pest is more likely to attack many age classes of a single species than to attack different species.

Nevertheless, there are important instances in which pure stands are more resistant to certain pests than stands of the same species mixed with highly susceptible ones. Pure stands of spruce are more resistant to the misnamed spruce budworm than spruce-fir stands that have overmature balsam firs on which the insect can thrive. It may be noted, however, that mixtures of hardwoods with spruce and fir are even more resistant.

The fact that the American chestnut grew in complex mixtures did not save it from the chestnut blight fungus. However, if it were still a major component in some eastern forests, there might be less difficulty with the gypsy moth, to which chestnut is unpalatable. The heteroecious stem rusts of conifers, or any other organism with alternate hosts, could not exist in absolutely pure stands. They are instead adaptations to the association of different species. The resistance of any kind of stand to its biotic enemies depends on the susceptibility to attack and vulnerability to damage of its individual components and not on doctrinaire generalizations. However, a mixed stand of both resistant and nonresistant species is bound to be more secure against injury than a pure stand of a nonresistant species, but less so than a pure stand of a resistant species. Sometimes the main value of mixed stands in this respect is the rather dreary consolation that the risk of losing whole stands all at once is reduced.

The spruce budworm, one of the most dangerous insects of the North American forest (Sanders et al., 1985), is an excellent example of an insect that is not merely a predator of the weak and also has a marked preference among species. The most typical form of this versatile insect is the scourge of the eastern portion of the spruce-fir forests, where massive outbreaks develop at intervals of several decades. It is poorly named because it is very dependent on and damaging to balsam fir. Spruces tend to be attacked when the supply of fir foliage has dwindled. The trees that are most susceptible to attack, but not extremely vulnerable to loss, are large balsam firs. These bear abundant staminate flower buds which are highly nutritious for young budworms. Susceptibility to attack is determined more by the proportion of mature balsam fir in whole forests than by the characteristics of individual trees or stands. It can be reduced to a limited extent by replacing large tracts of mature forests with an intermingled arrangement of stands in which the susceptible mature stands are diluted among younger and less susceptible stands. Vulnerability to the losses that are likely to occur after a stand is attacked can be somewhat reduced by presalvage of firs and spruces with poor live crown ratios.

As described in Chapter 20, large herbivores are as much pests of the forests as fires or introduced defoliating insects. The animal population of the forest must be managed just as that of the plants is; in fact, one cannot be managed without also managing the other.

Problems with Unnatural Conditions

Many, but not all, serious problems with biotic pests result from growing trees in habitats to which they are not adapted or from introducing exotic species of pests or hosts. Introduced pests are the most dangerous enemies of some species in almost every kind of forest. Sometimes the only good way to control them is by judicious introduction of the enemies of the pests that kept them in check in their native habitat. Introduced tree species that have left their natural enemies behind may flourish for a time and then suffer serious attack if their enemies catch up with them. They may also encounter serious difficulties with pests native to the new habitat.

Less spectacular difficulties arise when a species, native or exotic, is planted or is allowed to become established on a soil or site where it does not grow vigorously or does not grow in nature. The trees may develop well for some years and then fall victim to some pest that may be acting simply as one of the factors that determines the natural range of the species. Some of the most common cases of this sort develop when some species or strain adapted to a moist site is grown on a dry one but succumbs to a root disease or bark beetle attack after a drought year. This sort of difficulty can be reduced by timely thinning but is better avoided altogether by not allowing the ill-chosen species or strain to grow there in the first place.

Direct control with pesticides is far less important in forestry than in agriculture. Most commonly, it involves foliar applications of insecticides to combat defoliators or applications to bark to control bark beetles. Sometimes the tendency is to regard insecticides as perfect solutions for forest insect problems and even to heap scorn on indirect control measures. There is also the highly popular view that all pesticides are, at worst, dangerous and, at best, agencies that merely prolong pest outbreaks. The truth usually lies in the middle, but it is important to keep watch for the special cases in which extreme views are actually justified.

Other Aspects of Pest Management

Insecticides and fungicides are commonly used on shade trees and in nurseries, where the hosts are very crowded and the crop very valuable. It is very important to keep nursery seedlings from being infected with stem rusts, certain root rots, and other persistent disorders because infected plants seldom recover after planting.

Efforts to develop resistant strains of forest planting stock through genetic manipulations have played a more important role in dealing with fungus diseases than with insects. Most of this effort has involved difficult problems with fungi such as those causing stem rusts of conifers and the chestnut blight. Moderate success has been achieved in combatting the fusiform rust of loblolly and slash pine, as well as with some other diseases.

So many different kinds of modifications are aimed at reducing specific kinds of damage that it is hard to detect any general or consistent way in which they conflict with other objectives. However, these measures usually complicate harvesting and other operations. The indirect silvicultural measures of control are often the slowest to take effect, but they are the most enduring and automatic (Knight and Heikkenen, 1980). Where indirect measures work, they can usually be depended upon to do so even when no one remembers to do something at the right time.

DAMAGE CONTROL CUTTINGS

Certain kinds of cuttings are sometimes done in attempts either to anticipate damage or to prevent it. **Presalvage cutting** is designed to anticipate damage by removing highly vulnerable trees. **Sanitation cuttings** are more active measures designed to eliminate trees that have been attacked or appear in imminent danger of attack by dangerous insects and fungi in order to prevent these pests from spreading to other trees. **Salvage cuttings** are done to save the wood in dead or damaged trees. The opportunity to engage in such races with the damaging agencies depends mainly on whether stands are sufficiently accessible.

Presalvage Cuttings

The rational conduct of presalvage cuttings depends on identifying those trees that are likely to be lost and estimating the length of time they may be expected to endure.

If the main source of anticipated damage is wind, ice, or some other climatic agency, the mechanical structure of the trees and their position within the stand are the important criteria. Trees with asymmetrical crowns or previous injury from frozen precipitation are prone to additional damage from this source.

When biotic agencies or physiological factors such as those related to site factors are the causes of anticipated losses, the vigor of the trees is usually the criterion employed in presalvage cutting. Trees of relatively high vigor are less vulnerable if attacked because they are likely to endure the amount of loss of vital tissues that would kill trees of low vigor. Abundant pitch flow from trees of good vigor even actively resists attack by *Dendroctonus* bark beetles, which dictate much presalvage cutting. The presalvage of trees of low vigor is rendered all the more logical because they grow little in volume or value. A number of tree classifications have been developed as a guide to presalvage cutting (Hedden, Barras, and Coster, 1981). One of the best-known examples of such a classification is that developed by Keen for the interior ponderosa pine type of California and the Northwest (Miller and Keen, 1960). Vulnerability to loss from bark beetles is so closely correlated with growth rate that this classification is suitable for selecting trees for cutting even where risk of beetle attack is not the main consideration.

Keen's classification (Fig. 19.3) is based on the two major factors of age and crown vigor. There are four age classes and four vigor classes within each age class, making a total of 16 classes. The four age classes are termed: *1*—young, *2*—immature, *3*—mature, and *4*—overmature; they are grouped by relative maturity rather than by any definite age range. Actual age limits for the groups vary in different parts of the ponderosa pine region. Color and type of bark, total height, shape of top, characteristics of branches, and diameter are the chief external indications of maturity. The four crown vigor classes are: *A*—full vigor, *B*—good to fair vigor, *C*—fair to poor vigor, and *D*—very poor vigor. The size of the crown (length, width, and circumference), its density, and the shape of its top indicate the crown vigor and consequently the inherent capacity of the tree to grow and endure exposure to beetles.

Keen's classification was originally designed to provide a means of evaluating the risk that the trees left after selection cutting would be attacked by beetles during a cutting cycle of approximately 30 years. Although it still serves as a useful guide for many kinds of partial cutting and as an example of a tree classification, simpler means are now being used to determine which trees are in imminent danger of death. This is because it is now possible to make much lighter presalvage cuttings at much more frequent intervals.

Presalvage cutting has sometimes played an important role in putting decrepit old-growth stands under management, as on some forests in the western United States. The goal of developing the full range of age classes required for true sustained yield often requires that old, overmature stands be retained for decades until they are scheduled for replacement. The losses of high-value timber can be quite substantial during the waiting period, but these trees can be salvaged or forestalled by salvage and presalvage cuttings.

Sanitation Cuttings

Sanitation cuttings are often combined with salvage or presalvage cuttings. In fact, any cutting may be considered sanitation cutting to whatever extent it eliminates trees that are present or prospective sources of infection for insects or fungi that might attack other trees.

Sanitation cuttings are not worth conducting unless the removal of susceptible trees will actually interrupt the life cycle of the organisms sufficiently to reduce their spread to other trees. For example, the elimination of *Ribes* shrubs from stands of five-needled pines can be a moderately effective kind of sanitation cutting (or cleaning). However, the removal of pines already infected with the blister-rust fungus is not effective sanitation because the spores that carry *Cronartium ribicola* from pine to *Ribes* can travel so far that no practical advantage is gained. Sanitation cuttings are often applied to reduce the spread of *Dendroctonus* bark beetles in the southern pines, ponderosa pine, and Douglas-fir.

It should not be tacitly assumed that the vigor of trees is the sole criterion of their susceptibility to attack, even though it may be a good indicator of their capacity to endure damage. Some economic pests of the forest, notably insects, multiply most rapidly on vigorous hosts, just as grazing animals prefer forage from healthy plants. Fortunately, the vast majority of insects attracted to vigorous trees are well-adapted parasites that do not

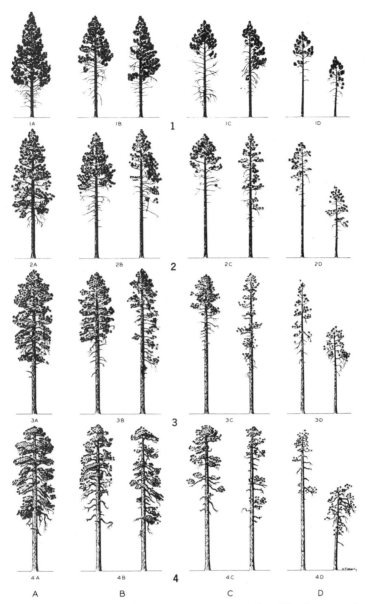

Figure 19.3 The Keen tree classification for ponderosa pine. The four age classes range from 1, the youngest, to 4, the oldest; the four crown vigor classes range from A, the most vigorous, to D, the poorest. *(Sketch by U.S. Forest Service.)*

endanger their own existence by killing their hosts; these insects rarely become cause for concern and may even escape notice.

Sometimes sanitation cuttings must be associated with special measures to provide additional assurance that the damaging organisms will not spread to the residual stand. If the pests are exclusively parasitic on living tissues, it is sufficient to kill the infested trees and leave them in the woods. If the insects or fungi involved are capable of multiplying as saprophytes in dead material, it may be best to utilize the wood even at a loss, provided that its transportation does not spread the infestation. This approach can be used to increased advantage against bark beetles if "trap" logs or trees are left temporarily to attract them and then hauled to mill or log pond before the emergence of the entrapped brood. It may also help to burn slash or treat stumps with insecticides in ways that directly kill the beetles.

Those species of heart-rotting fungi that spread as spores from conks on fallen trees are not easy to control if infected logs must be left in the woods. However, the felling of infected trees does reduce the distance over which the spores can travel, and the accelerated disintegration of the wood shortens the period of danger.

There are some agencies of damage against which sanitation cuttings are, for practical purposes, ineffective. Fungi, such as *Heterobasidum annosum*, that inhabit the soil and damage the roots of trees are not likely to be halted even by sanitation cuttings that are carried to the extent of removing the stumps. Some organisms spread so rapidly or over such long distances that sanitation cutting may be a meaningless gesture as far as the effective protection of the stand is concerned.

Sometimes the salvage of entire stands may be regarded as a desirable measure of sanitation, even if the damaged stands by themselves are not worth the effort. Stands that have been badly damaged by fire, wind, or similar agencies frequently support the development of large populations of bark beetles that can cause serious injury to adjacent stands. The expense of salvaging such foci of infection may be amply rewarded by reducing losses in adjacent stands.

In any kind of salvage or sanitation cutting in which the effects of insects or fungi are involved, there is no substitute for a thorough understanding of the life histories of the organisms involved. The details of a procedure useful in dealing with one kind of injury may be hopelessly ineffective or needlessly intensive as a means of reducing losses from other seemingly similar agencies of destruction.

Salvage Cuttings

When all else has failed, it may be desirable to salvage the losses. Salvage cuttings are made for the primary purpose of removing living or dead trees that are imminently threatened by mortality, damage, or loss from injurious agencies other than competition between trees.

In some cases, attacks by damaging agencies are so chronic that they rule silviculture, and so frequent damage control cuttings may be desirable. One example of this distressing state of affairs exists on the badly eroded, poorly aerated soils of the Piedmont Plateau where shortleaf pine is often lost to the little-leaf disease and root rots. Unfortunately, it is replaceable only with pines that are vulnerable to fusiform rust and bark beetles.

The recovery of timber values that might otherwise be lost is one important and expeditious silvicultural means of securing yields greater than those available from the managed forest. The objective is usually to utilize the injured trees to minimize financial loss. Salvage cuttings are not conducted unless the material taken out will at least pay for the expense of the operation, except in cases where real justification exists for true sanitation cuttings. Salvage operations are often unfeasible unless they are combined with the removal of healthy trees in other silvicultural operations such as thinnings or regeneration cuttings.

The immediate financial loss depends largely on the extent and distribution of damage. If trees die sporadically and at widely scattered places in a stand, they may become a total loss because of the impracticability of harvesting them. The loss may be small if the amount of damage is not great and is concentrated in time and space. When catastrophic losses have occurred over a wide area, the returns from salvage cutting are often reduced by the necessity of selling the products on a glutted market. The costs of logging are generally higher in salvage cutting than in operations where the trees to be removed have been chosen by intention rather than by accident.

Identifying the trees to be taken out in salvage cuttings is not usually a problem. Sometimes, however, the mortality caused by the attack of a damaging agency does not take place immediately. This is particularly true where surface fires have occurred because the main cause of mortality is the girdling that results from killing cambial tissues. As with other kinds of girdling, the top of the tree may remain alive until the stored materials in the roots are exhausted. It is usually a year or more before the majority of the mortality has occurred. By this time, those trees that were killed immediately have often deteriorated seriously. It is, therefore, advantageous to anticipate mortality before it has actually occurred. The predictions must be based on outward evidence of injury to the crown, roots, or stem.

Salvage cuttings should be completed as soon as possible after mortality or injury has occurred. Dead trees generally start to deteriorate rapidly during the first growing season after death, so it is usually advisable to get the trees out of the woods before insects and fungi have become active in the spring.

Unfortunately, the amount of salvageable material is often so great that it cannot be removed within a few months or a year, even if all other operations are suspended. Under such conditions, it is highly desirable to know how long the dead or damaged trees are likely to remain sufficiently sound to be worth salvaging. Entomological and pathological investigations have made this information available for a number of different species (Boyce, 1961). This type of knowledge makes it possible to conduct large salvage operations systematically and efficiently.

The amount of time allowable varies widely depending on the circumstances. The sapwood of virtually all species is highly perishable, but the heartwood of the most durable may remain sound for many years. Dead trees of small diameter become valueless long before large trees. Deterioration usually proceeds more rapidly on good sites than on poor. Differences in the rate of decay of various species are also significant. The basic objective should be to schedule salvage operations in different places in such order that the value of timber saved will be at a maximum. This does not necessarily mean that the most valuable material should be salvaged first; some less valuable wood may deteriorate much faster.

Harvesting of damaged timber is often more difficult and expensive than cutting in undamaged stands. This is especially true if large groups of trees have been broken off or uprooted by wind. In conducting salvage operations, the money lost because of expensive logging can easily exceed the potential values wasted through decay of unsalvaged timber. The extremes of both haste and procrastination should, therefore, be studiously avoided in salvage cutting.

Salvage cuttings will be necessary from time to time in any forest. In stands that are extraordinarily susceptible to injury, the time and place of harvesting operations may be dictated largely by damaging agencies. It is also a good policy to expect some mortality after even the most well-conducted partial cuttings. If full advantage is to be taken of opportunities for salvage of merchantable material, all parts of a forest must be made accessible by a good system of roads. The gains in production from salvage cutting and thinning can help justify the road network. Plans must always exist for action to be taken after episodes of major damage.

BIBLIOGRAPHY

Agee, J. K. 1993 . *Fire ecology of Pacific Northwest forests*. Island Press, Washington. 493 pp.

Anderson, R. L. 1973. A summary of white pine blister rust research in the Lake States. USFS Gen. Tech. Rept. NC-6. 12 p.

Anman, G. D., et al. 1977. Guidelines for reducing losses of lodgepole pine to the mountain pine beetle in unmanaged stands in the Rocky Mountains. USFS Gen. Tech. Rept. INT-36. 19 pp.

Baker, W. L. 1972. Eastern forest insects. *USDA Misc. Publ.* 1175. 642 pp.

Bega, R. V. (ed.). 1978. Diseases of Pacific Coast conifers. *USDA, Agr. Hbk.* 521. 206 pp.

Belanger, R. P., R. L. Hedden, and P. L. Lorio, Jr. 1993. Management strategies to reduce losses from the southern pine beetle. *SJAF*, 17:150–154.

Berg, N. H. 1989. Proceedings of the symposium on fire and watershed management. USFS Gen. Tech. Rept. PSW-109.

Berryman, A. A. (ed.). 1988. *Dynamics of forest insect populations: patterns, causes, implications*. Plenum, New York. 603 pp.

Berryman, A. A., G. D. Amman, and R. W. Stark (eds.). 1978. *Theory and practice of mountain pine beetle management in lodgepole pine forests*. Univ. of Idaho, Moscow. 220 pp.

Biswell, H. H. 1989. *Prescribed burning in California wildlands vegetation management*. University of California Press, Berkeley. 255 pp.

Black, H. C. (ed.). 1992. Silvicultural approaches to animal damage management in Pacific Northwest forests. USFS Gen. Tech. Rept. PNW-287. 422 pp.

Boyce, J. S. 1961. *Forest pathology*. 3rd ed. McGraw-Hill, New York. 572 pp.

Brookes, M. H., J. J. Colbert, R. G. Mitchell, and R. W. Stark (eds.). 1985. Managing trees and stands susceptible to western spruce budworm. *USDA Tech. Bul.* 1695. 111 pp.

Brookes, M. H., R. W. Stark, and R. W. Campbell (eds.). 1978. The Douglas-fir tussock moth: a synthesis. *USDA Tech. Bul.* 1585. 331 pp.

Brown, A. A., and K. P. Davis. 1973. *Forest fire: control and use*. 2nd ed. McGraw-Hill, New York. 686 pp.

Castello, J. D., D. J. Leopold, and P. J. Smallidge. 1995. Pathogens, patterns, and processes in forest ecosystems. *Bioscience*, 45(1):16–24.

Coulson, R. N., and J. A. Witter. 1984. *Forest entomology; ecology and management*. Wiley, New York. 675 pp.

Counts, M. P., and J. Grace (eds.). 1995. *Wind and trees*. Cambridge Univ. Press, Cambridge. 528 pp.

Dahlsten, D. L., R. Garcia, and H. Lorraine (eds.). 1989. *Eradication of exotic pests*. Yale Univ. Press, New Haven, Conn. 304 pp.

Dinus, R. J., and R. A. Schmidt (eds.). 1977. *Management of fusiform rust in southern pines, symposium proceedings, 1976*. Univ. of Florida, Gainesville. 163 pp.

Doane, C. C., and M. L. McManus. 1981. The gypsy moth: research toward integrated pest management. *USDA Tech. Bull.* 1584. 757 pp.

Filip, G. M., J. J. Colbert, C. A. Parks, and K. W. Seidel. 1989. Effects of thinning on volume growth of western larch infected with dwarf mistletoe in northeastern Oregon. *WJAF*, 4:143–145.

Filip, G. M., D. J. Goheen, D. W. Johnson, and J. H. Thompson. 1989. Precommercial thinning in a ponderosa pine stand affected by armillaria root disease: 20 years of growth and mortality in central Oregon. *WJAF*, 4:58–59.

Gottschalk, K. W. 1993. Silviculture guidelines for forest stands threatened by the gypsy moth. USFS Gen. Tech. Rept. NE-171. 50 pp.

Hard, J. S., and E. H. Holsten. 1985. Managing white and Lutz spruce stands in south-central Alaska for increased resistance to spruce beetle. USFS Gen. Tech. Rept. PNW-188. 21 pp.

Harris, A. W. 1989. Wind in the forests of southeast Alaska and guides for reducing damage. USFS Gen. Tech. Rept. PNW-25. 63 pp.

Hawksworth, F. G., and D. W. Johnson. 1989. Biology and management of dwarf mistletoe in the Rocky Mountains. USFS Gen. Tech. Rept. RM-169. 38 pp.

Hedden, R. L., S. J. Barras, and J. E. Coster (eds.). 1981. Hazard-rating systems in forest insect pest management: symposium proceedings, Athens, Georgia, 1980. USFS Gen. Tech. Rept. WO-27. 169 pp.

Hedden, R. L., R. P. Belanger, H. R. Powers, and T. Miller. 1991. Relation of Nantucket tip moth attack and fusiform rust infection in loblolly pine. *SJAF*, 15:204–208.

Hepting, G. H. (ed.). 1971. Diseases of forest and shade trees of the United States. *USDA, Agric. Hbk.* 386. 658 pp.

Johnson, W. T., and H. H. Lyon. 1976. *Insects that feed on trees and shrubs: an illustrated practical guide.* Cornell Univ. Press, Ithaca, N.Y. 464 pp.

Knight, F. B.and H. J. Heikkenen. 1980. *Principles of forest entomology.* 5th ed. McGraw-Hill, New York. 461 pp.

Krammes, J. C. (ed.). 1990. Effects of fire management of southwestern natural resources. USFS Gen. Tech. Rept. RM-191. 293 pp.

Manion, P. D. 1991. *Tree disease concepts.* 2nd ed. Prentice-Hall, Englewood Cliffs, N.J. 399 pp.

Martineau, R. 1984. *Insects harmful to forest trees.* Canadian Government Publishing Centre, Ottawa. 261 pp.

McLean, J. A., and S. M. Salom. 1989. Relative abundance of ambrosia beetles in an old-growth western hemlock/Pacific silver fir forest and adjacent harvesting areas. *WJAF*, 4:132–136.

Mergen, F. 1954. Mechanical aspects of wind-breakage and windfirmness. *Jour. For.* 52:119–125.

Miller, J. M., and F. P. Keen. 1960. Biology and control of the western pine beetle. *USDA Misc. Pub.* 800. 381 pp.

Mooney, H. A., T. M. Bennicksen, N. L. Christensen, J. E. Lotan, and W. A. Reiners (eds.). 1981. Fire regimes and ecosystem properties. USFS Gen. Tech. Rept. WO-26. 594 pp.

Moreno, J. M., and W. C. Oechel (eds.). 1994. *The role of fire in Mediterranean-type ecosystems.* Springer-Verlag, New York. 201 pp.

Powers, H. R., T. Miller, and R. P. Belanger. 1993. Management strategies to reduce losses from fusiform rust. *SJAF*, 17:146–149.

Pyne, S. J., P. L. Andrews, and R. D. Laven. 1996. *Introduction to wildland fire.* Wiley, New York. 450 pp.

Regelbrugge, J. C., and D. W. Smith. 1992. Postfire tree mortality in relation to wildfire severity in mixed oak forests in the Blue Ridge of Virginia. *NJAF*, 11:90–97.

Safranyik, L. (ed.). 1985. *Proceedings of the IUFRO conference on the role of the host in the population dynamics of forest insects.* Canadian Forestry Service, Victoria, BC. 240 pp.

Sanders, C. J., R. W. Stark, E. J. Mullins, and J. Murphy (eds.). 1985. *Recent advances in spruce budworm research. Proceedings of the CANUSA Spruce Budworm Research Symposium, Bangor, Me., Sept. 1984.* Canadian Forestry Service, Ottawa. 527 pp.

Sartwell, C., and R. E. Stevens. 1975. Mountain pine beetle in ponderosa pine, prospects for silvicultural control in second-growth stands. *J. For.*, 73:136–140.

Savill, P. S. 1983. Silviculture in windy climates. *For. Abst.*, 44:473–488.

Schmid, J., M. L. Thomas, and T. J. Rogers. 1981. Prescribed burning to increase mortality of pandora moth. USFS Res. Note RM-405. 3 pp.

Schmitt, D. M., D. G. Grimble, and J. L. Searcy (eds.). 1984. Managing the spruce budworm in eastern North America. *USDA, Agric. Hbk.* 620. 192 pp.

Schowalter, T. D., and G. M. Filip. 1993. *Beetle-pathogen interactions in conifer forests.* Academic Press, San Diego. 252 pp.

Shaw, G. C., III, and G. A. Kile. (eds.). 1991. Armillaria root disease. *USDA Agric. Hbk.* 691. 233 pp.

Sinclair, W. A., H. H. Lyon, and W. T. Johnson. 1987. *Diseases of trees and shrubs.* Cornell Univ. Press, Ithaca, N.Y. 574 pp.

Tainter, F. H., and F. A. Baker. 1996. *Principles of forest pathology.* Wiley, New York. 768 pp.

Thatcher, R. C., et al. (eds.). 1981. The southern pine beetle. *USDA Tech. Bull.* 1631. 266 pp.

Van Arsdel, E. P. 1961. Growing white pine in the Lake States to avoid blister rust. USFS Lake States For. Exp. Sta., Sta. Pap. 92. 11 p.

van der Kamp, B. J. 1991. Pathogens as agents of diversity in forested landscapes. *For. Chron.,* 67:353–354.

Walstad, J., S. R. Radosevich, and D. V. Sandberg. (eds.). 1990. *Natural and prescribed fire in Pacific Northwest forests.* Oregon State University Press, Corvallis. 317 pp.

Zobel, D. B., L. F. Roth, and G. M. Hawk. 1985. Ecology, pathology, and management of Port-Orford-cedar *(Chamaecyparis lawsoniana).* USFS Gen. Tech. Rept. PNW-184. 161 pp.

CHAPTER *20*

SILVICULTURAL MANAGEMENT OF WILDLIFE HABITAT

Forested ecosystems are the natural cover of only one-third of the earth's land surface, but they contain a large fraction of the plant, animal, fungal, and microbial species of the world. Because forested areas have been reduced in many regions by agricultural and urban development, it is important that management of forests for timber, watershed protection, and other purposes includes measures to maintain native species. Furthermore, public policy increasingly stipulates that this be done. Treatments of forest vegetation can also be designed specifically to improve habitats for species of particular concern.

The earliest silvicultural treatments were designed to maintain supplies of timber and fuelwood, but management of forests for wildlife populations has also had a lengthy history. Prescribed fire, tree cutting, and planting of preferred food species were used both by aboriginal peoples to improve subsistence hunting and by medieval game keepers to improve sport hunting and meat production for the upper classes of society. However, predator control and regulation of hunting were the main management tools used until fairly recently. A scientific approach to habitat management developed early in this century as the concepts of ecology were applied to traditional methods used for managing game species such as deer and grouse (Leopold, 1933). This ecological view tended to give more attention to the associated plant and animal species that formed habitats and food chains that supported game species. Recent management ideas have more explicitly included all organisms existing within an ecosystem; thus the terms and concepts of wildlife management are increasingly becoming associated with all of the native flora and fauna of a region (Hunter, 1990). However, concern for animals still tends to predominate in most wildlife management efforts.

Forest habitat management has several general kinds of objectives. In some cases,

483

the emphasis is on a single species; much of this kind of **species-oriented management** continues to focus on game species for sport hunting, but increasing interest is being shown in recreational viewing and photographing of animals. These two objectives often involve the same large mammal and bird species, but "watchable wildlife" includes a wider range of species than is considered for hunting. Other species become the objects of intensive habitat management because they are locally rare or are in danger of extinction; these may include organisms in any taxonomic group. A quite different objective of management is the **conservation of regional biodiversity**. This community-level approach aims to maintain not only the entire complement of native species in a locale, but also large enough populations of those species to preserve natural genetic diversity and allow natural selection and adaptation of the species to continue. The literature about forest habitat management for all these objectives is extensive; comprehensive sources of information include books by Hunter (1990), Patton (1992), Bookhout (1994), and Payne and Bryant (1994).

The species-level and community-level objectives require somewhat different approaches, but they are not necessarily incompatible. Species-oriented management clearly must be based on knowledge of the specific habitat requirements of a species. These are well known for some species, especially game animals for which detailed habitat management guidelines are available. In other cases (such as little-known species that have become rare), extensive research efforts are needed to identify and quantify habitat needs. The results of these studies can be found in the "recovery plans" for endangered species such as those for the northern spotted owl or the red-cockaded woodpecker. The species-oriented approach is sometimes expanded by defining **guilds** of animal species (groups of species that occupy similar niches, and therefore similar habitats). A variation on this idea is to focus on a single species not because of its exceptional importance, but because its population responses to habitat treatments may indicate the success of management for the guild as a whole. However, enough differences exist among the species within a guild that this has not always proven useful. It is impractical to expect to understand and quantify the habitat needs of all species inhabiting a forest area. Therefore, concerns for maintaining overall biodiversity must be met by providing a diversity of habitats.

In either case, habitat management is done principally by controlling **stand structure** (the ages, sizes, and density of trees within a stand) and **forest structure** (the sizes and spatial arrangement of stands within a forest). Stand and forest structure appears to be generally more important than tree species composition in providing for habitat, although particular species are sometimes important for certain food requirements. Silvicultural treatments can be applied most directly to creating particular *stand* structures for habitat purposes, just as is done to meet other objectives. The principles of designing *forest* structure can partly be drawn from traditional concepts of forest management for sustaining timber production, but additional ideas, some of which have formed the basis for the recently defined discipline of landscape ecology (Forman and Godron, 1986), also apply. In situations where individual animals range over very large areas, or when the maintenance of a sustainable population of a species requires a large area (even in cases where individuals have limited ranges), the spatial scale of wildlife management differs from that of timber management. In the latter situation, the landownership boundaries generally determine the extent of management interest. However, to achieve the goals of providing habitat for populations with large land requirements, stands should be treated in the context of the regional landscape, regardless of ownership lines. This presents one of the more challenging aspects of forest land management requiring economic, social, and political innovations to coordinate efforts. Under almost any circumstances, desirable patterns of landscape diversity represent long-term goals toward which foresters can work, but they are not patterns that can be created in a few years or even a few decades.

Habitat Elements

Species in a forest ecosystem are linked together in a food web through which the sun's energy is transferred from producers (plants) to herbivores, carnivores, insectivores, and decomposers. All animals require food, shelter, and water from their habitat. Manipulation of forest vegetation affects food supplies directly for herbivores and decomposers, and indirectly for other species. It affects shelter directly for all species, and may influence water availability in some cases. It is useful to subdivide habitat needs further into a set of habitat elements that can be controlled by silvicultural operations (Healy, 1987). Wildlife biologists associate the various habitat elements with a series of successional stages (DeGraaf et al., 1992; Thomas, 1979). One approach recognizes six stages: grass/forb, seedling/shrub, sapling/pole, intermediate-aged forest, mature forest, and old growth (Fig. 20.1). These stages are defined primarily by vegetation structure, but they are closely related to the developmental stages discussed in Chapter 2. The occurrence of each habitat

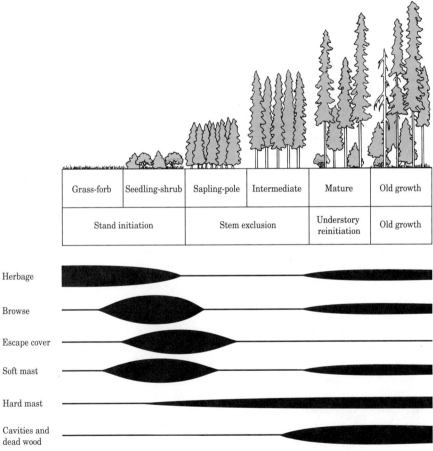

Figure 20.1   Successional stages used by wildlife biologists for describing changes in forest structure (Thomas, 1979), and their approximate relationship to developmental stages described by Oliver and Larson (1990). Relative abundance of different habitat elements is shown for each stage; these are meant only to indicate general trends, and do not necessarily apply to all forests types or wildlife species.

element in different developmental and structural stages, and the control of each element by silviculture, are considered next.

Food

Browse consists of the buds, twigs, and leaves of woody plants. It serves as a food source throughout the year for such species as deer, elk, moose, hares, beaver, and grouse, and is particularly important during the dormant season when other foods are unavailable. Twigs of hardwood species generally have higher nutrient content and are more palatable than those of conifers. The presence of hardwood browse is especially important where the dominant species are pines or spruces, which are rarely eaten; other conifers such as firs, hemlocks, and "cedars" (*Thuja, Chamaecyparis,* and *Juniperus*) produce more palatable browse. Browse is useful to most animals only within about 6 feet (2 m) of the ground, although grouse and moose feed at higher levels within the crowns. Thus availability of browse is greatest during the stand initiation stage, when shrubs and tree regeneration form dense stands.

As the canopy of a young stand rises, the edible twigs grow out of reach of most animals, and browse becomes nonexistent as the stand enters the sapling/pole stage. The persistence of habitat with plentiful browse is determined by the rate of height growth of regeneration. In areas with abundant precipitation, dense stands of regeneration tend to develop quickly, and height growth is rapid. In such areas in the Pacific Northwest and in eastern North America, browse production reaches a maximum 5 to 10 years after a disturbance and then declines rapidly. On drier areas such as in the Rocky Mountains, the stand initiation stage is prolonged as seedlings slowly colonize a disturbed area, and browse habitat may persist up to age 25. After closed canopies develop, browse production remains low until a woody understory becomes established again in the understory reinitiation stage, and then develops further in the gaps that occur in old-growth stands. However, the high-density browse that occurs during stand initiation generally does not redevelop in the understory.

The shrub and seedling growth response to any cutting of the overstory is proportional to the amount of canopy removed. Any harvest method that removes all or most of the overstory (clearcutting, heavy shelterwood cutting, selection cutting using large groups) will provide abundant browse. Thinnings can serve to increase the establishment of a woody understory, but they would have to be heavier than is usual for timber production to obtain an appreciable response. Thinning is more successful in producing browse habitat in stands of species with sparsely foliated crowns, as in pine stands in the southeastern United States. The usefulness of areas for browse production can be extended by periodically cutting back or burning young stands as they grow out of the stand initiation stage. This method will favor sprouting hardwood tree species and shrubs, but germination of new seedlings will occur as well. This technique is commonly used to maintain patches in the shrub/seedling stage in regions where the landscape is dominated by mature forests.

Herbaceous vegetation (or **herbage**) includes the leaves, stems, rhizomes, and roots of nonwoody plants. These are generally rich in nutrients and are highly palatable to many herbivores. The availability of herbaceous foods is similar to that of browse, being most abundant in the stand initiation stage. This kind of food is of such importance that wildlife biologists recognize a distinct grass/forb stage as the first following a major disturbance, even though it is often very brief. Herbage declines even sooner than browse, as tree seedlings and shrubs form a closed canopy and shade out the herbaceous plants. Thus it is difficult to maintain both good browse and good herbaceous foods in the same stand without repeated treatments. In many cases, skid trails and log landings may provide the only areas not dominated by woody plants. It is good to take advantage of them by sowing mixtures of seed of herbaceous plants especially chosen for their wildlife food value.

The grass/forb stage may last only one year in areas with prompt tree regeneration, but it can persist for many years in drier areas where woody plants are slow to become established. Herbage develops again in moderate amounts in the understory reinitiation and old-growth stages, but will be composed of shade-tolerant species that will differ from those of the grass/forb stage. Mowing can sometimes be used to prolong the herbaceous stage, but it will tend to favor sprouting woody plants if they had already become established prior to treatment. The most efficient method for maintaining herbaceous vegetation is the use of prescribed fires, either on a frequent basis or timed in the summer so that the roots of woody plants are killed. This can be accomplished either in open areas or beneath canopies. It is most easily done in areas with prolonged drought such as in ponderosa pine stands, where partially closed canopies and frequent light fires naturally maintain grass understories in mature stands. In the southeastern United States, fires are regularly used in the understory of pine stands to maintain either browse or herbage, depending on the frequency and season of burning (Stransky and Harlow, 1981). Winter burns timed at three- to five-year intervals are used to favor browse production for deer, but summer burns at one- to two-year intervals are used to produce herbage for quail and turkey (Fig. 20.2).

Mast is a term used for fruits and seeds of trees and shrubs; a distinction is made between **hard mast** (nuts and seeds) and **soft mast** (berries and other fleshy fruits). In boreal and temperate zones, soft mast is produced by early-successional species of many genera, including *Prunus*, *Sambucus*, and particularly *Rubus*. In later developmental stages, most of the berry production occurs on shade-tolerant understory shrubs, with relatively few overstory trees producing this kind of food. In contrast, many overstory tree species such as oaks, hickories, chestnut, and beech produce nuts that are among the most important foods in the diet of many animals; conifer seeds also contribute to these hard mast foods. This distinction cannot be made in the tropics, where both fleshy fruits and nuts are common in main canopy tree species.

The cycle of fruit production roughly coincides with patterns of browse production, being greatest in early-successional vegetation but declining rapidly as the tree canopy develops. During understory reinitiation, a different set of fruit-producing shrub species develops in the understory. In general, silvicultural treatments affect fruit production in ways similar to browse, but with a somewhat longer delay after cutting or burning. The importance of these foods for birds in particular have in the past led wildlife biologists to introduce exotic species of berry-producing shrubs such as multiflora rose and autumn-olive into North America, but it is generally preferable to work with native species.

Many long-lived main canopy tree species do not produce nuts until they are several decades old, especially if they grow in dense stands where crown expansion is restricted. As trees become sexually mature and crown sizes increase, large nut crops are produced and are widely used by both tree-dwelling and ground-dwelling animals. One advantage of nuts as a food is that they have a long shelf life, remaining usable for many months, either on the tree or the forest floor, or stored in caches by animals. Thus, they can be eaten during the dormant season, when other foods are in short supply. Animals that are highly dependent on nuts include small mammals (mice, chipmunks, voles, and squirrels), as well as turkeys and deer. The supply of hard mast is a limiting factor controlling winter survivorship in many of these species. Abundant nut crops are produced periodically (in "mast" years) by many of the mast-producing species, particularly the oaks, with very

Figure 20.2 Understory treatments in loblolly pine stands on the Santee Experimental Forest in South Carolina. *Top*: Untreated hardwood understory. *Bottom*: Summer fire used in foreground to produce herbaceous foods, and winter fire used in background to develop dense browse; both treatments also reduce the likelihood of dangerous crown fires by removing ladder fuels. These photographs were taken in areas adjacent to those of Figures 8.5 and 8.6, in which a different sequence of burning was used to establish understory pine regeneration.

low production during intervening years; cycles in animal populations are strongly affected by the intervals between mast years.

Stand density affects nut production in patterns similar to stem volume growth; individual-tree nut production is highest with heavy thinning, but stand-level production is little affected by thinning as long as large gaps are not made in the canopy. The above is true if all trees in a stand are mast species. Thinning mixed-species stands to favor mast-producing species will clearly increase nut production at the stand level; in doing so, it is preferable to maintain a diversity of mast species.

Shelter

Cover consists of any vegetation that shelters wildlife from predators (**escape** or **hiding cover**) or climatic extremes (**thermal cover**). All forest vegetation, living or dead, provides some benefits of shelter, but the high densities of woody stems that usually occur in young stands provide the best cover for many animal species. If an herbaceous stage occurs at the earliest part of stand initiation, escape cover will be poor or absent for most species. In fact, this condition provides important hunting habitat for raptors precisely because the cover for small mammals is intermittent. The dense shrub/seedling stage then becomes very important as cover for small mammals and birds, and once the height reaches about 6 feet (2 m), it is also useful to large mammals. The value of young stands for cover persists longer than for browse, forage, and soft mast, lasting into the sapling/pole stage. As stem density is reduced from natural mortality or thinning, the value for cover is reduced. When an understory of shrubs and shade-tolerant tree species develops in mature stands, escape and nesting cover will be present for some species of birds and small mammals.

Thermal cover coincides with escape cover for many animals. Any closed canopy will reduce solar heating during the day and reduce radiational cooling at night. But for large mammals in particular, the escape cover in young stands does not provide a sufficiently dense canopy for thermal cover purposes. Conifers have particular importance as thermal cover because they retain their sheltering effect during the winter, reducing wind-chill as well as radiational cooling and creating lower snow depths by intercepting much of the snowfall. This allows deer and elk to conserve energy, have better mobility, and more easily find browse and nuts on the forest floor. Maintaining nearly closed canopies in conifer stands is particularly important in forests that are dominated by deciduous species. In cold climates, deer concentrate in such areas, known as "deer yards."

Cavities in living and dead trees are of great importance for bird and mammal species as shelter for escape, sleeping, and rearing young. As stands develop, some trees begin to decline in vigor owing to competition, and some suffer injuries from snow, ice, wind, and fire. Poor vigor and broken branches allow invasion of decay insects and fungi, leading to the formation of decay cavities in living trees. Most mammal dens occur in cavities in living trees (**den trees**), but birds make more use of cavities in standing dead trees (**snags**). In addition to these decay cavities, woodpeckers excavate their own cavities, mainly in living trees infected with heart rot, which are subsequently used by other species.

Stands typically produce dead trees in the process of self-thinning, but trees dying from suppression in early stages are too small to provide cavities useful for any except the smallest animals. Smaller birds and mammals can use trees that are only 10 inches (25 cm) D.B.H., but many species require trees of at least 20 inches (50 cm). Larger snags tend to be used simultaneously by a number of birds and mammals. Thus even-aged stands do not form the most valuable den and snag trees until they reach the stage when large dominant trees decline in vigor and die. The density of cavities in old, even-aged stands may not differ greatly from that in old-growth stands in many forest types.

Scarcity of snags can limit the size of bird populations in intensively managed even-aged stands. The most direct approach to maintaining cavities throughout stand development is to retain a number of snags during thinning and during harvest of mature stands. Estimating the density of snags required to support maximum populations is complicated by the different minimum size requirements of bird species and their territorial behavior, which limits the usefulness of groups of snags for some species. Appropriate densities have been studied for a number of bird species (Tubbs et al., 1987; Thomas, 1979), and although some variations exist, fairly consistent guidelines of retaining two to five snags per acre (5 to 12 per ha) have been proposed for many forest types in North America (Hunter, 1990). Estimates of large den tree requirements for mammals are lower, ranging from 0.1 to 1 per acre (about 0.2 to 2 per ha). It may be difficult in practice to meet density goals during timber marking because of clumped distributions of cavity trees and random variation in the actual occurrence of cavities in trees of similar species and size. It may often be adequate to retain dead trees (unless it is hazardous to forest workers to do so) and den trees with actively used cavities. This will supply habitat needs but not result in a large proportion of living stand basal area being cull trees (Healy, 1987).

The practice of living retaining snag and den trees (often simply referred to as ''wildlife trees'') during harvest has become standard in many areas (Fig. 20.3), but it may only

Figure 20.3   Two methods of hastening the redevelopment of old-growth habitat characteristics following the harvest of Douglas-fir/western hemlock stands on the Willamette National Forest in Oregon. *Left*: Retention of individual old-growth elements, including fallen logs, snags, and scattered living trees. *Right*: Retention of 8 to 12 mature Douglas-fir per acre as reserve trees to grow through the next rotation. *(Photographs by U.S. Forest Service.)*

be a temporary solution to the problem in any one stand. Unless the snags are of large size, they may decay and fall before the surrounding stand is mature enough to produce a next set. If stands are managed on a rotation that does not produce large trees, a new set will never develop. Providing for future generations of snags can be accomplished by retaining not only current snags, but also an additional set of living trees, some with large branches or heart rot, which makes them likely to form cavities; these are just the sort of trees usually removed in partial cuttings. This would fit rather easily into selection system management, by allowing some trees of this kind to remain after they had reached the final diameter goal for timber production, thus providing a constant supply of den and snag trees.

With even-aged management, the retention of a set of living trees past final harvest is sometimes referred to as "green-tree retention" to distinguish it from snag retention (Franklin and Spies, 1991). This term may convey a rather specific practice in areas where clearcutting is the standard regeneration method used, such as in the Pacific Northwest, as depicted in Fig. 20.3. However, living trees are retained in a great variety of densities and spatial patterns in other areas, so it is not very useful as a general term. The key point is that a set of mature **reserve trees** are maintained well into, and possibly to the end of, the next rotation. In some cases, they may be reserved during clearcutting; in others, they may remain at the end of a series of cuttings that are part of the irregular shelterwood method. The idea of maintaining reserve trees is used in parts of Germany to produce a small number of very large trees of high timber quality in the next rotation; these are called "holdover trees" in this case. The general concept of retaining reserve trees from one rotation through the next can be adapted to the objectives of maintaining future den and snag tree, producing sawtimber trees, or a combination of both in the same stand.

These are solutions to the problem of providing cavities within a single managed stand, but they are not the only approach. These needs can also be met at the forest level without any particular silvicultural treatments, by maintaining a spatial arrangement of stands such that old stands with cavity trees are adjacent to young stands.

Under some circumstances, as for wood ducks, nesting boxes have been employed to substitute for natural cavities. Methods have also been developed to create snags by killing trees with herbicides in ways that make the decaying tree most useful to cavity nesters.

Dead wood in the form of logging slash and fallen trees and branches also serves an important habitat function. Cavities in large fallen logs provide the same kind of shelter as do those in standing trees, but accommodate species that do not fly or climb very high, such as red foxes, weasels, salamanders, and turtles. The branch-sized material that makes up the majority of logging slash also has value as escape cover for small animals, substituting for dense seedling cover during the earliest stages of stand development. Deep slash can inhibit the movement of large animals such as deer and elk, but this problem can be remedied by providing lanes cleared of slash throughout harvested areas.

Dead wood of all sizes, standing and down, also provides the main food source for fungi, insects, and other decomposer organisms. These in turn are the food of many animals higher on the food chain. Thus forgoing slash burning or other disposal methods will benefit many species, but this must be balanced against the need for protection against fire and insect pests that breed in the slash.

Stream Habitats

The vegetation structure in **riparian areas** surrounding streams, swamps, and pools affects the habitats of both aquatic and terrestrial animals. Aquatic food webs in small streams and vernal pools are based on the input of leaves and other detritus from the surrounding forest. Microbial decomposers and aquatic invertebrates (particularly insect larvae) feed on the detritus and become food for most fish and amphibian species. In larger streams and ponds with slower currents and more sunlight, the photosynthesis of algae and shoreline plants plays a more important role in the food web.

It is usually desirable to maintain shading over streams in order to keep the water in them cool, principally because the capacity of water to absorb oxygen and other gases *decreases* with warming. This is best done by retaining all or some trees that are tall enough and appropriately placed to shade the water. Such action is consistent with retaining filter strips of forest floor material and dense vegetation along the streams (see Chapter 18). Maintaining cool temperatures is most important in trout and salmon streams, because many of these species have rather limited tolerance for high water temperature and low dissolved oxygen content. In some situations, moderate opening of the canopy may improve fish habitat conditions by increasing photosynthetic levels in larger water bodies or by raising water temperatures in very cold, high-elevation streams, but in most cases maintaining a nearly complete canopy provides the best conditions.

A principal objective of maintaining dense riparian vegetation is to prevent particulate matter and dissolved nutrients from reaching streams. Paradoxically, some of the products of erosion are beneficial, although only when present in moderate levels close to that which occurs in unmanaged ecosystems. Some of the chemical nutrients that are leached into streams are necessary for the organisms that live there. The same is true of limited amounts of colloidal clay that provide base exchange capacity. However, excess sediment can cover gravel areas and fill in deeper pools, degrading habitats for both fish and insects, and excess nutrients may bring about the problems associated with eutrophication.

Streamside vegetation is also important as a source of large woody debris that falls into streams, creating deep pools with slower current. Pools provide escape and resting cover for fish by protecting them from predators and allowing them to maintain their position with low energy expenditure. However, fast-moving riffle areas provide the best habitat for insects, and therefore the best feeding habitat for fish. Optimum habitat for some species includes a fairly large proportion of stream area in pools, such as in mountain streams in Colorado where a pool:riffle ratio of 50:50 is considered most desirable for trout habitat (Moore et al., 1987). Pools tend to be in relatively short supply in smaller streams, and it may sometimes be desirable to install logs across streams to create additional ones. However, it would be carrying things to excess to use these benefits as an excuse for carelessly dumping tree-tops and other debris into streams. Excessive amounts of undecomposed organic matter in still ponded waters can reduce dissolved oxygen almost completely and render the waters essentially lifeless. It is better to avoid felling trees into streams or ponds in the first place, since pulling the slash or logs out can do damage to stream banks.

Terrestrial animals spend a disproportionate amount of time in riparian zones, largely because of a daily need for water, but also because the vegetation of these areas is often distinct from that of the surrounding forest. Such differences are minor along intermittent streams and vernal pools, but along larger streams, riparian vegetation may consist of entirely different tree, shrub, and herbaceous species. This contrast is greatest in coniferous forests in dry climates, where riparian zones often contain the only deciduous tree species in the region.

The key point about riparian vegetation management for terrestrial animals is that any habitat element will be more valuable if it exists near water rather than elsewhere. Maintaining a riparian area as mature forest provides cavities and dead wood and is compatible with the need for a filter strip and stream shading. The width of a special habitat management zone for terrestrial animals may be larger than is needed for the water-filtering function. This would depend on the prevalence of water bodies and the degree of contrast with nonriparian vegetation in the area. Harvesting patches in riparian areas can help to develop food and cover there, in addition to mature forest habitat. This is particularly important for deer wintering areas throughout the forest region dominated by northern hardwoods, spruce, and fir, where deer congregate in dense conifer stands near streams and wetlands. Providing patches of abundant hardwood browse near these deer yards improves the winter survivorship of deer herds.

Importance of Vertical Stand Structure

One stand attribute that is associated with wildlife species diversity is that of structural diversity in the vertical dimension. Much of the interest in vertical structure has come from studies of bird populations where it was found that crowns are partitioned among species for feeding and nesting at various levels, but any set of tree-dwelling animals will likely benefit. Epiphytic plants may also benefit, especially in moist climates where this life form is prevalent. High levels of vertical structural diversity are found in old-growth stands where canopy gaps have been created by the deaths of individual large trees. A continuous canopy occurs in some parts of these stands from the tops of the largest trees down to the released understory trees, regeneration, shrubs, and herbs growing in the gaps. This structure begins to develop as soon as old, even-aged stands enter the gap dynamics phase.

The lowest structural diversity is found in even-aged single-species stands, particularly those composed of a relatively shade-tolerant species in which most light is captured in one canopy layer, thus eliminating most understory vegetation. This kind of stand is typified by Norway spruce plantations maintained at high densities, but natural forests with low vertical diversity also occur, including nearly pure stands of lodgepole pine, jack pine, or black spruce which develop after hot fires. Vertical diversity in even-aged stands is greatest in species mixtures with pronounced vertical stratification among the crowns of the component tree species. This structural complexity is largely restricted to the tree and shrub layer, however, and lacks the dense seedling and herbaceous vegetation that grows in the gaps in uneven-aged stands.

The kind of vertical diversity found in old-growth stands can be mimicked in younger stands by the use of single-tree or small-group selection to create the equivalent of treefall gaps. However, results differ when the groups are larger in diameter than about twice the height of the main overstory trees (as discussed in Chapters 7 and 15). Cutting in larger groups essentially creates a number of small even-aged stands of different ages. This is more than just a fine point of terminology, because these larger groups create patches of early-successional vegetation which have greater similarity to that occurring in large stands than in small gaps.

A very different kind of vertical stratification occurs in two-aged stands in which younger vegetation becomes established beneath partially closed mature canopies. This occurs naturally after extensive but incomplete mortality from wind or fire and is a temporary condition. As these stands age, the rapid height growth of the regeneration relative to the mature stand obscures the distinct canopy layering. Some methods of creating this structure were described earlier in this chapter with regard to particular habitat elements; combining thinnings with prescribed burning to produce understory browse and herbage, or maintaining reserve trees after final harvest are ways of combining mature forest overstory with early successional vegetation. The creation of wooded pastures in medieval deer parks in Great Britain and Europe is an early example of this approach. It can be a useful way to mix timber production and wildlife habitat objectives, and usually is used to make up for the lack of one habitat type in an area. In some situations, this structure can improve habitat conditions. For example, browse tends to be more useful to deer in shelterwood areas than in clearcuts, because of the partial shelter provided. Although two-aged stands play an important part in habitat management, they do not contain the same kind of vertical structure found in stands of the kind created by selection management.

Importance of Modifying Forest Structure

There are limits to accomplishing habitat objectives by treating individual stands; much depends on management at the forest level. Habitat elements tend to occur either in the earliest stages of stand development or in old forests (Fig. 20.1). Herbage, browse, cover, and soft mast are in greatest abundance for most wildlife species during the stand initiation stage. A diverse insect community (providing food for many higher animals) also exists in young stands, attracted to the flowers, fruits, and distinct microclimate of this stage. Cavities, dead wood, hard mast, and a different set of insect species (mainly wood decomposers) occur in mature forests. Intermediate stages have much less value for most species but are of course a necessary step in the progress from young to old forest, in natural or managed forests. The most efficient way to provide a diversity of habitat structure is to maintain a distribution of stands of different ages across a forest landscape, such that all stages are represented at any time. This kind of age-class distribution is the goal of sustained-yield timber management (see Chapter 17).

It has been repeatedly suggested that ''good forest management (i.e., sustained-yield timber management) is automatically good wildlife management.'' The principal basis for this idea is the similarity in the need for a distribution of age classes. However, it is more accurate to say that the age-class balance of timber management can be taken as a starting point for the consideration of habitat management, with the question then being: what further factors or ideas should be considered for creating optimum habitat conditions? Some of these additional ideas include the effects of tree species composition, provision of old-growth habitat, sizes and spatial arrangement of stands, location of corridors, and the landscape context of the management area.

Tree Species Composition

Although vegetation structure is generally the single most important factor affecting habitat quality, in some situations tree species composition cannot be ignored. The most obvious cases are those in which an animal species is dependent on a particular tree for food, such as the dependence of ruffed grouse on aspen flower buds and catkins for winter food in some parts of its range (Gullion, 1984). But in other situations species composition has a broader effect on wildlife species. An important one is that in which native mixed-species stands have been replaced by plantations of a single species. The use of an exotic species would be an extreme case of reducing habitat value, because many native animal species would not be adapted to use it. However, a single native species grown in stands of simple structure across all site conditions could be nearly as limiting. The use of native tree species adapted to each site type would avoid many of these problems. In areas where plantation forestry is used on a large scale, maintaining a portion of each site type in native tree species (not necessarily reserved from timber management) represents an important compromise between meeting timber and wildlife habitat goals.

Old-Growth Habitat

Sustained-yield timber management will generally not provide old-growth habitat unless special accommodations are made. Old-growth habitat is distinctive because it includes particular habitat elements (den trees, snags, and fallen logs, all of large sizes) and particular stand structural characteristics (mature canopy with gaps and complex vertical canopy structure). This kind of structure was first described in detail for the Douglas-fir/western hemlock forests of the Pacific Northwest (Franklin and Spies, 1991), but similar structures with somewhat different dimensions have been described for forests elsewhere. This structure first develops naturally 150 to 250 years after a major disturbance in many forest types.

Some of these individual structural elements can be created in managed stands on normal timber rotations through retention of reserve trees, as previously described. An alternative is to manage a portion of stands on longer rotations, so that they pass the age at which old-growth characteristics develop, and then maintain that condition for a significant period. This extended-rotation approach can be used with either even-aged or uneven-aged stands (in the latter case, by setting a large final-diameter goal).

Maintaining a certain portion of old-growth stands in an essentially unmanaged condition is clearly another alternative. Such areas would be an important part of efforts to establish a system of nature reserves on all site types in each region. Old-growth stands are among the areas least disturbed by humans and therefore represent logical choices for the core areas for such reserves. Younger forests dominated by native species are also important for these conservation efforts, especially where old-growth stands are rare or entirely absent. Thinning or selection harvesting in younger stands may be useful to speed the development of old-growth structures, after which they would be left unmanaged. Reserves represent a community-level approach to preservation of natural processes, and provide habitats for the plant, animal, and microbial species for which habitat needs have not been determined, or for those species that have not even been identified. Other methods of creating old-growth structural features in managed stands are not a replacement for reserves, rather, they are a means of expanding elements of old-growth habitats beyond reserve areas.

Sizes and Spatial Arrangement of Stands

Sustained timber yield depends on a balanced stand age distribution, but not on the size and location of stands (although these can affect logging efficiency and therefore financial yield). Habitat conditions have a fairly complicated dependence on these factors. Many wildlife species are more common near **edges** where different plant communities meet. This phenomenon, known as the **edge effect**, was first described by Leopold (1933) for game animals. Most game species do not actually use the edge itself but need access to two or more habitats on a daily basis; thus they make much greater use of habitats near edges. For example, deer and elk need the youngest stages for browse and herbage, but require older forest for cover; the use of open areas by these animals decreases sharply with distance from escape or thermal cover. Ruffed grouse require the shrub/sapling stage for summer cover to raise their newly hatched broods, but sapling/pole stands are used for cover at other times, and older stands are needed for feeding on browse and mast from within the tree crowns. The importance for many species of having feeding areas adjacent to cover can be seen in the use of the word ''covert'' in wildlife management. This word originally referred to any area providing cover (Leopold, 1933) but has come to mean any area where at least three habitats meet in an edge (Hunter, 1990).

The initial interest in edges was for game purposes, but it was subsequently discovered that many nongame species are also edge-related. In some cases, there is an actual preference for the edge conditions rather than just a need for two or more habitats. Microclimate and vegetation change along a sharp gradient at an edge, creating conditions different from those found in either habitat. At edges between mature forest and low vegetation, differences in wind speed, humidity, temperature, and solar radiation extend well into the forest stand. Forest edges also have a marked vertical structure from the forest canopy top to the seedlings, shrubs, or grasses of the adjacent vegetation. The indigo bunting, a songbird that nests in the eastern United States, is an example of a species that makes use of the edge itself; the female generally builds a nest in dense shrubs, but the male defends the territory from a perch in tall trees. Some plants also prosper along edges, including vines that are relatively intolerant of shade but require large trees for support. In general, the diversity of species in many taxa has been found to be greater in edge habitats than in continuous forest vegetation.

Edges occur naturally along the perimeters of forest destruction caused by wind or fire and along abrupt transitions in site conditions. However, humans have added greatly to the prevalence of edges through agriculture and timber harvesting in many regions. Having recognized the importance of this habitat, wildlife biologists promoted the creation of edge as a basic goal of habitat management, by prescribed burning and cutting trees. This was the principal rule of habitat management for several decades after its introduction by Leopold. On forested areas where habitat management was the primary goal, rather artificial cutting patterns have been used to maximize edge, including harvesting in alternating strips or in checkerboard patterns. In areas where timber management was influenced by concern for wildlife, harvesting plans were modified by using a larger number of smaller cuts to achieve the desired harvest, by using stand shapes that were more oval than circular, by creating irregular edges, and by leaving islands of uncut trees within the cutting block. In addition, cutting areas were dispersed across the forest area to spread the influence of gaps across a larger area (Fig. 20.4). The fact that these edge-increasing modifications could be fairly easily incorporated into timber plans is another reason why timber management and habitat management have been regarded as compatible.

Although experience has shown that these practices are generally successful for benefiting edge species and increasing species diversity, wildlife biologists no longer believe that maximizing edge is a rule to be applied in all situations for all habitat purposes. It has been discovered that some species predictably decline in abundance when patchy edge habitat is created. These species require a minimum area of a single habitat and are referred to as **interior** species. The habitat requirements of these species were not recognized earlier, partly because of the concentration on game animals, but also because of the stress placed on quantitative measures of overall diversity, which generally show increases when edge is created. Increasing diversity at the local level may have an adverse impact on regional diversity, however. The problem is that when an edge-maximizing approach is used over a wide area, local diversity will in fact increase, but regional diversity may decline, as the same set of interior species is lost across the region. There are several potential reasons for the decline of forest interior species in patchy forests with plentiful edges: predators and parasites that live in nonforest habitats along edges may move into the stand to prey on forest species; edge-adapted species may be better competitors in the conditions near edges; or the remaining forest habitat simply may not be large enough to support the minimum social group of a species.

The best-known examples of interior species may be those such as the northern spotted owl that inhabit old-growth stands, but they are not the only ones. The large group of

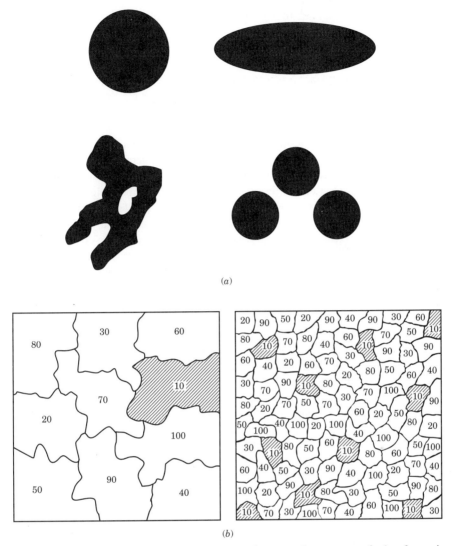

Figure 20.4 Modification of cutting patterns to increase the amount of edge for a given area harvested. *(a) Stand level:* Circular cuts have the shortest edge length; the edge can be increased moderately by using elongated shapes, but much greater increases can be achieved by creating a cluster of smaller cuts or making cuts with irregular edges that contain islands of residual trees. *(b) Forest level:* Both drawings represent 1,000-acre forests with balanced age-class distributions, as indicated by the age shown for each stand. The youngest stands are shaded, indicating the areas where distinct edge habitat exists. The use of 10-acre stands rather than 100-acre stands disperses edge habitat throughout the forest area.

songbird species known as neotropical migrants, which breed in North America but over-winter in Central and South America, have been the focus of considerable attention. Although some of these species are edge-inhabitants (such as the previously mentioned indigo bunting), others are forest interior species, and still others are shrubland or grassland interior species. Habitat management for these interior species requires the opposite of that for edge species; each vegetation age class should be created in the largest possible blocks with smooth edges.

Some of the most serious declines of interior species populations have been identified in the midwestern United States and in parts of the tropics (particularly in Brazil) where forest land has undergone fragmentation and reduction in overall extent as a result of the development of agricultural and suburban lands. Conservationists in these areas have advocated the approach of retaining forests in the largest blocks possible in order to reduce further fragmentation. The attention given to these cases has led to the possibility that minimizing forest fragmentation may become a rule to be universally applied in habitat management, just as creating edge had been in the past. It is useful to consider the difference between these situations and that of harvesting stands in a predominantly forested region. The ideas of edge, stand size, and spatial arrangement that have been incorporated into the theoretical framework of landscape ecology (Forman and Godron, 1986) are helpful in this regard. The key concept is that a landscape is made up of a **matrix** (the predominant vegetation type) in which **patches** of other vegetation exist, which may or may not be connected by **corridors** of vegetation similar to that in the patches.

The situations in which fragmentation has had the most serious effects on interior species are those in which agricultural or suburban development has produced a nonforest landscape matrix, with forest being reduced to patches. In contrast, harvesting trees generally produces grass/forb and shrub/seedling patches within a more mature forest matrix. The difference in the dominant vegetation of the landscapes is confounded with the amount of edge in these comparisons and makes it difficult to separate their effects on wildlife populations. However, it has been found that some forest interior species do not utilize mature forest near open areas if they are inhabiting forest fragments in a nonforest landscape matrix, but the same species will use the forest right up to the edge of an opening in the case of a forested landscape containing young stands regenerating after harvest.

Other differences also exist between these situations, an important one being the persistence of edge through time. Conversion of forests to agricultural fields or other uses, though not truly permanent, may last for centuries. In contrast, the goal of forest management is to produce regeneration that quickly develops into stands of large trees. Although it may take more than 100 years for these stands to reach the conditions needed for some interior species, many of the edge conditions disappear in a decade or two. Habitats may no longer exist for edge predators or parasites, and the vegetation may provide enough cover for interior species to travel across these areas to reach other mature forests. For example, the brown-headed cowbird has been implicated as one of the major problems reducing forest interior bird populations nesting in North America. Cowbirds feed mainly on seeds in grassy habitats, and lay eggs in the nests of other forest birds; the cowbird hatchlings then displace the young of the host species. This appears to be a problem mainly in forest fragments in agricultural and suburban areas, where cowbirds have abundant feeding habitat, rather than in continuously forested areas containing patches of young trees and shrubs. The important point is that the effects of edges created by forest fragmentation in some landscapes cannot be generalized to all situations (DeGraaf and Healy, 1990).

The situation in which cutting in forested areas is most similar to land conversion in its effects on forest interior species is that in which old-growth stands are harvested at a fairly rapid pace. In some forest types, the period before old-growth structure redevelops is so long that it can be considered a permanent change (at least on the same order of much agricultural land conversion). It is logical in such areas that the principle of minimizing fragmentation of the remaining old-growth has been emphasized (Franklin and Forman, 1987), but this again is not necessarily equivalent to the effects of forest management on a sustained-yield basis.

In all forest management, however, the different needs of interior and edge species do create a clear conflict between the two rules dealing with stand size and spatial arrangement: "create more edge" and "avoid fragmentation" (Hunter, 1990). In some cases, one or the other of these approaches may prevail, depending on the species of greatest concern. It may be possible to develop habitat for both kinds of species on the same area, if the management area is large enough. Patchy habitats providing edge could be restricted to one section, incorporating the ideas for maximizing edge previously discussed. Other sections could be managed in larger homogeneous blocks, arranged in such a way that mature forest had the least edge possible with younger forest. General approaches to allocating land to different management strategies have been outlined by Harris (1984) and Noss (1983). These include a central core of old-growth reserve surrounded by areas of successively more patchy, heterogeneous forest with increasing distance from the center.

An additional consideration is the appropriate stand size for habitat purposes. The realization that disturbances are a natural part of stand dynamics has led to the idea that creating stands that mimic the size and severity of natural disturbances occurring in a region would be beneficial to the native species adapted to those forests. The scale of disturbances range from large-scale stand-replacing hot fires to minor wind disturbances that create single-tree gaps. Although one kind may predominate in a region, most areas likely experience a range of disturbance effects. For example, large fires are the most apparent disturbance in the Northern Rockies, but these are infrequent, with low-intensity fires occurring in intervening years. This combination produces even-aged stands of various sizes, as well as some with partial residual overstories (Arno, 1980). This forest structure would be mimicked by a mixture of small and large clearcuts, incorporating heavy thinning or shelterwood cutting in some stands. In contrast, most New England forests are relatively free of large, stand-replacing fires. The death or windthrow of individual trees produces an old-growth structure at about age 150, but infrequent catastrophic windstorms destroy forests over large areas. Resulting stand structures range from a predominantly mature canopy with small gaps, to nearly complete destruction with only scattered residuals after major storm damage, to complete destruction when windstorm damage is followed by fire in the downed fuels (Foster, 1988). These structures could be mimicked by selection, irregular shelterwood, and clearcutting methods, respectively.

Not enough information is available for most regions to enable forest managers to precisely mimic natural disturbance processes, but it has been proposed as a general principle that silvicultural plans should incorporate a range of disturbance sizes across a landscape (Hunter, 1990; Harris, 1984). The important idea is that natural processes tend to leave a greater range of stand sizes and structures than occurs when a limited set of silvicultural methods is applied across a landscape.

In these considerations, it is clear that some species are dependent on mature forests in which single-tree gaps are the prevalent disturbance size. The question then arises about the other end of the scale: to what extent are some species dependent on large areas of early-successional stands? Small raptors such as the American kestrel appear to have the need for the largest areas of herbaceous and seedling vegetation in order to hunt for small mammals; a minimum of about 100 acres (40 ha) may be needed to support this species (Hunter, 1990). Young stands of this size are also important for providing habitat for populations of some warbler species, even though a pair may survive on smaller areas. It is unlikely that species would require stands resulting from harvest in larger contiguous blocks. Many species with large ranges can use heterogeneous habitats.

Corridors

In situations where patchy forest structure has been created, corridors of vegetation that connect patches of similar vegetation structure can help to minimize the negative effects on some wildlife species. They may be important at very different scales. In a regional context, maintaining forest vegetation in corridors that connect forest areas in an agricultural landscape allows the migration of plant and animal species over various time scales—daily, annual, or over many generations. These large corridors are beyond the scale of silvicultural planning; their most important attribute is that they remain forested in some form, rather than any particular management scheme that would be carried out within them. Such corridors are not considered beneficial in all cases; they can sometimes cause problems if they serve as a conduit for exotic species or diseases into previously unaffected areas.

Corridors connecting individual stands within a managed forest are of more direct importance in silviculture. The most important kind of corridor is often referred to as a **leave strip**—a strip of uncut or partially cut forest crossing through a harvested area. When cutting blocks are larger than about 100 acres (40 ha), the ability of wildlife to move between stands begins to be inhibited because of lack of cover, especially immediately after cutting. Leave strips are useful as long as they are wide enough to provide escape cover. A width of only 20 feet (6 m) is useful to some small animals, but corridors 10 times that width or more may be required for deer and elk. These larger widths are also needed if the objective is to provide food or other habitat elements within the corridor, rather than just escape cover for travel.

Riparian zones serve as the most logical locations for leave strips. These are preferred habitats anyway and form a natural network connecting various parts of the landscape. This use also is compatible with the need for filter strips around water bodies. Riparian leave strips have the greatest importance when large harvested areas are converted to plantations of simple structure with little habitat value; they serve to mitigate habitat effects that last through the rotation.

Corridors serving the reverse purpose of leave strips (harvested corridors connecting areas of early succession) are generally not necessary and may be detrimental by creating edge through mature stands. Most animal species using patches of open habitat will travel through older forest. Young vegetation corridors of limited length are sometimes valuable to connect small clearcuts within a cluster; sowing seed of herbaceous species on logging trails can generally provide for this when needed.

The Landscape Context of a Management Area

The success of many habitat management efforts depends on the forest conditions that lie beyond the boundaries of a particular forest owner. It may not be possible to create sufficient habitat for some species on an ownership of limited size if treatments produce only a small island in a regional landscape of unsuitable vegetation. Thus decisions about management objectives must consider the landscape and the ability of managers to affect forest conditions on large areas. This need is clearly of greatest importance when the objective is to maintain regional biodiversity.

The goal of such efforts is to create a diversity of stand sizes, ages, and structures across a landscape, as previously discussed. However, it is common for roughly synchronous waves of human activities (heavy timber cutting or abandonment of farmlands) to create a predominance of one forest age class across large areas. In considering the management of a relatively small ownership, often the best that can be accomplished is to provide habitats that are rare on the landscape as a whole. A more comprehensive design of landscape vegetation becomes possible even where a large number of owners are involved, if incentives, regulations, and cooperative enterprises among owners are developed (Irland, 1994; Oliver, 1992).

Examples of Application

Elk and Woodpecker Habitat in Eastern Oregon

Management of game species that require edge habitat has long been integrated with timber management, but concern for nongame species that inhabit forest interiors is being increasingly incorporated on the same lands. One example of this balancing effort, described by Thomas (1979), deals with timber production and habitat management for elk and pileated woodpeckers in the Blue Mountains in northeastern Oregon, in forests dominated by ponderosa pine, Douglas-fir, and grand fir. Elk require three kinds of habitat—forage areas, escape cover, and thermal cover. The elk habitat conditions provided at each stage of stand development are shown in Fig. 20.5, for a stand managed for sawtimber on a rotation of 120 years. Herbage and browse are produced during the first 10 years following cutting. Dense seedling and sapling stands form escape cover that lasts from age 10 to 20, ending with the reduction in stand density from precommercial thinning. Thermal cover develops at age 40 as the canopy closes and increases in height; it persists as long as the canopy exceeds 70 percent closure. Beyond that age, the stand shifts between providing foraging habitat and thermal cover as the stand is thinned and then the canopy recloses.

The key to habitat management for elk is to provide each of these three habitats in appropriate proportions and spatial arrangement. Both kinds of cover must occur within a short distance from forage areas; the appropriate scale to achieve this is at a stand size of about 25 acres (10 ha). These needs can be accommodated in a wide range of rotation lengths. If rotations are less than 50 years, thermal cover will not develop; if rotations extend beyond 140 years, the level of cutting becomes so low that early-successional forage and escape cover become scarce. However, the range of 50 to 140 years includes all the rotations likely to be used for timber production and demonstrates the general compatibility of habitat management for edge species and timber management.

Pileated woodpeckers use considerably different habitats. They are large birds that excavate nest cavities in trees that are at least 20 inches (50 cm) D.B.H. Their principal food consists of carpenter ants that inhabit trees or logs infected with heart rot. Stands develop trees suitable for nesting and feeding at an age of about 160 years. The territorial requirement for a pair of woodpeckers in this region is approximately 300 acres (120 ha) of mature forest. One approach for providing woodpecker habitat is to extend the rotation on approximately 25 percent of the forest to 240 years. These stands would have an old-growth structure for one-third of that rotation, which would give nearly 10 percent of the total forest as old-growth at any time, once a balanced age-class structure was achieved. These stands are best if they occur in large contiguous blocks; no special arrangement of stands is needed because woodpeckers do not need younger stands.

This approach would require modifications in management of the remaining 75 percent of the forest to maintain optimum conditions for elk, because of the overall reduction

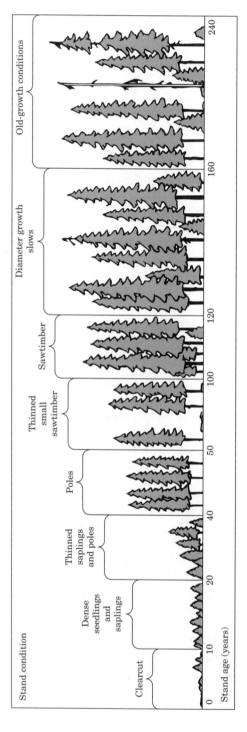

Figure 20.5 Diagram of habitat conditions in managed mixed-conifer stands in eastern Oregon, where extended rotations are used to provide old-growth stand structures. Elk habitat is provided in a 120-year rotation on 75 percent of the forest area, and pileated woodpecker habitat is provided on 25 percent of the area.

in harvest resulting from lengthening the rotation on 25 percent of the land. These modifications would include using prescribed fire or other means to delay regeneration on a portion of the harvested areas in order to maintain browse and herbage for an extended period, and altering the timing and intensity of precommercial thinning to lengthen the time that sapling/pole stands serve as escape cover (Smith and Long, 1987). This is only one possibility for balancing these habitats; other approaches include using old-growth reserves for woodpecker habitat or maintaining mature reserve trees after harvest to shorten the time before managed stands provide suitable old-growth conditions.

Habitat Diversity in Southern New England

A second example illustrates a situation in which the objective is the conservation of all native species in a region. The setting of this example is the rolling terrain of the Berkshire Mountains in western Massachusetts, where northern hardwoods (beech, sugar maple, and yellow birch) predominate, and northern red oak occurs only on warmer south-facing slopes, being at the northern extent of its range. Most of the land in the region was cleared for pasture and cropland in the eighteenth and nineteenth centuries, but was later abandoned. Nearly all the white pine stands that had invaded the abandoned fields were harvested, and forests are presently made up of mature hardwood stands that developed following the cutting. Only small stands of old-field pine remain. Most of the landscape is forested and is owned in small private woodlots, where only partial cutting is used. Harvests range from selection cutting to improvement thinning to high-grading, but they rarely remove a major part of the overstory canopy.

Management objectives on some of the lands in state ownership in this region are to create conditions favorable to overall species diversity. An analysis was conducted of the habitat requirements of all species of mammals, birds, amphibians, and reptiles native to the region (DeGraaf et al., 1992), and a goal was set for the appropriate balance of forest age classes to meet those requirements. The habitat that is most critically in short supply on the landscape are the grass/forb and seedling/shrub stages, both in small patches for edge species and in larger blocks for interior species that require extensive areas of the early-successional stages.

The following description represents the general silvicultural plans for one state management area of approximately 1000 acres (400 ha) in this landscape. Treatments are limited to those for which costs can be covered by harvesting revenues. Specific treatments include:

1. Creating clusters of three to five clearcuts, separated by strips of mature trees 50–100 ft (15–30 m) wide, where each cut is about 5 acres (2 ha), for a total area of 25 acres (10 ha). Some of these are converted to permanent openings by burning at three-year intervals to maintain vegetation in a mix of herbaceous and woody species; others are left to develop into sapling stands.

2. Producing a larger block of seedling/shrub habitat by regenerating stands of about 50 acres (20 ha). Mature stands containing red oak in the overstory are chosen in order to regenerate this most important mast species. Shelterwood cutting is used in these stands to establish oak advance regeneration; mature oaks are retained in the shelterwood overstory for mast production and as seed sources. The dense understory of beech, which dominates after harvesting in northern hardwood stands, is cut back and stump surfaces are treated with herbicide to prevent resprouting; this promotes the development of a high diversity of early-successional plant species.

3. Regenerating the few remaining white pine stands in order to maintain as much conifer winter cover as possible. This is accomplished by shelterwood cutting coupled with scarification and control of the understory hardwoods. Several removal cuttings are used to slowly release the pine advance regeneration in order to maintain the thermal cover value while assuring long-term dominance of pine on the site.

4. Developing vertical structure in some areas of mature forest by using selection cutting with small groups. However, this is of secondary importance, compared to creation of stands of young vegetation. Other areas of mature forest are to be left untreated.

This management plan is an example of a relatively new approach to active modification of landscapes in order to create the appropriate habitat diversity for all native wildlife species. It is based on a knowledge of species habitats and of the effects of silvicultural treatments on vegetation, but it cannot yet be considered a proven system. A proven system will be possible only after wildlife populations have been monitored for many decades on lands managed in this way.

Biodiversity Conservation in Boreal Forests

A final example of the application of habitat management principles is one in which concerns for biodiversity are combined with large-scale industrial timber management in boreal forests. Much of the boreal region of the world is well-suited to timber production because it contains vast conifer-dominated forests with few human inhabitants. Harsh climatic conditions cause natural levels of plant and animal diversity to be lower in boreal forests than in most other regions. Stands are dominated by one or a few tree species, consisting mainly of pine, spruce, fir, or larch. Hardwoods such as birch, willow, and poplar occur mainly in early-successional stages or in riparian areas. Forest industry has developed on a large scale in the countries of the boreal region, supplying the rest of the world with a substantial part of its industrial softwood.

As with other regions, timber harvesting in these forests was initially of an exploitative nature, but it was replaced by efforts to develop a sustainable timber economy. Although differences exist across this huge region, a common pattern in management is discernible. Silvicultural innovations have concentrated on the tasks of assuring prompt regeneration of conifers and mechanizing operations for economic efficiency. Silvicultural treatments often consist of a sequence of clearcutting, slash burning, mechanical site preparation, planting of improved seedlings of a single species, and removal of hardwood competition in young plantations. These techniques have been successful where they have been applied, but the result is the creation of ecosystems that are even simpler than those that existed prior to treatment.

This simplification is most evident with the higher plants, and the bird and mammal communities that depend on them, but much of the natural diversity of these ecosystems occurs in the insect and fungi communities, which are also affected. Many of those species are dependent on dead wood, particularly that of hardwoods because of the higher nutrient content; wood of deciduous species, living or dead, is in very short supply in intensively managed conifer stands.

Methods used to counteract these simplifying effects of silviculture have concentrated on saving mature living trees, snags, logs, and other important habitat elements during final harvesting (Fig. 20.6). These can be maintained during the process of conventional plantation establishment, but in some situations, a greater diversity in regeneration is also

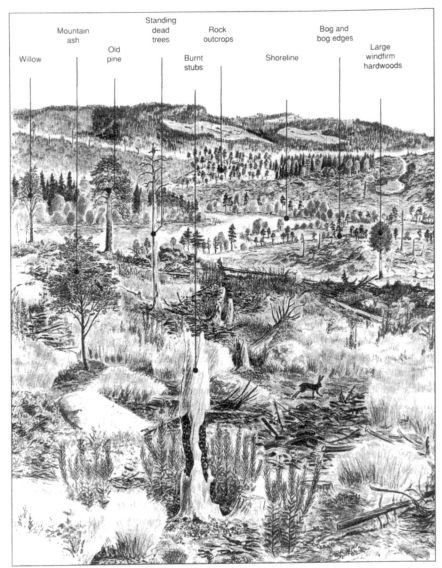

Figure 20.6 Diagram from a Swedish silviculture manual indicating important habitat elements to be retained during final harvest of conifer-dominated stands in boreal forests. *(From Hagner 1995).*

sought by altering planting and site preparation techniques. One form of this approach is to plant a conifer species at only about half the usual planting density, with site preparation being limited to those planting spots. This assures a minimum conifer crop on the site, which is supplemented by natural regeneration of both hardwood and conifer species that are allowed to develop between the planted trees.

Although it may be possible to use the selection system for conservation of some habitat elements, past attempts at selection cutting in boreal forests have frequently resulted in poor regeneration (Hagner, 1995). The use of mixed-species natural regeneration and the retention of snags, logs, and reserve trees in even-aged stands may be more suitable for habitat conservation, because the natural stand development processes in these forests are initiated by large hot fires.

Control of Wildlife Damage to Trees

This chapter has been devoted to ideas about manipulating forest vegetation to maintain or increase wildlife populations, but sometimes such populations become so large that they cause serious damage to stands. The most widespread problem results from deer and elk browsing on seedlings, but there are also situations in which small mammals damage seedlings by girdling them as they feed on the bark. Part of the solution involves favoring species such as spruce that are not palatable to browsing animals, as well as regulating the habitat of the animals to make it less favorable to them (Black, 1992).

Damage by deer and elk has long been such a serious impediment to regeneration in Europe, Australia, and New Zealand that foresters have resorted to fencing regeneration areas or using plastic tree shelters placed around individual seedlings. Chemical repellents are also used to spare seedlings from browsing. Such damage is of increasing concern in North America, and if other ways are not devised to control these animals, these expensive remedies may become more common here.

Another part of the solution is to recognize that large herbivores can become as much pests of the forests as are fires, root decay fungi, or defoliating insects. To meet many objectives, the animal population of the forest must be managed just like that of the plants. In fact, one cannot be managed without also managing the other. Hunting is an important tool of wildlife management for the same reason that cutting is in silviculture—each can be used as a means of mimicking natural processes that occur in forest ecosystems, in order to meet human objectives.

BIBLIOGRAPHY

Arno, S. F. 1980. Forest fire history in the northern Rockies. *J. For.*, 78(8): 450–465.

Black, H. C. (ed.). 1992. Silvicultural approaches to animal damage management in Pacific Northwest forests. USFS Gen. Tech. Rep. PNW-GTR-287. 422 pp.

Bookhout, T. A. (ed.). 1994. *Research and management techniques for wildlife and habitats*. The Wildlife Society, Bethesda, Md. 704 pp.

DeGraaf, R. M., and W. M. Healy. (eds.). 1990. *Is forest fragmentation an issue in the northeast?* USFS Gen. Tech. Rep. NE-140. 32 pp.

DeGraaf, R. M., M. Yamasaki, W. B. Leak, and J. W. Lanier. 1992. *New England wildlife: management of forested habitats*. USFS Gen. Tech. Rep. NE-144. 271 pp.

Forman, R.T.T., and M. Godron. 1986. *Landscape ecology*. Wiley, New York. 619 pp.

Foster, D. R. 1988. Species and stand response to catastrophic wind in central New England, U.S.A. *J. Ecol.*, 76: 135–151.

Franklin, J. F., and R.T.T. Forman. 1987. Creating landscape patterns by forest cutting: ecological consequences and principles. *Landscape Ecology*, 1: 5–18.

Franklin, J. F., and T. A. Spies. 1991. Composition, function, and structure of old-growth Douglas-fir forests. In: L. F. Ruggiero (ed.), *Wildlife and vegetation of unmanaged Douglas-fir forests*. USFS Gen. Tech. Rep. PNW-GTR-285. Pp. 71–80.

Gullion, G. W. 1984. *Managing northern forests for wildlife*. Minn. Agric. Expt. Sta., Misc. Journal Series Publ. No. 13,442. St. Paul, Minn. 72 pp.

Hagner, S. 1995. Silviculture in boreal forests. *Unasylva*, 46: 18–25.

Harris, L. D. 1984. *The fragmented forest: island biogeography theory and the preservation of biological diversity*. Univ. of Chicago Press, Chicago, Ill. 211 pp.

Healy, W. M. 1987. Habitat characteristics of uneven-aged stands. In: R. D. Nyland, (ed.), *Managing northern hardwoods, proceedings of a silvicultural symposium*. State Univ. of New York, Fac. of For. Misc. Pub. No. 13, Syracuse, N.Y. 430 p.

Hunter, M. L, Jr. 1990. *Wildlife, forests, and forestry: principles of managing forests for biological diversity*. Prentice Hall, Englewood Cliffs, N.J. 370 pp.

Irland, L. C. 1994. Getting to here from there: implementing ecosystem management on the ground. *J. For.*, 92(8):12–17.

Leopold, A. 1933. *Game management*. Scribner, New York. 481 pp.

Moore, R. L., R. L. Hoover, B. L. Melton, and D. J. Pfankuch. 1987. Aquatic and riparian wildlife. In: R. L. Hoover and P. L. Wills (eds.), *Managing forested lands for wildlife*. Colorado Div. of Wildlife, Denver, Colo. Pp. 261–301.

Noss, R. F. 1983. A regional landscape approach to maintain diversity. *Bioscience*, 33: 700–706.

Oliver, C. D. 1992. A landscape approach: achieving and maintaining biodiversity and economic productivity. *J. For.*, 90(9): 20–25.

Oliver, C. D., and B. C. Larson. 1990. *Forest stand dynamics*. McGraw-Hill, New York. 518 pp.

Patton, D. R. 1992. *Wildlife habitat relationships in forested ecosystems*. Timber Press, Portland, Oreg. 392 pp.

Payne, N. F., and F. C. Bryant. 1994. *Techniques for wildlife habitat management of uplands*. McGraw-Hill, New York. 500 pp.

Smith, F. W., and J. N. Long. 1987. Elk hiding and thermal cover guidelines in the context of lodgepole pine stand density. *WJAF* 2:6–10.

Stransky, J. J., and R. F. Harlow. 1981. Effects of fire on deer habitat in the southeast. In: G. W. Wood (ed.), *Prescribed fire and wildlife in southern forests*. Belle W. Baruch For. Sci. Inst., Univ. of Clemson, Georgetown, S.C. Pp. 135–142.

Thomas, J. W. 1979. *Wildlife habitats in managed forests: the Blue Mountains of Oregon and Washington*. USDA Agr. Hdbk. 553. 512 pp.

Tubbs, C. H., R. M. DeGraaf, M. Yamasaki, and W. M. Healy. 1987. *Guide to wildlife tree management in New England northern hardwoods*. USFS Gen. Tech. Rep. NE-118. 30 pp.

CHAPTER *21*

AGROFORESTRY

The basic human needs of food, shelter, and fuel are often supplied by separate systems of land management; building materials and fuelwood are harvested from forests, and food is produced from fields of herbaceous crops, fruit orchards, and livestock grazing in pastures. However, this has not always been the case and is not always so today. When trees are mixed on the same land with food crops or pasture for domestic animals, the term **agroforestry** is used to describe the management system. This term has been in use for only a few decades, but the practices that it describes are much older. Agroforestry was a traditional practice used throughout the world; it is a natural stage in the development of subsistence agriculture, often resulting from the partial clearing of natural forests and the desire to obtain a variety of products from a small plot of land. At least one agroforestry practice—the temporary growing of food crops between tree seedlings in young forest plantations—was part of organized silviculture since the early nineteenth century, but scientific agriculture and forestry have for the most part developed along separate lines. Only since the 1970s has there been widespread recognition that agroforestry has a role to play in modern silvicultural practice. Research and application of agroforestry practices have developed to the extent that proceedings of international symposia are available (Jarvis, 1991; MacDicken and Vergara, 1990), scientific journals are devoted solely to the topic, and a comprehensive overview of the field has been written (Nair, 1993).

Much of the impetus for the modern development of agroforestry practice has arisen from problems with the declining productivity of croplands in the humid tropics. The lack of seasonal fluctuation in climate, with high temperature and plentiful rainfall, allows biological and chemical processes to occur rapidly, with potentially high levels of plant productivity. However, these same factors of constantly favorable growing conditions also impose limitations on the ability to sustain high agricultural productivity on a given plot of land (Ewel, 1986). The deep weathering of soils and rapid leaching caused by heavy precipitation means that nutrients are soon depleted when forests are cleared and soils are tilled for agriculture, and they cannot readily be replaced by weathering of rocks. In addition, pest populations and weeds often build up to a greater extent than in temperate climates because there is no cessation of plant growth in harsh winter conditions.

The traditional system of agriculture used in such climates is called **shifting cultivation**. The natural forest vegetation is cut and burned, a crop such as maize or beans is grown for two or three cycles, and then the area is left fallow. Forest vegetation is allowed to develop, and soil nutrient stocks build up from atmospheric inputs and, in the case of nitrogen, from biological fixation. The native populations of soil organisms recover, and insects, fungi, and other pest species decline with the absence of their crop hosts. The developing tree canopy also shades out agricultural weeds. Food is grown on alternate plots until the fallow site has been restored to a condition suitable for clearing and cropping again. The ratio of fallow period to cropping period needed on a site varies with climate and soil conditions but is frequently on the order of 10:1. Thus this system requires a good deal of land.

Shifting cultivation is sustainable where human population densities are low. However, when the land base for food production is limited due to human population growth or the effects of land distribution policies, fallow periods must be shortened to the point where they no longer serve their purpose and agricultural yields decline.

The basic idea of agroforestry in such cases is to use the stabilizing aspects of the natural forest structure and function by mixing trees with herbaceous crops. The goal is to use the deep rooting and perennial nature of trees to maintain nutrient cycling and thus lengthen the time that crops can be grown on a given piece of land, ideally making agriculture permanently sustainable on a site.

Such problems are not restricted to rain forest climates, but extend to tropical savannas where seasonally heavy rainfalls and constantly warm temperatures create many of the same problems. Temperate forest ecosystems are not entirely different, but the effects are reduced in degree by differences in climate and soils. If chemical fertilizers and pest control measures were not available for large-scale commercial food production in most countries of the temperate region, the use of fallows or other agroforestry practices would likely become more important there. Indeed, the use of tree-dominated fallows was part of traditional subsistence agriculture in temperate as well as tropical regions.

Agroforestry practices of a quite different kind have been developed in semiarid regions of the world. Low precipitation restricts trees from developing a full canopy in these regions, and the natural vegetation is generally a mix of trees and herbaceous vegetation. Frequent fires and the grazing of large herbivores naturally adapted to these ecosystems cause grasses to dominate among the herbaceous species. Combining management of stands of trees for sawtimber and fuelwood with pasture for livestock is a logical approach to making maximum use of the structure of the natural vegetation.

Stages of Stand Development and Agroforestry

Any system that aims to combine food crops or pasture with trees must deal with the inherent problem of the competitive interactions that exist between the different plant lifeforms. As the tree layer increases in density, its canopy and roots dominate light, water, and nutrient resources and understory plant production declines. Competition for light between trees and understory is often the most critical factor because it is clearly one-sided. Many staple food crops are herbaceous plants that are low in stature, with relatively low shade tolerance, and cannot survive in forest understories. Even the more shade-tolerant shrub and vine crops decline in productivity with increasing overhead shade.

Because of this competition, agroforestry systems must take advantage of stages in forest stand development (described in Chapter 2) in which the tree canopy does not completely dominate site resources. A brief but highly productive period for crops occurs in the stand initiation stage when the tree seedlings and sprouts do not form a complete canopy. As a dense canopy forms in the stem exclusion stage, there is little opportunity for growing plants of low stature. Later, as the canopy rises in height in the understory reinitiation stage, growing space exists for crops in the diffuse light that reaches the ground level.

This pattern is illustrated by the intercropping practices used with plantations of coconut palms, described by Nair (1993). In the first eight years after planting, many shade-intolerant crops such as maize, rice, or peanuts can be grown between the young palms. The shade of the dense palm canopy from about age 8 to 25 precludes intercropping, but after that period, moderately shade-tolerant species can then be grown beneath the taller palm canopy, including cassava and cacao in the understory, and black pepper vines on the palm stems. The small rooting area of mature palms helps limit the below-ground competition with these crop plants.

Greater opportunity exists to grow crops beneath trees in dry climates, where canopies are not as dense, but it is necessary in nearly all climates to reduce tree density below that of natural stands in order to shift a substantial portion of the growing space to crops, even during the stand initiation or understory reinitiation stages. It is possible to maintain crops during the stem exclusion stage by thinning the overstory, essentially creating conditions for understory reinitiation at an artificially early age. In a sense, much of agroforestry can be understood as the manipulation of the stages of stand development so that plants of other lifeforms can be grown, although the complexities of carrying this out are enormous.

As with silvicultural systems, agroforestry systems must be developed for each set of ecological, economic, and social conditions. However, they consist of modifications and combinations of a relatively small set of practices. Classifications of agroforestry practices have recognized distinctions between those practices combining trees with food crops (herbs, vines, or shrubs), which are used in **agrisilvicultural systems**, and those combining trees with pasture and livestock, which are used in **silvopastoral systems**. There are also systems of **shelterbelts** and **hedgerows**, which are devised specifically to prevent erosion that occurs from tilling soil for crops. The basic principles of these practices are presented in this chapter, based largely upon the descriptions of Nair (1993).

Agrisilvicultural Practices

Multistory tree gardens are small plots that are characterized by a large variety of tree species, shrubs, and vines but with few or no herbaceous crops. This structure is often developed as a "homegarden" located around a farmhouse. The main purpose is the production of food for household consumption, so plots range from only a fraction of an acre to about 2 acres. The best-known examples of homegardens are those of Indonesian farms, but they are common in Central America, Africa, and other regions as well. The three or four vegetation layers may consist of large shade trees as an upper canopy and shade-tolerant tuber plants such as cassava, yam, or sweet potato in the lowest layer; intermediate levels are small trees or shrubs, producing fruit, nuts, spices, and condiments (Fig. 21.1).

A closely related practice is the creation of "mixed tree gardens," which are generally

Figure 21.1   Structure of a multi-story tree garden as used throughout the tropics. Fruits, nuts, medicines, condiments, and other products are harvested from different strata.

located on the commonly held land of a village rather than around individual farmhouses. The structure is very similar to that of homegardens, but the herbaceous crops are usually not included. There is more emphasis on fuelwood and timber harvesting, mixed with production of fruits, nuts, medicines, and spices.

Of all agrisilvicultural practices, multistory tree gardens bear the closest resemblance in structure and function to that of stratified forest stands. Much of the photosynthate is used for maintenance respiration and synthesis of woody tissues, so food production is rather low. However, there is low risk of catastrophic pest problems, and tight control exists over nutrient cycling and erosion. Tree gardens can stay in continuous production for centuries. The key benefit is the diversity of food, fuel, and medicine products derived from a small plot. Tree gardens are usually only one component of a farm system in which staple foods are grown in separate fields, often as monocultures.

A somewhat simpler but related practice is the management of **shade tree/crop combinations**. These are generally two-storied stands with one or more species of tall trees growing above an herbaceous, shrub, or small tree crop. In contrast to tree gardens, these are created for commercial crop production but are still used most commonly by small landowners. Examples of crops grown beneath shade trees include cacao, coffee, cotton, banana, peanuts, cassava, yams, and black pepper, all of which are at least moderately shade-tolerant. Shade trees are often chosen from among the native species of a region. It is necessary to thin the overstory trees as the stand develops, and in so doing, fuelwood, poles, and sawtimber are produced. Coconut palms are also widely used as the shade-tree component of these combinations, so an important food crop is also derived from the tree layer in this case (Nair, 1993).

The shade of the overstory canopy can be useful in reducing temperature and water loss in the crops below, but the shade trees often are more valuable for the role they play in nutrient cycling than for the shade they create. The deep roots of the trees take up nutrients that have been leached from the upper soil layers or have been weathered from rocks in deep strata, and thus are unavailable to the shallow roots of crops. A portion of the nutrients becomes available to crop plants through shade-tree litterfall. The overstory trees are sometimes pruned to reduce the shading effect and increase the rate of nutrient contribution of litter to the soil. Nitrogen-fixing trees are valuable as shade trees, because of the high nitrogen content of their foliage, but the litter decomposition rate and value

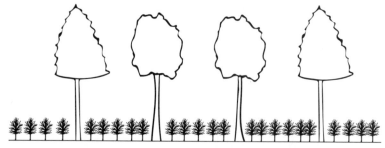

Figure 21.2   An example of a shade tree/crop combination used in Central America, in which the small tree crops of coffee and cacao are grown beneath valuable timber trees such as Spanish-cedar and laurél.

for timber or fuelwood are equally important in selecting species. Some of the tree species most widely used are legumes in the *Leucaena* and *Erythrina* genera.

An example of a shade tree/crop combination used in Central America consists of coffee or cacao grown beneath valuable timber trees (Fig. 21.2). In this system, shading may have some beneficial effect for the small tree crops in the dry season, but the main effect of the overstory shade is to decrease productivity during the long wet season. The value of the trees is in the maintenance of efficient nutrient cycling. When large commercial landowners grow coffee without shade trees, they compensate by adding chemical fertilizer, not shade cloth, to replace the effects of the overstory.

Another agrisilvicultural practice, known as **alley cropping**, shares the same objective of using trees for their nutrient cycling effects, but it is used for growing light-demanding herbaceous food crops. Crops are planted between hedgerows of trees that are maintained at shrub size by repeated pruning in order to reduce the effect of shading. The pruned branches from the hedgerow trees form a green mulch that is applied to the crops, improving the nutrient availability and organic matter content of the soil (Fig. 21.3).

Alley cropping has received considerable attention in agroforestry research, but trials have given variable results with respect to increasing crop yields. The shading of the trees is sometimes too great, even with pruning, and moisture competition between tree and crop roots is problematic on dry sites. In many of these cases, the competitive effects can outweigh the benefits to soil fertility, resulting in lower crop yields than those obtained without the hedgerows of trees. The success of this practice is generally determined by the yield of the herbaceous crop alone, because it is usually impossible to harvest much

Figure 21.3   The structure of an alley cropping system, in which maize is growing between tree hedges of a legume such as *Leucaena leucocephala*. The hedges are periodically pruned and the branches are used as mulch for the crops growing in the "alleys." Many other species combinations are used in the same arrangement.

from the hedgerows. It is often tempting to use the branches for livestock fodder, but this would defeat the nutrient management objectives of the system.

The poor results of many alley cropping trials caused by competition effects suggests that the use of a more direct modification of shifting agriculture, the idea of **improved fallows,** may be more successful in some conditions. Rather than mixing crops and trees spatially, crops are grown in monocultures for several cycles, followed by a fallow period in which the vegetation is managed to improve the efficiency of ecological restoration. By planting tree species with rapid growth, deep rooting, and nitrogen-fixing symbioses, it may be possible to restore the fertility of soil and shade out light-demanding weeds in a short time. In some soil and climatic conditions, fallow periods of two years or less may maintain soil fertility (Sanchez, 1995). Even with longer periods, the economic value of the fallow land can be maintained by planting tree species that produce fuelwood, fodder, or fruit or nut crops.

The temporary intercropping of herbaceous crops in the establishment phase of a forest stand, known as **taungya**, has somewhat more complicated economic objectives because the trees and crops usually have different owners. The system was devised in Southeast Asia in the nineteenth century to help establish government-owned teak plantations. Local farmers were allowed to take advantage of the vacant growing space between young teak trees and so gained usable farm land. In the process of weeding the crop plants, they also freed the tree seedlings of competition. The forest owners received the benefit of intensive weed control in their plantations, as well as any fees charged for the use of the land which helped offset establishment costs. The practice of taungya has proven very successful and is now a common method for establishing plantations throughout the tropics, with varying tree and crop ownership patterns.

Although the practice of taungya is temporary on any piece of land, it has served as a step in the development of more permanent agrisilvicultural systems. For example, taungya was initially used in Indonesia only during the first two to three years of teak plantation growth, when farmers grew maize and dryland rice crops. When government policies shifted to benefit local people more, taungya farmers were allowed to plant fruit and other multipurpose trees on 20 percent of the area, and carry them through the full rotation, thus creating a structure approaching a multistory tree garden (Kartasubrata and Wiersum, 1995).

The taungya approach is used on a limited basis in temperate forests and represents one of the few examples of agrisilviculture in North America. In eastern Canada, corn, soybeans, and other crops have been planted for the first few years in hybrid poplar plantations, which are grown on a 15-year rotation for pulpwood (Gold and Hanover, 1987). Considerable research has been done with the use of agroforestry to improve the financial aspects of establishing black walnut plantations for nut and timber production. Wheat or soybeans are grown between widely planted rows of walnut trees for the first 10 years; after the intensity of shading increases, livestock forage is produced for another decade. The effects of wide spacing of walnut trees are countered by pruning lower branches to improve timber quality. Nut harvesting begins at about age 15 and continues to the end of the timber rotation at age 60 to 80 (Garrett and Kurtz, 1983; Campbell, Lottes, and Dawson, 1991).

Silvopastoral Practices

Developing management methods that successfully combine livestock pasture with stands of trees presents even greater challenges than combining annual food crop production with trees. Not only does the shade from the tree canopy inhibit grasses and other forage plants, but also the animals may damage or destroy seedlings by browsing or trampling them. Indeed, some of the most complete kinds of deforestation have resulted from the continued pasturing of goats and other livestock in forests of semiarid regions. Yet, with careful management, some advantages can be gained by combining these two production methods on the same land.

Some agroforestry methods keep the animals separate from the trees in order to avoid the problems of interspecific plant competition and animal damage. These practices do not fit within the strict definition of silvopastoral, but they are closely related. They include the planting of small stands of **fodder trees**, which are coppiced on short rotations, with the cut foliage being carried to livestock in adjacent stalls or pastures. The same practice can be used when fodder trees are grown as **live fence posts** along pasture borders, but the stems are cut at a height of about 6 ft (2 m) (a practice known as **pollarding**) to keep the young sprouts out of reach of animals and to preserve the utility of the stem as a fence post. Legume species are frequent choices for fodder trees because their foliage and pods are rich in nitrogen; for this reason, stands of these fodder trees are sometimes called protein banks.

The central idea of silvopastoral systems is to produce timber or fuelwood from a stand of trees while developing productive pasture for livestock in the understory. This is one of the few agroforestry practices that has been developed on at least as large a scale in the temperate zone as in the tropics. Grazing domesticated animals in dry woodlands or savannas or in partially cleared forests in wetter climates was used throughout the world in the early history of agriculture (Long, 1993). Well-developed silvopastoral systems called **wood-pastures** were used in medieval Europe, and their origin is certainly much older than that (Rackham, 1986). The right of various groups to use the land was complex; often the landowner held the rights to the timber, but local farmers had government-granted rights or paid fees for pasturing their animals and pollarding nontimber trees for fuelwood, fence posts, and fodder. Animals were fenced out of areas only when new trees were being established. Some of these wood-pastures were contained within ''deer-parks'' in which both native and exotic deer species were fenced, and supplied with supplemental water and winter feed. This transitional stage between wildlife management and livestock production was an important form of meat production in medieval Europe (Rackham, 1986).

The main impetus for developing silvopastoral management remains similar to those ancient systems—to diversify production in a way that provides income from grazing fees during the growth of forest plantations. The forest land and the livestock still generally have different owners, which has at times been an impediment to the creation of efficient management schemes.

The opportunity to develop pasture vegetation during the various stages of forest stand development is similar to that for herbaceous crops. The most productive stage for shade-intolerant forage crops is during stand initiation; the closed canopy during stem exclusion precludes most herbaceous vegetation, and more shade-tolerant forage species develop during understory reinitiation. The main difference in this pattern is that the initiation stage, which is so useful for taungya cropping, is limited in value for pasture because of the problem of animals damaging tree seedlings. Thus older stands with open canopies in dry climates are the most logical kind to use for grazing. In the northwestern United States and western Canada, for example, it has long been a practice to graze cattle and sheep in the unimproved native grasses that grow beneath ponderosa pine and other conifers on the east side of the Cascade and Coastal Mountains; the main limitation to this practice is to restrict animals from regenerating stands. However, grazing and forestry are considered less compatible in the wetter climate on the west side of the mountains because stands form dense canopies that preclude herbaceous understories.

The practice of grazing livestock in older stands can be used in wet temperate or tropical regions only if stand density is reduced to a level below that which is normally used for timber production. This approach is used with widely spaced plantations of slash pine and longleaf pine in Florida (Lewis and Pearson, 1987) and of radiata pine in New Zealand and Australia (Knowles, 1991). It may be necessary to establish plantations in clumped or multiple-row configurations and to prune lower boles in order both to produce adequate forage yields and to develop acceptable wood volume production with good stem quality in these stands.

Grazing is also used in some situations as a silvicultural tool to release conifer seedlings from competing vegetation, because most conifers are less palatable than hardwoods and herbaceous plants (Fig. 21.4). This reduces the cost of weeding operations and is particularly useful on lands where it is not possible to use herbicides or prescribed fire. Considerable knowledge and care are required to use this technique without allowing damage to seedlings. The season and number of days of grazing must be controlled, which

Figure 21.4 Grazing being used to free ponderosa pine regeneration from competing weeds and brush in the Cascade Mountains of southern Oregon. Cattle must be moved frequently to avoid overgrazing and seedling damage. The stand is nearing the end of the stage in which sufficient forage is produced to support grazing. *(Photograph by S.D. Tesch.)*

requires frequent moving of the animals. It may be advantageous to sow seed of forage crops to prevent the growth of dense brush; sheep and cattle can maintain low herbaceous vegetation more efficiently, and are less likely to browse the conifers when much more palatable food is available. Livestock grazing is used for weed and brush control during stand establishment of Douglas-fir, ponderosa pine, and other conifers in Oregon, Washington, and British Columbia (Doescher, Tesch, and Alejandro-Castro, 1987), and of eucalypts in Brazil (Couto et al., 1994).

Shelterbelts and Contour Hedgerows

Some agroforestry practices are designed solely for the purpose of soil conservation, particularly in semiarid regions of the world, where wind erosion commonly accompanies farming. The planting of **shelterbelts** is an effective means of reducing such problems; rows of trees are established upwind of agricultural fields to reduce wind velocity and thus lower the potential for transport of soil particles. Wind reduction also decreases water loss from leaf and soil surfaces, increasing crop productivity. Shelterbelts are also used to protect livestock from temperature extremes caused by either hot or cold winds.

The beneficial effect of shelterbelts extends about 10 to 20 times tree height to the leeward side. The best structure is to have tall trees with a continuous vertical canopy. It helps to plant two or more rows of a mixture of tree and shrub species to create a stratified structure with the desired wall-like canopy structure. A single row of trees works poorly because gaps generally develop through which wind is channeled at high velocity.

Shelterbelts were established on a large scale in the prairie regions of North America, beginning in 1935, after the severe wind erosion that resulted from the droughts of the early 1930s; similar large-scale projects were carried out in China in the 1970s. Species used for shelterbelts often include drought-resistant trees and shrubs with a wide range of size at maturity, many of which are rarely used in silviculture. Species used in North America include hackberry, autumn-olive, green ash, poplars, boxelder, and eastern redcedar.

On moderate to steep slopes in wetter climates, excess runoff is the major cause of erosion on agricultural fields. Rows of trees or shrubs planted as **contour** or **barrier hedgerows** can serve much the same function as the filter strips that are retained along water bodies during silvicultural operations (see Chapter 19). Hedgerows are established at intervals along the contours of a slope, with crops being grown in the spaces between. This practice can also function as alley cropping, if the pruned branches are used as mulch for the crops, but it can also be used solely for stabilizing soil. These are so effective that they can replace the need for constructing terraces on moderately sloping land.

Selection of Tree Species for Agroforestry

Species used for silviculture are not necessarily the logical choice for agroforestry because trees play different roles in the two systems. Ideal species for many agroforestry practices are those that can serve the protective function of preventing nutrient leaching or soil loss, and at the same time provide a useful product of subsistence or commercial value. Species with these characteristics are called **multipurpose tree species**.

Multipurpose trees can be most useful for improved fallows, alley cropping, contour hedgerows, and shelterbelts. Here, the protective function is of greatest importance, but production of food or medicine, livestock fodder, wood suitable for burning or charcoal production, or poles or sawtimber at least partly compensates for the loss of cropland to the trees. These species can be used in most other agroforestry practices as well.

The concept of multipurpose trees has focused attention on lesser known trees that are not used for timber purposes. Much work has been done to identify appropriate species and provenances with desired characteristics (Gupta, 1993; Burley and von Carlowitz, 1984). Legumes, alders, and causarinas are highly desired for their nutrient contributions from symbiotic nitrogen fixation. However, different characteristics are required for use in different systems. For example, trees with rapid root growth are valuable for use in improved fallows, but they are likely to cause too much competition if used for hedgerows in alley cropping.

Efforts are being made to improve some tree species for use in agroforestry through controlled breeding. This task is made difficult because of the multiple functions desired of the species. In forest tree breeding, it has been found that concentrating on improvement of only a small number of characteristics gives the greatest gains. Improvement work has progressed the furthest with species and hybrids of *Leucaena*.

Some concern has been voiced that efforts to genetically improve multipurpose agro-forestry trees will result in the use of a few superior species as exotics throughout the tropics and subtropics, just as has been done in tropical timber plantations. As with sil-viculture, it is often better to use a wide diversity of native species adapted to local climates and site conditions. Promotion of exotic tree species in agroforestry may bring about the loss of local varieties similar to the loss of local food crops that occurred when improved crop varieties were distributed (Hughes, 1994). Other concerns are similar to those asso-ciated with the use of exotics in silviculture—that large areas may be affected by a disease or insect problem, or that some introduced species may invade native vegetation or farm-land, and become difficult to control.

BIBLIOGRAPHY

Adams, S. N. 1975. Sheep and cattle grazing in forests: a review. *J. of Appl. Ecol.*, 12:143–152.

Anderson, L. S., and F. L. Sinclair. 1993. Ecological interactions in agroforestry systems. *For. Abst.* 54:489–523.

Bandolin, T. H., and R. F. Fisher. 1991. Agroforestry systems in North America. *Agrofor. Syst.*, 16:95–118.

Burley, J., and P. von Carlowitz. 1984. *Multipurpose tree germplasm*. International Council for Research in Agroforestry, Nairobi, Kenya. 298 pp.

Campbell, G. E., G. J. Lottes, and J. O. Dawson. 1991. Design and development of agroforestry systems for Illinois, USA: silvicultural and economic considerations. *Agrofor. Syst.*, 13:203–224.

Couto, L., R. L. Roath, D. R. Betters, R. Garcia, and J. C. C. Almeida. 1994. Cattle and sheep in eucalypt plantations: a silvopastoral alternative in Minas Gerais, Brazil. *Agrofor. Syst.*, 28:173–185.

Doescher, P. S., S. D. Tesch, and M. Alejandro-Castro. 1987. Livestock grazing: a silvicultural tool for plantation establishment. *J. For.*, 85(10):29–37.

Ewel, J. J. 1986. Designing agricultural ecosystems for the humid tropics. *Ann. Rev. Ecol. Syst.*, 17:245–271.

Garrett, H. E., and W. B. Kurtz. 1983. Silvicultural and economic relationships of integrated forestry-farming with black walnut. *Agrofor. Syst.*, 1:245–256.

Gold, M. A., and J. W. Hanover. 1987. Agroforestry systems for the temperate zone. *Agrofor. Syst.*, 5:109–121.

Gupta, R. K. 1993. *Multipurpose trees for agroforestry and wasteland utilisation*. International Science Publishers, New York. 562 pp.

Hughes, C. E. 1994. Risks of species introductions in tropical forestry. Commonwealth For. Rev., 73:243–252.

Jarvis, P. G. (ed.). 1991. *Agroforestry: principles and practice*. Elsevier, Amsterdam. 356 pp.

Kartasubrata, J., and K. F. Wiersum. 1995. Traditions and recent advances in tropical silvicultural research in Indonesia. *Unasylva*, 46(2):30–35.

Knowles, R. L. 1991. New Zealand experience with silvopastoral systems: a review. *For. Ecol. and Manage.*, 45:251–268.

Lewis, C. E., and H. A. Pearson. 1987. Agroforestry using tame pastures under planted pines in the southeastern United States. In: H. L. Gholz (ed.), *Agroforestry: realities, possibilities and potentials*. Nijhoff, Dordrecht. Pp. 195–212.

Linnartz, N. E., and M. K. Johnson. (eds.). 1984. *Agroforestry in the southern United States*. 33rd Annual Forestry Symposium. Louisiana State Univ., Baton Rouge. 183 pp.

Long, A. J. 1993. Agroforestry in the temperate zone. In: *An introduction to agroforestry*. Kluwer, Dordrecht. Pp. 443–468.

MacDicken, K. G., and N. T. Vergara (eds.). 1990. *Agroforestry: classification and management*. Wiley, New York. 382 pp.

Nair, P.K.R. 1993. *An introduction to agroforestry*. Kluwer, Dordrecht. 499 pp.

Rackham, O. 1986. *The history of the countryside*. J. M. Dent, London. 445 pp.

Sanchez, P.A. 1995. Science in agroforestry. *Agrofor. Syst.*, 30:5–55.

Wiersum, K. F. 1982. Tree gardening and taungya on Java: examples of agroforestry techniques in the humid tropics. *Agrofor. Syst.* 1:53–70.

Young, A. 1989. *Agroforestry for soil conservation*. CAB International, Tucson. 280 pp.

COMMON AND SCIENTIFIC NAMES OF TREES AND SHRUBS MENTIONED IN THE TEXT

COMMON NAME	SCIENTIFIC NAME*
Alder, red	*Alnus rubra* Bong.
Ash, mountain (eucalypt)	*Eucalyptus regnans* F. Muell.
Ash, white	*Fraxinus americana* L.
green	*Fraxinus pennsylvanica* Marsh
tropical	*Fraxinus uhdei* (Wenzig) Lingelsh.
Aspen, bigtooth	*Populus grandidentata* Michx.
quaking	*Populus tremuloides* Michx.
Autumn-olive	*Eleaegnus umbellata* Thunb.
Bald-cypress	*Taxodium distichum* (L.) Rich.
Basswood, American	*Tilia americana* L.
Beech, American	*Fagus grandifolia* Ehrh.
European	*Fagus sylvatica* L.
Birch, black	*Betula lenta* L.
gray	*Betula populifolia* Marsh.
paper	*Betula papyrifera* Marsh.
yellow	*Betula alleghaniensis* Britton
Blackberries	*Rubus* spp.
Black-gum	*Nyssa sylvatica* Marsh.
Boxelder	*Acer negundo* L.
Blueberries	*Vaccinium* spp.

*Scientific names based on: E. L. Little, Jr., Checklist of United States trees. *USDA, Agr. Hbk.* 541, 1979, 375 pp; and D.J. Mabberley, *The Plant-book.* Cambridge Univ. Press, New York, 1987, 707 pp.

Buckeye, Ohio	*Aesculus glabra* Willd.
Cacao	*Theobroma cacao* L.
Catalpa	*Catalpa speciosa* Warder ex. Engelm.
Cecropia	*Cecropia* spp.
Cedar, Alaska yellow-	*Chamaecyparis nootkatensis* (D. Don.) Spach
Cedar, Spanish	*Cedrela odorata* L.
Cherry, black	*Prunus serotina* Ehrh.
choke	*Prunus virginiana* L.
pin	*Prunus pensylvanica* L. f.
Chestnut, American	*Castanea dentata* (Marsh.) Borkh.
Coconut palm	*Cocos nucifera* L.
Coffee	*Coffea* spp.
Cottonwood, black	*Populus trichocarpa* (Torr. & Gray) Brayshaw
eastern	*Populus deltoides* Bartr. (ex Marsh.)
Currant and gooseberry	*Ribes* spp.
Dogwood, flowering	*Cornus florida* L.
Douglas-fir	*Pseudotsuga menziesii* (Mirb.) Franco
Dipterocarps	*Dipterocarpus* spp.
Elms	*Ulmus* spp.
Eucalypts	*Eucalyptus* spp.
Figs	*Ficus* spp.
Fir, balsam	*Abies balsamea* (L.) Mill.
European silver	*Abies alba* (Mill.)
grand	*Abies grandis* (Dougl.) Lindl.
noble	*Abies procera* Rehd.
Pacific silver	*Abies amabilis* (Dougl.) Forbes
red	*Abies magnifica* A. Murr.
subalpine	*Abies lasciocarpa* (Hook.) Nutt.
white	*Abies concolor* (Gord. & Glend.) Lindl. (ex Hildebr.)
Gum, black	*Nyssa sylvatica* Marsh
Hackberries	*Celtis* spp.
Hemlock, eastern	*Tsuga canadensis* (L.) Carr.
mountain	*Tsuga mertensiana* (Bong.) Carr.
western	*Tsuga heterophylla* (Raf.) Sarg.
Hickory, pignut	*Carya glabra* (Mill.) Sweet
shagbark	*Carya ovata* (Mill.) K. Koch
water	*Carya aquatica* (Michx. f.) Nutt.
Hinoki	*Chamaecyparis obtusa* Endl.
Hop-hornbeam, eastern	*Ostrya virginiana* (Mill.) K. Koch.
Incense-cedar, California	*Libocedrus decurrens* Torr.
Juniper, western	*Juniperus occidentalis* Hook.
Larch, European	*Larix decidua* L.
Japanese	*Larix leptolepis* (Sieb. & Zucc.) Gord.
western	*Larix occidentalis* Nutt.
Locust, black	*Robinia pseudoacacia* L.
Madrone, Pacific	*Arbutus menziesii* Pursh.
Mahogany, Honduran	*Swietenia macrophylla* King.
Manzanita	*Arctostaphylos columbiana* Piper
Maple, big-leaf	*Acer macrophyllum* Pursh
red or soft	*Acer rubrum* L.
sugar	*Acer saccharum* Marsh.
silver	*Acer saccharinum* L.
vine	*Acer circinatum* Pursh

Oak, bear	*Quercus ilicifolia* Wangenh.
black	*Quercus velutina* Lam.
blue	*Quercus douglasii* Hook. & Arn.
bur	*Quercus macrocarpa* Michx.
canyon live	*Quercus chrysolepis* Liebm.
cherry-bark	*Quercus falcata* var. *pagodaefolia* Ell.
chestnut	*Quercus prinus* L.
live	*Quercus viriginiana* Mill.
northern red	*Quercus rubra* L.
overcup	*Quercus lyrata* Walt.
pin	*Quercus palustris* Muenchh.
post	*Quercus stellata* Wangenh.
scarlet	*Quercus coccinea* Muenchh.
southern red	*Quercus falcata* Michx.
white	*Quercus alba* L.
willow	*Quercus phellos* L.
Palmetto, saw	*Serenoa repens* (Bartr.) Small
Pecan	*Carya illinoensis* (Wagenh.) K. Koch
Pepper (vines), black	*Piper nigrum* L.
Persimmon	*Diospyros virginiana* L.
Pine, Aleppo	*Pinus halepensis* Mill.
bishop	*Pinus muricata* D. Don
Caribbean	*Pinus caribaea* Mor.
Digger	*Pinus sabiniana* Dougl.
Durango	*Pinus durangensis* Mart.
eastern white	*Pinus strobus* L.
jack	*Pinus banksiana* Lamb.
Jeffrey	*Pinus jeffreyi* Grev. & Balf.
jelecote	*Pinus patula* Schl. & Cham.
limber	*Pinus flexilis* James
loblolly	*Pinus taeda* L.
lodgepole	*Pinus contorta* Dougl. ex Loud.
longleaf	*Pinus palustris* Mill.
Monterey or radiata	*Pinus radiata* D. Don.
pinyon	*Pinus edulis* Englem.
pitch	*Pinus rigida* Mill.
pond	*Pinus serotina* Michx.
ponderosa	*Pinus ponderosa* Laws.
red or Norway	*Pinus resinosa* Ait.
sand	*Pinus clausa* (Chapm. ex Engelm.) Vasey ex Sarg.
Scotch	*Pinus sylvestris* L.
shortleaf	*Pinus echinata* Mill.
slash	*Pinus elliottii* Engelm.
sugar	*Pinus lambertiana* Dougl.
Virginia	*Pinus virginiana* Mill.
western white	*Pinus monticola* Dougl.
white-bark	*Pinus albicaulis* Engelm.
Poplars	*Populus* spp.
Raspberries	*Rubus* spp.
Red-cedar, eastern	*Juniperus virginiana* L.
western	*Thuja plicata* Donn
Redwood, coast	*Sequoia sempervirens* (D. Don.) Endl.
Rhododendron	*Rhododendron* spp.

Rubber-tree	*Hevea brasiliensis* (A. Juss.) Muell.
Sal	*Shorea robusta* Gaertn. f.
Sassafras	*Sassafras albidum* (Nutt.) Nees
Sequoia, big-tree or giant	*Sequoiadendron giganteum* (Lindl.) Buchholz
Silk-oak	*Grevillea robusta* Cunn.
Spruce, black	*Picea mariana* (Mill.) B.S.P.
blue	*Picea pungens* Engelm.
Engelmann	*Picea engelmannii* Parry
Norway	*Picea abies* (L.) Karst.
red	*Picea rubens* Sarg.
Sitka	*Picea sitchensis* (Bong.) Carr.
white	*Picea glauca* (Moench.) Voss.
Sugarberry	*Celtis laevigata* Willd.
Sugi	*Cryptomeria japonica* D. Don
Sweet-gum	*Liquidambar styraciflua* L.
Sycamores	*Platanus* spp.
Tamarack	*Larix* spp.
Tanoak	*Lithocarpus densiflorus* (Hook. & Arn.) Rehd.
Teak	*Tectona grandis* L.f.
Tupelo, water	*Nyssa aquatica* L.
Walnut, black	*Juglans nigra* L.
White-cedar, Atlantic	*Chamaecyparis thyoides* (L.) B.S.P.
northern	*Thuja occidentalis* L.
Willows	*Salix* spp.
Yellow-poplar	*Liriodendron tulipifera* L.
Yew, Pacific	*Taxus brevifolia* Nutt.

ABBREVIATIONS IN REFERENCE LISTS

Standard abbreviations are used in the names of serials and periodicals with the exception of those listed below. The abbreviations for names of regional experiment stations of the U. S. Forest Service are prefixes of the serial numbers of the publications.

AES   Agricultural Experiment Station
CJFR   Canadian Journal of Forest Research
FAO   Food and Agriculture Organization of the United Nations
FE&M   Forest Ecology and Management
INT   Intermountain Forest and Range Experiment Station
NE   Northeastern Forest Experiment Station
NC   North Central Forest Experiment Station
NJAF   Northern Journal of Applied Forestry
PNW   Pacific Northwest Forest and Range Experiment Station
PSW   Pacific Southwest Forest and Range Experiment Station
RM   Rocky Mountain Forest and Range Experiment Station
SE   Southeastern Forest Experiment Station
SJAF   Southern Journal of Applied Forestry
SO   Southern Forest Experiment Station
USDA   U.S. Department of Agriculture
USFS   Forest Service, U.S. Department of Agriculture
WJAF   Western Journal of Applied Forestry

INDEX

Note: Each page number entry may indicate the start of a long section dealing with the topic. Refer to Table of Contents for topics of chapters and sections.